ACCESSORY TO WAR

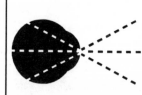

This Large Print Book carries the
Seal of Approval of N.A.V.H.

ACCESSORY TO WAR

THE UNSPOKEN ALLIANCE BETWEEN ASTROPHYSICS AND THE MILITARY

NEIL DEGRASSE TYSON AND AVIS LANG

THORNDIKE PRESS

A part of Gale, a Cengage Company

Farmington Hills, Mich • San Francisco • New York • Waterville, Maine
Meriden, Conn • Mason, Ohio • Chicago

LIBRARY OF CONGRESS CIP DATA ON FILE.
CATALOGUING IN PUBLICATION FOR THIS BOOK
IS AVAILABLE FROM THE LIBRARY OF CONGRESS

ISBN-13: 978-1-4328-6381-4 (hardcover)

Published in 2019 by arrangement with W. W. Norton & Company, Inc.

Printed in the United States of America
1 2 3 4 5 6 7 23 22 21 20 19

To everybody who has ever wondered
why astrophysicists have jobs at all

Do I not destroy my enemies when I make
them my friends?

— Abraham Lincoln

CONTENTS

PROLOGUE

In matters of battle, the role of science and technology often proves decisive, providing an asymmetric advantage whenever one side exploits this knowledge while the other side does not. The biologist, when enlisted for the war effort, may consider weaponizing bacteria and viruses; a rotting animal carcass catapulted over a castle wall during a siege may have been one of the first acts of biowarfare. The chemist, too, contributes — from the poisoned water-wells of antiquity, to mustard and chlorine gas during World War I, to defoliants and incendiary bombs in Vietnam and nerve agents in more contemporary conflicts. The physicist at war is an expert in matter, motion, and energy, and has one simple task: to take energy from here and put it over there. The strongest expressions of this role have been the atomic bombs of World War II and the more decisively deadly hydrogen fusion bombs that followed during the Cold War. Lastly, we have the engineer, who makes all things possible — enabling science to facilitate warfare.

The astrophysicist, however, does not make the missiles or the bombs. Astrophysicists make no weapons at all. Instead, we and the military happen to care about many of the same things: multispectral detection, ranging, tracking, imaging, high ground, nuclear fusion, access to space. The overlap is strong, and the knowledge flows in both directions. Astrophysicists as a community, like most academics, are overwhelmingly liberal and antiwar, yet we are curiously complicit in this alliance. *Accessory to War: The Unspoken Alliance Between Astrophysics and the Military* explores this relationship from the earliest times of celestial navigation in the service of conquest and hegemony to the latest exploitations of satellite-enabled warfare.

The idea for this book germinated in the early 2000s during my tour of duty serving on President George W. Bush's twelve-member Commission on the Future of the United States Aerospace Industry. That exposure to members of Congress, Air Force generals, captains of industry, and political advisors from both sides of the aisle was a baptism on the inner workings of science, technology, and power within the US government. My experiences led me to imagine what such encounters might have been like over the centuries in whatever country happened to be leading the world in cosmic discovery and in war.

Co-author Avis Lang is my longtime editor, from my days of contributing monthly essays to *Natural History* magazine. An art historian by training, Avis is a consummate researcher and an avid writer, with a deep interest in the universe. This book is a collaboration, a fusion of our talents. We each compensate for the weaknesses of the other. But the book got done because of Avis's sustained commitment to examining the role of science in society, as expressed in the printed word.

The reader will notice that in certain passages, such as here, first-person singular pronouns appear, primarily when I tell personal stories. But in no way does the occasional "I" or "my" deny Avis's co-authorship of every page in this book.

— Neil deGrasse Tyson and Avis Lang
New York City, January 2018

■ ■ ■ ■

SITUATIONAL
AWARENESS

■ ■ ■ ■

1
A TIME TO KILL

On February 10, 2009, two communications satellites — one Russian, the other American — smashed into each other five hundred miles above Siberia, at a closing speed of more than 25,000 miles an hour. Although the impetus for building their forerunners was war, this collision was a purely peacetime accident, the first of its kind. Someday, one of the hundreds of chunks of resulting debris might smash into another satellite or cripple a spaceship with people on board.

Down on the ground that same winter's day, the Dow Jones Industrial Average closed at 7888 — respectably above the decade's dip to 6440 in March 2009 but not much more than half its high of 14,198 in October 2007. In other news of the day, Muzak Holdings, the eponymous provider of elevator music, filed for bankruptcy; General Motors announced a cut of ten thousand white-collar jobs; federal investigators raided the offices of a Washington lobbying firm whose clients were major campaign contributors to the head of the House subcommittee on defense spending; the inflammatory Iranian president declared at a rally celebrating the thirtieth anniversary of his nation's Islamic Revolution that Iran was "ready to hold talks based on mutual respect and in a fair atmosphere"; and the brand-new American president's brand-new secretary of the Treasury presented a $2 trillion plan to lure speculators into buying the unstable American assets that had collapsed the global economy. Civil engineers announced that 70 percent of the salt applied to icy roads in the Twin Cities ended up in the watershed. An environmental physicist announced that a third of the top-selling laser printers form large numbers of lung-damaging ultrafine particles from vapors emitted when the printed image is heat-fused to the paper. Climatologists announced that the flowering ranges of almost a hundred plant species had crept uphill in Arizona's Santa Catalina Mountains over a twenty-year period, in lockstep with the rise in summer temperatures.

The world, in other words, was in flux and under threat, as it so often is.

Ten days later, an international group of distinguished economists, officials, and academics met under the auspices of Columbia University's Center on Capitalism and Society to discuss how the world might manage to emerge from its worse-than-usual financial crisis. The center's director, Nobel laureate in economics Edmund Phelps, argued that some financial re-regulation was called for but stressed that it must not "discourag[e] funding for investment in innovation in the non-financial business sector, which has been the main source of dynamism in the U.S. economy." What's the non-financial business sector? Military spending, medical equipment, aerospace, computers, Hollywood films, music, and more military spending. For Phelps, dynamism and innovation went hand in hand with capitalism — and with war. Asked by a BBC interviewer for a "big thought" on the crisis and whether it constituted "a permanent indictment of capitalism," he responded, "My big thought is, we desperately need capitalism in order to create interesting work to be done, for ordinary people — unless maybe we can go to war against Mars or something as an alternative."[1]

A vibrant economy, in other words, depends on at least one of the following: the profit motive, war on the ground, or war in space.

On September 14, 2009, just a few months after the satellite smashup and a few blocks from where the World Trade Center's Twin Towers had stood eight years and four days earlier, President Barack Obama spoke to Wall Street movers and shakers to mark the first anniversary of the collapse of Lehman Brothers, the investment firm whose bankruptcy is often presented as having triggered the avalanche of financial failures in 2008–2009. That same morning, China laid the cornerstone for its fourth space center on an island close to the equator — the latitude of choice for exploiting Earth's rotation speed, thereby minimizing the fuel necessary for a launch and maximizing the potential payload. By late 2014 construction was finished, well before the World Trade Center site would be fully rebuilt. An Associated Press reporter spoke of China's "soaring space ambitions" and, after presenting a daunting list of Chinese space achievements and ambitions, stated that "China says its space program is purely for peaceful ends, although its military background and Beijing's development of anti-satellite weapons have prompted some to question that."[2]

Much the same could be said of the background and backing of the lavishly funded space programs created by the Cold War superpowers.

Were he alive today, the seventeenth-century Dutch astronomer and mathematician Christiaan Huygens might tell us we'd be fools to think that ambitious undertakings in space can be achieved without massive military support. Back in the 1690s, as Huygens thought about life on Mars and the other planets then known to populate the night sky, he pondered how best to foster inventiveness. For him and his era, profit was a powerful incentive (capitalism was as yet unnamed), and conflict was a divinely endorsed stimulant of creativity:

> It has so pleased God to order the Earth . . . that this mixture of bad Men with good, and the Consequents of such a mixture as Misfortunes, Wars, Afflictions, Poverty and the like, were given us for this very good end, *viz.* the exercising our Wits and Sharpening our Inventions; by forcing us to provide for our own necessary defence against our Enemies.

Yes, waging war requires clever thinking and promotes technical innovation. Not controversial. But Huygens can't resist linking the absence of armed conflict with intellectual stagnation:

> And if Men were to lead their whole Lives in an undisturb'd continual Peace, in no fear of Poverty, no danger of War, I don't doubt they would live little better than Brutes, without all knowledge or enjoyment of those Advantages that make our Lives pass on with pleasure and profit. We should want the wonderful Art of Writing if its great use and necessity in Commerce and war had

not forc'd out the Invention. 'Tis to these we owe our Art of Sailing, our Art of Sowing, and most of those Discoveries of which we are Masters; and almost all the secrets in experimental Knowledge.[3]

So it's simple: no war equals no intellectual ferment. Arm in arm with trade, says Huygens, war has served as the catalyst for literacy, exploration, agriculture, and science.

Were Phelps and Huygens right? Must war and profit be what drive both civilization on Earth and the investigation of other worlds? History, including last week's history, makes it hard to answer no. Across the millennia, space studies and war planning have been business partners in the perennial quest of rulers to obtain and sustain power over others. Star charts, calendars, chronometers, telescopes, maps, compasses, rockets, satellites, drones — these were not inspirational civilian endeavors. Dominance was their goal; increase of knowledge was incidental.

But history needn't be destiny. Maybe the present calls for something different. Today we face "Enemies and Misfortunes" that Huygens never dreamed of. Surely the "exercising [of] our Wits" could be directed toward the betterment of all rather than the triumph of the few. Surely it's not too radical to suggest that capitalism won't have much to work with if several hundred million species vanish for lack of potable water, breathable air, or perhaps the aftereffects of a plummeting asteroid or an assault by cosmic rays.

Looking down at Earth from an orbiting spacecraft, a rational person could certainly feel that "necessary defence" may have more to do with the vulnerability of our beautiful blue planet, exposed to all the vicissitudes of the cosmos, than with the transient power of a single country's weapons, policymakers, nationalists, and ideologues, however virulent. From hundreds of kilometers above the surface of the globe, "Peace on Earth, Goodwill Toward Men" might sound less like a standard line on a Christmas card and more like an essential step toward a viable future, in which all of humankind cooperates in protecting Earth from the enemies among us and the threats above us.

On the chill evening of January 16, 1991, a thousand or so space scientists, myself included, tipped our wineglasses and schmoozed about our latest research projects at the closing banquet of the 177th semiannual meeting of the American Astronomical Society in Philadelphia. Sometime after the entrée but before the dessert, the organization's president, John Bahcall, stood up to announce that the United States was at war. Operation Desert Storm, the bombing blitz that launched the first US-led war in the

Persian Gulf, had begun at about half past six — the middle of the night in Baghdad. CNN journalists were reporting the aerial assault live and uncensored from the ninth floor of the Al-Rashid Hotel as the cloudless, starry desert skies filled with flashes of light. For the first time in warfare, America was showcasing its stealth bombers, virtually invisible to enemy radar and unseeable in the absence of moonlight. Not a cosmic accident. The attack was timed to coincide with the new Moon, the only phase not visible at any time of the day or night.

Our after-banquet speaker would not be coming, said Bahcall. No witticisms would accompany the coffee. The festivities would be cut short so that we could turn our attention to CNN or go home to be with our loved ones. The hall fell silent. The collective pall was no surprise. Fewer than twenty years had passed since the end of the Vietnam War, and memories of the US involvement in Southeast Asia still haunted many people in that room, myself included.

While most of my colleagues spent the rest of the night in Philadelphia glued to the tube, I strolled from the hotel alone to walk off some confused energy. Everywhere I went, TVs were tuned to CNN. Passing an auto repair shop, I shouted to a twenty-something mechanic working late, a fellow likely in kindergarten while Vietnam was becoming an

American nightmare, "Did you hear we went in?"

I'd expected to hear a word or two of regret. Instead, the guy gleefully shouted back, "Yup!" And with a fist pump in the air and a giddy pride I'd never before associated with warfare, he chanted, "Fuckin' A! We're at war!"

Probably I should have seen that coming, considering the patriotic enthusiasm so visible at Memorial Day parades and Fourth of July fireworks, with their backstory of war, bombs, rockets, and bloodshed. Like every other American, I'd sung the national anthem's soaring passage about rockets' red glare and bombs bursting in air. I was aware of the many wartime generals who subsequently became president, and the many public war memorials where statues portray, if not a solitary cannon, then one or more uniformed soldiers standing tall, standing brave, standing proud, occasionally astride a war horse, the statues' immobilized warriors brandishing the weapon of their time: saber, musket, carbine, assault rifle.

But none of those expressions of national pride and militarism meshed with my sense of armed conflict. I didn't understand how they fit together. That twenty-something grease monkey did, though. He was plugged into a primal passion that has energized so many wars across the millennia. Just not the

25

war I'd grown up with.

The US engagement in Vietnam, Laos, and Cambodia triggered a vehement antiwar movement, its strength and visibility without precedent and its numbers swelled by tens of thousands of returning American vets and active-duty GIs revolted by the war they helped wage. During the first few years following the 1973 peace accord and the departure of combat troops, the war's opponents may have expected that the US military budget would stage a retreat. Yet Office of Management and Budget figures show only a brief pause before a renewed escalation in spending — an escalation that became dramatic during the next administration.[4] Soon, promised the soon-to-be president, Ronald Reagan, it would be "morning again in America."[5] Reagan's first inaugural address, in 1981, officially heralded the era of ubiquitous heroism and insistent patriotism — heroes, whose "patriotism is quiet but deep," were to be met "every day . . . across a counter."[6] People hung the Stars and Stripes from their porches. Explicit signals of respect for the military and love of the homeland multiplied. Jingoism was in the air. Once again, war was glory.

Like the vast majority of my fellow astrophysicists, I recoil at the prospect of war — the death, the destruction, the disillusionment. My revulsion, like the patriotism of

26

Reagan's heroes, runs quiet but deep. In the early days of the Vietnam War, I heard the entire mainstream American political spectrum declare that we had to defeat communism because communism represented all that was evil and bad while we represented all that was God-fearing and good. Back then, I was old enough to listen but too young to understand. But by the time the lists and photos of dead GIs were being published weekly, I'd begun to have the occasional thought about world events, and to me the message came through loud and clear. Vietnamese were dying. Americans were dying. American soldiers were strafing rice paddies and villages. The images of suffering embedded themselves in my mind. Some lingered for decades.

Fast-forward to the summer of 2005, three decades after the end of the Vietnam War and days before my daughter's ninth birthday. Miranda is running from the shower to her room. She's naked, because she's accidentally left her bath towel there. As she scampers past me, arms extended from her sides and elbows slightly bent, time freezes. The 1972 Pulitzer Prize–winning photograph of a naked Vietnamese girl flashes into my mind. You know the one. She's escaping along a road after American jets have drenched her village in a firestorm of napalm.[7] She has the body development and proportions of an eight- or

nine-year-old girl. She has the body development and proportions of my daughter. In that
fleeting moment they were one and the same.

During the first Gulf War (1990–91), the
United States offered itself and its coalition
of willing nations as defenders of a helpless
Kuwait against an invading Iraq. As often as
not, demonstrators on the streets of America
were there to express well-mannered objections to the war rather than to denounce it
outright. The rage of the Vietnam era had
dissipated. Many antiwar activists adopted
the expedient stance of differentiating the
war from the warriors. Their placards were
likelier to display a slogan such as "SUPPORT
OUR TROOPS, BRING THEM HOME" THAN "NO
BLOOD FOR OIL." The song "When Johnny
Comes Marching Home Again," dating back
to the Civil War, made another hurrah. The
yellow ribbons of faithfulness and welcome
reappeared.

   A dozen years later, during the Iraq War,
the United States became the aggressor,
armed with upgraded space assets that provided an overwhelming asymmetric advantage. Weather satellites, spy satellites, military
communications satellites, and two dozen
Earth-orbiting GPS satellites charted and imaged the battlefield. Down on the ground,
young soldiers drove down danger-ridden
roads in armored vehicles. And because of

portable access to spaceborne assets, by and large they knew where their targets lay, how to get there, and what obstacles stood in the path. Meanwhile, anyone in America who publicly criticized the way the war was justified, funded, or conducted soon felt pressure to bracket their accusations with lavish declarations of support for the troops. Despite the pressure, hundreds of thousands of peace-minded US civilians, joined by hundreds of members of a fierce new generation of anti-war vets and by millions of Europeans, again put their bodies on the streets and their testimony on record to call for a swift end to the invasion.[8]

Congress, as usual, was not in the vanguard of the antiwar battalions. For more than half a century, it has neither exercised its constitutional right to declare war nor withheld funding to pursue a specific war. This time it simply voted on whether to give the president free rein to use US armed forces against Iraq "as he determines to be necessary and appropriate." In January 1991 — upholding the firm twentieth-century pattern of Democratic-controlled Congresses voting in support of war — a Congress with substantial Democratic majorities had voted 250–183 in the House and 52–47 in the Senate to authorize a Republican president to do as he saw fit with the troops.[9] Now, in October 2002, a more evenly split, but now Republican-controlled, Congress voted 296–133 in the House and 77–23 in the Senate to give another Republican president similar authorization. And so, ostensibly to avenge the horrors of September 11, 2001, we went to war to rid the world of Iraq's purported weapons of mass destruction and to liberate the citizens of Iraq from a tyrant who promulgated torture, repression, and poison-gas attacks on his own people but also, as it happens, supplied them with free university education, universal health care, paid maternity leave, and monthly allotments of flour, sugar, oil, milk, tea, and beans.[10]

The first few years after 9/11 were a fine time to be a mercenary, a military engineering firm, or a giant aerospace company. Vietnam felt very far away. Blackwater, Bechtel, Halliburton, KBR, and their brethren prospered; returns on one global aerospace and defense index rose nearly 90 percent, compared with a 60 percent rise for global equities.[11] At the mention of the words "terrorism" or "homeland security," liberal Democrats made common cause with conservative Republicans.

With the winding-down of the Cold War, the aerospace industry had undergone unrelenting shrinkage and consolidation. Seventy-five aerospace companies were in operation on the day Reagan was elected, merging into sixty-one by the time the Berlin Wall fell, and finally into just five titans — Lockheed Martin, Boeing, Raytheon, Northrop Grumman, and General Dynamics — by the time the Twin Towers crumbled into toxic dust. Some 600,000 scientific and technical jobs had vanished in just a dozen of those years, along with incalculable quantities of experience and intellectual capital.[12]

Terrorism to the rescue — if not for American S & T workers, then certainly for American industrialists. The rescue was ably assisted by the 2001 final report of the Commission to Assess United States National Security Space Management and Organization, better known as the Rumsfeld Space Commission after its aggressive chairman, who was about to begin serving as George W. Bush's secretary of defense. The report invokes vulnerabilities, hostile acts, attacks, deterrence, breakthrough technologies, space superiority, encouragement of the private sector, and prevention of a "Space Pearl Harbor" (a recurrent phrase). It calls for "power projection in, from and through space" to ensure that the United States "remain the world's leading space-faring nation," and declares that America must be able "to defend its space assets against hostile acts and to negate the hostile use of space against U.S. interests" — altogether a grandiose and open-ended agenda.[13] The report was published exactly eight months before 9/11, and while it mentions terrorism multiple times, Osama bin Laden is mentioned only once. The threat level of its pages, however, is consistently reddish orange.

One Rumsfeldian pillar of space management was missile defense, the much-questioned ballistic missile interception technology announced as a goal in 1983 by Ronald Reagan and quickly dubbed Star Wars. Under the budget for missile defense from 2001 to 2004 — George W. Bush's first term as president — Boeing's contracts doubled, Lockheed Martin's more than doubled, Raytheon's nearly tripled, and Northrop Grumman's quintupled.[14] At the same time, corporate aerospace contributions to both parties in election campaigns ranged in the tens of thousands of dollars, while the corporations' multiyear missile-defense contracts ranged in the billions — an enviable return on a modest investment.[15] The Defense Department's Star Wars budget, $5.8 billion in 2001, reached $9.1 billion in 2004. Early in its tenure, the Bush administration withdrew from the 1972 Anti-Ballistic Missile Treaty, thus escaping from international constraints on testing weapons technology in space and enabling the newly renamed Missile Defense Agency to execute its mandate.

The picture of overall military spending for 2001–2004 was as expansionary as the picture for Star Wars. The formal "budget authority" for national defense — the permission given to the Department of Defense, the Department of Energy, NASA, and other agencies to sign new contracts and place new orders — rose from $329 billion in 2001 to $491 billion in 2004. Meanwhile, America's military credit line plus preauthorized payments edged toward a trillion dollars a year, not to mention additional expenditures such as the off-the-record billions in shrink-wrapped packets of bills handed out in Baghdad.[16] Whether this spending increased US national security is a matter of debate.

People who care about politics — and about safety — barely intersect on a basic definition of security: national, global, or otherwise. According to the mission statement for the middle-of-the-road American Security Project, for instance:

> Gone are the days when a nation's security could be measured by bombers and battleships. Security in this new era requires harnessing all of America's strengths: the force of our diplomacy; the might of our military; the vigor and competitiveness of our economy; and the power of our ideals.[17]

A different spin comes from the left-of-center American Civil Liberties Union's National Security Project:

Our Constitution, laws, and values are the foundation of our strength and security. Yet, after the attacks of September 11, 2001, our government engaged in systematic policies of torture, targeted killing, indefinite detention, mass surveillance, and religious discrimination. It violated the law, eroded many of our most cherished values, and made us less free and less safe. . . . We work to ensure that the U.S. government renounces policies and practices that disregard due process, enshrine discrimination, and turn everyone into a suspect. We also seek accountability and redress for the victims of abuses perpetrated in the name of our national security. These are the ways to rebuild American moral authority and credibility both at home and abroad.[18]

The federal government's National Security Agency displays an extremely general motto on its home page — "Defending Our Nation. Securing the Future." Its Trump-era mission statement owes more to defense jargon than to political philosophy:

The National Security Agency/Central Security Service (NSA/ CSS) leads the U.S. Government in cryptology that encompasses both Signals Intelligence (SIGINT) and Information Assurance (IA) products and services, and enables Computer Network Operations (CNO) in order to gain a decision advantage for the Nation and our allies under all circumstances.[19]

Speaking of the National Security Agency, its most famous whistleblower, Edward Snowden, is far more sympathetic to the ACLU's vision than to that of his employers. Rather than invoking national security, to which he was quickly and widely accused of doing irreparable harm, he invokes the public interest — not the freedom of government to engage in massive, blanket surveillance of individuals for the ostensible goal of national security, but rather the right of individuals to know, debate, understand, and meaningfully consent to the actions of their government.[20]

Taking yet a different tack, the progressive Massachusetts-based National Priorities Project looks at national security in terms of the costs of various components and different viewpoints, noting that in 2016 US taxpayers shelled out, per hour, $57.52 million for the Department of Defense while spending $11.64 million on education and $2.95 million on the environment.[21]

Move outward from nation to globe. On the plane of raw human survival, scientists have cited overuse of antibiotics and the resultant increase in highly resistant microbes as a threat to national and ultimately global security, while the Pentagon, along with the

United Nations and scientists across the globe, has identified climate change as a parallel threat — a trigger for regional conflicts over freshwater, food, and refugees; a condition leading to drought, wildfires, and pandemics; and a cause of rising sea levels, which in turn would redraw coastlines and submerge low-lying countries.[22] The European Union contends that in the current era of "multifaceted, interrelated and transnational threats . . . the internal and external aspects of security are inextricably linked."[23]

By any definition, for any individual and any nation rich or poor, security — in the simplest sense of safety — is a central, if not the central, concern. Survival is merely step one. Beyond that, at the very least, are freedom from fear and freedom from want. On any scale — individual, familial, societal, national, or global — security also requires practices that are viable for the long term. In a technologically advanced world, an insufficiency of food, water, or education creates unviable, unsustainable conditions. Ultimately, security on the broadest scale is unachievable without an embrace of multilateral coexistence. From a couple of hundred miles up in space, after all, every nation is a landmass among landmasses — a collage, like the others, of green, brown, blue, and diminishing splashes of white — signaling the oneness of Earth and the inescapable togetherness of its inhabitants. It is a signal easily picked up by astronauts.[24]

In the years that followed 9/11, I was running New York's Hayden Planetarium, serving on a presidential commission charged with bolstering the prospects of the US aerospace industry, writing a monthly column for *Natural History* magazine, and scrambling to complete the unrealistic number of other projects I'd taken on. One of my newer commitments was serving on the board of the Colorado-based Space Foundation.

The 1983 charter of the Space Foundation, a not-for-profit advocacy group, has a noble ring:

> [T]o foster, develop, and promote, among the citizens of the United States of America and among other people of the world . . . a greater understanding and awareness . . . of the practical and theoretical utilization of space . . . for the benefit of civilization and the fostering of a peaceful and prosperous world.[25]

Two key pieces of the foundation's work, directed at anybody and everybody who conducts business in space, are the glossy, infocrammed annual publication *The Space Report: The Authoritative Guide to Global Space Activity* and the Space Foundation

Index of about thirty publicly traded companies. But the foundation's longest and liveliest commitment is its annual broad-spectrum conference: the giant, jam-packed, three-decade-old National Space Symposium.[26]

The first symposium I attended as a member of the board was the Space Foundation's nineteenth, held April 7–10, 2003. As usual, the venue was the venerable Broadmoor Hotel and Resort in Colorado Springs, with its acres of open, high-ceilinged display halls in which corporations, government agencies, branches of the military, and merchants display their aerospace wares in booths commonly staffed by attractive young women. Colorado Springs is a sunny, mid-sized, friendly city that happens to be home to a stunning battery of military entities, including the Peterson Air Force Base, the Schriever Air Force Base, the Cheyenne Mountain Air Force Station, the North American Aerospace Defense Command (NORAD), Fort Carson, the US Air Force Academy, the US Northern Command, the Air Force Space Command, the US Army Space and Missile Defense Command/ Army Forces Strategic Command, the Missile Defense Integration and Operations Center, the Joint Functional Component Command for Integrated Missile Defense, the 21st Space Wing, the 50th Space Wing, the 302nd Airlift Wing, the 310th Space Wing, and the National Security

Space Institute. It also hosts the offices or headquarters of more than a hundred aerospace and defense contractors, including giants like Ball Aerospace and Technologies, Boeing, Lockheed Martin, Northrop Grumman, and Raytheon. The region further boasts three universities with graduate programs in space sciences, and, not surprisingly, hosts the headquarters of the Space Foundation — all this in a state that ranks twenty-second in population but bobs annually between first and third in total aerospace activity.

Just three weeks before the start of the 2003 conference, the second President Bush had announced from the Oval Office the "decapitation attack" that launched Operation Iraqi Freedom, assuring the world that it would not be "a campaign of half measures" and that "no outcome except victory" would be acceptable.[27]

Typically, registrants at the National Space Symposium include Air Force generals, corporate executives, heads of space centers, and administrators of NASA and other government agencies. You'll also find engineers, entrepreneurs, inventors, investors, flyboys, space weapons traders, communications specialists, space tourism mavens, and the occasional astrophysicist, as well as selected members of Congress, representatives of state government, and diplomats and scientists from the ever-growing international community of spacefaring nations. There are students. There are teachers. Most of the registrants are men. That year, many of the five thousand people gathered at the Broadmoor had some professional link to Operation Iraqi Freedom. The symposium's organizers, in fact, had worried that the long list of military officers slated to give plenary talks would be called to war, precluding any trip to the symposium and any talk about wars yet to come. Yet they showed up in higher numbers than ever: 20 percent higher than the year before.[28] Rather than thinking of the several-day symposium as a place that would take them away from their space business, everybody presumed it would be the best place in the world to conduct it — and they were right.

Anybody who needs to hear about US space assets, or state-of-the-art communications, or the future of war; any general who needs to know how corporate R & D might influence the warfighter's vision of spaceborne weapons; any industry manager who needs to know what's in the latest vision statement drawn up by military strategists: they're all there, in the same place at the same time. Although academic scientists are a far less prominent part of the mix, it has long been clear to me that the space research my colleagues and I conduct plugs firmly and fundamentally into the nation's military might.

But not everyone on the Broadmoor's turf was enthusiastic about US military control of space. On a brief stroll from the lovely grand hotel to the brand-new conference center the first morning of the symposium, I found myself facing a dozen protesters denouncing the conference as a weapons bazaar. I'm not a fan of war. I'm somebody who imagined the naked, napalmed Vietnamese girl child running from the bathroom of my apartment. And yet, face-to-face with the protestors that day in Colorado Springs, in a shift of heart and mind I could not have foreseen, I suddenly felt I was confronting "them."

Yes, Boeing makes thermal-kill antimissile systems. Yes, Lockheed Martin makes laser-guided missiles, Northrop Grumman makes kinetic-kill missile interceptors, Raytheon makes cruise missiles, and General Dynamics makes the guidance and weapon-control systems for nuclear-armed ballistic missiles. They all make weapons that break stuff and kill people. Some are ground-based; some are aircraft-based; others are space-based. Yes, in most directions you cared to turn at the National Space Symposium in 2003, space-inspired arms trading was going on. But to me the conference was primarily about peaceful things — cosmic things — and so I wasn't ready to paint the entire enterprise as evil just because it facilitated a bit of arms trading on the side. I told myself that accountability lay with the voters and their elected officials, not with the corporations.

Treating my new vision as though it was a long-held personal conviction, I inwardly labeled the protesters as politically naive, as well as ungrateful to the defenders of the freedoms they took for granted. With a tinge of indignation, I stepped across their phalanx and walked into the events center.

The banquet hall, repurposed daily for all plenary talks, is so large you can barely see the speaker's podium from the back rows. The ceiling is high, the thousands of red-upholstered chairs sturdy, the red-flowered azure carpeting thick. The backdrop of the stage looks like the cockpit of a spacecraft. Jumbo video screens hang along each side of the room, about halfway to the back row, so that each of the thousands of attendees can get a close view of the speakers and panelists.

General Lance W. Lord, the tall, calm, affable man who headed the Air Force Space Command, delivered the keynote address. "If you don't have a dream, you can't have a dream come true," he declared, managing to echo both *South Pacific* and the Cold War. "If you're not in space, you are not in the race."[29] Offstage, General Lord offered an avuncular handclasp and an engaging friendliness, at odds with both his central-casting name and my Vietnam-era stereotype of the warmongering commander.

During a break in the proceedings, I opened my laptop and started perusing my email. But my mind, like the minds of everyone around me, was preoccupied with the war. The battle for Baghdad began on April 5.[30] The first American transport plane carrying troops and equipment landed at the Baghdad airport on April 6. American troops took over Saddam Hussein's main presidential palace on April 7, the opening day of the symposium. A gigantic Milstar satellite, the fifth of a constellation of five satellites geared for military communications, had been launched that very morning from Cape Canaveral, applauded at liftoff by General Lord. The American campaign of "shock and awe" seemed to be succeeding. And hanging in the sky was a different sort of Moon from the one that witnessed the start of the Gulf War. This phase was visible, a waxing crescent Moon. The coalition forces did not require the cover of a dark, moonless night, because the assault on Baghdad did not rely on airborne stealth. Much of the assault depended on infantry, tanks, and armored personnel carriers, invading by ground.

Suddenly the informational PowerPoint slides that ordinarily fill the hall's large display screens during breaks were replaced by war coverage from CNN: Operation Iraqi Freedom, live and in color. Intense fighting was taking place in the center of Baghdad. The office of the news agency Al Jazeera had been bombed. The Palestine Hotel, preferred habitat of the international media, had been bombed. Tank-buster jets were hitting Iraqi positions on a bridge over the Tigris River. Helicopter gunships were pummeling a compound thought to be used by the Republican Guard. British troops were gaining control of Basra, Iraq's second-largest city. On-screen, reporters and anchors and spokespersons and generals described the weaponry and announced the names of the corporations that manufactured it — the same names highlighted in the symposium's display booths and printed on the badges of the people surrounding me. And every time a corporation was identified as the producer of a particular instrument of destruction, its employees and executives in the audience broke into applause.

Up to that point I'd felt okay. But now I was anguished. Once again, America was invading the soil of a sovereign nation that hadn't attacked us. In video games you're expected to cheer when you destroy your virtual targets and proceed to the next level. But it's hard to accept that kind of behavior when your targets are real. People die when a Boeing B-1B Lancer drops a quartet of

GBU-31 bombs on a Baghdad restaurant that Saddam Hussein has reportedly entered. People are killed when a barrage of Lockheed Martin AGM-114 Hellfire missiles strikes a convoy suspected to include Saddam Hussein.

Blinking back tears and fighting to keep my composure, I thought about leaving the conference. I began to choreograph my resignation from the board of the Space Foundation. But at the same time I felt I couldn't just walk out of the sanctum of war and put my head in the sand. It's better to see than not to see, I said to myself. It's better to know than not to know, better to understand than not to understand. Then and there I grasped the unattractive, undeniable fact that without the Space Symposium, without the many symposia like it, without all its predecessors and counterparts across culture and time, without the power sought by its participants — both for themselves and for the nations they represent — and without the tandem investments in technology fostered by that quest for power, there would be no astronomy, no astrophysics, no astronauts, no exploration of the solar system, and barely any comprehension of the cosmos.

So I stayed put, and decided to explore other ways to reconcile my emotions with the histories, contradictions, priorities, and possibilities inherent in that day.

The universe is both the ultimate frontier and the highest of high grounds. Shared by both space scientists and space warriors, it's a laboratory for one and a battlefield for the other. The explorer wants to understand it; the soldier wants to dominate it. But without the right technology — which is more or less the same technology for both parties — nobody can get to it, operate in it, scrutinize it, dominate it, or use it to their advantage and someone else's disadvantage. Absent that technology, neither side can achieve its ends. In the words of the Rumsfeld Commission report, "The U.S. will not remain the world's leading space-faring nation by relying on yesterday's technology to meet today's requirements at tomorrow's prices."[31] The technology all sides seek is both cutting-edge and potentially dual-use.

So whether you're an astrophysicist peaceably seeking a good look at Saturn's rings or an army general aggressively seeking high-resolution satellite information about a bunker inside a mountain, you're dependent on the same pool of engineers. Some of them work or consult for corporations; some of them teach at universities; some of them do both. Most of the contracts they work under, as well as most of the contracts for celebrated space-science projects, are funded with taxpayer dollars. NASA is a major funder of space research in academia, and most of the leading corporations contract to NASA in one way or another. The contract could come from a range of sources: Air Force Space Command, the Defense Advanced Research Projects Agency, the Air Force Research Laboratory, the National Reconnaissance Office, the National Aeronautics and Space Administration, a private company such as SpaceX. Doesn't matter much. Whoever puts up the contract, both space scientists and space warriors will utilize the results.

Where do these aerospace engineers, astrophysicists, physicists, and computer geniuses come from, and where do they end up working? How do those agencies and companies attract them? Probably not by talking up their missile-defense contracts. These people want to do science. They want to do space.

Look at a recent *Space Report* from the Space Foundation or a *Science and Engineering Indicators* from the National Science Foundation, and you'll see some eye-popping statistical trends: Knowledge-based and tech-intensive industries in the United States account for almost 40 percent of gross domestic product, the highest in the developed world, rendering this country's economy deeply dependent on a highly educated workforce. Employment in science and engineering as a percentage of total employment doubled between 1960 and 2013, and immigrants formed a substantial part of that doubling. While the century-long average of US residents who are immigrants is about 10 percent, 33 percent of all the US Nobel Prizes in the sciences have gone to immigrants. Almost half the workers with doctorates in the physical sciences, and more than half with doctorates in computer science, mathematics, or engineering, were born outside the United States. Meanwhile, other countries — most notably China and India — are now far outpacing the United States in the number of first university degrees awarded in these fields, and the fraction of foreign-born US graduate students will diminish, since many of those other countries have been assiduously building their own academic infrastructure. At the rate we're going, the United States will soon cease to be a prime destination of aspiring young scientists from "developing" nations — further starving our dreams of reclaimed greatness. What will happen when the ramifications of both xenophobia and the contraction of public support for higher education play out over the medium to long term? And simply tallying the numbers of students in graduate school doesn't capture the full challenge. When it comes specifically to space, the civilian US workforce has been shrinking every year for the past decade or so, even as global space activity has been soaring. Japan's civilian space workforce has increased two-thirds from its low point in 2008, and the European Union's has increased by a third.[32]

No matter the absolute numbers in the United States, the PhD astrophysicist has few problems finding work. Astrophysicists are expert coders and trained problem solvers. We are fluent in multiple computer languages and comfortable with the analysis of large quantities of data, besides having a facility with mathematics that exceeds what most job descriptions demand. Those who don't become professors or educators get snatched up by Wall Street, by NASA, by the US Department of Energy, by any of several branches of the US Department of Defense, or by the information technology or aerospace industries — out-earning their academic counterparts at every step.

One company they might decide to join is Northrop Grumman. Its legacy runs deep. In the 1960s, the Grumman part of that merger built the module that delivered all moonwalking astronauts to the lunar surface. And from 2012 through 2016, Northrop Grumman fared significantly better than the S&P 500 Index. Government contracts, primarily from America's Department of Defense and intelligence agencies, are its bread and butter. During the three-year period 2014–2016, the

US government accounted for five-sixths of its total sales. Most of the remaining sixth comprised foreign military sales, contracted through the US government. Although Northrop Grumman characterizes itself as a "leading global security company" and foregrounds its military work — "We provide products, systems and solutions in autonomous systems; cyber; command, control, communications and computers, intelligence, surveillance, and reconnaissance (C4ISR); strike; and logistics and modernization" — it is also the prime contractor for NASA's James Webb Space Telescope, a state-of-the-art, state-of-the-science infrared observatory designed to orbit the Sun a million miles from Earth as, alongside other goals, it tracks the birth of galaxies in the early universe. Conceived in 1996 as the follow-on to Hubble, the Webb telescope, named for the administrator of NASA during much of the Apollo era, represents a total cost of about $9 billion — about $375 million per year when spread from birth to launch. Feels like a lot, but it's less than 2 percent of Northrop Grumman's total annual sales.

No doubt a freshly minted astrophysicist would be thrilled to work on the Webb telescope itself or on the company's petal-shaped Starshade screen, which is designed to fly thousands of miles in front of any space telescope, blocking the light of various stars so that their planetary systems can be studied. Plus, working in industry means a higher salary than a university can offer. So the attraction is strong. But, having joined the 67,000 other employees at Northrop Grumman, our starry-eyed scientist might instead be deployed to military-related aerospace projects: radar arrays, multi-spectral hi-res imaging, ballistic missile defense, high-energy lasers, EHF (extremely high frequency) protected communications systems, space-based infrared surveillance, maybe even stealth bombers.[33] Space exploration may pull in the talent, but war pays the bills.

On April 14, 2003, the day the US military established a seemingly firm hold on the hometown of Saddam Hussein, the director of operations for the Joint Chiefs of Staff said, "I would anticipate the major combat engagements are over." Operation Iraqi Freedom, a twenty-seven-day siege, was ending. Two weeks later, clad in a strappy flight suit and speaking from the flight deck of the aircraft carrier USS *Abraham Lincoln* at sea near San Diego, California, President Bush echoed the Joint Chiefs' assessment:

In the Battle of Iraq, the United States and our allies have prevailed. . . . In this battle, we have fought for the cause of liberty, and for the peace of the world. . . . With new

51

tactics and precision weapons, we can achieve military objectives without directing violence against civilians. . . . The war on terror is not over, yet it is not endless. We do not know the day of final victory, but we have seen the turning of the tide. No act of the terrorists will change our purpose, or weaken our resolve, or alter their fate. Their cause is lost. Free nations will press on to victory.

Other nations in history have fought in foreign lands and remained to occupy and exploit. Americans, following a battle, want nothing more than to return home.[34]

As we now know, the terrorists' cause was not lost, and the United States had not prevailed. Much violence was yet to be directed against civilians. Nor was returning home on that year's, or that decade's, agenda. But occupation was.

Three Aprils later, in 2006, as the third anniversary of Bush's premature declaration approached, I attended another National Space Symposium at the Broadmoor. Conditions in Iraq were dire: civil strife, almost-daily insurgent attacks, the infrastructure still in wreckage, the prospects for a stable and effective national government still teetering despite an encouraging turnout for national elections some two months earlier. Many American troops were on their third or fourth tour of duty.[35] Some 100,000 well-paid private contractors — supplied by firms such as Halliburton and Blackwater USA — were working alongside active-duty troops at the rate of almost 1:1.[36] US military deaths had exceeded 2,000; Iraqi civilian deaths had exceeded 30,000 by some accounts and 600,000 by others. The official number of US military wounded was nearing 20,000.[37] A half dozen retired generals were about to come out publicly against the Bush administration's handling of the war. And the money was flowing: the direct costs of war in 2006 approached $10 billion a month, while the full costs of war exceeded $15 billion a month.[38]

Business boomed at the symposium, with more than seven thousand participants and more than a hundred companies and space-related organizations showcasing their wares, sprawled across far more square footage than they had occupied at the exhibit center the previous year. The theme was "One Industry — Go for Launch!" a phrase that suggests the meshing of military, scientific, technological, corporate, and political interests into a unified command structure. A post-conference news release featured a rave review from Elliot Pulham, the Space Foundation's late president and CEO: "There was an excitement and sense of industry unity in the air. . . . With the civil, commercial, national security and entrepreneurial space community all converging from across the nation and around the world, there was a feeling that humanity is poised to take the next bold steps in the greatest adventure of all."[39] Excitement, industry unity, humanity poised for boldness and adventure. But was all of humanity equally poised?

Three news releases that appeared in my in-box later that April point to some answers. One was a policy bulletin from the American Institute of Physics about a hearing in late March of the Science, State, Justice, and Commerce Subcommittee of the US House of Representatives. The bulletin summarizes statements by Michael Griffin, then the NASA administrator, on the difficulties of apportioning available funds in a way that would satisfy everyone, and quotes his contention that the capability to put people into space is one of the things that "define a nation as a superpower."[40] In other words, mastery in space strongly correlates with mastery on Earth.

You don't have to be head of NASA to hold that view, of course; policymakers in the world's most populous nation think that way too. In 2003, China became the third nation to send humans into space all on its own. By late 2005 China was regarded as the likeliest imminent superpower by two-thirds of the American electorate.[41] And during the National Space Symposium of April 2006 at the Broadmoor, Griffin's counterpart at the China National Space Administration delivered a quite stunning list of his agency's achievements, projects, and goals to a huge and extremely attentive crowd that looked like the very embodiment of "industry unity" — lots of blue uniforms and good suits, with a scattering of low-fashion engineers — except that this crowd seemed to be radiating consternation rather than excitement.

"Industry unity," by the way, is simply a suaver way of referring to the emergent space-industrial complex, akin to the oft-referenced military-industrial complex named by General Dwight D. Eisenhower in his presidential exit speech. Consider, then, the second of my April 2006 emails. In this one, Michael Griffin announces the appointment of Simon P. "Pete" Worden, a retired US Air Force brigadier general and research professor of astronomy at the University of Arizona, as the next director of NASA's Ames Research Center in California.[42] Worden is pedigreed: a PhD in astronomy; a past director of the Air Force Space Command; a commander of the 50th Space Wing of the Air Force Space Command; an official with the Strategic Defense Initiative Organization, home of "Star Wars" missile defense; and director of the 1994 Clementine lunar probe, a collaboration between NASA and the

Department of Defense. This is industry unity rolled into a single person: seamless transitions and frictionless flow between power and war and space science.

But there may not be enough Pete Wordens out there to keep the pipeline pumped. Consider the third of my April emails worth sharing. This one, a news release from the University of Arizona distributed through the news portal of the American Astronomical Society, summarized a survey of US-based planetary scientists' opinions concerning priorities for exploration of the solar system. More than a thousand people responded — fully half the community — and their view was overwhelmingly that research and analysis are far more important than missions. Also included was commentary from the director of UA's Lunar and Planetary Laboratory, who was worried about something even more basic than basic research. Based on trend lines in American demography, he contended, "the real issue is that half of the American workforce will be retired 10 years from now."[43] Half of the workforce means half the astrophysicists, half the accountants, half the pharmacists, teachers, carpenters, journalists, bartenders, fishermen, auto mechanics, apple growers, rocket engineers, *everybody*. It represents a staggering breadth and depth of expertise.

So, either the United States ponies up to

encourage, educate, support, and utilize current and future scientists, or else US science evaporates, along with all the jobs, breakthroughs, space missions, discoveries, power, and money that flow from it. You can already see that evaporation if you look at the job losses. The termination of the space shuttle program had a huge impact: from a high of 32,000 in the 1990s, the shuttle workforce dropped to 6,000 by mid-2011 (though some people were fortunate enough to be reassigned). More generally, in terms of the early years of our no-longer-new century, core employment across America's space industry dropped from a high of 266,700 in 2006 to 216,300 in early 2016. That's a 19 percent drop in the course of a decade when, according to the Bureau of Labor Statistics, the total non-farm workforce rose by 6 percent (despite the slide/drop/slow recovery during 2008–2010). The US space employment situation looks even grimmer when contrasted to that in Europe and Japan during the same period.[44]

Today, while private, for-profit US companies perfect a space taxi that may replace the shuttle program, our uneasy partner Russia ferries America's astronauts to and from the International Space Station for a steep fee: about $71 million per seat for a round trip through 2016, increasing to $82 million under the next contract.[45] Since Russia is, for the time being, the only game in town, that price increase is not a shock. Just supply and demand at work.

Today, unless they're lucky enough to have been hired there or to have European collaborators, American particle physicists wistfully gaze across the Atlantic Ocean and over the Alps at the Large Hadron Collider near Geneva, Switzerland — the most powerful accelerator ever built, in which controlled conditions that rival the earliest high-energy moments of the Big Bang have yielded evidence of the long-sought subatomic particle called the Higgs boson. They're wistful because Europe's collider is only a fifth as powerful as America's Superconducting Super Collider would have been, had Congress not cut the entire project in 1993, a few short years after peace broke out between the United States and the Soviet Union. A tale worth telling.

In the 1970s, astrophysicists came to realize that the dense, hot conditions of the infant universe, fourteen billion years in the past, could be recreated within a particle accelerator. The higher the energy attained by the accelerator, the closer scientists could get to the Big Bang itself.

The key to attaining ever-higher energies is to generate ever-stronger magnetic fields, which accelerate charged particles to stupendously high speeds. The ring of the accelerator becomes a particle racetrack. Slam the particles into one another from opposite directions, and brand-new particles are birthed — some predicted, others unimagined. By the 1980s, the introduction of superconducting materials enabled accelerators to generate significantly stronger magnetic fields and thus even more wildly energetic collisions.

Currently the US Department of Energy controls seventeen national science laboratories. Often aligned with universities, some have particle accelerators, each more powerful than its predecessors. Notable labs on this list include Stanford's SLAC National Accelerator Lab in California and the Fermi National Accelerator Lab in Illinois, as well as Lawrence Berkeley National Lab, managed by the University of California; Oak Ridge National Lab, managed by the University of Tennessee; and Brookhaven National Lab in New York, associated with Stony Brook University. These institutions employ bevies of engineers and teams of high-energy particle physicists hunting for the fundamental structure of matter.

Has the United States funded and engaged in this kind of research purely for the sake of discovery? Hardly. Most American accelerators were built during the Cold War, when the particle physicist was a vital resource for increasing the lethality of nuclear weapons.

That's how astrophysics — specifically astro-particle physics, a branch of cosmology — became an auxiliary beneficiary of Cold War science priorities. Astrophysics and the military are conjoined whether their shared buoy bobs up or down in the tidewaters of politics.

In the fall of 1987, past the midway point of his second term, President Ronald Reagan approved construction of the Superconducting Super Collider (SSC), which the chair of the House Science, Space, and Technology Committee called "the biggest public works project in the history of the United States." Fifty-four miles in circumference, the SSC needed a big enough state with large swaths of relatively unpopulated areas whose deep-down geology would tolerate tunneling. Texas — specifically, the town of Waxahachie, situated above the geological formation known as the Austin Chalk — won the eight-state competition. At twenty times the energy of any previous or planned collider in the world, the SSC would be an engineering marvel, assuring American leadership in particle physics for decades to come. And with an initial price tag of $4.4 billion, it would be the most expensive accelerator ever built.[46]

Two years later, the Berlin Wall came down; two years after that, the Soviet Union dissolved. Cold War funding enthusiasm evaporated. By February 1993 the US General Accounting Office had prepared a document for Congress titled simply "Super Collider Is Over Budget and Behind Schedule."[47] In June 1993, project managers were called in front of the House Subcommittee on Oversight and Investigations, not to defend the value of the collider in terms of its contributions to the frontier of physics but, much more important to the members of the subcommittee, to defend it against detailed charges of mismanagement.[48] In peacetime, cost overruns and poor administration were seen as fatal blows to the project rather than as the normal speed bumps of creating something never before created.

Congress's October 1993 decision to cancel funding for the SSC, two years after construction began, did not explicitly say, "We won the Cold War, so we don't need physicists and their expensive toys anymore." Rather, the grounds cited were cost overruns and shifting national priorities. Besides, at a comparable price tag, Texas was getting the new space station, headquartered at NASA's Johnson Space Center in Houston. Two major, congressionally approved projects in one state during peacetime was a hard sell.[49]

Pushed aside during all this oversight were the cosmologists — hidden casualties of peace. Our species' understanding of the greatest explosion ever, the event that created the universe itself, was thwarted because a

half-century-long standoff that held humanity hostage to the world's most explosive
weapons had ended.

Just because America pulls the plug on a
scientific project doesn't mean research, planning, and hoping come to a screeching halt
elsewhere in the world. Other nations, developed and developing, have begun to pick up
where the United States has left off. Leading
the way is China, contributing more than 31
percent between 2000 and 2015 to the
growth of worldwide research and development spending, while the United States
contributed 19 percent.[50]

Buffed by the ambitions, creativity, and
war-inspired innovations of the twentieth
century, America's high-gloss technological
and scientific self-image blinds us to the
Dorian Gray reality of our times. Hierarchies
have been reordered before. It's happened in
art, in commerce, in exploration, in sports.
Why shouldn't it happen in space? Some
commentators presume it's already happened, and that henceforth America will aim
no higher than the creation and aggressive
marketing of minor consumer products that
replace similar, and perfectly satisfactory,
consumer products. "America may be losing
a competitive edge in many enterprises, from
cars to space," riffed National Public Radio
host Scott Simon in the summer of 2010,
"but as long as we can devise a five-bladed,

mineral-oil-saturated razor, we face the future well-shaved."[51]

That the United States could someday be a secondary, supplicant nation, begging for a seat at Europe's or China's decision table, is not the America that most Americans want. To the patriot, the thought is repellent.[52] To the policymaker, it's frightening. To the student, it's deflating. The February 2001 report of the Hart–Rudman Commission on national security, for instance, minces no words on the matter:

Second only to a weapon of mass destruction detonating in an American city, we can think of nothing more dangerous than a failure to manage properly science, technology, and education for the common good over the next quarter century. . . . America's international reputation, and therefore a significant aspect of its global influence, depends on its reputation for excellence in these areas. U.S. performance is not keeping up with its reputation. Other countries are striving hard, and with discipline they will outstrip us.

This is not a matter merely of national pride or international image. It is an issue of fundamental importance to national security. . . . Complacency with our current achievement of national wealth and international power will put all of this at risk.[53]

Half a decade later, the *Competitiveness Index: Where America Stands,* a twenty-year overview published by the Council on Competitiveness, also sounded the alarm. While noting America's position as the world's largest economy and its responsibility for one-third of global economic growth from 1986 to 2005, among other achievements, the report marshals reams of statistics showing that the future might not be so bright as the past. "America Still Leads the World in Science and Technology, But That Lead Is Narrowing," reads one heading. Underneath it, bar graphs track the two-decade decline in the US share of total global activity in many categories, ranging from degrees awarded in science and engineering — a ten-point drop in bachelor's degrees, a thirty-point drop in doctoral degrees — to the twelve-point drop in scientific researchers.[54]

As for the prospect of enduring American dominance in space, Joan Johnson-Freese, professor of national security studies at the US Naval War College, foresees only uncertainty. "There is no magic solution, no sudden discovery of warp drives or phaser beams or ion cannons, that will get us to such a secure future."[55]

No question, America's space program has had some major recent successes. But so have the space programs of China, India, Canada, and South Korea. The European Space Agency, Russia's Roskosmos, and Japan's JAXA are central to the collective space endeavor. Space programs have been operating for several decades in Azerbaijan, Bulgaria, Egypt, Israel, Indonesia, North Korea, Pakistan, Peru, Turkey, Uruguay, and most countries in Western Europe. Bahrain, Bolivia, Costa Rica, Mexico, New Zealand, Poland, South Africa, Turkmenistan, and the United Arab Emirates joined the list during the second decade of the current century. Australia and Sri Lanka will soon join as well. Altogether there are now more than seventy government-run space agencies. Some four dozen countries operate satellites. More than a dozen have launch facilities.

Intensive and successful, the Chinese space program is readily comparable to those of the United States and the Soviet Union in their better years. On January 11, 2007, when it sent a kinetic-kill vehicle more than eight hundred kilometers into space to destroy one of its own aged weather satellites in a direct hit, China in effect announced its status as a space power with potentially lethal capabilities. It could now deny another country freedom of operation in space.

The hit put tens of thousands of long-lived fragments into high Earth orbit, adding to the already considerable dangers posed by debris previously generated by other countries, notably ours. China was roundly criticized by other spacefaring nations for making such a mess; twelve days later, its foreign ministry declared that the action "was not directed at any country and does not constitute a threat to any country." Hmm. That's a little like saying the Soviet Union's launch of the world's first satellite, Sputnik, in October 1957 was not a threat — even though Sputnik's booster rocket was an intercontinental ballistic missile, even though Cold Warriors had been thirsting for a space-based reconnaissance vehicle since the end of World War II, even though postwar Soviet rocket research had been focusing on the delivery of a nuclear bomb across the Pacific, and even though Sputnik's peacefully pulsing radio transmitter was sitting where a nuclear warhead would otherwise have been.

Of course, among the many implications of China's successful kill, one seemed unmissable: a US spy satellite or a bit of US missile-defense hardware orbiting at the same altitude could just as easily have been the target. General T. Michael Moseley, the US Air Force Chief of Staff, called the Chinese achievement a "strategically dislocating event." If you can hit a six-foot-long object at five hundred miles, he said, you can "certainly hit something out beyond 20,000 miles. It's just a physics problem."[56]

Since then, space has only become more crowded, militarized, and globalized. More dislocation seems inevitable, and more co-operation essential.

Nearly three thousand years ago, architects, stonemasons, sculptors, and slaves built a breathtaking palace complex for the Assyrian ruler Ashurnasirpal II in the ancient city of Kalhu, some three hundred kilometers north of modern-day Baghdad. Wall panels from the Northwest Palace now hang in the British Museum in London and the Metropolitan Museum of Art in New York City. Carved in bas-relief on the panels are muscled archers, charging chariots, stricken lions, supplicants bearing tribute, and other iterations of victory. Across the center of each panel in the British Museum runs a cuneiform text, the so-called Standard Inscription of Ashurnasirpal, proclaiming the ruler's invincibility:

> . . . the great King, the mighty King, King of Assyria; the valiant man, who acts with the support of Ashur, his lord, and has no equal among the princes of the four quarters of the world; . . . the King who makes those who are not subject to him submissive; who has subjugated all mankind; the mighty warrior who treads on the neck of his enemies, tramples down all foes, and shatters the forces of the proud; the King who acts with

the support of the great gods, and whose hand has conquered all lands, who has subjugated all the mountains and received their tribute, taking hostages and establishing his power over all countries . . .[57]

Invincibility does have its limits, however. Ashurnasirpal II's kin ruled northern Mesopotamia for two centuries. The Assyrian empire and the great palaces at Kalhu, later called Nimrud, endured for a century more. Today Nimrud's glory exists only within the walls of Western museums. In 2007, with the onset of the US "troop surge," the Iraqi city of Mosul, Nimrud's nearest neighbor, became a place where gunmen would fire on a wedding procession and nine unidentified bodies could be delivered to the morgue in a single day. In 2014, with the advance of the Islamic State of Iraq and Greater Syria, Mosul became a ruin from which the inhabitants had to flee and where the US-trained Iraqi army scattered like powder in the wind. By July 2017, when Iraq's prime minister arrived in the broken city to declare victory over ISIS, *New York Times* headlines were announcing, "Civilians Emerge from Mosul's Rubble Starving, Injured and Traumatized" and "Basic Infrastructure Repair in Mosul Will Cost Over $1 Billion: U.N." And the archaeological treasures that ISIS had not already looted had mostly been smashed to bits.

The praise of Ashurnasirpal was praise of empire. The inscription paints him as "an uncommanded commander," to borrow a phrase from John Horace Parry, the distinguished British historian of European empire. In 1971 Parry noted that in the second half of the twentieth century, "some major western states, notably the United States of America, whose political traditions included a profound suspicion of imperialism, found themselves drawn, with much misgiving, into widespread ventures and responsibilities of a quasi-imperial kind."[58] Yet as the century drew to a close, many political thinkers and commanders relinquished their misgivings, in effect dropping the "quasi-" from "imperial." They acquired the habit of proclaiming America's capacity to subjugate its foes on Earth and broadcasting its intentions to suppress them in space. Parry might have applied the word "dominion," the original meaning of *imperium*.

Today it is mostly fantasists who praise empire, and video gamers who hunger for it. The late American political essayist and novelist Gore Vidal, a witty lefty patrician, made American empire a prime target, evidenced in titles such as *The Last Empire* (2001) and *Imperial America: Reflections of the United States of Amnesia* (2004). The late political scientist Chalmers Johnson, an East Asia specialist who gave us the useful term "blowback," is another writer whose works evinced disillusionment with the course of US foreign policy: *The Sorrows of Empire* (2004), *Dismantling the Empire* (2010). J. M. Coetzee, a South African writer who won the Nobel Prize for Literature in 2003, presented an especially tragic view of empire in his 1980 novel *Waiting for the Barbarians*. Near the end of the novel the main character, a discredited petty functionary who once oversaw a nameless walled outpost, accuses empire of preoccupation with a single thought: "how not to end, how not to die, how to prolong its era."[59]

In a less literary portrayal of the workings of empire, Pulitzer Prize–winning journalist Ron Suskind recounted his meeting in 2002 with a senior advisor to George W. Bush, who upbraided the author for a recent article and then dismissed his work in a way that, as Suskind realized only later, captured "the very heart of the Bush presidency":

> The aide said that guys like me were "in what we call the reality-based community," which he defined as people who "believe that solutions emerge from your judicious study of discernible reality." I nodded and murmured something about enlightenment

principles and empiricism. He cut me off. "That's not the way the world really works anymore," he continued. "We're an empire now, and when we act, we create our own reality. And while you're studying that reality — judiciously, as you will — we'll act again, creating other new realities, which you can study too, and that's how things will sort out. We're history's actors . . . and you, all of you, will be left to just study what we do."[60]

Despite the bluster, Americans might be going the way of the Assyrians, not to speak of the Romans or the Maya or the Ottomans. By the end of the twenty-first century's opening decade, references to the disappearing American empire had become commonplace in platforms across much of the political spectrum. *New York Times* oped columnist Maureen Dowd colorfully observed on October 11, 2008, following the Dow's worst week of its life till then: "With modernity crumbling, our thoughts turn to antiquity. The decline and fall of the American Empire echoes the experience of the Romans, who also tumbled into the trap of becoming overleveraged empire hussies."[61] Long before the word "empire" began popping up in the *Times,* journalists, scholars, Baghdad bureau chiefs, former CIA higher-ups, counterterrorism experts, historians, and political commentators of every stripe began to sprinkle their writing and their titles with it, occasionally linking it with "hubris." So pervasive had these terms become that in mid-2008, a Yahoo! stock blog depicted investors as wondering "if the dollar's swoon signals a much longer-term displacing of the global American empire" and proposed "Portfolio Adjustments for the End of the American Empire."[62]

What does it take to build an empire? What resources must be consumed to sustain one? Why do some people hunger for power while others shun it? What nations, if any, must be inculcated or invaded by other nations to achieve security — real or imagined? Which people, if any, must be placated or silenced to prevent uprisings? What must get broken and who must get killed to achieve these ends?

Centuries ago, Christiaan Huygens contended that war and commerce had "forc'd out . . . most of those Discoveries of which we are Masters; and almost all the secrets in experimental Knowledge." With only rare exceptions, history shows that while strategy and bravery can win a battle, the frontiers of science and technology must be exploited to win a war. Though the night sky itself is the quintessential frontier, the astrophysicist neither declares war nor makes international enemies. Countries manage that with no help from scientists. Yet for every empire the world has known, skywatchers have been in attendance, offering arcane cosmic knowledge that has been enabled by, and has also reinforced, the power wielded by leaders — leaders who sought the highest ground and judged, once again, that it was time to kill.

2
STAR POWER

Throughout much of history, knowledge of the heavens informed the rhythms of life and the mastery of territory. Astronomy moved arm in arm with agriculture, trade, migration, empire, and war. It created and marked time; it registered place on Earth. It was both a sacred mystery and a blue-chip stock. Astronomers wielded power and served the powerful.

Millennia before anybody had drawn usable maps of the continents, people memorized imagined maps of the sky. Long before there were astrolabes or sextants or precision portable clocks to establish distance, latitude, and longitude, people gauged their position with no tools but their eyes and the sky. To go where no one had gone before, to know how long it took to get there, and to return there if you liked what you found, you needed guides. The sky was a good one, especially if your path lay across uncharted ocean, unstable dunes, sweeping grasslands, or barren tundra. Heaven itself was both compass and clock, direction-finder and time-keeper. For many, it was also ultimate cause, crystal ball, and the home of deities — astronomy, astrology, history, folklore, religion, psychology, and poetry rolled into one. Knowing the rhythms of the sky was a means to knowing the character and fate of all things.

It's anybody's guess when and where a community chronicler, or maybe an insomniac, first decided to track the cycles of change in the Moon's illuminated disk, or the alternate lengthening and shortening of the Sun's arc across the sky, or the periodic comings and goings of Venus. Such tracking would have predated the first stone tools. Maybe an antecedent of *Homo sapiens* was the first to do it. Whoever it was and whenever it happened, that signaled the birth of astronomy, a source of both wonder and power for our nascent species.

Consider units of time. If the Sun never set and the Moon never waned, our measures of time might be grounded solely in biology — the beating heart, circadian rhythms, menstruation — because "periodicity is part of who we are."[1] But the Sun does set, and the Moon predictably waxes and wanes. Transitions recur endlessly in the skies above. Celestial cycles offer themselves as a natural measure of time in units we care about.

Earth's early cultures, population centers, and central governments required official methods of organizing time, especially when they needed to plan ahead. Sacrifices, festivals, planting, harvesting, tax collection, daily work shifts, and daily prayers took place at predictable intervals. In Upper Egypt, farmers needed to know when the dazzling Dog Star, Sirius — the brightest star in the night sky — would appear in the dawn sky just before the rising Sun, because that was when the Nile, too, would be rising. Hunters, gatherers, herders, and nomads also required advance planning: their lives depended on knowing when the regional waterholes would dry up, when the cattle or gazelles or bison would give birth and the eggs of the mallee hen could be stolen, when to visit the wild strawberry patches and when to dig up the yams. It was useful to know how many days' travel were needed to reach the nearest oasis. It was useful to monitor fertility. Everybody needed ways to track the passing days.

More than twenty thousand years ago, people made notches in animal bones and painted rows of dots on the walls of caves to mark the days of a lunar cycle.[2] But no round number of lunar cycles matches the duration of the solar year, a discrepancy that gave rise to continual fussing with calendars. Several early cultures followed a twelve-month year; some added the occasional thirteenth month or five-day bloc to keep things on track. Discrepancies notwithstanding, sometime around the middle of the fifth millennium BC the Egyptians counted the correct round number of whole days in a year. They also devised a 365-day solar calendar that began with the rising of Sirius on July 19, 4236 BC — possibly the earliest secure date in history.[3]

Unlike the solar day, the lunar month, the Earth year, or the other celestial cycles that our ancestors could observe, subunits of time such as the hour, the minute, and the second are a matter of cultural and mathematical taste. Sociologically, they suggest the emergence of oversight, labor, standardization, and penalty: slaves and prisoners on construction gangs, priests reciting prayers at fixed intervals, sentries posted for a fixed watch — and, more recently, trains running on time, workers punching in, and spacecraft systems synchronized for launch. On a more personal level, they suggest practicalities and annoyances such as waiting for the bread to finish baking or your mate to return home. Enter the clock, whether based on a moving shadow (the obelisk or sundial), flowing water (the clepsydra), an advancing gear, a swinging pendulum, or a transitioning electron in an atom of cesium.

The Sumerians divided the day into twelfths, and each twelfth into thirtieths. The

Egyptians divided both day and night into twelfths: voilà, the twenty-four-hour day. The Babylonians came up with the fraction-friendly sixty-minute hour and sixty-second minute. But not all units of time are as practical as the minute or the month. Plato, for instance, wrote of the "perfect year," the period necessary for all the planets to return to their initial configuration. A scheme devised by the ancient Hindus employs even vaster units, such as the *kalpa,* the length of a single day or single night in the lifetime of Brahma, who dreams the universe into existence each time he sleeps. When he awakes, the universe begins anew; 4.32 billion years later, when he next falls asleep, it vanishes. The Maya, too, formulated an overview of time based on attenuated cycles of creation; the most recent cycle expressed through their complicated "long count" began on August 12, 3114 BC.[4] Nor did such imaginative conceptions cease in the modern world. A mystical quasi-mentor of Adolf Hitler's, for instance, foretold that the 730-year "cosmic week" beginning in 1920 would, because of Jupiter's entrance into Pisces, bring about the millennarian triumph of blond Christians under the wise and genial rule of aristocrats, priests, and führers.[5]

Besides marking time, there was the challenge of mapping the sky. If Heaven was the fount of fortune and disaster, prudence demanded that the stars and the constellations they trace be demarcated and monitored. Some early Chinese astronomers divided the sky into the Five Palaces; others divided it into the Nine Fields or the twelve Earthly Branches or the twenty-eight Lunar Mansions. Early Mesopotamian astronomers divided the eastern horizon into the paths of three gods, with sixty fixed stars and constellations rising within the paths; later Mesopotamian (Babylonian) astronomers divided the sky into twelve segments, each associated with a constellation and each enclosing thirty degrees of the Sun's yearlong path across the sky — forming the now-classic twelve constellations of the Western zodiac.

Inevitably, references to the cosmos show up in the art and architecture of antiquity. Cuneiform tablets inscribed five thousand years ago in Mesopotamia mention the Bull (Taurus), the Lion (Leo), and the Scorpion (Scorpio). A tablet inscribed almost four thousand years ago in the Mesopotamian city of Nineveh lists the apparitions of Venus during the reign of King Ammisaduqa. The arched ceiling of a first-century BC Han dynasty tomb unearthed on the campus of Jiaotong University in Xi'an, China, presents a painted diagram of the heavens showing the Sun and the Moon surrounded by a circular band filled with symbolic figures representing the twenty-eight "lunar lodges" that mark the path of the Moon.[6]

Scattered across our planet are enormous stone temple ruins and looming stone monuments whose structure reveals well-established knowledge of sky patterns. In the ancient world, architecture, owing in part to the expense, labor, and time necessary for its construction, was the very embodiment of state and religious power. Among the oldest undisputed monuments with a celestial tinge are the fourth millennium BC stone "passage tombs" of County Meath, Ireland: low burial mounds where, at the winter solstice, sunlight streams through an opening above the entrance and illuminates a long passageway leading to a large chamber.[7]

Doorways and sight lines of massive stoneworks — whose manyton components were, in some cases, quarried, transported, shaped, and positioned without the aid of metal tools — align, perhaps not precisely but still convincingly, with the rising or setting Sun at the spring equinox or winter solstice, the setting full Moon at the summer solstice, the cardinal directions, or the apparitions of a planet or the never-setting Pole Star. The slew of far-flung examples include the pyramids at Giza, stone circles throughout the British Isles, roofed temple complexes in Malta, octagons in the Basque region, the Caracol at Chichén Itzá, the Templo Mayor in Mexico City, and the Thirteen Towers at Chankillo,

Peru, which consists of a row of towers strung across a ridge plus two observation structures, one to the west and one to the east. Other, more modest constructions embody the same principles: at Nabta Playa in southern Egypt, two upright stone "gates" in a small circle of sandstone slabs, akin to a small Stonehenge, align with what would have been the position of the rising Sun at summer solstice.[8]

In fits and starts, astronomy became a science. During the first millennium BC, the astronomers of Mesopotamia and China — in the service of hereditary rulers, warrior-kings, and eminent priests — compiled systematic records of what happened before their eyes and developed systems and even instruments for predicting what would happen in the future. About fifteen hundred Late Babylonian clay tablets, in the form of diaries chronicling routine observations, have been found to date. Spanning eight centuries, the tablets list such things as lunar eclipses, weather conditions, intervals between moonrise and sunrise and between sunset and moonset at different times of each month, and the changing positions of the planets in relation to thirty-one reference stars. By about 500 BC, Babylonian astronomers had devised mathematical ways of predicting the dates of new and full Moons. The world's earliest known record of a series of solar eclipses, between about 720 and 480 BC, comes from China. By 200 BC, Chinese court astronomers had begun to chronicle most celestial phenomena visible to the unaided eye, both cyclic and episodic, whether or not they understood what they saw: auroras, comets, meteors, sunspots, novas, and super-novas, as well as the paths of planets month by month. The presumed relationship be-tween the unfolding universe above and the affairs of state below rendered this record-keeping a guarded activity. In today's par-lance, it was classified research.[9]

When I was a postdoc at Princeton Univer-sity in the early 1990s, a graduate student specializing in ancient Chinese culture stopped by my office with a query about a certain historical date. Sometime around 1950 BC — he couldn't pinpoint the year — major events had taken place in China, and he suspected that some kind of sky event had preceded them. He was right.

Whipping out my planetarium sky-search software, I discovered that February 26, 1953 BC, corresponded with the tightest conjunc-tion of planets ever witnessed by civilization: Mercury, Venus, Mars, and Saturn gathered on the sky within half the area of your pinky fingernail held at arm's length (half a degree), with Jupiter two finger-widths away (four and a half degrees), creating a conjunction of all five known planets. Four days later, the very thin, waning crescent Moon would join the jamboree. All six objects were now nicely contained within the top-to-bottom area of your fist at arm's length (ten degrees). Other spaceniks with equal access to computational tools would independently discover this alignment.

Although uncertainties abound when you're trying to date events from early history, it turns out that 1953 BC just may coincide with the founding of the Xia Dynasty by its first ruler, Yu, of whom it was recorded in the *Xiaojing Gouming Jue:* "At the time of Yu the planets were stacked like strung pearls." More important, the first-century BC *Hong Fan Zhuan* ("Account of the Great Plan"), now lost, declared that a new calendar began on a spring morning in about 2000 BC during a five-planet conjunction with the new Moon. All of which makes the February 1953 BC conjunction a convincing candidate for the start date of what became the modern Chinese calendar.[10]

While the Chinese were occupied in observing and recording the behavior of objects, the Greeks were expanding astronomy's reach, making it both more conceptual, more practical, and more accessible. Empowered by geometry, they began to measure and map the universe as no civilization had done before. Triangulation, an idea set down in

Euclid's *Elements* (ca. 300 BC) as a pure mathematical statement, proved useful for estimating the distance between Earth and the Sun. Several centuries after *Elements* hit the market, an expert toolmaker — perhaps on the island of Rhodes, likely in collaboration with an astronomer — built a sophisticated calendar/ astronomical computer/ almanac/planetarium known today as the Antikythera Mechanism, perhaps the most-debated scientific object from the ancient world.

Alexander Jones, a classicist and historian of the mathematical sciences, proposes that the Antikythera Mechanism be called a cosmochronicon. Found along with other high-end cargo in a large Mediterranean shipwreck at a depth of 180 feet, and fitted with dozens of bronze gear-wheels, a hand crank, multiple dials, and multiple inscriptions, it was a shoebox-sized creation that could calculate the phases of the Moon, the changing longitudes of the Sun, Moon, and planets, the timing of eclipses, solstices, and equinoxes, and several long-term time cycles. Investigators derive its date — most likely first century BC and certainly no later than first century AD — from such factors as the Hellenistic-era vocabulary and lettering in the inscriptions, the state of astronomical knowledge incorporated in the object, and the scores of coins found nearby in the wreck. Though almost shocking in its sophistication, the Mechanism does have several known antecedents. It also has a cultural context: astronomy was treated as a topic suitable for popularization (think *Cosmos* and *Star Talk* rather than the guarded cosmic secrets of ancient imperial China), and both public and private spaces in the Mediterranean world were liberally sprinkled with astronomy-related objects, such as sundials large and small, armillary spheres, star globes, and stone tablets called parapegma, which had movable pegs that fit into holes beside each numbered day and served as public almanacs. The Antikythera Mechanism, whose complex inner workings have recently been revealed through X-ray computer tomography (CT) and whose surface details have become more legible through reflectance imaging, strikingly exemplifies the Greek concept of "uniformly flowing time that could be measured on instruments."[11]

Physics, too, now came to the fore. Ever since the second century BC, writers have been recounting the story of the Greek mathematician and military inventor Archimedes, who, they say, devised a "burning mirror" in about the year 213 to redirect and focus the Sun's rays onto a fleet of Roman ships anchored in the harbor of Syracuse, thereby, in the words of Lucian, "set[ting] ablaze the triremes of the enemy through art." But even before Archimedes did (or didn't do) it, mathematicians and engineers had begun to consider what a workable burning mirror would look like. The earliest detailed analyses concluded it would have to be concave, perhaps parabolic, and made up of an array of at least two dozen hinged, movable mirrors rather than just one. Presumably the mirrors would be large and cast of polished bronze. To this day, mechanical engineers, teenage science-fair types, and TV crews stage the occasional simulation of Archimedes's endeavor, some resulting in out-and-out failure, some in qualified success.[12]

Despite astronomy's growing practicality, celestial events could still provide a potent magical kick. Sometimes they even swayed the course of history. Rulers could be dethroned because of a comet or a supernova. Battles were launched, won, lost, or abandoned because of an eclipse. The day Odysseus rejoined his waiting and presumably widowed wife and slaughtered the hordes of suitors who had been hanging out at his house may well have coincided with a noontime eclipse in 1178 BC.[13] And Herodotus — the fifth-century BC war historian, travel writer, and investigative reporter — recounts the effect of an eclipse during the sixth year of battle between the Lydians and the Medes. The participants, he writes, were so shocked to see "day on a sudden changed into night" that both sides stopped fighting and started negotiating.[14] Modern eclipse calculations, based on celestial mechanics, yield a precise date for that armistice: May 28, 585 BC, at about 7:30 PM. While the time of an ancient event is often uncertain, its location is typically well-documented. For this reason, total solar eclipses have served as a type of laboratory, permitting a comparison between where you would have expected to see a given eclipse, based on the assumption that Earth's rotation rate has been constant over the millennia, and where the eclipse was actually observed on Earth. That these turn out to be two different locations on our planet's surface offers incontrovertible evidence that Earth's rotation rate has been slowing down, primarily due to friction from oceanic tides sloshing on our continental shelves. In modern times, this phenomenon is well-known and well-measured, which has led to the occasional addition of a "leap second" to the calendar.

While many ancient writers, no strangers to war, discussed the military advantage afforded by astronomy, Socrates discounts it. In Plato's *Republic,* written twenty-four centuries ago (before Archimedes and his mirrors), Socrates and Glaucon debate which branches of knowledge would be useful to the rulers of Athens. Socrates contends, in Book 7, that the most valuable branches "have a double use, military and philosophical" and that a firm command of arithmetic and geometry is essential both for war and for the soul. Glaucon replies that astronomy — by which he means the observation of the seasons, months, and years — is as useful to the general as to the farmer or the sailor, but Socrates doesn't agree. To him, astronomy is too wedded to observation, too dependent on the senses, and therefore antithetical to noble philosophy.

Two centuries later, in the section of his *Histories* titled "On the Art of a Commander,"[15] the Greek politician-historian Polybius ranks astronomy up alongside geometry. Elaborating on the importance of knowing the movements and positions of the Sun, the Moon, and the constellations of the zodiac, he writes:

It is time, indeed, which rules all human action and especially the affairs of war. So that a general must be familiar with the dates of the summer and winter solstices, and the equinoxes, and with the rate of increase and decrease of days and nights between these; for by no other means can he compute correctly the distances he will be able to traverse either by sea or land. He must also be acquainted with the subdivisions of day and night so as to know when to sound the [reveille] and to be on the march; for it is impossible to obtain a happy end unless the beginning is happily timed.[16]

Neglect these matters, warns Polybius, and you'll make a mess of things. Bad timing is fatal. To prove his point he cites a number of examples, including a precipitous decision taken on August 27, 413 BC, during the siege of Syracuse, a major campaign of the Peloponnesian War (though not the siege of Syracuse in which Archimedes is said to have deployed his mirrors):

> Nicias, again, the Athenian general, could have saved the army before Syracuse, and had fixed on the proper hour of the night to withdraw into a position of safety unobserved by the enemy; but on an eclipse of the Moon taking place he was struck with terror as if it foreboded some calamity, and deferred his departure. The consequence of this was that when he abandoned his camp on the following night, the enemy had divined his intention, and both the army and the generals were made prisoners by the Syracusans. Yet had he only inquired from men acquainted with astronomy so far from throwing away his opportunity owing to such an occurrence, he could have utilized the ignorance of the enemy.[17]

It's hard to avoid lunar eclipses. When they occur, on average every couple of years, they last hours, and the entire half of Earth facing the Moon will witness them. That's because, unlike a total solar eclipse — an event that takes place on Earth's surface — a total lunar eclipse takes place in space, with the full Moon entering Earth's shadow. In fact, in ancient Greece and Rome, the intelligentsia already understood that, as Alan Bowen, a historian of the exact sciences in the classical world, puts it, "the antidote to the fear induced in the ignorant at the occurrence of an eclipse is learning that eclipses take place in the regular course of nature and are not omens or signs from the gods."[18]

During his fourth voyage to the New World, Christopher Columbus decided that an upcoming lunar eclipse would be a good way to threaten the locals on Hispaniola, who, because they produced almost no excess food, had been unable to supply Columbus with enough provisions to ensure that his crew remained loyal to him. He warned the locals that God, punisher of evildoers, would make the Moon disappear if they did not hand over more food. He even specified when this would happen. Divine wrath aside, Columbus — being familiar with recently compiled eclipse tables — knew that astronomy would back up his threat. February 29, 1504, would be the night. The eighteenth-century British historian Edward Drake relates the incident:

[K]nowing that there would be an eclipse of the moon within three days, [Columbus] sent an Indian, who spoke Spanish, to assemble the [community] on an affair of the utmost importance to their welfare: being met on the day preceding the eclipse, the Indian told them that the Christians believed in God who made heaven and earth [and who] was angry with them for not supplying his distressed servants with provisions, and would therefore chastise them with famine and other calamities, and, as a proof that what he told them was true; they should, that very night, observe the moon rising with a bloody aspect, as a warning of the punishment God would inflict upon them.

[T]he eclipse beginning as soon as the moon was up, and the darkness continuing to increase, it put them in such a consternation, that they hastened to [entreat] the Admiral that he would pray to God to be no longer angry with them, and they would bring as much provisions as he would have occasion for.

[R]etiring to his cabin, [Columbus] shut himself up till the eclipse was at its heighth, when he came out and told them he had prayed for them, promising they would be good . . . ; whereupon God had forgiven them, and they should see the moon by degrees, recover her usual appearance.[19]

Fourteen centuries before Columbus, Ptolemy had supplied the mathematics needed to compute the timing, magnitude, and duration of eclipses. Nevertheless, to the unlettered, they continued to seem special and portentous. In fact, every event and feature in the heavens, whether special or ordinary, has long been regarded as relevant to or even directly causal for the affairs of humans on Earth, if only its meaning could be divined.

Enter astrology.

For the Mesopotamians, astrology and astronomy were more or less the same thing. For the emperors of ancient China, as for the ancient Greeks, astrology and astronomy were intertwined. The skies spoke; the skywatcher listened and translated. Copernicus did astrology; Tycho Brahe did astrology; the great Galileo did astrology. Johannes Kepler, though critical of many aspects of astrology and aware of its cynical use, cast hundreds of horoscopes. In 1601, just after becoming imperial mathematician to the Holy Roman Emperor Rudolf II, Kepler published a treatise entitled *Concerning the More Certain Fundamentals of Astrology;* a quarter century later he served as astrologer to General Albrecht von Wallenstein.[20]

The modern distinction between astronomy and astrology was, at one time, fuzzy and irrelevant, just like the distinction between alchemy and chemistry or between magic and medicine. Promoting the propitious, avoiding calamity, and forecasting death suggest astrological interpretation. Yet prediction, which is an offshoot of analysis, can be exacting and scientific, given the right practitioner. Accurate observation of the skies, combined with a grasp of physics and a cartography of the cosmos, is the cornerstone of both.

Claudius Ptolemy, a renowned second-century AD Alexandrian mathematician, addressed all of the above. Besides writing astronomy's formidable founding document, the *Almagest,* he wrote the influential compilation of the geographical and cartographical knowledge of his day, the *Geographike Hyphegesis,* and the equally influential astrology opus, the *Tetrabiblos.* He begins *Tetrabiblos* by asserting a link between sky and Earth and the dual nature of sky studies:

Of the means of prediction through astronomy . . . , two are the most important and valid. One, which is first both in order and in effectiveness, is that whereby we apprehend the aspects of the movements of sun, moon, and stars in relation to each other and to the earth, as they occur from time to time; the second is that in which by means of the natural character of these aspects themselves we investigate the changes which they bring about in that which they surround.[21]

Ptolemy did not doubt that the cosmos was a unified, harmonious system (the Greek word *kosmos* means "order" as well as "world") or that the celestial affected the terrestrial. He traced a natural progression from heavenly configurations within and among the zodiacal constellations to the differing strengths of their influence on different sectors of Earth, to the general temperaments of persons born in those sectors, to the particular temperaments of persons born at particular times when particular influences were prominent. The sky was the seal that stamped the wax.[22]

By calculating where everything was, is, and will be on the sky, the astrologer could assign cause — preferably before, but sometimes after, the effect. Excesses of the body, flaws and felicities of character, distresses of the soul, and disruptions of society and nature could be traced to a source. Jupiter and Venus were temperate and moistening, hence fertile, active, and beneficent; Saturn and Mars were chilling and drying, hence destructive. Leo the Lion, the Sun, Mars, Saturn, and Jupiter embodied masculinity; Virgo the Virgin, the Moon, and Venus embodied femininity. Europe, the northwestern quadrant of the known world, was familiar with Leo, Aries the Ram, and Sagittarius the centaur Archer and was governed by Jupiter and Mars; therefore, wrote Ptolemy, the men there were warlike, commanding, clean, fond of liberty, and indifferent to women. The inhabitants of Britain and Germany, he added, were especially fierce because of their greater familiarity with Aries and Mars.[23]

Your horoscope (*hora,* "hour"; *skopos,* "watcher"), which derived from the Sun's location among the stars at the time of your birth, delineated your basic individual tendencies. In addition, the changing skies triggered changing effects. Knowing your own as well as the planets' proclivities, you could calmly prepare yourself for what lay ahead and, if necessary, rein in your worst tendencies so as to reduce your risk. Events and cities were subject to similar influences: while celestial configurations predisposed an individual to violence or acquiescence, and a polity to peace or endless strife,[24] they could also signal a coming shipwreck, earthquake, or robbery, or suggest the most advantageous timing for a marriage, a coronation, a prayer, or an invasion.[25]

The influence of classical astrology lasted for centuries. Astrologers cast horoscopes not only for Jesus and Pope Urban VIII but also for the fate of Florence and Rome and for the belligerent parties in World War I. They predicted or retroactively explained the assassination of monarchs, the success of empires, the rise of religions, even the end of history.[26] Not that everybody thought Ptolemy's legacy was a good thing. Astrologers, contended the critics, usurped power that rightly belonged to others; horoscopes were too persuasive. Even Ptolemy himself had reservations.[27] Hardly had the ink of *Tetrabiblos* dried than astrologers began to be expelled from Rome. The practice of astrology was restricted or even banned by the emperors Augustus, Diocletian, Theodosius, and Justinian. Saint Augustine said it was untenable to propose that the stars, whose power derived from God, could cause evil. Martin Luther pointed out that numerous astrologers had predicted a Great Flood of 1524, which didn't happen, whereas none predicted the massive Peasants' Revolt of 1524–25, which did. Urban VIII, whose death had been misforecast for 1630 by a renowned abbot-astrologer, issued a papal bull against astrologers in 1631.[28]

But even vocal opponents of astrology sometimes played it safe: Francesco Guicciardini, a politician in Renaissance Florence who ridiculed the widespread tendency to remember astrologers' successes while forgetting their far more numerous mistakes, had his own horoscope cast by a murderer. Nor did astrology vanish with the rise of rationalism or the increasing interest in observational astronomy or the spread of the telescope, despite the sudden appearance of brilliant new stars (supernovas) in 1572 and 1604, not to mention Galileo's discoveries in 1609 and 1610 of the mountains and craters of the Moon, the four largest satellites of Jupiter, and the two seeming companions of Saturn — all of which made astrology's foundations tremble. Suddenly the map of the sky, and consequently the analysis of celestial influences, had to be revised. William Herschel's discovery in 1781 that Uranus was a planet further confounded the profession.

Beliefs, however, have a strong grip. Although many educated Europeans came to reject celestial determinism in the lives of individuals, many continued to embrace the idea that the stars and planets affect the more general course of nature. Diplomats advised a limited reliance on astrology rather than an all-out rejection of it, especially in times of war.[29] The English philosopher-scientist Francis Bacon discounted the doctrine of horoscopes and felt astrology to be "so full of superstition, that scarce anything sound can be discovered in it," yet declared that purifying it was preferable to discarding it altogether.[30] Britain's first Astronomer Royal, John Flamsteed, trod lightly on astrologers' toes: referring to the rare triad of Saturn–Jupiter conjunctions in 1682–83, he wrote that while astrologers had "affrighted" the "Common People" with "fearful Predictions of direful events . . . , the more Judicious are desirous to know how often and at what time their Conjunctions happen."[31] Galileo, too, was caught between the old and the new. He himself drew up horoscopes for his friends, his daughters, his patrons, himself. The dedication of his paradigm-destroying *Sidereus Nuncius,* published in March 1610, includes a panegyric to Jupiter and to his patron Cosimo II de' Medici, whose horoscope Galileo slightly rigged to make that regal planet maximally dominant.[32]

Astrological prognostication kept its hold well into the seventeenth century. After enduring a few body blows in the eighteenth century, it gained ground in the nineteenth, survived the twentieth, and is alive and well across the globe in the twenty-first, especially among those with limited science literacy.[33] Few people, including those who hold power, are immune to the suspicion that astrology might have something to offer or to swear by.

Take America. For most of the past thirty years, the fraction of the US population that embraces astrology has held steady at one-fourth but is now growing (about the same fraction believes in reincarnation, while twice as many have had what they call a "mystical experience"). While Ronald Reagan was president, he and his wife Nancy consulted a Vassar-grad astrologer who prescribed the timing (sometimes right down to the second) of presidential election debates, the announcement of Justice Anthony Kennedy's appointment to the Supreme Court, press conferences, takeoffs of Air Force One, State of the Union addresses, and much else. Right after 9/11, a "prophecy" ostensibly written by the illustrious, obscurantist sixteenth-century astrologer and seer Nostradamus raced across the Internet, further terrifying masses of already terrified Americans and further priming them for retaliation. In fact, the quatrain was an intentional fabrication, written for a twentieth-century student essay: "Two brothers torn apart by Chaos, / while the fortress endures, / the great leader will succumb, / The third big war will begin when the big city is burning." Embellishments soon included "On the 11th day of the 9 month, / two metal birds will crash into two tall statues" and "In the city of york there will be a great collapse." In 2004, after several years of intensive fear-promotion by media and public officials alike, the most popular search term on AOL was "horoscope." On the night of September 4, 2008, as a vast-screen video introduced John McCain to the party faithful at the Republican National Convention, to the accompaniment of swelling music and applause, a sonorous voiceover proclaimed: "The stars are aligned; change will come."[34]

Or take India, where astrology is Vedic rather than Ptolemaic, and the Moon figures more prominently than the Sun in horoscopes. Today, as in the past, few Hindus marry without consulting — and obeying — an astrologer. As the diplomat, journalist, and writer Khushwant Singh put it: "Astronomical harmony was the one guarantee of happiness." On November 27, 2003, twelve thousand couples got married in New Delhi because Jupiter's "planetary mischief" would be kept at bay that night. In late October and early November 2006, Delhi was again awash in weddings, this time because even couples whose discordant horoscopes would normally rule out a marriage were being assured of a happy outcome. But matters of matrimony are far from the only domain where astrology rules: political candidates' nomination papers are filed, and the winners are sworn in, at astrologically opportune times. In 2001, with the Bharatiya Janata Party in power, publicly funded Indian universities were urged to offer courses in Vedic astrology. Many Indian scientists and academics vilified the policy — "For our Government to send satellites into space yet permit astrology to be taught using public funds is too great a contradiction to bear further mention" — but it was upheld by the Supreme Court of India.[35]

In his panoramic 2007 novel *Sacred Games* (now a Netflix series), Indian-American writer and computer geek Vikram Chandra offers an extraordinary depiction of astrology wedded to annihilation. One of his characters is a guru whose goal is to engineer the nuclear annihilation of the city so as to start Earth's cycle of time and life afresh. Speaking to a gang kingpin who has become his disciple, he says,

Think of life itself. Do you think it has no violence in it? Life feeds on life, Ganesh. And the beginning of life is violence. Do you know where our energy comes from? The sun, you say. Everything depends on the sun. We live because of the sun. But the sun is not a peaceful place. It is a place of unbelievable violence. It is one huge explosion, a chain of explosions. When the violence ceases, the sun dies, and we die. . . . Have not holy men fought before? Have they not urged warriors to battle? Does spiritual advancement mean that you should not take up weapons when confronted by evil? . . . We must resist this so-called peace which emasculates spirituality and makes it weak.[36]

Money, an overwhelmingly obvious source of power, is another matter in which astrologers are hard at work. Read the business press, and you're likely to run across a quote from the economist John Kenneth Galbraith: "The only function of economic forecasting is to make astrology look respectable." But plenty of people seem to feel that money is power and that astrology is control, and that if you put the two together, you've got control over money. The Gilded Age banking magnate John Pierpont Morgan reputedly said, "Millionaires don't hire astrologers, but billionaires do." Both Morgan and Seymour Cromwell, president of the New York Stock Exchange from 1921 to 1924, consulted a high-profile astrologer named Evangeline Adams, who received her clients in a suite above Carnegie Hall.[37] More recent financial astrologers' mottoes and book titles may give skeptics pause ("Market timing by planetary cycles and technical analysis"; *Planetary Harmonics of Speculative Markets*), but investors, fund managers, bankers, and corporate executives still seek their advice. Statistically, of course, amid all the misses there's the occasional hit: one astrologer forecast that the stock market would plummet in October 1987; another forecast that gold would hit $487 an ounce in 2005. Horoscopes can be, and are, drawn up for a bond, a Treasury bill, a company, or even a stock exchange, based on the hour of its first offering or its incorporation or the start of trading.[38]

Astrologers were no doubt heartened by the findings of two business-school professors, to the effect that across the full history of the Dow Jones Industrial Average, the S & P 500, the NYSE, and the NASDAQ, stock returns have been as much as 8 percent higher (about double) for the fifteen days around the new Moon than for the fifteen days around the full Moon. Elsewhere the "lunar cycle effect" has been even more pronounced: for the final three decades of the twentieth century, in stock exchanges around the world, the returns were as much as 10 percent higher.[39] Meanwhile, the half month surrounding the new Moon generates, on average, the same gravity and the same tidal forces as the half month that surrounds the full Moon.

One "classical scientific astrologer" and financial commentator has stressed planetary transits and oppositions rather than the phases of the Moon. Posting his analyses in the late summer of 2007 amid the deepening credit crunch, the flood of home foreclosures and bank failures, and the ubiquitous (though widely ignored) signs of imminent global economic meltdown, Theodore White warned of the bursting housing bubble, contending that Jupiter helped inflate values and that Saturn had begun to deflate them. "Saturn's long transit of Virgo (26 months) and another four months by retrograde in the year 2010, takes place in a sign ruled by Mercury," he wrote. "This transit will have a devastating, nearly depressing, effect on those severely affected by the downturn in the housing market nationwide." In addition, Saturn would be "rising in sunrise diurnal charts, and will be Lord of the months of October & November, with continuing strong influences into December 2007." The latter's transit near the South Lunar Node pointed to the subprime mortgage crisis "crystalliz[ing] into a major call for regulation throughout the economic climate of the United States."[40]

After-the-fact revelations are easy to come by when you consult the nearly limitless number of cyclic phenomena. It's not hard to find one that matches your needs or expectations. There's the eleven-year sunspot cycle, the twenty-six-month cycle of Earth and Mars in space, the 18.6-year cycle of lunar eclipses. There's also the yearly cycle of months: in 1907, 1929, 1987, and 2008, the stock market sustained huge hits in October. Other Octobers saw pullbacks. Does that mean the "October effect" is a real thing? No. But if significant numbers of buyers and sellers believe that cosmic forces will bring down the market, a sell-off will follow, thereby fulfilling their prediction. Plus, keep in mind all the *failed* predictions.

The prosecution of war is at least as durable and hard-nosed a pursuit as the acquisition of a fortune, and certain real-life warriors have been as interested in astrology as were the rulers of Mesopotamia and ancient China. Nazi Germany offers a stunning case study, chronicled in detail by Ellic Howe, a writer, historian, and expert forger who during World War II worked for a British agency called the Political Warfare Executive.[41]

Interest in astrology rose rapidly in defeated, inflated Germany following World War I, writes Howe — more rapidly than in the rest of Europe. A graphologist-journalist named Elsbeth Ebertin was fast becoming a well-paid, widely read professional astrologer, and in the spring of 1923 a follower of Adolf Hitler's, hoping to learn about his leader's horoscope, sent Ebertin the rising politician's date of birth (though not the exact hour, a crucial detail). Ebertin decided to publish the horoscope in the 1924 edition of her annual almanac, *A Glance into the Future.* She did not name Hitler, but she didn't have to:

A man of action born on 20 April 1889, with Sun in 29° Aries at the time of his birth, can expose himself to personal danger by excessively uncautious action and could very likely trigger off an uncontrollable crisis. His constellations show that this man is to be taken very seriously indeed; he is destined to play a "Führer-role" in future battles. It seems that the man I have in mind, with this strong Aries influence, is destined to *sacrifice himself for the German nation,* also to face up to all circumstances with audacity and courage, even when it is a matter of *life and death,* and to give an impulse, which will burst forth quite suddenly, to a German Freedom Movement. But I will not anticipate destiny.[42]

Ebertin's prognosis, calculated on the assumption of a noonday birth, appeared in July 1923. In November, Hitler participated in what could easily qualify as an "excessively uncautious action": the Beer Hall Putsch. By the time he landed in jail for his part in the putsch, Ebertin had learned that he'd been born at 6:30 PM. No matter. Astrology's star was rising in Germany, aglow with swiftly multiplying societies, publishers, manuals, conferences, and adherents of every sort. More than a hundred *Herren Doktoren* — philosophers, paleontologists, physicians, even an astronomer who worked on ballistics problems and possibly the dreaded V-2 rocket — publicly joined their ranks. As Howe put it, "In Germany between the two wars there were more astrologers per square mile than anywhere else in the world."[43] Popular as it was, astrology also spawned powerful opponents.

With Hitler's appointment as chancellor of the Third Reich on January 30, 1933, his horoscope became a matter of wider interest. Aiming to justify various characterizations of the Führer, some astrologers even "corrected" the hour of his birth, putting the Sun in Taurus rather than in Aries and, in a few cases, questioning his capacities. The authorities saw that as a line in the sand. In the spring of 1934 the Berlin police banned most forms of astrological activity, and by the end of the year the Reich Ministry for Propaganda and Public Enlightenment, headed by Dr. Paul Joseph Goebbels, had silenced public astrological speculation concerning the fortunes of the Third Reich and the horoscopes of prominent Nazis. Astrological literature, both popular and abstruse, was confiscated from publishers and booksellers. Homes were searched, persons arrested. The last major annual astrological conference took place in 1936. One after another, periodicals ceased publication during 1937 and 1938.[44]

Late in the afternoon on May 10, 1941 — coincidentally, a few hours before the most horrific night of the London Blitz — the mentally unstable Rudolf Hess, occupant of the third highest leadership position in the Third Reich and, like so many of his countrymen, something of an aficionado of astrology, climbed into a Messerschmitt-110 fighter plane and headed for Scotland. He had secretly decided to embark on a peculiar, unvetted mission of peace: trying to convince British high officials to accept German supremacy in Europe and thereby save their country from further devastation. Practicalities soon intervened: the plane's fuel supply wasn't sufficient for the trip, so he had to bail out, leaving the plane to crash onto a farmer's field near Glasgow while he himself parachuted down, breaking his ankle and ending up in a British military hospital. Reich officials, needing to produce an explanation for the surprise flight that would somehow satisfy not only the German people but also the rest of the world, decided to blame a combination of insanity and astrology. Rumors flew around Europe; the London *Times* posited that Hess was Hitler's secret personal astrologer. Propagandists on both sides went into overdrive.[45] Within a couple of days the Gestapo arrested and questioned several astrologers; within the month they arrested hundreds more, primarily those who belonged to astrological societies and who had published their analyses, along with many more people involved in activities tinged with the occult. On June 24 public lectures and performances involving astrology, clairvoyance, telepathy, and other esoteric practices were thenceforth forbidden. On October 3 the ban was extended to the print media. Some astrologers ended up in concentration camps.

Yet despite the considerable censorship, astrology and the occult flourished behind both closed and open doors, supported in part by Goebbels. On November 22, 1939, at one of his almost-daily ministerial conferences (convened for the attendees to assent, not confer), he decreed that a psy-ops leaflet based on Nostradamus's prophesying of the far future be swiftly prepared for dissemination in France.[46] In 1940 the Propaganda Ministry hired Karl Ernst Krafft, a fervent, statistics-minded Swiss astrologer, to annotate selections from Nostradamus.[47] In 1942–43, chastened by a few months of incarceration following the Hess affair, Krafft and another notable, though more pragmatic, astrologer named F. G. Goerner were conscripted to spend their days excerpting Nostradamus and preparing the horoscopes of Allied generals. Other recently arrested astrologers, along with astronomers, mathematicians, and psychics, were recruited for the Pendulum Institute, where, during the spring of 1942, under the directorship of a captain of the German navy, the professional staff assiduously swung pendulums over maps of the Atlantic Ocean, searching for the positions of enemy ships.[48]

As the Reich's fortunes wavered, the drafting of prophecies and the private study of horoscopes mounted.[49] Publicly, prediction became the general order of the day — at least while it remained useful — and radio was the propagandists' preferred medium. From September 1942 through March 1943, the second winter of the campaign in Russia, nearly one in every eight items in the Ger-

man news bulletins was an explicit prediction.[50] Refugee German intellectuals described the Nazis' approach to prediction and
prophecy:

Belligerent governments invariably predict
victory. The stakes are high, and the public
is naturally anxious. Not to predict is to
encourage suspicion and to destroy confidence. To predict ultimate failure is morally
to surrender. Thus propagandists predict
victory, for it is the only thing they can
do. . . . Above all, the Leader is forced to
prophesy to demonstrate his charismatic
gift. . . .
   Reassurance was given by initial victories.
As time went on, however, the propagan-
dist . . . found it convenient to deal with the
increased tension of the German people, by
an increased use of predictions. [There is a
greater] need to predict in times of distress
rather than in times of comfort. For a long
time predictions took the place of good
news. [But w]hen . . . Russia's force remained unbroken, the policy was suddenly
changed, and prediction became rare. It
was at this time that Goebbels began telling
the German people in so many words that
this was a world in which one could not
predict and that the war was simply "the
riddle of riddles.["][51]

The specifically astrological form of prediction, however, retained its appeal. From the late 1930s onward, rumors about a Hitler–astrology connection multiplied. One astrologically sophisticated writer — Louis de Wohl, a part-Jewish Berliner who got himself out of Germany in 1935 and wished to survive in style in London and New York — found astrology a convenient way to facilitate his survival, and so he let it be known that Krafft was Hitler's personal astrologer. The president of Columbia University soon announced that Hitler had a team of five astrologers. The London *Evening Standard* named Elsbeth Ebertin the Führer's favorite astrologer.[52]

In fact, neither Hitler nor most of his closest Nazi colleagues[53] turned to astrologers for advice on what to do when, even though the Nazis' angry nationalism and ardent racism put them on the same side of the fence as many people who embraced not only the political mission of a racially pure, redemptive Aryan future but also the fairytale vision of a golden Aryan past, full of spiritualism, folk identity, cosmic mysteries, and astrological constructs. Nevertheless, as Goebbels put it, "crazy times call for crazy measures," and the Third Reich's waning weeks must have been intensely crazy, not least because its leaders had not yet grasped that their nanosecond of supremacy had already come and gone.[54] And those were the weeks when Hitler turned to prophecy.

From the April 1945 diary of Hitler's minister of finance, Count Lutz Schwerin von Krosigk (a former Rhodes Scholar at Oxford), we learn that around midmonth, Goebbels and Hitler decided that the time had come to examine two horoscopes: that of the Führer himself, which had been cast in 1933, and that of Greater Germany, cast in 1918. The revelations must have been thrilling. As the diarist writes,

Both horoscopes had unanimously predicted the outbreak of war in 1939, the victories till 1941, and then the series of defeats culminating in the worst disasters in the early months of 1945, especially the first half of April. Then there was to be an overwhelming victory for us in the second half of April, stagnation till August, and in August peace. After the peace there would be a difficult time for Germany for three years; but from 1948 she would rise to greatness again. . . . [N]ow I am eagerly awaiting the second half of April.[55]

Early on Friday the thirteenth of April 1945, the Reich's state secretary rang up the finance minister to announce that President Roosevelt had died the previous day. "We felt the wings of the Angel of History rustle through the room," Schwerin von Krosigk records. "Could this be the long-desired change of fortune?" Goebbels thought so. When a reporter told him the news, he called for a bottle of the best champagne and telephoned Hitler to say that just such a turning point had been "written in the stars." Goebbels was ecstatic.[56]

Less than four weeks later, the Nazis surrendered.

To proponents of National Socialism, the discovery of icy little Pluto in 1930 seemed pregnant with implications. Astrologers swiftly integrated Pluto into their horoscopes, and in 1935, two years after Hitler became chancellor of the Third Reich, a German astrologer, Fritz Brunhübner, published a brief but detailed book on the newcomer, *Der neue Planet Pluto.* According to Brunhübner, Pluto is "the end of the old world and the ascent of a new spiritual epoch." It is "a malefic in the greatest form," "the planet bringing death," "the instigator of the turn in world events." Its "destiny is to clean up the old and to march before the new era in a new form."[57]

But the creepiest connection he makes between Pluto and Hitler's Germany is the following:

Moreover, I believe Pluto to be the planet of National Socialism and the Third Reich. Adolf Hitler and almost all of the leading men now in the government, also the Nazi Party, and the horoscope of the Third Reich (January 30, 1933, the day of Potsdam, the Reichstag elections of March 5th and November 12th, 1933) show — besides a very dominant Uranus — a strong Pluto.

It has to be like this. Pluto is the planet of the turning-point. The National Socialistic movement, in the horoscope of which Pluto is elevated above all other planets, brought about, according to the laws of Pluto, a reversal in German history. And what tells the horoscope of Adolf Hitler? At that moment, when Reich President Hindenburg handed Adolf Hitler the fate of the German people, transiting Pluto stood in the Zenith, tied to the most important places of the radix horoscope . . . a trial of strength, a seizure of power, a turning-point, a crisis.[58]

"Turning-points" keep turning. At the war's end, the Allies dissolved and banned the National Socialist German Workers Party, and Germany itself now deems the performance of the Nazi salute a criminal offense. In the decades following the war, astronomers found that Pluto is smaller than not only our own Moon but six other moons in our solar system as well, and the International Astronomical Union no longer classifies Pluto as a true planet. The search for sources of sky power, conquest, and "new eras" must turn elsewhere.

3
SEA POWER

An expanded, ethnically purified Germany — Grossdeutschland — was the vision that drove the Nazis. The lands they intended to conquer had long since been explored, settled, and fought over, their latitudes and longitudes established, their terrain mapped, their rivers traced, their inhabitants identified and named. No such vision impelled the first courageous, curious, or desperate peoples who walked up the Rift Valley, rowed and sailed into uncharted tracts of the Pacific Ocean, or rode horses through unknown wastes of the Taklimakan Desert. They had no idea what they were in for.

Yet by forty thousand years ago, bands of anatomically modern humans had trekked as far as Sri Lanka and the east coast of China and transported themselves across the sea from Africa to somewhere in Southeast Asia to the then-continent of Sahul, a fusion of Australia and New Guinea.[1] Early explorers, gatherers, exiles, sea drifters, traders, and raiders had neither compasses nor maps. Geography and navigation were nascent practices. On land, travelers could follow a river, a mountain pass, or an animal path; on the sea they could try to stay within sight of land but had to avoid the subsurface, rocky perils of hugging the shore.

Oceangoing wayfinders catalogued and memorized landmarks. They consulted clouds, winds, and sounds for additional clues. They became familiar with swells and currents, phosphorescence, tides, the implications of floating palm fronds and coconut husks, the plants and fishes that dwelled at different depths, variations in the color of the water and in the smell and taste of samples drawn from sediment below the boat.

The flight of a bird could reliably indicate land beyond the horizon. A mariner might carry a caged "shore-sighting" raven, booby, or frigate bird on board, freeing it periodically to see whether it would return to the dry safety of the boat or head off toward the preferable safety of land. Genesis 8:11 tells us that Noah sent forth a dove, which returned with an olive sprig in its beak. Ancient Polynesians, seeing the long-tailed cuckoo migrate southwestward each year, would have realized it was heading for unseen terra firma, because the cuckoo is a landlubber. Taking their cue from the cuckoo and steering southwestward in their twin-hulled voyaging canoes, the Polynesians came upon New Zealand. Medieval Irish monks saw vast, honking flocks of geese head northward every spring from the Shannon estuary and return every autumn; sailing north in their curraghs, they came upon Iceland. Columbus, sighting pelicans en route to what he expected would be the Indies, noted in his log that this bird does not venture more than twenty leagues from land.[2]

Once out on the open sea, however, mariners could depend on the sky to tell them where they were. Aside from announcing seasons, the proximity of land, and the weather, the sky signaled location and direction: where the boat was and where it should be heading. In other words, the sky transformed wayfinding into navigation, the "haven-finding art"[3] — valued greatly in Europe by the late sixteenth century, when a mathematics-minded instrument maker in Antwerp penned this definition:

This art is divided into two, namely common navigation and grand navigation. . . . The whole science of common navigation is nothing more than knowing perfectly by sight all capes, ports, and rivers, how they appear from the sea, what distance lies between them, and what is the course from one to another; also in knowing the bearing of the moon on which high and low tides occur, the ebb and flow of the waters, the depth, and the nature of the bottom. . . . Grand navigation, on the other hand, employs, besides the above-mentioned practices, several other very ingenious rules and instruments derived from the art of Astronomy and Cosmography.[4]

A century later, John Seller, Hydrographer in Ordinary to several British kings — the nation's official surveyor of rivers, lakes, and seas — described navigation as "guiding the ship in her Course through the Immense Ocean to any part of the known World; which cannot be done unless it be determined in what place the Ship is at all times."[5] And indeed, by his day, the immensity of both Ocean and World were well known. Travel books, both factual and fanciful, were perennial best sellers. Owing to a potent combination of astronomy, mathematics, cartography, literacy, weaponry, instrumentation, navigation, and intimidation, Ocean and World had been discovered, explored, charted, inventoried, fictionalized, bought, sold, colonized, grabbed, planted, harvested, and mined, and many millions of the residents forcibly Christianized or enslaved.

But there's a backstory.

To determine the precise location of his ship, the early navigator needed reliable objects against which to compare his position. But even on a given stretch of sea, a feature that was there in the spring might be absent in the fall. And because the navigator was moving rather than stationary, sailing rather than standing, the reliability factor changed month to month, week to week, even day to day.

Since Earth goes around the Sun once a year, a stargazer looking upward from the same rooftop once a month at the same time of night sees a sky that has shifted westward one-twelfth of 360 degrees, or thirty degrees, from the previous month's sky. Early astronomers tracked this cycle carefully. *Shangshu,* or *The Book of History,* written in China in the first millennium BC, states that Taurus rises in the east in the Sixth Month (of the Chinese year), reaches its zenith in the Eighth Month, and sets in the west in the Tenth Month — all, implicitly, at the same hour of night. *Kitab al-Fawa'id fi usul al-bahr wa-l-qawa'id,* or the *Book of Useful Information on the Principles and Rules of Navigation,* compiled in the fifteenth century AD in what is now the United Arab Emirates, states that the bright star Canopus sets due west at dawn on the 40th day (of the Islamic year) and rises due east at dawn on the 222nd day.[6]

Another way to think about this cycle is that, day after day, decade after decade, a stargazer will see the same stars rise at the same point on the horizon — but they will rise four minutes earlier each day. Now, add a much smaller but very real factor to the daily four-minute and monthly thirty-degree changes: the wobble of Earth's tilted axis of rotation, at the rate of one full revolution every 25,700 years. Discovered by the ancients, that wobble — called the precession of the equinoxes — has the effect across the centuries of shifting the stars' positions relative to the month of the year. It also affects the North Star. In Homer's time, that star, which today we call Polaris, stood a dozen degrees from the North Pole; in Columbus's time it stood three and a half degrees away; in Sputnik's time, it stood right near the pole. By about AD 15,000, as Earth keeps wobbling like a top, Polaris will sit forty-five degrees away.[7]

When you're sailing the high seas, the slow, centuries-long shifting of Polaris is irrelevant. But not knowing north from east can be fatal. Direction is key. Fortunately, the Sun's appearance, disappearance, and midday shadows, as well as the paths taken by other stars and the places from which winds of different character blow, are archetypes of directionality. For example, a bright star named Alnilam — corresponding to the center of Orion's Belt — rises due east and sets due west. As for finding north in the Northern Hemisphere, you could look more or less to where

Ursa Major, the Great Bear — with its seven bright stars, the Big Dipper[8] — wheels around an axis, neither ascending, culminating, nor descending. The reputedly blind bard Homer, though confused about the northern nighttime sky, knew that stellar navigation would have been important to any voyager, and so he writes that Odysseus, wishing to return home, was instructed by the nymph Calypso to keep rightward of the Great Bear, the constellation that "alone of them all never takes a bath in the Ocean."[9]

Indo-European languages have long distinguished Orient (rising/ east) from Occident (setting/west). The Greeks differentiated sunrise and sunset at the solstices from those at the equinoxes, creating six directions from two. The Vikings, sailing from Scandinavia into the sea, differentiated landward from oceanward: land-south and land-north were easterly directions; out-south and out-north were westerly. For early navigators in low latitudes such as the Mediterranean and the Arabian Sea, the points at which the Sun rose and set were useful direction markers year-round, whereas for the Vikings, who lived at high latitudes, those points changed too drastically from month to month to be helpful. The closer a mariner sailed to the North Pole, the harder it was to gain his bearings from the Sun or the stars, and the more he had to rely on winds, birds, and tides, though he could consult Polaris as a reliable rough indicator of north. Pacific Islanders took another tack. Voyaging across Oceania, they steered their course by *kavengas,* or star paths: arcs described by the successive risings or settings of a series of familiar stars. These arcs would guide them from one known island to another.[10]

Three, four, and five millennia ago, large numbers of slow, big-bellied merchant ships crisscrossed the Old World waterways, carrying both luxuries and necessities.[11] But merchants had neither the seas nor the harbors to themselves. By 2400 BC Egyptian armies were being ferried to what is now the coast of Lebanon. By 2000 BC the first true maritime power of the Mediterranean — the Minoans, inhabitants of the island of Crete — had built itself a navy. By 1300 BC fleets of marauding northerners were seizing ships and blockading naval bases that the pharaoh Thutmose III had established along the Lebanese coast.

From the earliest centuries of maritime commerce, writes the historian Lionel Casson, "the freighter had to share the seas with the man-of-war."[12] Piracy, plunder, and the taking of slaves increased in direct proportion to trade, travel, and the taking of land. Seaborne raids on both vessels and coastal settlements became commonplace; sea battles

123

increased in scale and complexity. Meanwhile, the hunger for foreign goods mounted. Athens' "Achilles' heel," exploited in war by both Sparta and Macedon, was its dependence on grain shipped from Egypt, Sicily, and southern Russia.

An amazing range and quantity of cargo was carried across the seas in ancient times. In the third millennium BC, South Asian gold, ivory, carnelian, and lapis lazuli, Lebanese cedar, and Omani and Cypriot copper changed hands at the eastern Mediterranean port of Byblos, at the Persian Gulf port of Bahrain, and at the mouths of the Indus River. Frankincense and myrrh from the Horn of Africa were ferried up the Red Sea to Egypt; lapis lazuli from Harappan settlements in the Indus Valley also made its way to Egypt. Fragments of Indian teak appear in the ruins of the Sumerian city of Ur; Minoan craftsmen worked amber from the Baltic; Mycenean jars arrived at the palace of the pharaoh Akhenaten; Chinese silk was woven into the hair of Egyptian mummies; Sri Lankan cinnamon bark scented the women of Arabia; Zimbabwean gold crossed the Indian Ocean long before Europeans staked claims in southern Africa; Han Chinese rulers had such need of war horses that they imported the beasts both by land and by sea. Each year, freighters transported hundreds of tons of wheat, olive oil, marble, and herb-laden fish sauce to Athens, to Rome, to Alexandria. Local versions of fermented shrimp paste, a staple condiment in the cuisines of Southeast Asia, made their way across the South China Sea. A single merchant vessel wrecked in the first century BC near Albenga, on the Italian coast between Genoa and Monaco, held between 11,000 and 13,500 amphorae of wine.[13]

The Bronze Age made tin a prized commodity. Generally an alloy of copper and tin, bronze was a brilliant invention, a strong, corrosion-resistant material that could be cast at relatively low temperatures into weapons, ritual vessels, ornaments, statues, and tools. The intimidating rams at the prows of the warships that kept the seas open for the merchant ships were made of bronze. But since copper and tin are rarely found in the same patch of Earth's crust, long-distance trade was essential to their union. And since tin could fetch many times the price of copper,[14] it was definitely worth a trader's time and effort.

By the eighth century BC, the search for tin, as well as for silver and gold, had taken the Phoenicians through the Pillars of Hercules and the Strait of Gibraltar at the western exit from the Mediterranean and onward to the Atlantic side of the Iberian peninsula, to an area called Tartessos.[15] Some tin could be extracted locally there, but much more was transported overland from major sources farther north, including Cornwall, the south-western tip of Britain, to which Herodotus seems to have been referring when he wrote in the mid-fifth century BC of the "Tin Islands, whence the tin comes which we use." To Herodotus, those places, unglimpsed by him and everyone he'd ever met, were "the ends of the earth." One reason none of his acquaintances had seen those sources of tin firsthand was that the navy of Carthage, the strong North African colony planted by the Phoenicians, had blockaded the Strait of Gibraltar. Nevertheless, little more than a century after Herodotus wrote those words, a daring Greek from Massalia named Pytheas may well have made his way to the Atlantic Ocean, the tin works of Cornwall, and much else besides.[16]

Massalia (Marseille) was the colony of a colony, one of many Greek and Phoenician maritime cities that sprouted across and beyond the full breadth of the Mediterranean from the early to the middle of the first millennium BC. During those centuries, the founding of colonies and the forging of trade routes proceeded hand in hand with the development of warships and the establishment of navies.[17] Alongside all that commerce and conflict, inquiry and learning flourished as well. Interchange took place on every coast; information poured in from every direction. Anaximander, a resident of the thriving Greek city of Miletus, drew the first map of the inhabited regions of Earth. Soon afterward, Hecataeus of Miletus improved upon Anaximander's map and produced a comprehensive geography of the known world: a doughnut-shaped collage of land-masses, a flat map of a flat Earth, with the Medi-terranean (literally, "Middle Earth") at its heart and the continuous Ocean at its outer boundary. Soon after that, a globe-trotting mathematician-astronomer named Eudoxus of Cnidus wrote his own work of geography and also devised a model of planetary motion, presenting it as an inter-connected system of twenty-seven spheres, each of which rotated on an axis that passed through the center of Earth.

Pytheas thus came of age in a cosmopolitan, contentious, intellectually active world that grew larger, more acquisitive, and more fact-hungry by the day. How he got past the Pillars of Hercules is much debated; that he did so is generally accepted, as is the contention that he saw Cornwall and followed the west coast of Britain north to the Orkney Islands, with a stopover at the Isle of Man. What some scholars do dispute is whether Pytheas himself then voyaged six days north to a place the ancients called Thule (which might be Iceland) and thence almost up to the Arctic

Circle.[18]

Let us be believers. Let us say Pytheas did all the things his advocates say he did. So, during his voyage, besides searching out tin, he periodically measured the height of the Sun; recorded the shadows cast by a gnomon at various locations; gasped at the prodigious tides of the Pentland Firth; counted the number of islands in the Orkney group; and took note of the houses, crops, and beverages of the communities he visited. In Thule, at the outskirts of the Arctic Circle, he witnessed extreme phenomena: "the place where the sun lies down [and] straightaway rises again" and the "Congealed Sea" lying one day's journey from land, a region "where neither earth was in existence by itself nor sea nor vapor, but instead a sort of mixture of these . . . [where] the earth and the sea and all things are together suspended, . . . existing in a form impassable by foot or ship." From Thule he traveled east in search of amber and then south, completing his circumnavigation of Prettanikē (whence "Britannia") and masterfully approximating its rough perimeter as the equivalent of 4,400 modern miles.[19] Upon returning to Massalia he wrote a *periplus* ("sailing around"), a treatise called *On the Ocean,* of which not a single copy survives — only respectful paraphrases and skeptical dismissals.[20] Pytheas wasn't the first Mediterranean mariner who reputedly entered the North Atlantic; he was just more adventurous and science-minded than his predecessors.

Traditionally, the conceptual universes of navigators and scholars did not overlap. Seafarers had little truck with scientists' determinations, nor did scientists with seafarers' findings. But Pytheas's data were used by astronomers and geographers for centuries after his death and became as useful to plunderers and conquerors as to merchants and diplomats. Hipparchus — the mathematician-astronomer who developed the framework of degrees, parallels, and meridians that is still used to describe latitude and longitude — translated into degrees of latitude Pytheas's careful measurements of gnomon shadows, hours of daylight, solar altitudes, and distances traveled. That's how we know that Pytheas placed Massalia at 43° 3' N (he was only a fourth of a degree off) and paused on his northward journey at 48° 40' N (northwestern Brittany, likely the island of Ushant in the English Channel), 54° 14' N (the Isle of Man), 58° 13' N (the island of Lewis in the Outer Hebrides), about 61° (the Shetlands), and about 66° (northern Iceland).[21] Hipparchus, no mean authority himself, invoked the authority of Pytheas when correcting other scientists' blunders:

Indeed, concerning the north pole, Eudoxus . . . certainly doesn't know what he is talking about when he says, "There is a certain star remaining always at the same place; this star is the pole of the cosmos," since no single star lies at the pole, but an empty place [instead], near which lie three stars. The spot marking the pole, aided by these [stars] encloses a figure very nearly resembling a quadrilateral — exactly, in fact, as Pytheas the Massaliote says.[22]

Ambitious early wayfinders reputedly also went in the other direction: south. One voyage, a several-year clockwise circumnavigation of Africa undertaken by Phoenician mariners in about 600 BC, was initiated at the behest of the military-minded Egyptian king Necho II. A century or more later, Hanno, the king of Carthage, took a counterclockwise route with many thousands of colonists and great numbers of vessels. How far did those voyages get? Hard to say for sure.[23]

Hipparchus's 360-degree system of latitude and longitude, and the calculations it made possible, gave a big boost to the sciences of geography, cartography, and astronomy. The terms "latitude" and "longitude" derive from the Greek for "breadth" and "length," respectively, denoting a binary directionality in early maps of the known world. But the difference between the two goes very deep. The American historian Dava Sobel describes it this way:

> The zero-degree parallel of latitude is fixed by the laws of nature, while the zero-degree meridian of longitude shifts like the sands of time. This difference makes finding latitude child's play, and turns the determination of longitude, especially at sea, into an adult dilemma — one that stumped the wisest minds of the world for the better part of human history.[24]

Though Polaris was not yet in place to serve as a convenient North Star, the Greeks understood that if the same star or stars barely skimmed the horizon in two different cities, those cities lay at the same latitude. Latitude could be calculated from the highest altitude reached by certain catalogued stars. One that could be consulted alone or as part of a pair was Canopus, a bright southern star known in Arabic as Suhail. Eudoxus knew that Canopus–Suhail could barely be seen in Rhodes but hit an angle of 7 1/2 degrees in Alexandria. The medieval Arab navigator-poet Ahmad ibn Mājid, who measured both in degrees and in *isba* (the width of the knuckle of the middle finger, held at arm's length against the horizon), advised his readers that when the star Aldebaran reached its highest ascent, the angle of Suhail would be six degrees in Sindabūr (present-day Goa) and 7 3/4 *isba* at Cape Madraka in present-day Oman. "The best method of measuring latitudes is using Suhail," wrote ibn Mājid, "and another like this will never be seen throughout all eternity."[25]

Eternity is a long time. In fact, it would take fewer than a thousand years for Polaris to elbow Suhail aside and become the present century's best tool for finding latitude. Everywhere north of the equator, the elevation of Polaris above the horizon will currently locate you within a degree of your actual latitude on Earth.

The bright southern star Sulbār (also called Achernar, Arabic for "end of the river") served as another reference point for calculations of latitude; according to ibn Mājid, the *mu'allim* ("navigators") who spent weeks crossing the Indian Ocean in tandem with the monsoons relied heavily on it:

> By your life, had it not been for Sulbār, the pilots
> Of the fig, the date and the betel would never have been guided.
> No instrument which they use over Madwara is like it
> As a guide . . .[26]

Ibn Mājid's praise of naked-eye observations notwithstanding, specialized instruments have long proved valuable in navigation. Some, including the quadrant and the astrolabe, began life as land-based aids, devised by astronomers and mathematicians, and were later simplified to make them seaworthy.

Foremost among readily available length-measuring devices, of course, are the independently movable parts of the human body: fingers, hands, arms, the landlubber's striding foot. In the 1150s an Icelander who had just visited the Holy Land declared that there a man could determine the altitude of the Pole Star by lying on the ground, putting his fist upon his raised knee, and raising his thumb from his fist. A Venetian sailing in the 1450s for the Portuguese crown described the Pole Star's altitude at a certain location along the coast of West Africa as "the height of a man above the sea." In the 1950s a commodore in Britain's Royal Navy still felt free to declare that even a modern navigator might approximate the altitude of a star by consulting the span of his wrist (eight degrees) or his hand (eighteen degrees) held at arm's length.[27] Today, too, any amateur astronomer knows that a fist held at arm's length spans ten degrees on the sky. This system works because people with big hands tend to have longer arms, preserving the standardized angles of measurement.

Early Indian Ocean navigators consulted the width of a typical knuckle but also the *kamal.* In its most streamlined form the *kamal* is a card-shaped piece of wood through whose center passes a piece of string, knotted at equal intervals representing units of latitude. One end of the string is held between your teeth and the other in a hand. You mark the altitude by pulling the string taut, parallel to the ground, and moving the card with the other hand until its top aligns with the target star, Polaris, and its bottom aligns with the horizon; the resulting number of knots sitting between your teeth and the card translates into the latitude. Used across the Indian Ocean until well into the nineteenth century, the *kamal* was deployed again in the late twentieth century on a recreated voyage from Oman to China, and its effectiveness reconfirmed. Marco Polo mentions that Chinese navigators relied on a similar instrument, the *qianxingban,* "polar star–aiming plates," which is a series of different-sized plates held at arm's length in such a way as to align top and bottom with star and horizon, respectively. The choice of plate depended on the altitude of the star. A millennium earlier, the Chinese estimated latitude with the *liangtianchi,* "star-measuring ruler."[28]

The real sea change in navigation, however,

came with the rapid spread of the seemingly magical magnetic compass. Now one could have an instant sense of direction, clouds or no clouds, stars or Sun, day or night.

Many countries have claimed authorship or at least awareness of crucial components of the compass. Both the ancient Greeks and the ancient Chinese saw that certain brownish stones drew iron to them; the "lode" in lodestone, the magnetic form of the iron-rich mineral magnetite, is Old English for "way." Some scholars confidently apply the term "compass" to objects that began to be used by Chinese navigators around AD 500 as sea routes to Japan were established, and the first mention of a south-pointing shipboard needle appears in a Chinese navigational text written in AD 1100. A resident of Amalfi, a southern Italian maritime power in the twelfth century, has traditionally been credited with the invention of the north-pointing mariner's compass, and a contemporary chronicler described medieval Amalfi itself as famous for showing sailors the paths of the sea and sky. The first Arab text to mention a compass, written in the thirteenth century, calls the instrument by its Italian name. To some historians, the fact that the Chinese referred to south-seeking needles and the Italians to north-seeking ones suggests the likelihood of independent invention.[29]

Whatever the origins, compasses worked,

and the way they worked was well understood in the Mediterranean by 1200, when a French writer described in detail how to rely on a compass to navigate by "the star that never moves":

> This is the star that the sailors watch whenever they can, for by it they keep course. [W]hile all the other stars wheel round, this stands fixed and motionless. By the virtue of the magnet-stone they practice an art which cannot lie. Taking this ugly dark stone, to which iron will attach itself of its own accord, they find the right point on it which they touch with a needle. Then they lay the needle in a straw and simply place it in water, where the straw makes it float. Its point then turns exactly to the star. There is never any doubt about it; it will never deceive. When the sea is dark and misty, so that neither star nor Moon can be seen, they put a light beside the needle, and then they know their way. Its point is toward the star, so that the sailor knows how to steer. It is an art that never fails.[30]

In other words, float a magnetic iron needle by attaching it to something buoyant, and it will invariably come to rest along Earth's north–south magnetic axis, with its point aimed north.

Soon came the pivoting compass needle

and the compass's essential partner: a radial diagram called a compass card or wind rose, divided into as many as sixty-four directions. At sea, a direction meant a wind, and each wind bore a name. With the aid of this new technology, the literate, numerate ship's pilot sailing the Mediterranean or the Black Sea in the early thirteenth century could confidently determine not only when he was heading Tramontane (north) or Ostro (south), Greco (northeast) or Sirocco (southeast), but also when he was heading Tramontane quarter Greco — and he could keep to his course by dead reckoning, a mostly reliable technique based on knowing the relative positions of one's departure point and specific or general destination, as well as the direction of the vessel's movement and the distance traveled, both of which are determined at regular intervals. The pilot's sons and grandsons, continuing in the family profession, would have even more aids: scaled marine charts and a pilot book filled with detailed sailing directions.

Imagine you're the captain of a Venetian vessel in the year 1320. You've just brought in a load of grain from Egypt, and you're now heading to the east coast of Spain with some prized Sardinian cheeses and another ship-load of grain, this one picked up at Constantinople. On the return trip you'll carry Spanish wool. Your masters and the navies they support have thus far kept Portuguese ships out of your path, and neither the Black Death nor the mounted Ottoman Turks have yet arrived on European soil. Your exceptional nephew at the University of Bologna has told you of two groundbreaking books that he says are relevant to your profession, though you have no intention of reading them: Fibonacci's *Book of Calculation* and Sacrobosco's *On the Sphere of the World.* The former includes a reader-friendly introduction to Hindu-Arabic numerals, including the indispensable 0; the latter is your era's preferred text for Astronomy 101. What you do read closely, and keep with you on board, is a hand-lettered copy of *Lo Compasso da Navigare,* which directs you on a clockwise circuit of the Mediterranean, and a *Toleta de Marteloio,* a series of tables showing you how to correct your course as you tack with the wind. You also have an exquisite portolan chart of the whole sea, showing distances, harbors, and major landmarks; it is carefully scaled and even signed by an illustrious Jewish cartographer from Majorca. On your oak table sit a pair of silver dividers and a silver ruler to work the chart. Your ship's compass, with its freely pivoting needle and its attached compass card, is safely housed inside a circular metal box; your sandglasses (you keep some spares) were blown in Venice.[31] Thanks to all this state-of-the-art equipment, you can figure out which direction your ship is heading; you can monitor the advancing and waning of night and thus the hours of the watch; and you know the distance to the nearest port, how many days' sailing it will take to get there, and what to look for as you approach it. Unlike your Arab, Indian, Polynesian, and Chinese counterparts, you extract almost no information from the stars, and since you stay within the confines of your home sea, the Mediterranean, you find little cause to heed latitudes and even less to ponder longitude.

But the known world had already reached well beyond the Mediterranean, and change was coming fast. Centuries had elapsed since the Vikings had begun to ship dried cod to Britain, since Icelanders had sojourned on Vinland (Newfoundland), since Polynesians had settled on New Zealand, and since the Chinese had crossed the Arabian Sea and learned that certain inhabitants of East Africa drank fresh ox blood mixed with milk. A handful of recent European-drawn maps had begun to show the southern half of Africa, which Ptolemy had hinted, more than a millennium earlier, extended well below the equator. Plutarch knew Africa was circumnavigable, and Alexander the Great knew it could be reached by sea from the mouth of the Euphrates. But in the interim it had, so to speak, gone missing. By the end of the thirteenth century, a Venetian had escorted a Mongol princess from the South China Sea to the Persian Gulf, and a Genoese had built a castle in the Canary Islands. By the end of the fourteenth century, Arab and Indian traders had established themselves along the East African coast as far south as present-day Mozambique. At the beginning of the fifteenth century, a heavily armed fleet of more than three hundred Chinese vessels, commanded by the formidable eunuch admiral Zheng He and carrying nearly 28,000 soldiers and half a dozen astrologers, had sailed forth to impress and intimidate China's southern neighbors by a lavish display of both Ming treasures and military might.[32] Last but certainly not least from the perspective of Western Europe, the Portuguese had begun to sail the Atlantic far and wide.

Portugal's Prince Henry the Navigator, born in 1394, dedicated himself to the discovery of Africa's "River of Gold," to the erasure of Islam, to the gathering of slaves and pepper, and, according to the court chronicler of his day, to the fulfillment of his horoscope — the "inclination of the heavenly wheels" that inclined him to conquer new lands:

His ascendant was Aries, which is the house of Mars and exaltation of the sun, and his lord is in the XIth house, in the company of the sun. And because the said Mars was in Aquarius, which is the house of Saturn and in the mansion of Hope, it is signified that the Lord [Henry] should toil at high and mighty conquests, especially in seeking out things that were hidden from other men and secret, according to the nature of Saturn, in whose house he is. And the fact of his being accompanied by the sun, as I said, and the sun being in the house of Jupiter, signified that all his traffic and his conquests would be loyally carried out.[33]

There are many rational, strategic reasons why one might invoke the universe in the name of conquest. You might want to stage a nighttime attack at the time of the new Moon, giving you maximum darkness, as was done at the start of Operation Desert Storm in 1991. You might need to carefully monitor lunar tides during a naval invasion, to ensure that your ships don't run aground in shallow waters. You might decide to invade during a phase of high auroral activity, which will muck up the other side's radio communication. The reasons ascribed to Prince Henry, anchored in the pseudoscience of astrology, are neither rational nor strategic.

In his own day, it was well understood that Prince Henry — governor of the wealthy Order of Christ, which had replaced the wealthy Knights Templar of earlier centuries — was undertaking a crusade, the quintessential fusion of war, profit-seeking, exploration, and the imposition of foreign ideas. Writing in the twentieth century about space exploration, the American journalist William E. Burrows called it a "drive that was defined and tempered by politics and competition on every level," which could easily be said as well of Prince Henry's program. As Burrows says, "Exploration was always done for the wrong reasons. But it was done."[34] What he means, of course, is that whether or not the explorer is an explorer, exploration is hardly ever motivated by the desire to explore. Part the curtains of curiosity, and you'll find individuals hungry for political, cultural, or economic dominion funding the expedition.

Henry and his actual navigators could not have done their venturing without astronomy, a fact made explicit in the florid decoration of noble works of Portuguese architecture executed during and shortly after his lifetime. The exuberantly sculptured windowframes and archways, the floor mosaics and ceiling paintings of the vast Convento de Cristo and elsewhere repeatedly present the astronomer's armillary sphere linked with the Crusader's cross as well as with the vegetation of exotic lands. Historian Jorge Cañizares-Esguerra captures the association between astronomical knowledge and conquest embodied by Prince Henry and succeeding generations of Iberian colonizers: "[T]he cosmographer as knight, or the knight as cosmographer, was a hallmark of the Portuguese and Spanish fifteenth- and sixteenth-century colonial expansion." The gathering of knowledge, he argues, was "an expansion of crusading virtues." An influential mid-sixteenth-century book by a royal cosmographer, *Arte de Navegar,* presented ships' pilots as "the new knights, whose horses were their vessels and whose swords and shields were their compasses, charts, cross-staffs, and astrolabes."[35]

Prince Henry's first conquest was Ceuta, a Mediterranean town in what is now Morocco identified with the southerly Pillar of Hercules and piled high with African goods of great beauty and value. Under Henry's sponsorship and direction, many Atlantic islands, including the Azores, Canaries, and Madeiras, were tilled and grazed. Navigators in the prince's employ mastered a route around fearsome Cape Bojador, far out to sea and down past the cape's winds and currents, and rounded the westernmost point of Africa, reaching as far as Sierra Leone. Along the way, the captains recorded star altitudes at notable capes, islands, and river mouths, which astronomers back in Portugal translated into tables of latitude. During Henry's fiftieth year of fulfilling destiny, his brother,

143

the king of Portugal, granted him monopolistic rights over the lands discovered and all enslavable persons therein. Henry's death in 1460 barely interrupted the travels of the Portuguese. In 1473 Lopes Gonçalves crossed the equator; in 1488 Bartolomeu Dias rounded the "cape of storms" at the southern tip of Africa; in 1498 Vasco da Gama reached southern India by sea; in 1500, eight years after Columbus's first crossing of the Atlantic, Pedro Álvares Cabral arrived in Brazil. Their aims, like those of Henry and of the conquistadors who followed them, were "to serve God and His Majesty, to give light to those who were in darkness, and to grow rich, as all men desire to do."[36] In short, they were laying the cornerstones of empire.

Not that trade hadn't already made many people rich, as well as building a global economy across the Old World. Consider this: the Muslim soldiers who fought in the Middle East against the Crusaders wore chain mail from the Caucasus and wielded steel swords smelted in South Asia from sub-Saharan iron. As the Ottoman caliphs extracted taxes and the Chinese emperors extracted tribute (and invented paper money), merchants transported product from market to port and port to market. Much intercontinental trade was pan-Asian, private, and conducted by diasporas of blood, dialect, or faith: Jews, Hindus, Muslims, Armenians, Lebanese, Fujianese,

Gujaratis. The Indian Ocean was the cross-roads of a network of trading centers that stretched for thousands of miles, and local lords from the East China Sea to the eastern shores of Africa generally permitted merchants of every stripe and origin to enter their ports. But medieval Middle Eastern and Asian trade networks, however sprawling, were not colonial empires. The Muslim caliphate settled for the collection of taxes hefty enough to fund the army that safeguarded the highways, which allowed for the commerce that yielded life's luxuries. And the Chinese state, which grew plenty of sugar and other tropical delights within its own borders, had little cause to put money, personnel, and effort into creating colonies overseas.[37]

The Portuguese, on the other hand, representing king, country, and God, sought both control and colonies. Good ships and new-fangled guns gave them the advantage as they revived the practices of building forts, blocking trade routes, claiming trade monopolies, boarding foreign ships, and generally seeking to rule the waves and harbors. A key part of their program was to find routes free of Ottoman control and thus free of Ottoman tax collectors.[38]

Venturing far across the Atlantic and Indian oceans in the fifteenth century, the Portuguese had need of knowledge and instruments more elaborate than those used by the average captain crisscrossing the Mediterranean or exploring the east coast of Africa or taking soundings for depth, sampling bottom silt, and monitoring the tides in the fogbound English Channel or the Baltic. Portugal's sailors may have been wary of unfamiliar concepts and new techniques, but Portugal's ocean-bound navigators had little choice other than to use the mounting numbers of charts, pilot books, and mathematical rules. Far more than their predecessors, they consulted the stars as well as the compass. Having mastered the art of dead reckoning, they continually checked their bearing against the height of the Sun or the Pole Star as measured with a quadrant or mariner's astrolabe and used arithmetic and geometry to recalculate that bearing when winds and currents threatened to send the ship off course. Instructions for the use of the quadrant cautioned that the Pole Star was not entirely stationary and should be observed only when its two Guards lay east–west. Compilations of tables, known as ephemerides, listed the predicted daily position of the sky's major occupants. Tables of midday solar altitude in various cities helped navigators sail to the correct latitude and then keep to it as their ship headed due east or west — known as running down the latitude.[39]

The drive to amass navigational information that would grant an advantage over one's seagoing adversaries kept accelerating, and the stakes kept rising. Faith, glory, and commerce, writes the historian Emilia Viotti da Costa, were the driving motives. The pope himself pronounced Portugal's African project to be a "just war," and three papal bulls concerning it were issued during Prince Henry's last ten years of life. The first, in 1452, proclaimed the right of the Portuguese king to attack and enslave non-Christians and to confiscate their goods and lands. The second, in 1455, specified that this right applied to Africans from Morocco down to Cape Bojador, who

> had lived in perdition of soul and body: of their souls in that they were yet pagans without the custom of reasonable beings . . . and worst of all through the great ignorance that was in them, in that they had no understanding of good, but only knew how to live in a bestial sloth.

Translation: If you despise the way certain people live, you're entitled to take everything they own, and you're officially free to use force to do so.[40]

A quarter century after Henry's death, King João II of Portugal (*o Príncipe Perfeito,* "the Perfect Prince") picked up where his uncle had left off. In 1484 he called together a group of savants from across Europe to work out rules for calculating latitude based on direct observation of the Sun's midday altitude. Their findings were published in a comprehensive manual of navigation called the *Regimento do astrolabio e do quadrante.* It includes a list of latitudes spanning the territory from Lisbon to the equator, nearly all of which are correct to within half a degree. It even includes a translation of Sacrobosco's *On the Sphere of the World.* Word was getting around again: the world was not flat. Geographers began to wrap their maps around a globe, while astronomer-astrologers kept themselves busy refining the coordinates of naturally occurring astronomical objects and phenomena.[41]

Besides seeking slaves, converts, and knowledge, the expeditions of the fifteenth century — the early part of what the maritime historian J. H. Parry calls the Age of Reconnaissance — sought gems and precious metals, spices and medicaments, good land for growing sugarcane, grapes, coffee, and tobacco, new fishing grounds, new pastures for sheep, new sources of timber of a suitable size for masts and palaces.[42] But as each ship's hold filled with the products of faraway places, and each ship's captain returned to tell his tales to eager listeners, it became increasingly obvious that all this adventuring, conquering, colonizing, commodifying, and profiteering also required that every captain become expert at determining the exact location of his vessel, his destination, and his home.

Fifteenth-century navigation was still a huge challenge. Few navigators could master the contents of the *Regimento*. There was no widely accepted north–south baseline against which to gauge distances east or west. There were no seaworthy chronometers and nothing approaching an odometer or speedometer. Quadrants and astrolabes, which depended on a stable gravity vector to stay vertical, were ill-suited to rough open seas. The compass needle had to be periodically remagnetized.

And the problems didn't end there. Mariners suspected magnetic variation but had no reliable means of isolating it, so they fiddled with the findings of their compasses in unhelpful ways. Without international standards of measurement, they used conflicting units of the mile, league, stade, and degree, so they ended up assigning varying equivalents to the distances noted in ancient literature. Their old-fashioned planar maritime charts suffered not only from a lack of up-to-date information but also from a disregard for Earth's roundness. That roundness produces a (then mysterious) convergence of meridians as you approach the Poles, which means that heading sixty leagues east along the equator will take you to a different meridian than if you headed sixty leagues east along the Tropic of Cancer. Yet as late as the late seventeenth century, a navigator could rely on a plane chart, lose his ship, and nonetheless become a Fellow of the Royal Society.[43]

As for matters of hunger and health, a well-stocked expedition might carry enough pickled pork, salt fish, ship's biscuit, cheese, onions, and dried beans to fill the seamen's stomachs, and enough wine to give each man a liter and a half a day, but the casks of water soon turned foul, and scurvy took a heavy toll.[44]

Despite all the handicaps, with every voyage Portugal's mariners and travelers added to empirical knowledge of what lay where in the western and eastern oceans, and what was seen when in the sky both above and below the equator. With every passing year, their findings and narratives revealed more errors in the maps and coordinates accompanying the much-read Latin translation of Ptolemy's second-century AD Greek *Geographike* that had come out in the first decade of the fifteenth century, and each round of errors triggered another round of updated maps and geographical treatises.

At the respectable age of forty-one, having already sailed the Atlantic as far north as Iceland and as far south as Ghana,[45] Christopher Columbus headed west from the

Canary Islands on August 3, 1492, expecting that he and his fleet of three ships would come upon Japan within several weeks, at a distance of some four thousand kilometers, and would subsequently reach the fabled Indies. The monarchs of Portugal, Spain, France, and England, and likely also the city-states of Genoa and Venice, had all turned down his proposition at least once. But after having second, third, and fourth thoughts on the matter, and after convening a group of experts — who saw that Columbus had used the wrong version of the mile to calculate the circumference of the round Earth, and had therefore come up with the wrong distance to his destination — Isabella I and Ferdinand II, already the rulers of Castile, León, Aragon, Majorca, Minorca, Sardinia, Sicily, and elsewhere, finally gave him the go-ahead "to discover and subdue some Islands and Continent in the ocean" on their behalf.[46]

That Columbus and his three-boat crew of ninety were not the first Europeans to cross the Atlantic does not diminish the ambitiousness of their agenda, the extent of their navigational challenges, or the magnitude of their eventual impact — no matter the errors of calculation or the failure to reach their intended destination. Nearly every man aboard was a seaman; none were soldiers, and weapons were few. Although he later complained that "neither reason, nor mathematics, nor maps were of any use to me," Columbus had consulted maps, charts, globes, books, and instruments, especially the compass. He read Marco Polo's *Travels* and the soon-to-be Pope Pius II's *Historia rerum ubique gestarum,* based on Ptolemy's *Geography.* He read and copied a June 1474 letter to the king of Portugal from the Italian cosmographer Paolo dal Pozzo Toscanelli, who declared that the shortest way to get from Lisbon to China was to head west, across a nearly empty Atlantic, rather than to go around Africa, and that the distance in a straight line was nearly one-third the circumference of the Earth. He and perhaps his honored cartographer brother Bartholomew read and extensively annotated Pierre d'Ailly's cosmography *Imago Mundi.* Like many people of their time, whether learned or merely literate, they both almost certainly read the hugely popular *Travels of Sir John Mandeville,* a mid-fourteenth-century mishmash of fact, fiction, and faith. They studied recent world maps that suggested the possibility of heading west to arrive at the East. Bartholomew's own maps of the late 1480s, in fact, show that the brothers may have altered and invented a more attractive geography so as to more effectively persuade their royal patrons to fund the Indies expedition.[47]

Columbus consulted aids to reconnaissance

willingly. The masters and pilots of his ships would have resisted both reading and calculation; their expertise came from hands-on, hard-won experience maneuvering real vessels within sight of real eastern Atlantic coastlines. Yet even if every man aboard had been a mathematician and literary scholar, of what use were charts, handbooks, and sailing directions in unknown waters? Columbus therefore turned to dead reckoning, the Pole Star, and the compass.[48]

Star and compass read differently in different locations, however. Season, time of day, and latitude affect the former; magnetic variation affects the latter, as Columbus found, much to his distress: "The needles declined north-west a full point. In the morning the needles were true. The star appears to change its position, not the needles."[49] Moreover, knowing only the relative "easting" and "northing" would get a navigator (or land surveyor) only so far. If he wished to record exactly where an impassable carpet of seaweed, a luxuriant pearl fishery, or a fortuitous defensive promontory lay, he had to know exactly how far east and north he was — and what he was east and north of. A sophisticated navigator might know how to calculate his craft's geometric relation to the sky's salient inhabitants, but to note a position in such a way as to be unambiguously and automatically understood, he needed a standard point of reference — two, in fact. He needed coordinates, a grid, a graticule with both an equator and a prime meridian at right angles to it.

With its main parallel and its prime meridian crossing at the Aegean island of Rhodes, Eratosthenes's ancient world map had a grid that Hipparchus found arbitrary. Ptolemy's map, with its prime meridian passing through the westernmost known islands in the Atlantic, had a more astronomically inspired grid. The maps of Columbus's day — made for scholars and kings, and treated as classified information — had something of a grid, while the marine charts — made for sailors — had none. On the earliest surviving terrestrial globe, Martin Behaim's "Erdapfel" ("Earth Apple"), completed in 1492, there's a minimal grid that includes the equator, the tropics, and one meridian.[50] When the New World entered the picture, the question of paral-lels and meridians got messier.

When grabbing land, who is entitled to what and who gets to decide are not minor issues. For the crowns of Portugal, Spain, and Christendom, the choice of deciders was obvious: themselves. After all, the actual inhabitants of all that attractive New World real estate "lived in perdition of soul and body," were filled with "great ignorance," and "had no understanding of good," so why ask them?[51] In 1493 the pope issued the first of a new spate of papal bulls meant to regulate the explorers' land seizures, giving Spain the lion's share of everything. Predictably, Portugal was unhappy. As a result, in 1494 Spain and Portugal, both Catholic nations, negotiated and signed the Treaty of Tordesillas, which was reinforced by a papal bull a dozen years later. The treaty fundamentally split the Western world in two: everything east of a north–south line 370 leagues west of the Cape Verde islands would belong to Portugal, and everything west of the line would belong to Spain. In 1529, in the complementary Treaty of Saragossa, the duo split the other side of the world along a line 297.5 leagues or seventeen degrees east of the Moluccas — the so-called Spice Islands, home of the prized clove tree. Portugal ended up with about 191 degrees worth of the world's girth, and Spain with about 169 degrees. So the conflicts carried on.

The dividing lines that Spain, Portugal, and the pope adopted had nothing to do with astronomy or mathematics or the science of geography. They were territorial markers, battle lines, private fences, announcements of yours and mine. Neither treaty served up a universal prime meridian. Meanwhile, Iberian expeditions proceeded apace.

In September 1522 a Portuguese navigator named Juan Sebastián del Cano — who, as part of an expedition of five ships and almost three hundred crew under Ferdinand Magellan's command, had sailed from the southern Spanish port of Sanlúcar three years earlier — returned to Spain with eighteen men (sans Magellan himself, who had been killed in battle) on the sole surviving ship, the *Victoria*. Those eighteen had thus completed the first circumnavigation of the globe. Along the way, Magellan's men inadvertently discovered the international dateline — or rather, they discovered the need for one. Antonio Pigafetta, an Italian nobleman/knight who joined the expedition as a volunteer, served as an occasional diplomat, and kept an account of "all the things that had occurred day by day during our voyage," described "the mistake of one day which our men discovered" at their last port of call in Portugal before returning home to Spain:

And we charged our men in the boat that, when they were ashore, they should ask what day it was. They were answered that to the Portuguese it was Thursday, at which they were much amazed, for to us it was Wednesday, and we knew not how we had fallen into error. For every day I, being always in health, had written down each day without any intermission. But, as we were told since, there had been no mistake, for we had always made our voyage westward and had returned to the same place of departure as the sun, wherefore the long voyage had brought the gain of twenty-four hours, as is clearly seen.[52]

Three and a half centuries later, the international date line, along with the corresponding prime meridian, would be formally established at the International Meridian Conference in Washington, DC. The date line would be a line running from North Pole to South Pole, crossing the middle of the Pacific Ocean exactly halfway around the world, 180 degrees, from the prime meridian at zero degrees longitude. The prime meridian itself would pass pole-to-pole right through the Royal Observatory Greenwich, near London.

Though as yet unaffected by prime meridians and missing days, fifteenth-century Portuguese marine charts began to look a little more like maps: while still planar, many show a meridian marked off with latitudes and drawn north–south through Cape St. Vincent, Portugal's southernmost headland. Soon the maps, if not the charts, would show Earth's landmasses and coastlines in reasonable proportion and detail. Often the cartography is embellished with references to ownership and allegiance: national flags, coats of arms, religious iconography.[53]

Cartography helped conceptualize and

display the "theater of the world," and the stage itself was steadily increasing. A map was the preeminent portable expression of geographical and cosmographical understanding. During the sixteenth century, as noted by the British historical geographer Denis Cosgrove, "the scale and wonder of global diversity — physiographic, climatic, biotic, ethnographic — overwhelmed the European episteme." Complicit in European oceanic expansion, the map made the case for world citizenship at the same time as it smoothed the way for Western dreams of subjugation and empire. While the cartographer/cosmographer would likely have been a humanist, cosmopolitan scholar who embraced religious tolerance, his Iberian royal patrons would have been bent on aggrandizement and religious hegemony.[54]

In 1569 one of those humanists, the Flemish cartographer Gerardus Mercator, produced a world map, a *mappa mundi,* which he called a "new and augmented description of Earth corrected for the use of navigation," with meridians, parallels, and sea routes all projected onto a huge rectangle comprising twenty-four separate sheets of paper. Meanwhile, Iberia persevered in its search for cartographical consensus, going so far as to issue questionnaires to its ships' pilots in hopes of determining the latitudes and longitudes of its New World conquests.[55]

In the final decade of the sixteenth century, as ships were being enlarged and redesigned to carry more and heavier gunnery, and their captains were learning to excel in both fighting and navigation, a British mathematician-astronomer-cartographer at Cambridge named Edward Wright applied himself to mastering Mercator's maps and producing practical charts suitable for seamen.[56] Other nation-states followed suit, developing their guns and fleets and cultivating cartographers so that they could dispute Spain's and Portugal's claims to territory and amass enough funding to bypass Genoese and Venetian financiers.[57] The ships of the Indian Ocean's traditional navigators could not keep up. By the close of the seventeenth century, Europeans had sailed to, trod upon, confronted the inhabitants of, and extracted products and persons from nearly every landmass on Earth — mapping as they went.

Astronomy and the natural sciences were indispensable to Europe's voracious seafaring empire builders. "Eighteenth-century monarchs," writes historian Joyce E. Chaplin,

> dispatched men of science to the far ends of the earth in order to claim, not only sovereignty over land and sea, but cultural supremacy through the exercise of learning and the gathering of knowledge on the far side of the world. These goals came to-

gether perfectly in Captain James Cook's three voyages into the Pacific Ocean.[58]

Funded by Britain's Royal Society, the first of Cook's voyages was timed to coincide with a rare event: the 1769 transit of Venus across the face of the Sun, visible only from the South Pacific. One of the biggest scientific unknowns of the day was the physical size of the solar system. Although astronomers had figured out planetary separations in units of Earth–Sun distance, they did not know the Earth–Sun distance itself. But if observers in more than one location, separated by known distances, could precisely time the duration of Venus's transit, through triangulation they would be able to calculate the distance between Earth and the Sun and thus deduce the distances to all other planets in the solar system.

The transit made a good top story for the voyage, but it wasn't Captain Cook's only directive. After heading for newly discovered Tahiti and setting up an observatory there to monitor the transit, Cook and his crew of eighty-five — plus ten civilians, including four artists and one astronomer[59] — were to find and chart other islands in the area and, most important, discover *Terra Australis Incognita,* a mythical continent lurking in the southern reaches of the globe. If they didn't find Terra Australis, they were to search for and explore other lands instead. In other words, their other job was to augment existing maps.[60]

But to what ends?

Like a calendar, a map — though shaped by scientific thinking — is a statement of political and social power. Writing shortly after World War II, the British historian of navigation E. G. R. Taylor remarked that "during the European wars of the 18th century it was discovered that an accurate map is a weapon of war. And so it remains." Forty years and several wars later, the British-born historian of cartography J. Brian Harley offered a postmodernist articulation of a similar idea, stressing Foucault's idea of power-knowledge: "cartography is primarily a form of political discourse concerned with the acquisition and maintenance of power." David Turnbull pointed out that maps "connect the territory with the social order" and therefore, quoting Pierre Bourdieu, "naturalise the arbitrary." Novelist Vikram Chandra has also weighed in on the meaning of maps: "A map is a kind of conquest, the precursor to all other conquests. . . . [O]ne kind of knowledge can conceal another. Information nests inside information."[61] And if the "knowledge space" embodied in a map is indispensable to warmakers and other practitioners of power, the map is nearly worthless in peacetime unless its measurements and delineations arise from knowledge that is both shared and internationally binding. For the monarch, the navigator, the admiral, and the general, imperfect cartography was a hazard.

During his first voyage to the South Pacific, James Cook carefully charted Australia's east coast and promptly claimed ownership of it on behalf of the British crown. Within two decades, Great Britain had established a penal colony at Sydney Cove: the Colony of New South Wales. Convicts, some in leg irons, some in chain gangs, became the workforce of British colonization across Australia. Britain wasn't the only power interested in obtaining a precise picture of Australia's coastline. The Dutch — seeking spices that could help fund their military operations against Spain — had already spent the previous century and a half charting the north, south, and west coasts. The French, too, explored and charted the landmasses of the South Pacific. One thing is certain, though: absent the hegemonistic agenda of the British empire builders, nobody would have measured the 1769 transit of Venus.

Before the International Meridian Conference of October 1884, and even for decades afterward, the world was in a muddle in the matter of determining time and place.

Time had long served as the marker of distance, if not place. The unit used by the ancient Greeks to measure distances on land was the "day's journey," on the open sea the "day's sail." Medieval English mariners were advised to "go south a glass or two" — that is, to sail southward for the time necessary to drain a sandglass.[62] Medieval Arab navigators marked distances traveled in *zam*s, three hours' sailing. Even today, in a car-loving location like Los Angeles, locals will tell you that LAX is thirty minutes from the Staples Center.

Scientists turned units of time into units of angular measure: the degree divided into the minute and the second. For everyone else, units of every sort remained a local matter, subject to great variation. The distance represented by, for instance, the ancient Greek stade (the length of one footrace, which gave rise to the stadium) varied so much from region to region that it could barely serve travelers as a unit of length, and so the conquering Romans replaced it with the mile. Meanwhile, the width of a seaman's middle finger held at arm's reach, whether he be fat or thin, marked a span of two degrees.[63]

Yet place remained elusive, as did the problem of finding the longitude — indispensable when ascertaining place. From Hipparchus in the second century BC to Kepler, Galileo, Newton, and other luminaries from the sixteenth to the eighteenth centuries, nobody was able to figure out how to achieve it with precision. This involved devising a rigorous system and good instrumentation with which to measure it; choosing a widely acceptable zero point, or meridian, from which to start measuring it; and convincing everybody else to adopt both measurement and meridian. In fact, "finding the longitude" became slang for the pursuit of a task either insanely difficult or just plain absurd.[64]

But difficulty did not obviate necessity, and the founding of France's Royal Academy of Sciences and the Paris Observatory during the reign of Louis XIV, and Britain's Royal Observatory Greenwich during the reign of Charles II, had a lot to do with the need to resolve the issue. The better-known sea lanes were filling up with massive vessels, laden with cargo and cannon. Merchants were pursuing their fortunes, kings were pursuing their empires, and privateers and pirates were pursuing everyone. In the absence of a precision system of longitude, it was not only courageous but also foolhardy, greedy, and suicidal to seek new routes to new places. And so, in March 1675, a twenty-eight-year-old ordained deacon, John Flamsteed, was chosen the first Astronomer Royal of Britain and charged "forthwith to apply himself with the most exact care and diligence to rectifying the tables of motions of the heavens, and the places of the fixed stars, so as to find out the so-much desired longitude of places for the perfecting [of] the art of navigation."[65]

Among the many spots used by philosophers and astronomers over the centuries to mark the meridian for zero degrees longitude were Ferro, in the Canary Islands; Ujjain, in the Indian state of Madhya Pradesh; the "agonic line" (a line along which true north and magnetic north coincide, but not forever) that passed through the Azores; the Paris Observatory; the Royal Observatory at Greenwich; the White House; and the Church of the Holy Sepulcher in Jerusalem. Among the proposed yardsticks by which to ascertain degrees east and west of zero were an eclipse of the Moon or the Sun; an eclipse of Jupiter's four Galilean satellites; an occultation of a star by the Moon; an excellent compass, impervious to variations in terrestrial magnetism; and the joint efforts of an excellent clock, a fleet of gunships, and a fleet of vessels equipped for a sound-and-light show.[66]

If you relied on an astronomical event, you would consult precise and exhaustive ephemerides for a known meridian and then compare them with your own observations, performed wherever you happened to be, reckoning fifteen degrees of longitude for each hour of time difference, since twenty-four hours' worth of fifteen degrees equals a full 360-degree turn of the Earth.

Easier said than done.

For one thing, ephemerides were still inexact. For another, you'd need a long, powerful telescope — and how would you keep such an unwieldy object unmarred by salt air, and steady on a heaving ship? Having faced these difficulties in 1764 when trying to observe Jupiter's satellites at sea, Reverend Nevil Maskelyne, author of the *British Mariner's Guide* and the first *Nautical Almanac,* opined, "I am afraid the complete Management of a Telescope on Shipboard will always remain among the Desiderata."[67]

Surely a reliable portable timepiece would be a better solution. It would "enabl[e] mariners," writes Dava Sobel, "to carry the homeport time with them, like a barrel of water or a side of beef." The rub was reliability. In 1500, even a fine clock sitting firmly on solid ground would generally accumulate an error of ten or fifteen minutes with each passing day. But that didn't faze Regnier Gemma Frisius, a Dutch mathematician who proposed that a good clock, set to the exact moment a ship left the dock, could serve as a stable point of comparison for the local time as ascertained at sea by Sun, star, or other means — assuming that the clock's exactness could be preserved despite the moisture, cold, heat, salt, gravity, and tumult.[68] Quite a task. Not until 1759, after thirty years of effort, did a provincial English craftsman named John Harrison manage to implement Gemma's proposal.

Harrison undertook the project not out of enthusiasm for a challenge or concern for his shipwrecked countrymen but because in the summer of 1714 the British parliament had, in desperation, put up a series of substantial cash prizes for a solution to the longitude problem. Spain had been the first to offer a prize, in 1598; Portugal, Venice, and Holland had followed suit — but to no avail, which is why France and Britain soon turned to the founding of scientific academies, the building of observatories, and the luring of Europe's name-brand astronomers, still to no avail. Throughout the seventeenth century, neither wrecks nor rewards led to longitudinal certitude. Meanwhile, empire building accelerated and maritime tragedies multiplied.

Then, in 1707, Britain suffered an especially horrible wreck: a fleet of Her Majesty's warships under the command of Admiral Sir Cloudesley Shovell (spellings vary) foundered on the Scilly Isles, causing the loss of four vessels and the death of two thousand men. A coalition of dismayed ship's captains, naval commanders, and London merchants soon petitioned the government to offer "due Encouragement" so that "some persons would offer themselves" to the task of "Discovery of the Longitude." The method and mechanism were unspecified. Parliament consulted Newton, Halley, and other notable scientists, drafted the Longitude Act, and set up the Board of Longitude to vet proposals and results. The guidelines were clear: £20,000 for accuracy within a margin of error of half a degree, £15,000 for two-thirds of a degree, and £10,000 for one degree. The accuracy would be assessed on a voyage between the homeland and the West Indies aboard a British ship. Since such a voyage would take six weeks, any mechanism that lost or gained more than two minutes — the time equivalent of half a degree — over the course of the journey could not fetch the top prize.[69] Sounds strict, until you consider that being off by half a degree is like heading for Times Square in the heart of Manhattan but ending up across the Hudson River in Plainfield, New Jersey, or telling your navigator you want to go to Fort Worth, Texas, but getting dropped in Dallas.

John Harrison fashioned not just one but several chronometers, whose accuracy exceeded the most stringent demand of the Longitude Act. The first, completed in 1735 and known as H-1, was an intricate brass tabletop contrivance that ran on springs, wheels, rods, and balances; the fourth, H-4, completed in 1759, was an exquisite outsize watch that lay supine in a cushioned box and ran on diamonds and rubies. Of the latter, its maker declared, "I think I may make bold to say, that there is neither any other Mechanical or Mathematical thing in the World that is more beautiful or curious in texture than this my watch or Timekeeper for the Longitude."[70]

Unswayed by H-4's beauty, powerful members of the Board of Longitude — fervent advocates of finding the longitude by comparing the Moon's observed angular distance from major stars with the distances listed in continually updated tables compiled by the world's top astronomers — fought for decades against Harrison's receiving the money and recognition that were his due. Instead they kept presenting him with dribbles of interim funding, new conditions, new insults, and eventually the outright confiscation of his creations by his most dedicated enemy: Reverend Nevil Maskelyne, a lunar-distance partisan and now Britain's Astronomer Royal. King George III (the same monarch whose "injuries and usurpations" are enumerated in America's Declaration of Independence) finally stepped into the fray on behalf of the elderly clockmaker in 1772, and the next year Parliament decided in his favor. Never, however, did the unrelenting Board of Longitude itself award Harrison the top prize, and never did he receive the full £20,000 to which he was entitled.[71]

What Harrison did receive, however, was vindication from James Cook, who carried an exact replica of H-4 on his second voyage to the Pacific, in 1772–75. As valuable to navigation as a sharp-eyed person scanning the waters from a ship's bow, Harrison's chronometer endowed the word "watch" with new meaning. This timepiece, wrote Cook, "exceeded the expectations of its most zealous advocate and . . . has been our faithful guide through all vicissitudes of climates." He referred to it as "our trusty friend the Watch," "our never failing guide, the Watch," and asserted that "indeed our error (in Longitude) can never be great, so long as we have so good a guide as [the] Watch."[72] With its help, he crossed the Antarctic Circle, conclusively disproved the existence of a massive southern continent extending well north of Antarctica, claimed some chilly islands for Britain, and charted regions of the South Pacific so accurately that twentieth-century mariners continued to depend on his findings.

John Harrison died in 1776, but even before he was laid to rest, a skilled assistant had begun to make knock-offs of H-4: the cheaper and less functional K-2 and K-3. The race for an affordable chronometer was on. Within a decade or so, competition among chronometer designers had become almost as fierce as the original race to discover the longitude. In the service of both commerce and conquest, on behalf of the East India Company as well as the Royal Navy, ship's captains spent their own money to buy, not just one, but often several chronometers, so that they could be cross-checked with one another. Smaller and cheaper versions of Harrison's invention became essential equipment. The navy kept a stash at Portsmouth. In 1737 there was one lone marine timekeeper in existence; in 1815 there were about five thousand. HMS *Beagle,* whose task in 1831–36 was to register a circle of longitudes around the Earth, carried twenty-two chronometers — in addition to a then-unknown twenty-something naturalist named Charles Darwin.[73]

But until 1884 the world was unable to agree on an official Earth-wide midnight hour at which an Earth-wide day would begin at an agreed-upon place, so there was no recognized zero point from which geographic eastward and westward would originate. Preferences regarding the designation of zero degrees longitude depended more on nationality, religion, and patriotism than on the obvious utility of having a common international standard for time and place. Astronomers at the Royal Observatory Greenwich had long obtained and maintained precise data on celestial coordinates for the stars passing overhead — coordinates based on a prime meridian that traversed the site of their own telescope. Early eighteenth-century Europeans tended to use the Paris Observatory as their zero-degree reference for longitude on land; nineteenth-century Europeans tended to use the Greenwich observatory for longitude at sea.[74] By the late nineteenth century, ships' captains, railway magnates, armies, navies, astronomers, geographers, and hydrographers could wait no longer for complete consistency. Agreement had to happen.

At long last, compelled by an act of the US Congress, a conference took place in 1884 at the State Department. Twenty-five nations sent representatives. Sixteen sent diplomats rather than scientists, signaling lack of serious intent. One of the earliest resolutions had to do with whether a group of invited astronomers, representing the broad interests of science, would be free to contribute their thoughts to the discussion as they saw fit. It failed.[75] After enduring the first several sessions, the reporter for the weekly journal *Science* complained, "The time has been mostly taken up with political diplomacy and sentiment." An irritated British representative, Lieutenant General R. Strachey, encountering resistance to the prospect of full international agreement and precision, declared that "longitude was longitude, and as a geographer he must repudiate the idea of first-class longitudes for astronomical purposes and second or third rate geographical longitudes." An equally irritated American representative, the astronomer Lewis Rutherfurd, pointed out that "the delegates must have studied the matter before coming here; and that no one would be likely to come unless he knew, or thought he knew, something about the matter."[76] Altogether a contentious affair — a forerunner of early twenty-first-century climate conferences.

In the end, on October 22, 1884, the assembled delegates bowed to the inevitable and acknowledged the benefits of adopting "a single prime meridian for all nations, in place of the multiplicity of initial meridians which now exist." They agreed it would bisect the base of a very special telescope at the Greenwich observatory. Henceforth there would be a "universal" day "to begin for all the world at the moment of mean midnight of the initial meridian" and that "the astronomical and nautical days will be arranged everywhere to begin" at that same moment.[77] Not until 1911, however, did France officially accede to the Greenwich-based meridian.

Far into the foreseeable future, even as the continents drift and national borders are redrawn by force or by justice, Earth's hard-won coordinate system of latitude and longitude will persist as a frame of reference. But not for everyone and not for all purposes. One century after the International Meridian Conference of 1884, the sky-and-telescope-based Greenwich prime meridian lost its overarching authority to a more refined meridian, based on Earth's global gravitational field and established by pulses of laser light aimed at reflective satellites. Because of the uneven distribution of mass in Earth's crust and mantle, if you head straight down from the original prime meridian you will not intersect with the center of our planet. But if you follow the upstart, "geodetic" meridian — 102 meters east of Greenwich's traditional prime meridian — you'll pass right through Earth's exact center of mass.

The US Department of Defense had been working on a geodetic meridian since early in the Cold War. By the 1980s, new techniques and greater quantities of data made it possible for Earth and space scientists to agree on a viable, internationally consistent system that, having been adopted by the US Defense Mapping Agency in 1984 and incorporated into America's GPS constellation, has become the global standard for satellite navigation and the basis for Universal Time.[78] Once again, in a pattern as old as civilization, stars and bars joined hands for the sake of ever greater exactitude — exploiting each other's needs, passively and actively achieving each other's ends.

4
ARMING THE EYE

Unassisted seeing is a weak way to engage with the glories of the universe. Without optical aids to bridge all those physically unbridgeable distances, we can't come close to knowing what's out there. Human beings need huge amounts of help just to recognize what takes place in the visible cosmos, let alone the multitudinous events happening in nonvisible bands of light.

On its own, the human eye is a good but not great detector, capable of resolving the visual data in only about one-sixtieth of one degree of a complete, 360-degree circle. With its embarrassingly narrow wave-length range of 400–700 billionths of a meter, the human retina detects a mere sliver of the electromagnetic spectrum. That sliver has been assigned a self-evident name: visible light. If you think of light as a wave traveling through space, the wavelength is simply the distance between two consecutive crests. Clock how many crests pass per second, and you get the frequency. Whatever the speed of the passing wave, the shorter the wavelength, the higher the wave's frequency.

The full electromagnetic spectrum may extend forever in both directions: toward ever longer wavelengths, perhaps limited by the size of the universe itself, and toward ever smaller wavelengths, perhaps limited by quantum physics. Currently we have the technology to detect wavelengths ranging from less than a hundred-billionth of a meter (high-frequency gamma rays) to many hundreds of kilometers (extremely low-frequency radio waves). That's a factor of quadrillions.

Millennia ago, if you wanted to look up at the sky or across a broad valley, you might have used a long sighting tube to focus your attention and reduce glare, as did Aristotle and probably his predecessors. But no empty tube, however long — and whether cast in gold by an ancient Assyrian metalworker, carved from jade by an ancient Chinese artisan, or fastened to an armillary sphere by a mathematically astute medieval pope[1] — would improve your physiological capacity to detect the planet Neptune or assess the size of a rival army or navy massing on a distant shore.

Put a couple of lenses in the tube, though, and you've got yourself an optical telescope.

A tool for augmenting the senses, the telescope enables you both to *detect* things too faint to see and to *resolve* detail where your eyes would otherwise fail you. First, it shows you that an object exists; then, by revealing the object's shape, motion, and color, it hints at what the object might be. The telescope's task is to collect at a distance as much visual information as it can, and feed it to your brain via your eyes.

Whether you're scanning the enemy or scanning the skies, every bit of information delivered by your telescope rides a beam of light. Structurally, a telescope is little more than a bucket for catching photons. Whether your goal is detection or resolution, the bigger the diameter of your light bucket, the more photons you'll catch. The collection area increases as the square of the diameter. So if you triple the diameter, you increase the telescope's power of detection nine-fold. Resolution depends on the diameter of your telescope divided by the wavelength of light you're observing. To maximize resolution, you want a bucket that's much, much wider than your chosen wavelength. For visible light, with its wavelengths measured in hundreds of nanometers, a several-meter-wide bucket does swimmingly. And just as the wine lover wants a wineglass to be so thin that it is nearly absent as a boundary between lips and wine, the astrophysicist wants the design limitations of the telescope, the susceptibilities of the human observer, and the distortions of Earth's atmosphere to be as absent as possible from the data.

Assistance in seeing at a distance arrived just four centuries ago, in the form of a pair of cookie-sized lenses firmly fixed inside a tube and presented by a spectacle maker named Hans Lipperhey in September 1608 — right in the middle of the Catholic–Protestant conflict known as the Eighty Years' War — to Prince Maurice of Nassau, commander-in-chief of the armed forces of the United Provinces of the Netherlands. This tube was the first historically substantiated, honest-to-god telescope, although allusions to earlier ones abound. Within half a year Galileo had learned of Lipperhey's indispensable instrument and had built a better one of his own design.

Early telescopes gathered little light, and their images of distant objects, whether celestial or terrestrial, were blurry, distorted, and faint. The lenses were small and thick, made of imperfect glass, imperfectly curved and polished. Back in the days when, despite the panegyrics of the early writers, a telescope delivered barely more data than would an ordinary pair of opera glasses, its achievements were usually described in terms of magnification rather than resolution. Galileo's very first telescope — a lead tube with two store-bought spectacle lenses, assembled early in the summer of 1609 — made objects appear three times closer. As with the arithmetic that applies to the collecting area of telescopes, when we square the factor of three we get objects that are nine times larger than they would appear with the unassisted eye. By late autumn, Galileo had built himself a telescope in which objects appeared sixty times larger.[2]

Of course, seventeenth-century astronomers didn't know how bad their telescopes were. All they knew was how good their telescopes were compared with human vision. So they did manage to discover some marvelous things. In the summer of 1609 the English astronomer Thomas Harriot, scientific aide to Sir Walter Raleigh, saw the crescent Moon sufficiently well to sketch a few of its surface features: the earliest known portrayal of the Moon as seen through a spyglass.[3] That fall, Galileo, with a much better telescope at his disposal, saw and drew our Moon's mountains and craters, as well as other "very great and wonderful sights": a quartet of moons circling Jupiter; extra stars in the Orion nebula and the Pleiades cluster; and a pair of intermittent companions close to Saturn. Half a century later, looking through an even bigger, better telescope, Christiaan Huygens observed that Saturn's companions were actually two arcs of a ring. A mere twenty years later, through a still better telescope, Giovanni Cassini picked out two concentric rings, separated by a gap.

During the millennia before aerial bombing, the sky was the domain of air, light, rain, wind, and deities. There was no reason to imagine that military dangers could be circumvented by looking up. Armies advanced in waves on the ground. The notion that the sky is a place to be monitored for protection from human adversaries is a twentieth-century perversion. Monitoring the faraway terrestrial landscape, however, was a long-held dream of generals, opticians, navigators, and surveyors alike.

It so happens that when Hans Lipperhey arrived in The Hague in September 1608 to present what his letter of introduction called "a certain device by means of which all things at a very great distance can be seen as if they were nearby," intense peace negotiations were taking place, and the city was swarming with delegations of diplomats. The French were mediating between the Dutch representatives and their Spanish/Belgian adversaries, and both sides were internally divided about the wisdom of continuing to fight. Into the middle of all this walked the nice man from Middelburg with his optical invention, seeking promises of a patent and a pension. Not only did he get what he wanted, but his invention seems to have played an astonishing bit part in the negotiations.

According to an insider's account of the invention's wondrous capabilities, written in early October, just days after the commander-in-chief of Spain's armed forces left The Hague, "From the tower of The Hague, one clearly sees, with the said glasses, the clock of Delft and the windows of the church of Leiden, despite the fact that these cities are distant from The Hague one-and-a-half, and three-and-a-half hours by road, respectively." So impressed was the Dutch parliament with Lipperhey's device that they sent the instrument to Prince Maurice, saying that "with these glasses they would see the tricks of the enemy." The Spanish commander-in-chief, equally impressed, told Maurice's kinsman Prince Henry, "From now on I could no longer be safe, for you will see me from afar." To which Henry replied, "We shall forbid our men to shoot at you."

The writer then elaborates on the instrument's potential:

The said glasses are very useful in sieges and similar occasions, for from a mile and more away one can detect all things as distinctly as if they were very close to us. And even the stars which ordinarily are invisible to our sight and our eyes, because of their smallness and the weakness of our

181

sight, can be seen by means of this instrument.[4]

From birth, the telescope represented the convergence of war and astronomy. It was obviously a dual-use instrument. Any courtier could see that it would revolutionize both intelligence-gathering and skywatching. Which is why Lipperhey got his money, Prince Maurice got his "glasses," and Spain signed the Twelve Years' Truce with the young Dutch nation on April 9, 1609.

The Vatican, too, was made aware of the worldly implications of the invention. In a letter written to Cardinal Scipione Borghese just before the signing of the truce, the archbishop of Rhodes spends a full three paragraphs describing Maurice's new possession and announcing that a similar item is on its way to His Holiness by the next post. The Spanish commander, writes the archbishop, thought that Maurice "had procured this instrument in order in time of war to reconnoitre from a distance, or observe places he might want to besiege, or sites of encampments, or enemy forces on the march, or similar situations that might be turned to his advantage." Having himself tried out one such instrument and been quite impressed by what was visible ten miles away, the archbishop is certain it will "provide much diversion [and] pleasure" to his superiors.[5]

Not five months later, in late August 1609, Galileo Galilei — who described himself as "Florentine patrician and public mathematician of the University of Padua" — ascended Saint Mark's Campanile in the Republic of Venice, accompanied by the republic's senators, to demonstrate his own significantly improved spyglass. Having fulfilled his mission, he donated the spyglass to the senate and (successfully) petitioned the doge, the republic's chief magistrate, for his patronage. Other entrepreneur-inventors had been busy working up and demonstrating versions of this desirable new instrument, of which word had spread far and wide. One such inventor even seems to have petitioned the senators of Venice before Galileo did. But Galileo had a powerful Venetian connection who smoothed the way for him and conceivably even let him examine his competitor's version.[6]

Accompanying Galileo's donation was a written sales pitch to the doge:

Galileo Galilei, a most humble servant of Your Serene Highness, . . . appears now before You with a new contrivance of glasses, drawn from the most recondite speculations of perspective, which renders visible objects so close to the eye and represents them so distinctly that those that are distant, for example, 9 miles appear as though they were only 1 mile distant. This is a thing of inestimable benefit for all transactions and undertakings, maritime or terrestrial, allowing us at sea to discover at a much greater distance than usual the hulls and sails of the enemy, so that for 2 hours and more we can detect him before he detects us and, distinguishing the number and kind of the vessels, judge his force, in order to prepare for chase, combat, or flight; and likewise, allowing us on land to look inside the fortresses, billets, and defenses of the enemy from some prominence, although far away, or also in open campaign to see and to distinguish in detail, to our very great advantage, all his movements and preparations; besides many other benefits, clearly manifest to all judicious persons.[7]

What could be more militarily useful to a seventeenth-century seafaring republic than the capacity to monitor enemy vessels? Indeed, few things could be more useful to any sort of republic, in any century, than the capacity to monitor the enemy's movements anywhere: land, sea, air, space, or online. Eventually, satellites — descendants of the spyglass — would enable them to do so.

In the year 1267, more than three centuries before Hans Lipperhey put two lenses in a tube and betook himself to the nearest general, a scholarly Franciscan friar named Roger Bacon sent Pope Clement IV a hefty scientific treatise. Some of his thoughts were ahead of their time:

> [W]e can so shape transparent bodies, and arrange them in such a way with respect to our sight and objects of vision, that the rays will be refracted and bent in any direction we desire, and under any angle we wish we shall see the object near or at a distance. . . . Thus a small army might appear very large, and situated at a distance might appear close at hand, and the reverse. So also we might cause the sun, moon, and stars in appearance to descend here below, and similarly to appear above the heads of our enemies. . . .

No one, possibly including Bacon himself, followed through on his suggestion, whether because Bacon's concept was too spooky for its day or because glassmakers were not yet up to the task or because learned gentlemen had little interest in practical matters. By the sixteenth century, though, his writings had been dusted off and his ideas revivified.

One revivifier was the learned Oxford mathematician, astronomer, and all-round science maven John Dee, who owned at least one of Bacon's works. In his "very fruitfull praeface" to a 1570 English translation of

Euclid's *Elements of Geometry,* Dee told his readers that anyone wishing to "make true report . . . of the numbers and Summes, of footemen or horsemen, in the Enemyes or-dring" would "wonderfully helpe him selfe, by perspective Glasses." Less than a decade later, in a book titled *Inventions or devices. Very necessary for all generalles and captaines, or leaders of men, as well by sea as by land,* one William Bourne wrote that a pair of properly positioned lenses "is very necessary in diverse respects, as the viewing of an army of men, and such other like causes." And in 1589, in his best seller *Natural Magick,* Gio-vanbattista Della Porta spoke of the ancient "Glass" of an Egyptian king, "whereby for six hundred miles he saw the enemies ships com-ing."[8]

Earth's actual horizon sits no farther away than a few dozen miles from any observer, but exaggerations of the distance to the enemy are surely forgivable, given the perco-lating anticipation of what glass lenses would one day do. Little wonder, then, that when Lipperhey demonstrated his optical aid in The Hague in the autumn of 1608 before a group of military men, ministers, and media-tors, they instantly understood its military utility. The chief French negotiator lost no time in procuring two for the French court. By the following spring, not only had Galileo learned of the invention, but the archduke of

Austria and the pope each owned a telescope, foot-long spyglasses were for sale on the streets of Paris and Milan, and peace had been declared between Catholic Spain and Protestant Netherlands.[9]

The truce lasted until 1621. With the resumption of war, the Spanish commander-in-chief Ambrogio Spinola resumed his command. The siege and surrender of the fortress city of Breda in 1624–25 brought death to Prince Maurice and transient victory to Spinola, who is depicted graciously receiving the key to the city in an imposing canvas by the Spanish court painter Diego Velázquez. Grasped in Spinola's gloved left hand and positioned near the focal point of the painting, as if to emphasize its role in the victory, is a spyglass nearly two feet long.[10]

Few people in history have arrived at the end of life without witnessing war, and seventeenth-century Europeans were no exception. What distinguishes their chapter is the unprecedented level of commercialization and bureaucratization of the pursuit. European entrepreneurs, merchants, and rulers devoted huge amounts of money and effort to the improvement of weapons and the institutionalization of paid standing armies that numbered in the tens of thousands. Many of Europe's best scientists and inventors — in addition to considering questions related to commerce, mining, and marine transport — addressed themselves to matters directly or indirectly related to military technology: explosives, ballistics, velocity, air resistance, impact, innovative armaments, new methods of timekeeping, new means of surveying, and, of course, a raft of new sighting instruments.[11] In the words of the seventeenth-century Irish optics expert William Molyneux, to use a sighting instrument was, in effect, to arm oneself:

Tis manifest by Experiments, that the ordinary Power of Man's Eye extends no farther than perceiving what subtends an Angle of about a Minute, or something less. But when an Eye is armed with a Telescope, it may discern an Angle less than a Second.[12]

Nor were bards and scriveners immune to war fever. In England, beset in the 1600s by fifty-five years of actual warfare and many additional years of almost warfare, writers invented a slew of military metaphors. While the Royal Navy prepared for war with the Dutch by ordering hundreds of cannons and thousands of hand grenades,[13] the poet Samuel Butler composed a long satire on the astronomers of the Royal Society observing the full Moon, in which he depicts his subjects as hungry for cosmic conquest:

And now the lofty tube, the scale
With which they heav'n itself assail,
Was mounted full against the Moon;
And all stood ready to fall on,
Impatient who should have the honour
To plant an ensign first upon her.[14]

Think of the nascent telescope as an emblem of an entire society readying itself for expansion, not of the mind but of the wallet, the jewel box, the dinner table, and the wardrobe. Merchants were on the lookout for opportunities. Armies and navies were on the go. Getting a good view of not merely the heavens but also the hills, forests, ports, palaces, and sea lanes was becoming strategically necessary.

Within a century of its invention, the telescope came in many models: some with mirrors, some with two lenses, some with three lenses, some meant to be propped up on stands, some small enough to be carried in the pocket and certainly in the hand, some whose tubes were as big as a building, some whose widely separated components were suspended in the air without benefit of a tube.[15] Some of the earliest versions were binoculars, including three commissioned from Hans Lipperhey by the Dutch government and delivered in good working order by February 1609.

Although none of the seventeenth-century

telescopes measured up to those of later centuries, and not everyone could master the knack of seeing through them, the telescope and its cousin the binocular nonetheless brimmed with potential, both astronomical and military. But the breadth of astronomical possibilities surfaced only gradually and incidentally. Passable wide-field astronomical telescopes barely existed until the second half of the century, by which time Isaac Newton's nemesis, the talented British scientist Robert Hooke, would have good cause to speculate that "there may be yet invented several other helps for the eye, as much exceeding those already found, as those do the bare eye, such as by which we may perhaps be able to discover *living Creatures* in the Moon, or other Planets [italics in original]." At first, however, rather than gazing at an uncharted section of sky to see what they could see and thus make discoveries of their own, skywatchers generally aimed their telescopes upward just to look at some of Galileo's discoveries: the four major satellites of Jupiter, the textured surface of the Moon, the two "servants" of Saturn "who help him walk and never leave his side" (Saturn's barely resolved rings, extending to both sides of the planet itself, second largest in the solar system).[16]

So no, astronomical discovery was hardly the main agenda. The earliest telescopes were regarded primarily as aids to reconnaissance

— as terrestrial devices, meant to be turned toward the sea rather than the sky and to be used in the daytime rather than at night. Marketed to the Chiang Kai-sheks and Benito Mussolinis of their era, not to the Carl Sagans and Stephen Hawkings, the best of these instruments would have been the prized possession of only a select few senior officers.

Until the 1630s and 1640s the lightweight, portable, two-lensed Galilean spyglass — the lens near the eye concave, the one nearer the object convex — had the market pretty much to itself. Though relatively small and fuzzy, its images at least arrived at the eye right side up. As later described by Johannes Kepler, an alternative version — with two convex lenses — offered a larger field of view, but the image arrived upside down. For non-hurried astronomers studying space, where there is no up or down, an inverted image is not a crippling disadvantage. But for generals and admirals, who usually do their reconnaissance under pressures of time and position, rapid readability is crucial, whether on battlefield, battlements, deck, or promontory.

Despite all the advance publicity that attended the spyglass, a few military planners did remain blind to its benefits.[17] By and large, though, evidence from around the globe shows that terrestrial telescopes soon figured in a variety of military situations, especially surveillance and reconnaissance.

In 1615, for instance, following the Dutch East India Company's sinking of the six-hundred-ton Spanish vessel *Santa Ana* near Lima, Peru, a Spanish captain who had been taken prisoner and held until the Dutch reached Acapulco, Mexico, reported to Mexican officials that while the Dutch were lying at anchor near a Peruvian port, they had "sighted a vessel through some tubes they carry, by means of which they can see more than six leagues," or about twenty miles. In 1620 the English colonial governor of Bermuda reported having spent several hours looking through his "perspective glasse" from the vantage point of Warwick Fort, monitoring the approach of a strange ship. In 1626, before entering the harbor of Havana, the commander of a fleet from the West India Company relied on his own *anteojo de larga vista* to survey the situation. Lookouts aboard the many Dutch vessels plying the seas decade after decade from Java and New Amsterdam to South Africa and South America used their telescopes to scan the horizon for privateers. In seventeenth-century Japan, when Christianity was banned and Christian missionaries were understood to be agents of European colonialism, foreign merchant vessels were permitted to enter the country through only two ports, one of which was Nagasaki, where coastal watch stations were equipped with telescopes for scanning the waters.[18]

On land, commanders who had telescopes found they could now exercise a degree of control over fronts several miles wide rather than dashing around for a close-up view of small segments of the battlefield. By the mid-eighteenth century, Frederick the Great, the formidable king of Prussia — who greatly valued detailed maps but believed that "when we can make use of our own eyes, we ought never to trust to those of other people" — took to having his camp set up on an overlook, where he could use his own telescope at his own convenience.[19] Meanwhile, four thousand miles to the west, an educated Virginian named George Washington was ordering telescopes from London to help him in his work as a public surveyor and mapmaker and in his efforts to make sure that Virginian veterans of the French and Indian Wars got the "bounty lands" they had been promised.[20]

In colonial America, almost anyone who wanted a telescope — or, indeed, almost any scientific instrument — would have ordered it from London or Paris. Many of the men shelling out pounds sterling for a telescope constructed by the esteemed English instrument makers John and Peter Dollond had been born or educated in Great Britain or identified with British interests in the colonies. Others were rebels, members of the Continental Congress, officers of the Continental Army, signers of the Declaration of Independence. In 1776 at King's College (forerunner of New York City's Columbia University), for instance, while the president, most of the faculty, and half the students declared themselves loyalists, the college's Irish-born librarian and tutor Robert Harpur, an astronomer, joined the rebels.

Gradually the study of astronomy, geography, mathematics, and physics in the colonies gained ground. Usefulness — "an *Inclination* join'd with an *Ability* to serve Mankind, one's Country, Friends and Family," as Benjamin Franklin put it[21] — became a major goal of education and scientific inquiry. In 1743 in Philadelphia, Franklin and fellow enquirers founded the American Philosophical Society, dedicated to the pursuit of "all philosophical Experiments that let Light into the Nature of Things, tend to increase the Power of Man over Matter, and multiply the Conveniences or Pleasures of Life." Four decades later in Massachusetts, a similarly enquiring crew founded the American Academy of Arts and Sciences (whose seal features Minerva, the Roman goddess of both war and wisdom) to "advance the interest, honour, dignity, and happiness of a free, independent, and virtuous people." Franklin, Washington, and other founding fathers soon joined their ranks.[22]

And if all that doesn't make you yearn for yesteryear, consider this: the fourth US presidential election, in 1800, pitted the serving president of the American Philosophical Society against the serving president of the American Academy of Arts and Sciences.

Before that, of course, there had to be a first presidential election, which was preceded by George Washington's taking command of the Continental Army in 1775. One of Washington's early initiatives was the collection of military equipment for use in the field. Clothing and tents were big concerns; spyglasses for his officers were another. As the campaign for control of New York drew near, he also set his sights on getting a powerful telescope through which to observe British camps on Long Island and British ships in the Hudson River. The only one he knew of anywhere in the colonies was at King's College.

New Yorkers were pleased to cooperate. The records of the New-York Convention of August 1776 — one month after the Second Continental Congress met in Philadelphia and ratified the Declaration of Independence — include the following resolution:

Whereas his Excellency General *Washington* is in want of the use of a good Telescope; and whereas a good Telescope is absolutely necessary for the Commander-in-Chief of the Continental Army, to discover the arrangements and operations of the enemy:

Resolved, That the Chairman of the General Committee of the City of *New-York,* with such other members of that Committee as he may think proper, take and deliver to His Excellency General *Washington,* for his use, the Telescope which belongs to, and is part of the apparatus of the College of New-York.

No. 2. Resolved, That the Convention of this State of New-York will indemnify the governours of the College at New-York, for any injury, loss, or damage, that may happen to the Telescope belonging to the said College.[23]

By August 7, the instrument had been delivered to Washington's headquarters in New York City. Soon afterward, Washington wrote to Brigadier General George Clinton (who would soon become the governor of New York State and eventually Jefferson's and Madison's vice president): "By intelligence received and movements observed of the enemy, we have the greatest reason to believe a general attack will be made in the course of a few days."

Of course, the mere possession of a telescope to provide intelligence is no guarantee of victory. At the end of August, the British routed the Revolutionary Army on Long

Island, and the remaining soldiers escaped to Manhattan Island in the dead of night. On September 5 Washington wrote to Major General William Heath, advising him how to conduct his operations under the dangerous conditions at hand:

> As everything in a manner depends upon obtaining intelligence of the enemy's motions, I do most earnestly entreat you and General Clinton to exert yourselves to accomplish this most desirable end. Leave no stone unturned, nor do not stick at expense to bring this to pass. . . .
>
> Keep, besides this precaution, constant lookouts (with good glasses [that is, spyglasses]) on some commanding heights that look well on to the other shore (and especially into the bays, where boats can be concealed), that they may observe, more particularly in the evening, if there be any uncommon movements. . . . I should much approve of small harassing parties, stealing, as it were, over in the night, as they might keep the enemy alarmed, and more than probably bring off a prisoner, from whom some valuable intelligence may be obtained.[24]

The advice was excellent (the CIA and various writers deem George Washington a first-rate intelligence chief and spymaster), but the results were mixed. By mid-November the British and their mercenaries had taken over all of Manhattan, and Washington's forces had retreated to New Jersey. By mid-December the much-defeated Revolutionary Army was running out of resources, soldiers, time, and morale. Nevertheless, five thousand or so hungry men and a handful of women under Washington's command, many of them sick, some of them barefoot, reached the Pennsylvania side of the Delaware River before the worst of winter set in. To make the journey, they had seized every heavy wooden cargo boat they could lay their hands on. Soon the remnants of a couple of other divisions joined them.

On the windy, sleeting night of December 25, 1776, more than two thousand soldiers made it back to the New Jersey side of the river. At dawn they took the enemy by surprise at Trenton. It was a remarkable turnaround. Emanuel Leutze's heroic painting *Washington Crossing the Delaware* — honoring both the imminent victory and the newborn nation — depicts a line of rowboats stretching almost to the horizon, with Washington standing tall and determined in the foremost boat, his right leg planted on the bow as the multi-ethnic crew of revolutionaries struggles with poles and oars in the ice-choked river and light begins to flood the morning sky. At the commander's left side hangs a saber; in his right hand is a telescope.[25]

By the late eighteenth century, the telescope had a recognized role in the waging of war. No first-rate tactician would have engaged the enemy without one. The option of a collapsible midsection — a refinement of the standard hollow tube with its lenses at each end — increased the instrument's portability. One British optical firm advertised its refracting telescope as having "been favoured with the Approbation of the best Judges in Theory, as well as those Gentlemen whose naval, or military Capacity, has made them more than ordinary conversant with the Use of it."[26] Today, writers of history sprinkle their accounts of bygone battles with references to a long-dead colonel, general, ship captain, or concerned citizen watching in alarm through a telescope as a forest of masts materializes on the horizon, or slowly pivoting a telescope across a landscape like a swiveling gun, or peering through a telescope and muttering, or decisively collapsing his telescope shut once he's seen what he needed to see.

The telescope also serves as the key prop in a tallish tale involving England's most famous admiral, the one-eyed, one-armed Horatio Nelson. In 1801, during the Battle of Copenhagen — fresh from having prevented Napoleon from executing his plans for Egypt and

India — Nelson served as second-in-command to Admiral Sir Hyde Parker. Their goal was to break up a Northern European free-trade/ free-passage alliance that Britain saw as overly advantageous to France. Parker and Nelson and their fleet were dispatched to convince Denmark, by whatever methods they could muster, to withdraw from the alliance. Parker (who favored caution and negotiation) positioned his ships to the north of Copenhagen and sent Nelson (who favored intimidation to the point of annihilation) and his ships to attack from the south. The battle was fierce and the smoke was thick, but Nelson did not back down. Two hours into the attack, when Parker's flagship signaled that the British bombardment should stop, Nelson raised his spyglass to his blind right eye and announced that he simply didn't see the signal. Despite his cautiousness, Parker was killed in the battle, while the confrontational Nelson prevailed. The Danes signed a truce, and the phrase "turn a blind eye" entered the language.

But the portable terrestrial telescope, even in the hands of a brilliant commander, could not by itself revolutionize war. George Washington, for instance, valued spies even more than he valued spyglasses, as evidenced by a letter of July 10, 1779, in which he writes to a brigadier general, "Single men in the night will be more likely to ascertain facts than the best glasses in the day."[27] A telescope could facilitate the gathering of data about nearby enemy forces, one's own nearby forces, local terrain, local weather, and local roads. It could facilitate the rejection of some tactics and the adoption of others. But achieving victory in even a single battle remained as daunting, multifaceted, cumbersome, and diffuse a process as ever. A commander with a spyglass might observe an advance contingent of enemy cavalry beyond the next hill or across the river and swiftly devise a method to kill them all, but it would be his lieutenants' and soldiers' job to implement his orders. And if the deadliest weapons are effective only when fired at close range, and available cavalry are scarce, the telescope can contribute almost nothing to strategic planning and very little to tactics in the moment. In terms of "who ordered whom to do what, when, by what means, on the basis of what information, what for, and to what effect," as the military historian Martin van Creveld describes the parameters of military command, the telescope could play only a narrow role. Some prominent historians of war and technology, in fact, accord the telescope no military role whatsoever for its first century or two of life.[28]

For starters, consider how land war was waged in seventeenth-century and much of eighteenth-century Europe. Good information was hard to come by, and rapid communication was unknown. Decent roads were few and far between. Maps of any kind were a rarity; countrywide maps that represented towns, roads, and distances at a proper scale were rarer still; and maps that represented the topography were nonexistent. General information about the inhabitants, customs, and features of foreign territories came from a few books, a few newspapers, unreliable censuses, the accounts of pilgrims and merchants and diplomats. Information of greater tactical relevance might come from soldier-spies or from the statements of deserters, prisoners, or peasants; the spy, disguised as a laborer or servant, might enter an enemy camp in the company of a peasant selling turnips or textiles. To ensure the peasant's usefulness, a member of his household might have been taken hostage.[29] Most information not derived from the commander's own observations traveled to and fro no faster than the fastest horses could gallop. The same speed held for the commander's orders. Quick decisions premised on up-to-the-minute information were unknown; spur-of-the-moment commands, if ever issued, were unlikely to be executed. Most commands were verbal rather than written, although, before issuing them, the commander may have had to send reports to the king and wait a couple of weeks for instructions.

Whether his army was tramping across fields or besieging a fortress, the commander's biggest headaches were securing enough bread, brew, and meat to sustain his troops, providing his many mercenaries with a paycheck and shelter, making sure the horses were fed and watered, and getting hold of enough weapons and ammunition. Thanks to the innovations of Prince Maurice, the troops' everyday activities, when they weren't looting the locals or actually doing battle, came to include drill, marching, and ditch-digging. Van Creveld offers a capsule description of the enterprise: "Well into the eighteenth century, battle and warfare were all but identical . . . war apart from battle being almost indistinguishable from a somewhat violent form of tourism accompanied by large-scale robbery."[30]

Portable firearms were relatively new. For every battle in an open field, three or four sieges took place at the walls of Europe's fortresses, from which a siege cannon sitting on a carriage would fire heavy iron roundshot at the target.[31] With the thousands and sometimes tens of thousands of shouting soldiers rushing in and out of formation, the noise of firearms and cannons, the clouds of smoke from exploding gunpowder and flaming siege towers and incendiary devices hurled over the battlements, even a first-rate telescope would have limited impact on the outcome.

At sea, a telescope could be more useful. All-weather maritime commerce had been expanding all around Europe since the fourteenth century, and, absent armed protection either on board or sailing alongside, no laden merchant vessel or convoy could expect to reach its destination unmolested. Distant voyages had increased in popularity as hunger for coffee, gold, spices, sugar, slaves, tobacco, tea, textiles, and tax revenues mounted. Most naval battles took place near coastlines, at close range, from which one of the era's newly large, fully rigged men-o'-war sailing ships would fire its shot from a hundred or more giant cannons. Ships were wooden, and more vulnerable to impact and flame than a fortress wall would be. Because of the closeness of the confrontations and the more limited number of hiding places for a convoy of large ships, a telescope could come in handier than it did on land — if and when the commander was lucky enough to get a break in the fog, smoke, fire, cannonades, and tumult.

Notwithstanding the many limitations displayed during its first century and a half in battle, the telescope did enable some reconnaissance and did still promise military advantage. Inventors were far from giving up on it.

Combine a good late-eighteenth-century telescope, a signaling system based on readily visible elements, a capacious code, and a series of relay stations that stretches from county to county, and you get the "optical telegraph," a signally useful military innovation and the most advanced communications technology of the early nineteenth century. Never mind that it was superseded by the electric telegraph at midcentury. Before it went extinct, local versions of the optical (sometimes called the aerial) telegraph had been built from Stockholm to Sydney, from Curaçao to the Crimea. Some bankers used the telegraph to get a jump on stock quotes. Originally, though, it was meant for admirals and generals.

Sending urgent messages by means of a moving relay, whether of runners or mounted couriers, is a time-honored way to communicate at a distance. Twenty-five centuries ago, for instance, Darius the Great set up a relay of men whose shouts could carry great distances. A vast array of other visual and acoustic tricks can transcend time and space: bonfires, smoke, and torches; flags, mirrors, and polished shields; trumpets, drums, animal horns, and seashells. Using extremely simple, prearranged visual codes, especially torch codes, for the most common wartime contingencies dates back some twenty-five centuries as well. As the second-century BC

Greek historian Polybius writes in *The Histories*, "It is evident to all that in every matter, and especially in warfare, the power of acting at the right time contributes very much to the success of enterprises, and fire-signals are the most efficient of all the devices which aid us to do this." Trouble is, he notes,

> it is impossible to agree beforehand about things of which one cannot be aware before they happen. And this is the vital matter; for how can anyone consider how to render assistance if he does not know how many of the enemy have arrived, or where? And how can anyone be of good cheer or the reverse, or in fact think of anything at all, if he does not understand how many ships or how much corn has arrived from the allies?[32]

Clearly the next step, says Polybius, is to develop a far more powerful, flexible visual code that could capture the gist of an important message. To the great thinkers of his era, the obvious choice was a code based on the alphabet, though still conveyed via torches. And how best to view the distant signal fires? Through the still-empty sighting tubes of the time.

Two thousand years later, and less than a century after those tubes began to be occupied by lenses, John Wilkins — soon to become master of Trinity College, Cambridge

— published a treatise titled *Mercury: or the Secret and Swift Messenger. Shewing, How a Man may with Privacy and Speed communicate his Thoughts to a Friend at any distance* (1641). In it, he describes a form of coding, and how coded messages could be cleverly encrypted and conveyed via torch signals. Less than half a century later, in a 1684 lecture to the Royal Society titled "Shewing a way how to communicate one's Mind at great distances," the brilliant Robert Hooke proposed a marriage between the ancients' optical telegraph, the modern telescope, and the changeable billboard.

Hooke outlined a system with multiple stations, each equipped with a telescope and each located in a high, isolated spot, well above the obscuring fog of a typical British morning, "to convey Intelligence from any one high and eminent Place, to any other that lies in Sight of it, tho' 30 or 40 Miles distant, in as short a Time almost, as a Man can write what he would have sent." He even mentioned "Cruptography." Along with what would now be called control codes, the system would use twenty-four large symbols made of lightweight wood, hoisted in quick succession via pulley to the top of a high pole.[33]

In the waning years of the eighteenth century, in part spurred by the image quality attainable through the newest telescopes, a number of inventors experimented with long-distance communication. They tried synchronization by banging on pots or by flipping from large black surfaces to white ones. They tried smoke, fire, pendulums, shutters, windmills, synchronized clocks, and sliding panels. Among those inventors were the five brothers Chappe, descendants of a French baron and, as of late 1789, unemployed because of the Revolution.

On March 24, 1792, Claude Chappe, priest and physics buff, and the most committed and persistent of the brothers, addressed the French legislature in a bid to gain government support for an official demonstration of their optical telegraph, the *tachygraphe:*[34]

I have come to offer to the National Assembly the tribute of a discovery that I believe to be useful to the public cause. . . . I can, in twenty minutes, transmit over a distance of eight to ten miles, the following, or any other similar phrase: "Lukner has left for Mons to besiege that city. Bender is advancing for its defense. The two generals are present. Tomorrow the battle will start." These same phrases are communicated in twenty-four minutes over a distance twice that of before; in thirty-three minutes they cover fifty miles.[35]

While the proposal languished in a series of committees, France was declared a republic, Louis XVI was beheaded, the republic declared war against its monarchist neighbors, and Chappe's experiments were twice destroyed by citizens suspicious that the apparatus would be used to get in touch with enemies of the state. Finally, success. On July 12, 1793 (a day before the radical doctor-journalist Jean-Paul Marat, vocal proponent of the guillotine, was stabbed to death in his bathtub), Claude Chappe, in the presence of members of the legislature, issued a two-sentence message from a tower near Paris. Eleven minutes later, one of his brothers received it — that is to say, saw it through his telescope — at a tower sixteen miles away. Chappe had handily bested the time and distance of his own original estimate. On July 26 (a day before the radical lawyer-philosopher Robespierre was elected to the powerful Committee of Public Safety), Chappe was given military rank and the title of telegraph engineer. On August 4 the Committee of Public Safety ordered that construction begin on a two-hundred-kilometer telegraph line between Paris and the northern city of Lille. The project was placed under the authority of the minister of war.

There were to be eighteen high towers in all. The coded message would be carried atop a pole by one long, movable bar and two smaller, hinged bars attached at either end — three lines that could be swiftly manipulated from below by wires, pulleys, and rods. Of the ninety-eight signals that could be configured by the three bars, six were reserved for special instructions. The remaining ninety-two conveyed the message through a pair of signals. The first one directed the telescope operator to the page number in the accompanying codebook, in which each page listed ninety-two words or phrases. The second signal directed the operator to the item number on the page. All told, a compendium of almost 8,500 message bits.[36]

Enthusiasm ran rampant. The 1797 *Encyclopaedia Britannica* presented the telegraph as a bringer of peace: "The capitals of distant nations might be united by chains of [telegraph] posts and the settling of those disputes which at present take up years or months might then be accomplished in as many hours." Napoleon himself embraced the optical telegraph with both arms. Here was a man who wanted everything to be done yesterday and wanted to be everywhere at once. Via the royal mail service, he reckoned, information could move only about twice as fast as it had in Julius Caesar's time; as one historian of France puts it, "the fastest communication could be no faster than a mounted rider or a sail before the wind."[37] Not only was that much too slow by Napoleon's standards, but, owing to mail seizures by the likes of Admiral Nelson, sending a letter was no guarantee of its arrival. The optical telegraph, on the other hand, promised both instantaneity and lack of interference.

One news flash that had to be disseminated as swiftly and widely as possible was the coup d'état that began on 18 Brumaire of the French Republic's year VIII (November 9, 1799). An official copy of the dispatch, in flowing script on official letterhead, survives. "Bonaparte is named Commandant of Paris," it declares. "All is calm and happy."[38] The letterhead itself is worth a good look. Kneeling at the base of a towering stone pyramid topped by the Chappe signal bars is the messenger god Mercury, about to finish engraving on the pyramid a line from Virgil's *Aeneid:* "HIS EGO NEC METAS RERUM NEC TEMPORA PONO." Add the next few words of the famous quotation (the words of Jupiter, king of the gods), and it sums up the aims of both Chappe and Napoleon: "For them I set no boundaries of things or time; I give empire without end."[39]

The optical telegraph has been called the first practical telecommunications system, the first nationwide data network, the first internet. Claude Chappe himself has been called the first telecom mogul. By the late 1700s, however, electricity had become the darling of experimenters — fueled in part by Benjamin Franklin's internationally read 1751 treatise *Experiments and Observations on Electricity* — and by the 1830s inventors had begun to experiment with the electric telegraph. Wasting no time, in the 1840s France began to replace its optical system with an electric one. In early September 1855, during the Crimean War, news of the fall of Sevastopol arrived via the Chappe telegraph; shortly afterward, the network fell silent.

But the idea of an optical system was not yet dead. You could still use one to overlook the battlefield, monitor the enemy's approach, or evade enemy forces under close conditions — if and only if your system was low-tech and portable, the signaler was in the receiver's line of sight, the signals weren't swallowed by the smoke of battle, the weather cooperated, and your enemy didn't have a similar system or couldn't decipher the sender's code. That list of qualifiers may sound impossibly long, but on a few occasions during the American Civil War all or most of them were fulfilled. These were the occasions when Signal Corps officers, standing guard and observing through their telescopes, together with the flagmen who communicated the officers' warnings and requests, influenced the course of battle.

By 1862, America had three separate entities responsible for military communications:

one organized by a US Army surgeon named Albert J. Myer, the second organized by a West Point graduate named Edward Porter Alexander, and the third a wartime expedient called the United States Military Telegraph, which relied primarily on professional civilian operators and on the electric telegraph lines owned by private companies.[40] The result was turf battles, conflicting loyalties, mistrust, and espionage.

Myer, a Northerner, was the right person at the right time. Having worked at the New York State Telegraph Company, he was familiar with the new, electric technology as well as the basic concepts of coding. He'd already adapted a popular telegraph code for use as a sign language: spelling out words letter by letter, in a binary code, by tapping on a nearby surface. Upon joining the Army, he readapted the code so that each letter could be communicated with a single flag by a single signaler and seen by a distant observer peering through a telescope.

In 1856 Myer pitched his system to the secretary of war, Jefferson Davis of Mississippi, who did not pursue it. A few years later, a new secretary of war, along with a committee headed by another Southerner, Robert E. Lee of Virginia, gave Myer the go-ahead to borrow some personnel and run some tests. The most assiduous of the borrowed assistants was a third Southerner, the Georgia-born second lieutenant Edward Porter Alexander. The tests went better than expected, and in the spring of 1860 Congress made Myer the first-ever US Army signal officer.

Deployed to New Mexico in late 1860 to help stamp out Navajo resistance to westward expansion onto Native lands, Myer and his signalers did both reconnaissance and communication. It would not be the first time that technological innovations would help displace a resident population.

Soon came secession and civil war. In February 1861, Jefferson Davis became provisional president of the newborn Confederate States of America. In April Confederate forces fired on Fort Sumter. Myer was ordered east in May; in June he began training Union signal officers and flagmen. In July, during the First Battle of Bull Run — while Myer and twenty members of the 26th Pennsylvania Infantry were tangled up in a tree with a reconnaissance balloon — Myer's former collaborator Edward Alexander, now a captain in the Confederate Army, made brilliant use of his own telescope and of Myer's code to warn his side of the approaching Union troops.[41] In August, Myer became chief signal officer of the Army of the Potomac. Less than a year later the Confederate congress voted to create a full-fledged signal corps; a year after that, the Federal congress voted to do the same.

Myer's system was both simpler and slower than the Chappes'. More important, because it was thoroughly portable, it could be used for communications to and from the battlefield.[42] It was also affordable and flexible. But all parties to a communication had to be on the same page. They needed a common textbook, and in early 1864 Myer published the first of his many editions. The text leaves little to guesswork: it even tells the reader to hold binoculars with two hands when looking through them.[43]

The signaler, called a wigwagger, positions himself on a hilltop, tower, isolated tree, ship's masthead, or anywhere else that commands a good view. Holding a large flag attached to a pole, he starts off with his arms in a vertical position. He briskly sweeps the flag down to the right to signal "1," returns it to the vertical, and sweeps down to the left to signal "2." Four swoops at most take care of the entire alphabet. A single forward swoop signals the end of a word, two the end of a sentence, three the end of a message.[44] A choice of flags — white, red, and black, each with a contrasting center — makes the motions visible during the day in any environment.

The lookout was the signaler's superior. He carried the optical aids and assessed the circumstances, using binoculars for reading signals less than five miles away and a portable telescope for greater distances. To avoid detection, standard Signal Corps collapsible telescopes were camouflaged: Myer describes the four-jointed draw as "bronzed black, in order that there may be neither glitter to attract the enemy, nor glare to disturb the eye of the observer."[45] Sometimes one officer acted as both signaler and lookout. Sometimes the optical aid minus the flags provided the main advantage: observing while unobserved. At well-concealed signal stations, the lookouts sometimes tracked enemy movements with their telescopes while the flagmen raised nary a flag, because wigwagging would announce the station's position — and that would be the end of the advantage.

Myer makes absolutely clear that the telescope is a precious thing:

> Telescopes ought never to be allowed to fall into the hands of the enemy. Officers, on dangerous stations, should conceal their glasses when not in use. When a glass is to be hidden for precaution, the object-lens, or one joint of the telescope, should be hidden separately from the body of the telescope. A single joint or one lens is so small an object, that it can be concealed almost beyond the possibility of discovery. If an officer is in danger of capture, and there are no means of concealment, the telescope-glasses must be shattered or rendered

worthless rather than surrendered.[46]

Both North and South used the same basic binary signaling system. As a result, both sides could read at least some of the other side's messages, even when the codes were re-enciphered. Signal duty brought much criticism, few medals, and a disproportionate chance of death.[47] Yet signalers and encipherers on both sides showed remarkable ingenuity and steadfastness, and certain battles might have gone differently were it not for the officers perched in trees, cupolas, and the hundred-foot towers built expressly to give them the high ground.[48]

Consider Gettysburg, the southern Pennsylvania battleground where some fifty thousand soldiers lost their lives in the first three days of July 1863. By the last week of June, a dozen signal officers were installed near the Maryland–Pennsylvania border, watching for the advance of the Army of Northern Virginia. By the morning of June 30 it was evident to the Northerners that the Confederate columns — nearly the whole of Lee's army — were converging on Gettysburg. The Southern generals did not expect to be met by a massive Union force.[49]

On July 1, as he moved from steeple to cupola in Gettysburg, a Union signal officer named Aaron B. Jerome alerted his commanding general that he had detected the rebels close by. Short of men, the commander could muster only two brigades along the road to intercept them. Within hours, Jerome signaled details of the Confederates' progress to a colleague on a nearby hill: "Over a division of the rebels is making a flank movement on our right; the line extends over a mile, and is advancing, skirmishing. There is nothing but cavalry to oppose them."[50]

That day, the Confederates took Gettysburg. However, Union signalers managed to reach Little Round Top, a now-famous hill alternately occupied and abandoned by Union troops during the next two days. By noon on July 2, Lieutenant Jerome, again in the thick of things, sent this message from Little Round Top to headquarters: "The rebels are in force, and our skirmishers give way. One mile west of Round Top signal station, the woods are full of them."[51] Numerous though they were, the Confederate forces were constrained by having to avoid being seen by Union signalers. Eventually, despite heavy fire, Union forces took Little Round Top. As Myer's erstwhile protégé Edward Alexander, by now a brigadier general who served as the rebels' artillery commander at Gettysburg, later complained, "That wretched little signal station upon Round Top that day caused one of our divisions to lose over two hours and probably delayed our assault nearly that long."[52]

On July 3, intense Confederate fire from the foot of the exposed hill made it impossible for Union wigwaggers to use their flags, so they sent out their orderlies on horseback every few minutes to deliver messages to headquarters.[53] Signalers at other stations around Gettysburg showed their determination in other ways. One captain stationed near Cemetery Hill stayed behind after all the other Union officers and troops had been forced away, taking the signal equipment with them. Undeterred, under fire, and needing to send a few important messages, he quickly cut himself a pole and attached a bedsheet to it to serve as the flag.[54]

The next morning, the Confederates began to withdraw, their mission undermined in part by the wigwaggers' resourcefulness.

Communication had long been a weak link in the structure of command. Myer and his signalers helped change that — for a while. Generals didn't stop using scouts and spies and couriers. Nor did they stop looking through their own telescopes to get firsthand information. The rapidly improving electric telegraph soon erased the need for the optical telegraph. But credit should be given where credit is due. In melding the historical practices of enciphering and aerial signaling with the rapidly improving craft of telescope manufacture, Myer's simple method linked widely dispersed as well as vulnerably close commanders and troops, enabling not just the rapid exchange of information but also rapid intervention.

After the Civil War, as the paradigm of national security shifted from conquest to prevention of loss of life and property, the US Army Signal Corps — still run by Myer — began to perform the work of a national weather service. Among its innovations were daily weather bulletins, telegraphed across the country to be displayed at local post offices, and the daily publication of an international weather map. Scientific collaboration became a key feature of its work. Myer's successor established the corps's Scientific and Study Division, sought input from consultants such as Alexander Graham Bell and the astronomer Samuel Langley, and sponsored a textbook on meteorology. The metamorphosis of the corps's identity and activity from wartime to peacetime is a case study in adaptability.[55]

Besides becoming the world's weatherman after the Civil War, the US Army Signal Corps helped launch many other practices that are now integral to military operations: combat photography, airborne radiotelephones, photoreconnaissance and aerial mapping, communications satellites, even (with the help of Wilbur Wright) military flights. During World War I the corps took responsibility for combat and surveillance photography both foreign and domestic, on the ground and in the air, producing tens of thousands of stills and hundreds of thousands of feet of motion pictures.[56] As communications historian Joseph W. Slade wrote, by the end of the twentieth century the Signal Corps had turned into "Ma Bell with guns."[57] Telescopes and binoculars, reconnaissance aircraft, bombs, satellites, and telecommunications: the intersection of war and astrophysics is neatly embodied in the corps's evolving duties.

Speaking of the erstwhile North American telephone conglomerate Ma Bell: during World War I its parent company at the time, AT&T, supplied its chief engineer to the Signal Corps Officers' Reserve Corps.[58] Since then, corporate giants have become integral to every war effort. The envisioning, anticipation, and implementation of war have in fact spawned some of these corporations and multiplied the profits of others. Today there are no standardized armaments without manufacturers, no inventions without patents, no stockpiles without suppliers — global webs of interdependence, benefit, and responsibility. The elimination of a single supplier, the sudden unavailability of a single product, can cripple a country or help shift the course of a war.

■ ■ ■ ■

Like so many sectors of what is now a global industrial marketplace, the precision optics industry began with a scattering of assiduous, independent-minded individuals. A barrister hobbyist, for instance, working alone in a gentleman's laboratory in Essex, discovered a major principle of refractive lens design — how to minimize the spurious appearance of color in the image — but sought no recognition for it. He was simply solving an intriguing puzzle for his own pleasure.[59]

The curve of a lens determines the angle at which the light rays will bend as they pass through it, and thus the distance over which they come to a focus or diverge. If the curve bulges out, like a beer belly, the lens is convex and will bring the rays to a focus. If the curve sinks inward, like a cupped palm, the lens is concave and will force the rays to diverge. If one side is flat and the other curved, the lens is called either plano-convex or plano-concave. And if both sides are curved, you've got a double convex or double concave lens.

The color problem in lens optics derives from a simple feature of angled glass. A triangular prism, by design, splits white light into its component colors, with each color emerging from the other side of the prism at a slightly different angle from all the others.

A double convex lens — a crucial feature of telescopes — is not very different from two prisms cemented to each other at their base. While it doesn't produce such extreme coloristic effects as two pure prisms, this lens will focus different colors of light at different distances within the telescope tube, creating unwanted colorful aberrations unless corrective lenses are added to the system. The thicker the double convex lens, the shorter the telescope tube can be, but the more severe the problem becomes. Reflecting telescopes create no such problem, because all colors of light reflect at the same angle.

The beginning of the end for color problems came in 1758, when two things happened. The first was that a mathematically inclined, London-based ex–silk weaver named John Dollond published in *Philosophical Transactions* an account of his experiments with lens sandwiches formed of two different kinds of glass — crown and flint — which exhibit different refractive qualities. The second was that John Dollond applied for a British patent for his sandwich, calling it the Achromatic Lens, "whereby the errors arising from the different refrangibility of light, as well as those which are produced by the spherical surfaces of the glasses, are perfectly corrected."[60]

By rights that patent (of only fourteen years' duration) should have belonged to the barrister Chester Moor Hall. But he hadn't sought it, and Dollond had. The following decade, John Dollond's son Peter added a third lens, eliminating residual aberrations and creating the perfect club sandwich. Never again would a telescope have to be fifty feet long to yield clear, crisp results. Soon the seamen of the Royal Navy began to call a telescope a "dollond,"[61] and the progeny of the Dollonds' dollonds became standard field equipment for warfighters on the move. George Washington and Napoleon alike (not to mention Captain Cook, Frederick the Great, a long list of British royals, the father of Wolfgang Mozart, and untold others) would have been lost without J Dollond & Son or, subsequently, P & J Dollond Instrument Makers, foremost suppliers of a variety of precision optics for most of the eighteenth and much of the nineteenth century.

Neither the Dollonds nor Britain held the field unchallenged. In 1846 a thirty-year-old technologist-optician named Carl Friedrich Zeiss opened a workshop in the small town of Jena, Germany, that soon became the dominant corporate power in the optics industry. And just before the Civil War the American company Alvan Clark & Sons set up shop in Massachusetts. Most American observatories built in the second half of the nineteenth century, when enthusiasm for astronomy was on the rise, relied on one or more of the Clarks' superbly hand-crafted telescopes, and during the war itself the company sold the US Navy nearly two hundred expensive spyglasses.[62]

One item required by all manufacturers of precision optics was fine, clear, homogeneous optical glass — blank slabs ready to be ground and polished by exacting craftsmen such as Alvan Clark, who finished them with his bare thumbs rather than resort to an insufficiently soft cloth.[63]

A material at least as ancient as the pharaohs, glass is made mostly from molten sand, cooled in such a way that it bypasses the crystallization phase. But optical glass is a far cry from the glass used for bottles and beads, and no pharaoh's workshop could have produced it. Nor, centuries later, was it an easy sideline for the producers of window glass, though some of them made the attempt. As the American astrophysicist Heber D. Curtis wrote at the close of World War I, it is "a substance which differs from ordinary glass almost as much as does the diamond from graphite."[64] (A year later, Curtis would bet on the wrong horse in a highly publicized debate about whether the Milky Way was the entire universe or whether the spiral fuzzy objects seen dotting the night sky were other galaxies, rendering the actual universe vastly larger than previously imagined.)

Quality optical glass requires vast quantities of fuel and highly controllable furnaces. It needs melting pots that won't contaminate the fiery brew, and it must be stirred well. It needs the right flux to draw out impurities. Bubbles, veins, strains, and cloudy patches must be prevented from forming during cooling. If the goal is to vary the refractive effects in different parts of the spectrum, various substances may be added: lead, barium, boron, sodium, silver, uranium, mercury, arsenic. Above all, optical glass must be utterly transparent and uniform.[65]

Fine optical glass blanks of a decent size were hard to come by until well into the nineteenth century, and instrument makers paid dearly for them.[66] Dollond had come up with a lens design that promised excellent astronomical telescopes, but the promise was infrequently fulfilled. A design is only a recipe. If you don't have avocados, you can't make guacamole.

For decades, just two companies — Chance Brothers of Birmingham, England, and Parra Mantois et Cie. of Paris — satisfied most of Europe's appetite for optical glass. In the early 1880s the spotlight switched to Jena, where Carl Zeiss and two university-trained scientists had formed a legendary industrial collaboration. The senior scientist was the physicist Ernst Abbe, who had made major contributions to the mathematics of optics —

having determined, for example, that the resolution of a telescope or microscope is limited by the size of the instrument and the wavelength of the light it brings to focus — and was already collaborating with Zeiss in the manufacture of advanced microscopes. The junior scientist was a young PhD chemist named Otto Schott, whose dissertation topic was the fabrication of glass. No longer could trial-and-error craftsmanship suffice. Apprentices now needed the input of academics, and the optician himself attended university lectures whenever he could.

Together these men expanded Carl Zeiss's already impressive optical workshop and also set up Schott & Associates Glass Technology Laboratory. Shortly after Zeiss's death in 1888, Abbe formed the Carl Zeiss Foundation, which today owns Carl Zeiss AG and Schott AG and thus is responsible for the awesome star projector — the Zeiss Mark IX — that rises up out of the floor of the Space Theater in New York City's Hayden Planetarium.[67] Among the early Zeiss/Schott corporate conquests were the perfection of low-expansion borosilicate glass (what the rest of us call Pyrex), the apochromatic lens (a significant advance on the achromatic lens, bringing all wavelengths into focus in the same plane), and the mass-produced prismatic binocular. By the eve of World War I, Zeiss was the preferred supplier of most "optical munitions" — one-person observation devices that included binoculars, rangefinders, panoramic artillery sights, and submarine periscopes.[68] But Zeiss was producing fine nonmilitary equipment as well: astrophysicists sought its new-generation large refracting telescopes, photographers sought its cameras, all sorts of people sought its microscopes. In June 1914 the many departments of the Zeiss works in Jena employed more than five thousand people.[69] (In June 1945, by the way, US occupation forces removed seventy-seven Zeiss scientists and executives from Jena — which sits squarely in Germany's east — and took them to the southwest, where they set up a Zeiss subsidiary in Oberkochen. Cold War politics intervened in 1953, when the government of East Germany cut off contact between the eastern and western branches. In 1991, soon after Germany's reunification, Zeiss reunited as well.[70])

Despite the many advances made by Zeiss, Abbe, and Schott, size remained a challenge. The curved metallic surface of a reflecting telescope's polished mirror could not be shaped perfectly. For those who sought ever-larger glass lenses in the nineteenth century, Alvan Clark had seemed a godsend. But the refracting telescope's glass lens had problems of its own. Hand-and-thumb craftsmanship is hardly mass production, and the continuing paucity of fine optical glass limited the quality of a telescope's optics. Most important, the sheer weight of a large glass lens, which must be held in place only at its perimeter, posed a severe engineering challenge.

Fortunately for astrophysicists, the germ of a better solution was already available. In 1835 the German chemist Justus von Liebig had introduced the silvered-glass mirror. Made by depositing a thin layer of silver vapor on one side of a slab of polished glass, it offered an excellent image and soon became a fixture of every bourgeois household. Two decades later, Jean-Bernard-Léon Foucault (the pendulum fellow), in collaboration with the Paris Observatory's official optician, improved upon this technique by adding a subsequent phase: localized repolishing to correct errors of form. This enabled Foucault to make ever-larger reflecting telescopes, culminating in an eighty-centimeter telescope installed in the Marseille Observatory in 1864.[71]

Today the largest telescopes in the world are all reflectors, and all of them use a mirror with a vapor-deposited metal coating on one polished glass surface. While the lens of the largest extant refracting telescope is one meter across, the mirror of the largest reflecting telescope is more than ten meters across. Others in the works approach forty meters in diameter. Hardly anything limits the size of the mirror, because it is mounted from the back. As a result, since the end of the nineteenth century, the reflector has been the astrophysicist's instrument of choice.

The military solution, however, lay elsewhere. For nearly the entire nineteenth century, military planners and artillerymen alike fretted far less than astronomers about the limited availability of fine optical glass. An infantry rifle that could be fired effectively at a target more than a mile away was not yet on the market.[72] Gunners did not rely on barrel-mounted spotting scopes. Civil War cannons were fired point blank in the general direction of a nearby visible enemy; battling Northerners and Southerners estimated distances strictly by eye and aimed their guns with the aid of spirit levels and plumb lines, hoping to overwhelm the enemy with a barrage of shot. "The gunners sighted their fieldpieces hastily and banged away, trusting to hit some vital spot," writes Lieutenant Colonel F. E. Wright in a historical overview produced in 1921 for the Ordnance Department of the US Army.

By 1914, gunners equipped with optical munitions were able to attack unseen targets fifty thousand yards away, targets whose positions had been calculated on a map. Optical aids had become indispensable. The gunner who lacked them, says the colonel, "is almost helpless in the presence of the enemy; he can not see to aim properly . . . and his firing serves little purpose." The manufacture of optical glass had become "a singularly important key industry."[73] Writing in 1919, Heber Curtis was equally forceful: "When we pass from the needs of peace to the requirements of a nation waging modern scientific war, optical glass changes from a mere essential of the observatory or the laboratory to an element nearly as indispensable as the high explosive." Or, to use a phrase of economic historian Stephen Sambrook, "no gunnery without glass."[74]

So, you might think that by the eve of World War I, every Western nation-state with an industrial base and a habit of waging war would have funded the building of factories to make their very own optical glass and optical munitions, that they would have stockpiled raw materials, fuel, and finished products, ensured an adequate workforce of skilled personnel, and signed the treaties that would guarantee a steady supply of optics to their armies and navies. But no. They hadn't.

Among their other failings, key countries of the Entente now relied heavily on a single factory for a great deal of their optical glass: Schott & Associates Glass Technology Laboratory, located well within the borders of what was soon to become enemy territory.[75]

The UK was Schott's top importer of optical glass; the USA was second.[76] The details of Schott's manufacture were proprietary information. Despite the recent spate of wars and despite warnings from informed individuals,[77] the West's large nation-states — whose kings and parliaments had for four centuries been putting 30 or 50 or sometimes 70 percent of their annual budgets into war and armaments[78] — had directed inadequate attention and money toward securing local production during wartime.

Inevitably, crunch time came.

Suddenly countries were scrambling to fill urgent needs, not only for optics but also for photographic chemicals, pharmaceuticals, synthetic dyes, high explosives — much of which had previously been imported from Germany, duty free. Nor was the cutoff of imports the only difficulty. Vast armies, new industries, new materials, and new practices had to be created almost from scratch. The war effort required bombs, mass-produced vacuum tubes, carrier pigeons, ammonia, pilots' clothing, unprecedented numbers of airplane motors, the airplanes themselves. From 1903 through 1916 a mere thousand planes, none intended for combat, had been built in the United States, and yet in late May 1917 the US government was asked to come up with two thousand planes and four thousand engines a month, as well as five thousand pilots and fifty thousand mechanics within a year.[79] Near-instantaneous demand for optical glass and optical munitions reached comparable levels. The only solution was intensive cooperation among industrialists, scientists, diplomats, patent lawyers, military brass, procurement officers, and the factory floor.

As for Britain, the military's prewar demands could be satisfied by a few flourishing British manufacturers. The Royal Navy had been a patron of homegrown precision optics companies since the 1890s, followed within a decade by the British Army. Barr and Stroud Ltd, which started in 1888 as a casual collaboration between a professor of engineering and a professor of physics, had by 1897 become the world's sole manufacturer of rangefinders. Soon it was supplying them to Japan and every major European power except Germany. Between 1903 and 1914 it pulled in £750,000 in foreign contracts and £450,000 in Royal Navy and War Office contracts.[80]

But with the onset of war, existing channels of glass supply had to be realigned or relinquished. Three British optical-munitions manufacturers, specializing in three different instruments, had become almost entirely dependent on French-supplied glass. Starting in 1909, Chance Brothers of Birmingham, primarily a maker of window glass, had been investigating the secrets of making the optical varieties, and in August 1914 its monthly output was a thousand pounds of the good stuff. Not even close to enough. Within a year, the War Office required a monthly output of seventeen thousand pounds, and British glassmakers were being hamstrung by their dependence on imported raw materials, some of which came from — you guessed it — Germany.

In mid-1915 Chance Brothers and the Optical Munitions and Glassware Department of the Ministry of Munitions (whose first director was a lecturer in physics, an expert on optics generally and range-finders specifically, and a former Examiner of Patents, thus embodying the modern alliance of science and war with industry) finally agreed on the terms of a public–private partnership.[81] The government would supply money and procure scientific input, and Chance would maintain adequate facilities and personnel and would achieve specified outputs; after the war, Chance would become a monopoly supplier to the military but could continue to use the facility for ordinary commercial production. It was a win-win situation. By war's end, the company was producing more than ten tons — comprising seventy different types — of optical glass a month.[82]

Germany's path from prewar to postwar was

more dramatic. A formidable prewar exporter not only of superb glass and optics but of steel, chemicals, and electrical goods, Germany had swiftly gained export ground on cotton-and-coal Britain since the 1890s, raising British fears of being overtaken and undercut. In 1897, the year Barr and Stroud set up the world's only rangefinder factory, Britain was the world's top exporter at $1.4 billion, the United States was a close second at $1.2 billion, and Germany a lagging third at $865 million. By 1913, while Britain's exports had doubled, Germany's had more than tripled.[83]

War and its blockades, followed by defeat, armistice, and the Treaty of Versailles, should have decisively halted Germany's race toward the top. Under the terms of the treaty, signed in June 1919, every business enterprise engaged in "the manufacture, preparation, storage or design of arms, munitions, or any war material whatever" was to be closed. Both import into and export from Germany of "arms, munitions and war material of every kind" was to be "strictly forbidden." Aside from specified allowable quotas, all German armaments, munitions, and war material, including "aiming apparatus" and the "component parts" of various guns (both of which lie within the bailiwick of optics), were to be swiftly "surrendered to the Governments of the Principal Allied and Associated Powers to be destroyed or rendered useless."[84]

Ah, but what is "war material"? That question kept the members of the Inter-Allied Military Control Commission (IAMCC), the Treaty's disarmament inspector-overseers, awake at night and drawing up lists all day.[85] As the exasperated British brigadier general who served as second-in-command on the IAMCC's Armaments Subcommission later wrote,

The thing defies definition. Is a field-kitchen war material? Or a field ambulance? Or a motor-lorry? All three are capable of civilian use. When are you to "call a spade a spade," and when should you call it an entrenching tool? How are you to distinguish between war explosives and "commercial" explosives? The dynamite which serves to blast a quarry is as useful to the sapper in war as to the quarryman in peace. . . .

Our categories of war material grew and grew until they filled scores of pages of print. The species and sub-species extended to hundreds of articles. The list of "optical" war material, from periscopes to range-finders, alone ran to fifty-two items. "Signaling material" was almost equally multitudinous. In both cases many of the incriminated articles, such as field-glasses, telephones, and wireless apparatus, were unquestionably ambiguous in character, equally susceptible of use for war and for peace.[86]

Brigadier General J. H. Morgan and his fellow overseers found it equally frustrating to decide which factories to close. While the war had decimated France's industrial capacity, it had left Germany's largely intact. More than 7,500 engineering, electrical, and chemical factories had been tasked with the production of war material; as of the war's end, claimed Germany, most were "reconverted" to production for civilian purposes. Obligated by the treaty to permit Germany to continue production of the stipulated quantities of armaments, and feeling pressured not to constrain Germany's capacity to pay reparations, the overseers ended up deciding to "spare every factory and workshop which could establish . . . its re-conversion. The result was that Germany was left with every lathe that ever turned a shell" — and, though General Morgan doesn't mention it, probably every grinder that had ever polished a periscope lens, including those at Zeiss. Plus, he and his colleagues discovered "in due course, that vast stocks of arms which never appeared in the official returns made to us by the German Government were being concealed all over Germany."[87]

Thus the halt was more like a brief interruption. In 1913 Zeiss was among Germany's largest business endeavors, with total assets triple those of Schott; together, these twin companies were safely ensconced among the top hundred. By then, Zeiss was not simply a German enterprise but an international conglomerate; it not only exported its products but managed a web of foreign sales agencies, foreign manufacturing licenses (including some held by Bausch and Lomb Optical Company, founded in upstate New York by German immigrants), and foreign factories (including a very lucrative one near London). Schott, with a simpler business model, nonetheless exported more than half its total glass production and about a quarter of its optical glass before the war. The outcome of the war did cramp their international style — Zeiss's London factory was sold in 1918 for a mere £10,000, for instance, and Schott's exports to the UK in 1920–21 were a mere one percent rather than the steady prewar 5 or 6 percent. But by the mid-1920s, despite treaty restrictions and increased tariffs, both Zeiss and Schott were again making deals with British companies.[88]

More important, in Jena they went deep into R & D and were soon pushing the limits of optical technology again, with high-profile civilian achievements alongside those of use to the military.[89] In 1925 the world's first planetarium opened in Munich, equipped with the world's first star projector, designed and built by Zeiss. In 1930 America's first planetarium opened in Chicago, again with a Zeiss projector. And in 1933, as audiences were being enraptured by the sight of the stars in Zeiss-equipped planetarium domes from Stockholm to Rome to Moscow, a heavily remilitarized Germany made clear its displeasure at disarmament by finally withdrawing altogether from the League of Nations, that high-minded pioneering world association brought into being by the Treaty of Versailles.

What about America's part in the wartime production of optical glass? Before the United States joined the war, its imports of optical glass cost about half a million dollars a year.[90] Bausch and Lomb, the major domestic producer (of which Zeiss had bought a 25 percent share), made barely one ton of optical glass a month. Yet, upon joining, America was expected to supply one ton of optical glass a day to the Allies. While US citizens loaned their binoculars to the military, the country's glassmakers geared up.[91] Again, the transformation resulted from a public–private partnership, but unlike Britain's piecemeal approach, based around voluntary cooperation, the American solution was top-down and carefully focused.

By late spring 1917 the Council of National Defense had dispatched silicate scientists (silica being the main component of common sand, the main component of glass) from the Carnegie Institution's Geophysical Laboratory to the nation's glass factories. The US Army Ordnance Department made the scientist in charge, F. E. Wright, a lieutenant colonel. Consequently, the Army itself was, as Wright put it later, "the court of last appeal," which he found "a useful lever" in wartime conditions. So the Army ran the show, the scientists obeyed, and the factories ramped up production as fast as they could, with the assistance (and coercion) of other government agencies. Given the strict controls and tight deadlines, the experts opted for basics and high volume — just six types of glass, adequate for most instruments — rather than range, innovation, and top quality. In September 1917 US factories produced more than five tons of optical glass, in December more than twenty tons. In 1918 the total US production of "satisfactory" glass for optical munitions amounted to nearly three hundred tons, two-thirds of which came from Bausch and Lomb.[92]

In World War I, unlike its successor, the air was not initially a strategic battlefield. Space was decades away from becoming a site for surveillance and reconnaissance. Radio and aircraft were still rudimentary. The intimate alliance between astrophysics and the military would not be forged until just before the next world war.

The modern, Western offspring of astronomy, astrophysics is not even a century and a half old. Its midwives were two nineteenth-century technological innovations. The more widely known of the two, photography — literally "light drawing" — stemmed from a welter of investigations into the image-forming proclivities of light. The lesser-known and more specialized innovation, spectroscopy — which separates light into its component colors, yielding heaps of information about its source — derived from the prismatic study of the Sun's spectrum and the discovery that every substance radiates a characteristic and unique combination of colors. Jointly, photography and spectroscopy empowered the astronomer to record and analyze whatever light the available telescopes could gather from the sky.

The inception of photography during the 1830s and 1840s changed the ground rules of representation and the concept of evidence. Astronomers had long needed a convincing way to record their observations. In the seventeenth and eighteenth centuries they could talk about, write about, compose anagrams about, or draw what they saw. Their audience had to trust in their honor and take their word. Drawings were the best anybody could do, but they have inherent limitations.

As long as a human hand holding a pencil is recording the photons, the record is susceptible to error: human beings, especially sleepy ones with eyestrain and variable artistic skill, are not reliable recorders. On occasion, Galileo circumvented the problem by using symbols. In *Sidereus Nuncius* ("The Starry Messenger"), rushed to publication in February 1610, his drawings of the movements of Jupiter and its largest moons consist simply of a large circle and several dots; his drawings of stars are either six-pointed asterisks (small or medium-sized) or six-pointed cookie-cutter stars with a dot in the middle.[93]

Finally, in the mid-nineteenth century, a presumptively unbiased recording device came to the rescue: the camera. By employing one of the multifarious new light-drawing techniques, you could record the terrestrial and celestial worlds with minimal interference from eye, hand, brain, or personality. Your quirks and limitations would fade to irrelevancy, whether you used a silver-plated, highly polished sheet of copper exposed to iodine vapors and mercury fumes or a glass plate coated with a gelatin concoction.

One of photography's inventors, Louis-Jacques-Mandé Daguerre, and many of its first commentators were concerned primarily with art, specifically painting, which they thought would either be facilitated or nullified by the miraculous mechanical invention.

One writer hailed the daguerreotype as "equally valuable to art as the power-loom and steam-engine to manufactures, and the drill and steam-plough to agriculture."[94] Others contended that photography heralded the death of painting. Soon photography would, in fact, unshackle artists from any remaining obligation to capture visual reality, thus clearing a broad path for modernist painters such as Gauguin, van Gogh, and Picasso, not to mention early art photographers such as Julia Margaret Cameron. While scientists embraced photography as a tool to gather data and remove the observer's impression of a scene, artists embraced it as another good reason to convey subjective impressions, internally generated visions, or the essence of their medium.

Among photography's pioneers and proponents were several high-profile scientists. William Henry Fox Talbot, inventor of the light-sensitive paper negative in 1834–35, was a Royal Society gold medalist in mathematics and a Fellow of the Royal Astronomical Society.[95] Another Englishman, Sir John Frederick William Herschel, president of the Royal Astronomical Society, coined the word "photography" in 1839. He also coined the word "snapshot" in 1860, introduced the photographic usage of the words "positive" and "negative," discovered that sodium hyposulfite — "hypo" for short — could be used as a photographic fixative (rendering the emulsion no longer sensitive to light), made the acquaintance of Fox Talbot, corresponded with Daguerre, and, all in all, threw himself so early and so thoroughly into the new endeavor of drawing with light that he practically ranks as one of photography's inventors.

Even more influential than Sir John Herschel during the early months of photography's official existence was the French astronomer and physicist François Arago, director of the Paris Observatory, perpetual secretary of the French Academy of Sciences, and, following the Revolutions of 1848, the provisional government's colonial minister as well as its minister of war. He was also a great publicist. On January 7, 1839, acting as Daguerre's spokesperson and scientific advocate, Arago announced the invention of the daguerreotype at the Academy. It was a thrilling moment for science, art, commerce, national heritage, and much else besides. "Monsieur Daguerre," said Arago, "has discovered special surfaces on which an optical image will leave a perfect imprint — surfaces on which every feature of the object is visually reproduced, down to the most minute details, with incredible exactitude and subtlety."[96]

Arago also asserted that the new technique was "bound to furnish physicists and astronomers with extremely valuable methods of investigation." Together with two noted physicists of his day, Arago himself had tried but failed to make an image of the Moon by projecting moonlight onto a screen coated with silver chloride. Now, at the urging of several members of the academy, Daguerre had managed to "cast an image of the Moon, formed by a very ordinary lens, onto one of his specially prepared surfaces, where it left an obvious white imprint," and had thus become "the first to produce a perceptible chemical change with the help of the luminous rays of Earth's satellite."[97] To contemporary eyes, the image is unimpressive; to mid-nineteenth-century eyes, it was mind-blowing. Anyone who knew anything about chemistry or physics now rushed to attempt *un daguerréotype.*

In early July, on behalf of a commission charged with assessing the wisdom of granting Daguerre a lifetime government pension in exchange for permitting France to present the discovery to the world, Arago reported to the Chamber of Deputies that the daguerreotype would rank with the telescope and the microscope in its potential range of applications:

We do not hesitate to say that the reagents discovered by M. Daguerre will accelerate the progress of one of the sciences, which most honors the human spirit. With its aid the physicist will be able henceforth to

245

proceed to the determination of absolute intensities; he will compare the various lights by their relative effects. If needs be, this same photographic plate will give him the impressions of the dazzling rays of the sun, of the rays of the moon which are three hundred thousand times weaker, or of the rays of the stars.[98]

By August 19, Daguerre's pension was a done deal, and Arago announced the details of the process. Every aspiring daguerreotypist could now just follow directions.[99]

The first impressive daguerreotype of a celestial object dates to early 1840. It was a portrait of the Moon one inch in diameter, the product of a twenty-minute exposure from the roof of a building in New York City, made by a physician–chemist named John William Draper. In 1845, by exposing a silvered plate for a mere sixtieth of a second, two French physicists — Léon Foucault and Armand-Hippolyte-Louis Fizeau — produced a respectable image of the Sun. In 1850 two Bostonians — John Adams Whipple, a professional photographer, and William Cranch Bond, first director of the Harvard College Observatory — daguerreotyped Vega, the sixth brightest star in the nighttime sky, by exposing their plate for a hundred seconds. The next year, another professional photographer, Johann Julius

Friedrich Berkowski, in collaboration with the director of the Royal Observatory in Königsberg, Prussia, used an exposure of eighty-four seconds to daguerreotype a total solar eclipse. Astrophotography was well and truly under way.

Meanwhile, inventive individuals were hard at work making photography more user-friendly. Within a few years, the one-off daguerreotype positive would be a relic, replaced by a glass plate coated with a light-sensitive emulsion that yielded a negative, thereby ushering in a new era of reproducibility. In 1880 hand-craftsmanship gave way to mechanization when the Eastman Dry Plate and Film Company opened in Rochester, New York. By the end of the decade, photography had become an essential tool of the astronomer's trade.[100]

Compared with photography, spectroscopy — the other midwife of astrophysics — might seem an arcane development. No populist fanfare or breathless newspaper accounts attended its birth.

As soon as telescopes became standard equipment, piles of people began to spend gobs of time finding dim blips of light, mapping their positions, estimating their brightness and colors, and adding them to the ballooning catalogue of stars, nebulas, and comets. The task was limitless. But no sky map says much about the stuff of which the stars are made, or about their life cycles or their motions. For that you need to know their chemistry and understand their physics. That's where spectroscopy comes in.

Every element, every molecule — each atom of calcium or sodium, each molecule of methane or ammonia, no matter where it exists in the universe — absorbs and emits light in a unique way. That's because each electron in a calcium atom, and each electron bond between atoms in a methane molecule, makes the same wiggles and jiggles as its counterpart in every other calcium atom or methane molecule, and each of those wiggles absorbs or emits the same amount of energy. That energy announces itself to the universe as a specific wavelength of light. Combine all the wiggles of all the electrons, and you've got the atom's or molecule's electromagnetic signature, its very own rainbow. Spectroscopy is how astrophysicists capture and interpret that rainbow.

Spectroscopy's prehistory begins with Isaac Newton in 1666, when he showed, using prisms, that a visible beam of "white" sunlight harbors a continuous spectrum of seven visible colors, which he named Red, Orange, Yellow, Green, Blue, Indigo, and Violet (playfully known to students as ROY G. BIV). For the next couple of centuries, investigators on several continents followed his lead. In 1752 a Scot named Thomas Melvill found that when he burned a chunk of sea salt (think sodium) and passed the firelight through a slit onto a prism, it yielded a striking bright yellow line; two and a half centuries later, sodium would be the active ingredient in yellow-tinged sodium vapor streetlights.[101] In 1785 a Pennsylvanian named David Rittenhouse devised a way to produce spectra with something other than a prism: a screen made of stretched hairs, densely packed in parallel lines and arranged to provide a series of slits that could disperse a beam of light into its constituent wavelengths. In 1802 an Englishman named William Hyde Wollaston found that the Sun's spectrum includes not only the seven colors that met Newton's eye but also seven dark lines or gaps amid the colors. It was now evident that visible light held a lot of hidden information, reinforcing the prior two years' discoveries of infrared and ultraviolet, which had shown that light itself could be hidden from human view.

A dozen years later, Joseph von Fraunhofer, a German physicist and top-notch glassmaker who had committed himself to producing the most distortion-free telescope lenses money could buy, made a major breakthrough in examining the spectrum of the Sun. He decided to place a prism in front of a lens and look at sunlight that had passed through both intermediaries. What he saw in 1814 were hundreds more dark spectral lines than Wollaston had seen in 1802. In experiments with different types of glass over the next couple of years, the lines always appeared in the same places on the spectrum. Today tens of thousands of these "Fraunhofer lines" are known to exist in the solar spectrum. They're dark because the light that would otherwise show up at those specific wavelengths is being absorbed by the lower-temperature, outermost layers of the Sun. By contrast, certain bright lines that show up in the spectra of flames from laboratory experiments are the result of those specific wavelengths being emitted, rather than absorbed.

Not only did Fraunhofer assiduously map the solar spectrum; he also noticed that the position of two bright yellow lines in the spectrum of a sodium flame matched the position of two prominent dark lines in the solar spectrum. Moreover, he saw that the spectrum of the Sun matched the spectra of sunlight reflected from the planets but that the Sun and the other bright stars in the sky each had its own spectral signature. By some people's standards, he made the first true spectroscope.[102]

Light was a topic of hot debate and cutting-edge research, and its fundamental nature remained elusive for much of the nineteenth century. Was it made of corpuscles, as Newton had argued, or of waves? Was it propagated through a ubiquitous, flexible, invisible medium? At what speed did it travel? Was it related to electricity? To magnetism? At mid-nineteenth century, spectroscopy didn't yet exist as a specialty, but it soon would — thanks largely to the collaboration of two professors at the University of Heidelberg, the physicist Gustav Kirchhoff and the chemist Robert Bunsen (who, by the way, improved but did not invent the Bunsen burner). In the late 1850s they began to devote themselves to

> a common work which doesn't let us sleep. . . . [A] means has been found to determine the composition of the sun and fixed stars with the same accuracy as we determine sulfuric acid, chlorine, etc., with our chemical reagents. Substances on the earth can be determined by this method just as easily as on the sun.[103]

In 1859 Bunsen and Kirchhoff devised a way to superimpose the spectrum of a beam of light from a sodium vapor lamp on the spectrum of a beam of sunlight, thereby confirming Fraunhofer's suspicion of a connection between two of his dark lines and the two bright-yellow sodium lines, forever linking the lab chemist's table with the matter that occupies the farthest reaches of the cosmos. Over the next few years, by burning various substances on their Bunsen burner and passing the light through a spectroscope of their own design, they methodically mapped the patterns made by known elements, discovered several new ones, and enabled their students and other investigators to discover still more.

One person who probably rolled over in his fairly fresh grave when Bunsen and Kirchhoff began to publish their findings in 1860 was the French philosopher Auguste Comte, who in 1835, in the second volume of his six-volume *Course on Positive Philosophy,* boneheadedly declared the impossibility of gleaning any chemical information, or more than limited physical information, about the stars:

We understand the possibility of determining their shapes, their distances, their sizes and their movements; whereas we would never know how to study by any means their chemical composition, or their mineralogical structure, and, even more so, the nature of any organized beings that might live on their surface. . . . I persist in the opinion that every notion of the true mean temperatures of the stars will necessarily always be concealed from us.[104]

Had Comte been correct, astrophysics would not exist. But shortly after the publication of volume two of his magnum opus, spectroscopic revelations about Earth's cosmic neighborhood began to multiply. Soon spectra would be not merely detected but also photographed, despite the challenge of grabbing enough photons of any given wavelength so that a line would actually register in the emulsion. Astrophotographs would capture previously unseen and unimagined attributes of distant celestial bodies. Thirty years before being discovered on Earth, helium would be discovered in the Sun's spectrum and named for the Greek sun god, Helios. By 1887 — four decades after two Bostonians daguerreotyped the star Vega in a hundred seconds — two French brothers, Paul-Pierre and Matthieu-Prosper Henry, took only twenty seconds to photograph a star ten thousand times dimmer.[105]

The Astrographic Congress of April 1887, convened by the French Academy of Sciences and attended by scientists from nineteen countries, marked the official marriage of photography and astronomy.[106] During their eleven days in Paris, the delegates agreed to undertake a two-pronged international effort to use a standard instrument and standard methodology not only to map the sky photographically but also to precisely catalogue the two million brightest stars — a significant goal, given that the average unaided eye sees not much more than six thousand. The instrument of choice was one developed by the brothers Henry. The very next year, an American astronomer/physicist/aircraft pioneer, Samuel P. Langley, published a book titled *The New Astronomy* — although not everyone saluted the idea of newness. As one hidebound nineteenth-century astrophysicist wrote, "The new astronomy, unlike the old astronomy to which we are indebted for skill in the navigation of the seas, the calculation of the tides, and the daily regulation of time, can lay no claim to afford us material help in the routine of daily life."[107]

The new astronomy needed a new journal and a new organization. In 1895 *The Astrophysical Journal, an International Review of Spectroscopy and Astronomical Physics* published its first issue. Four years later, the various subspecies of skywatchers came together to form the Astronomical and Astrophysical Society of America. Under truncated titles — *The Astrophysical Journal* and the American Astronomical Society — both the journal and the organization still thrive.

Today's astrophysicists have at our disposal individual telescopes that collect seventy thousand times more light than Galileo's first attempts at a spyglass, and spectrometers that can reveal hydrogen in a galaxy that dates back to the first billion years after the Big Bang. We're also armed with an abundance of auxiliary tools and tactics: adaptive optics, digital detectors, supercomputers, devices for masking the overwhelming brilliance of a host star so that nearby planets can be detected, methods for separating signal from noise. But no matter the innovations, no matter the complexity of the technology, the twenty-first-century astrophysicist's fundamental challenge remains the same as Galileo's: to collect the maximum amount of light from extremely dim and distant objects, and then extract from that light as much information as possible. It's how the contemporary astrophysicist — and the contemporary warfighter — wants to use the light that makes all the difference.

Astrophysicists deduce nearly everything we know of the contents and behavior of the universe from the analysis of light. Most of the cosmic objects and events we observe materialized long ago, and so their attenuated light arrives here on Earth after delays that stretch up to thirteen billion years. Since the observable universe now spans nearly 900,000 billion billion kilometers, and the actual universe is vastly larger than that, astrophysicists are proximity-challenged. Most of the objects of our affection lie forever out of reach and are, at best, barely visible from Earth. They don't grow in a laboratory, they release stupendous energy, and they're immune to manipulation. For the most part, they are accessible only by night. We can't easily visit them in their natural habitat, and, beyond our solar system, it's not yet possible to touch (or contaminate) them. Though smitten by the cosmos, we have no choice but to embrace it from multiple degrees of separation: when we want to know the motions of a star, we examine not the star itself, not an image of the star, not even the spectrum derived from the light recorded in an image of the star, but rather the shifts in the patterns in the spectrum derived from the light recorded in an image of the star. A convoluted consummation.

So astrophysicists have learned to be lateral thinkers, to come up with indirect solutions. True, scientists in general are skillful problem solvers. Physicists can build a better vacuum chamber or a bigger particle accelerator. Chemists can purify their ingredients, change the temperature, try out a novel catalyst. Biologists can experiment on organisms born and bred in the lab. Physicians can question their patients. Animal behaviorists can spend hours watching clans of their favorite creatures. Geologists can scrutinize a hillside ravine or dig up sample rocks. But astrophysicists need to find another way, never forgetting that we're the passive party in a singularly one-sided relationship.

Down here in our labs and offices, though, we become somewhat more aggressive, owing to our mutually advantageous alliance with the military. Many significant advances in our understanding of the cosmos are by-products of government investment in the apparatus of warfare, and many innovative instruments of destruction are by-products of advances in astrophysics.

As a group, astrophysicists don't embrace a military approach to problem solving. Rarely do you find an astrophysicist thinking, I'll do *a* or *b* so that it will someday help the military, or, I hope the military does *x* or *y* so that someday it will help me. The connection is more fundamental, more deeply embedded in the nature of the astrophysicist's domain and the capabilities of the astrophysicist's tools. Space — our turf — is the new high ground, the new command post, the new military force multiplier, the new locus of control, although in fact it's not very new. Space has been politicized and militarized from the opening moments of the race to reach it.

The recurrent interconnections between sky work and war work have not gone unnoticed by either space scientists or space policy analysts. In his 1981 book *Cosmic Discovery,* Martin Harwit, director of the Smithsonian's National Air and Space Museum from 1987 to 1995,[108] profiles five turning points in the history of astronomy — the telescope, the birth of cosmic-ray astronomy, the birth of radio astronomy, the birth of X-ray astronomy, and finally the then-recent discovery of distant gamma-ray bursts. Only the account of radio astronomy's early days includes no reference to military involvement. Harwit further points out that the discoveries of new phenomena often involved equipment originally designed for use by the military. British political scientist Michael J. Sheehan puts forth a related position in his 2007 book *The International Politics of Space:* "Space has always been militarised. Military considerations were at the heart of the original efforts to enter space and have remained so to the present day."[109]

Much has been written about the making of the atomic bomb. The relationship between physics and war is clear: the ruler and the general want to threaten or obliterate targets; destruction requires energy; the physicist is the expert on matter, motion, and energy. It's the physicist who invents the bomb. But to destroy a target, you have to locate it precisely, identify it accurately, and track it as it moves. That's where astrophysics comes in. Neither protagonists nor accomplices, astrophysicists are accessories to war. We don't design the bombs. We don't make the bombs. We don't calculate the damage a bomb will wreak. Instead, we calculate how stars in our galaxy self-destruct through thermonuclear explosions — calculations that may prove helpful to those who do design and make thermonuclear bombs.

Our utility is broad. We understand trajectories and orbits, and so we're key to the launching of both spacecraft and space weapons. We're skilled at the art and science of image analysis, especially at the limits of detection — a suite of techniques indispensable to the selection of targets and the interpretation of elusive evidence. We understand reflectivity and absorptivity, and so we've laid the groundwork for an entire industry of stealth matériel. We can distinguish an asteroid from a spy satellite by studying the differing wavelengths of light that they absorb and reflect. We know, by their light, which molecules inhabit which celestial bodies, and so we could spot an alien intrusion if one were suddenly to appear. We recognize the multi-spectral light signatures of naturally occurring collisions, explosions, impacts, magnetic storms, shock waves, and sonic booms, and we can differentiate them from dangers and catastrophes induced by a living agent.

But whether an astrophysicist's work is done at the behest of the military or for the sake of science, the tools are the same. The techniques are the same.

After a swift yet peaceful journey of tens, hundreds, or thousands of light-years, the sharp pinpoint of light from a distant star reaches Earth's lower atmosphere. A fraction of a second later, skygazers with telescopes see it as a fuzzy, jiggling blob, while naked-eye skygazers see it as a pleasantly twinkling, distant jewel. Back in 1704, Sir Isaac Newton was already worried that twinkling would hamper astronomers of the future:

> If the Theory of making Telescopes could at length be fully brought into Practice, yet there would be certain Bounds beyond which Telescopes could not perform. For the Air through which we look upon the Stars, is in a perpetual Tremor; as may be seen by the . . . twinkling of the fix'd Stars.[110]

Newton went on to suggest that a mountaintop might be a good place to put a telescope, and he was right. But even given optimal placement, the atmosphere may not cooperate. Robert W. Duffner, an optics historian at the Air Force Research Laboratory in New Mexico, describes looking at stars through the atmosphere as akin to looking through the frosted glass of a shower stall: you see shapes but no detail.[111]

What happens when a star twinkles? The atmosphere is a tapestry of air patches with different temperatures and densities, and thus different optical properties. Each time a light wave crosses from one patch to another, it bends a little and slightly shifts direction. The scene resembles the fate of pond ripples moving across an untidy ridge of stones, which disturb the smooth shape of each wave crest before it reaches the shore. Under the influence of our undulating atmosphere, a star's image will not only drift to and fro but will also change in brightness from one moment to the next. A time-lapse photograph will record a smeared circular blob; your eyes will record a twinkling star. In fully turbulent air, the patches are small and numerous, causing the star to twinkle ferociously.

What's needed is a way to compensate for how the varying patches of atmosphere disrupt the starlight. This amounts to reconstructing our pristine pond ripple after it has passed over the rocks. To do this, you'll have to record the light hundreds of times per second. Each time, you'll need sufficient light to simultaneously track and correct for any ongoing atmospheric changes. To facilitate the corrections, you'll need a luminous "guide star" for comparison, close enough to your target object to be influenced by the same patch of atmosphere at the same time. Alas, such stars are few and far between, unlikely to sit conveniently nearby on the sky. The solution? Create an artificial star. Send a powerful laser beam high above the stratosphere, where turbulence is minimal and there's a continually replenished supply of sodium atoms left behind by vaporized meteors. Excite some of the sodium atoms so that they radiate back at you, and position this now-luminous spot exactly where it will serve you best.

Before the 1990s, anybody seeking high-resolution images of a star field or galaxy on a twinkle-ridden night had two obvious options. Plan A: Close the telescope dome and go to bed. Plan B: Raise several billion dollars, build a new telescope, launch it into orbit above the layers of atmospheric disturbance, and observe the universe from there. In 1990 Plan B gave us the Hubble Space Telescope, which offered a leap in resolution from the ground-based telescopes of its day as impressive as the leap from the unaided eye to Galileo's first telescope.

But now there's a less obvious remedy to the twinkling problem. Welcome to the field of adaptive optics. This innovation uses laser guide stars and deformable mirror surfaces to correct for the unwanted twinkling caused by Earth's atmosphere. A matrix of push-pull pistons affixed to the back of a deformable telescope mirror continually adjusts the exact shape in such a way as to correct for the transient atmospheric turbulence, canceling out the patch-to-patch, moment-to-moment atmospheric variations. All adaptive optics systems also include a second, non-segmented mirror to monitor and correct the wandering of the image due to larger-scale motions in our atmosphere. Rounding out the system components, adaptive optics uses beam splitters, interferometers, monitoring cameras, and of course specialized software. The whole contraption is costly and complex. It's also strikingly effective, enabling the sharpness of ground-based images to rival that of images taken from space.[112]

Did civilian astrophysicists make adaptive optics a reality? No. But that was not for want of trying. From the 1950s onward, astrophysicists developed concepts and potential solutions. But while they were still focusing on possibilities, the US Department of Defense was secretly achieving results — through classified research funded and conducted from the late 1960s through the late 1980s by organizations such as the Defense Advanced Research Projects Agency, the Air Force Research Laboratory and Phillips Laboratory at Kirtland Air Force Base in New Mexico, the Air Force Maui Optical Site, Itek Optical Systems near Hanscom Air Force Base in Massachusetts, the Air Force's Rome Air Development Center in New York, MIT's Lincoln Laboratory, the Visibility Laboratory at Scripps Institution of Oceanography, and the Strategic Defense Initiative. Additional expertise came from the top-secret national-security science advisory group called the Jasons. Formed in 1960 and comprising Mac-

Arthur geniuses, Nobel laureates, and prominent academic physicists, the Jasons provide the military with bleeding-edge ideas for how to wage war, end war, and prevent war. From their earliest years of summertime meetings, there have always been a few Jasons whose specialty is the cosmos.[113]

It was a Jason who thought up adaptive optics, and it was not until May 27, 1991, that the details of the research were made public. Addressing a packed room at the 178th meeting of the American Astronomical Society that afternoon, Robert Fugate, technical director of the USAF's Starfire Optical Range at Kirtland AFB, began his presentation by saying, "Ladies and gentlemen, I am here to tell you that laser guide star adaptive optics works!" Two images of the binary star 53 Ursa Major proved his point. One showed the stellar duo smeared into a single blob of light by the effects of atmospheric turbulence; the other, a beneficiary of adaptive optics, showed two distinctly separate, glowing objects. In that moment, Fugate had declassified adaptive optics. Space scientists could now take it to their own next level.[114]

The Pentagon's interest in clearer seeing was consistent with the centuries-long military desire for more accurate information, and its interest in laser beacons meshed with the equally long-standing desire for new kinds of weaponry. Cold War thinking dominated US policy during the two decades of groundbreaking adaptive optics research before declassification. Not only did the intelligence community seek sharp images of newly launched enemy satellites, incoming enemy missiles, and troublesome space debris for the sake of space situational awareness; warfighters sought ways of directing powerful lasers at those missiles and satellites for the sake of destroying them.

In the early 1970s, sharpening the images could only be done through post-detection digital cleanup of short-exposure films, with deeply unsatisfactory results. Reliance on photographs, scanners, and mainframe computers meant delays of a day or more when measuring wavefronts. The military needed much better technology to provide instant information, and they were prepared to pay for it. The first adaptive optics system for a large telescope was installed in 1982 on the Air Force's satellite tracker at Mount Haleakala on Maui. By then, on the laser front, the military had already seen considerable progress toward controlling and maximizing the intensity of the beam. Building on earlier research, in 1975 the Air Force began to transform an aged Boeing KC-135A into the Airborne Laser Laboratory, which in 1983 succeeded in shooting down a series of air-to-air missiles and ground-launched drones. The use of lasers in airborne antimissile defense held promise. Ronald Reagan's 1983 public announcement of the Strategic Defense Initiative — Star Wars — promised yet more.[115]

With declassification, the divergent goals and tasks of the warfighters and the space scientists came into focus. British-born electrical engineer John W. Hardy, who in 1972 developed the first successful image-compensation system to use adaptive optics, described the "vast disparity" in his 1998 book *Adaptive Optics for Astronomical Telescopes:*

> Equipment for military applications must work reliably under the worst conditions, and must produce a specified level of performance[, which] usually requires advancing the state of the art, an expensive proposition. Astronomers, on the other hand, usually work in good [observing] conditions and are able to exploit small improvements in technology that allow more information to be extracted from their observations. . . .
>
> The defense community must continually push the limits of technology to keep ahead of assumed adversaries; it usually takes some time for the value of new technology to be appreciated and to be applied to scientific work.[116]

"Some time" in this instance was less than a decade. By the end of the 1990s, space scientists were already benefiting from the new technology. And today almost every giant ground-based visible-light telescope incorporates a version of this corrective system. Unlike other cases, in which research progresses via the resonance of ideas, adaptive optics was a baton pass from the warfighter to the astrophysicist.

If the capacity to monitor an enemy's movements has always been necessary to military success, what could be more useful to a twenty-first-century spacefaring superpower than the capacity to monitor not only our entire planet but also the surrounding envelope of space? Since time immemorial, it's been obvious that defense is enhanced by surveillance and reconnaissance, which are enhanced by gaining the high ground. Having gained it, you may then be able to keep it and control it.

In 1958, while still a senator, Lyndon B. Johnson called space control "the ultimate position":

There is something more important than any ultimate weapon. That is the ultimate position — the position of total control over Earth that lies somewhere out in space. That is . . . the distant future, though not so distant as we may have thought. Whoever gains that ultimate position gains control, total control, over the Earth, for the purposes of tyranny or for the service of freedom.[117]

Given the perennial patterns of unrest in human history, the prospect of any single nation having total control over Earth is unlikely to engender universal confidence. As President Kennedy said, in a famous speech delivered to a joint session of Congress in May 1961, just six weeks after the Soviet Union's Yuri Gagarin became the first person to orbit Earth, "No one can predict with certainty what the ultimate meaning will be of mastery of space."[118] What's certain is that if the past behavior of nations is any indication of the future behavior of nations, such mastery will not be wholly benign.

Benign or otherwise, monitoring to achieve even partial control is standard operating procedure. The US military uses the term "situational awareness" for the product of its varied forms of monitoring. This awareness is achieved through intelligence, surveillance, and reconnaissance — ISR, a modern abbreviation for the age-old challenge of knowing what the enemy is up to. Hand in hand with ISR is C3I: command, control, communication, and intelligence. Whatever the acronym, it's clear that neither rulers nor warfighters can make sensible decisions in defense of the nation if they can't quickly muster the facts.

That's where satellites come in, because nothing provides more hard facts today than the many hundreds of navigation, remote-sensing (also called Earth-observation), and weather satellites that now circle Earth 24/7.[119]

Take America's Global Positioning System, GPS — two dozen satellites in orbit at about 12,500 miles above Earth, more than fifty times higher than ordinary low-Earth-orbit satellites. You use it to navigate to a cousin's new house ten miles from nowhere for Thanksgiving dinner; geologists use it to chart earthquake fault zones in western India; conservation biologists use it to track the tagged grizzly bear population in Alberta, Canada; and people looking for immediate sex use it to triangulate on potential partners within range of their own location. GPS is everybody's handy helper. You probably wouldn't guess that it was created for the US Department of Defense and is controlled by the Air Force Space Command. Civilians can use GPS, but the navigation data they're given is less precise than what is supplied to military interests. People in other countries use it too, but there's no iron-clad guarantee of their having permanent access irrespective of changes in the political situation.

Then there are (and were) the satellites in

America's Defense Support Program, the Defense Meteorological Satellite Program, the Defense Satellite Communications System, the Missile Defense Alarm System, the Space-Based Infrared System, the Military Strategic and Tactical Relay system, the Galactic Radiation and Background program — all the various classified and declassified satellites whose ISR capabilities our multiple defense agencies rely on. Military satellites have been around for half a century, sent aloft shortly after the Soviet Union alarmed the United States by putting the first artificial satellite, Sputnik 1, into orbit on October 4, 1957. From the early days of spaceflight, ISR formed a major chunk of the agenda: America's Corona missions, beginning in August 1960, and the Soviet Union's Zenit missions, beginning in April 1962, were Cold War spies that took hundreds of thousands of photographs — although both programs were given a civilian, scientific face and a different name for public consumption.[120]

The high-altitude cameras of today's Earth-observation satellites are useful for the planning of roads and the monitoring of hurricanes, for locating ancient ruins swallowed up by sand or jungle, and for routing disaster assistance to villages cut off by fires, floods, landslides, or earthquakes. Most of them are mounted on satellites that orbit our planet somewhere between two hundred miles and twenty-two thousand miles overhead. The same (or similar) cameras that are used to surveil dwindling forests and shrinking glaciers can be used to surveil adversaries.

Most satellites, in fact, are "dual use." And if, as Joan Johnson-Freese of the US Naval War College points out, dual use covers both civilian/military and defensive/offensive uses, then "space technology is at least 95 percent dual use."[121]

India, for example, has a satellite called TES, Technology Experiment Satellite, which has orbited at an altitude of about 350 miles since late 2001. Asked whether TES's optical camera, sharp enough for one-meter resolution of Earth's surface, was intended for spying, the chairman of the Indian Space Research Organisation responded: "It will be for civilian use consistent with our security concerns. . . . All earth observation satellites look at the earth. Whether you call it earth observation or spying, it is a matter of interpretation." If one hi-res remote-sensing satellite is good, two are even better. In the spring of 2009 India's space agency launched RISAT-2, an Israeli-built satellite with all-weather, round-the-clock radar sensing, suitable for monitoring both crops and borders. On the question of its uses, the *Times of India* quoted a senior Indian space official as saying, "It will be primarily used for defence and surveillance. The satellite also has good application in the area of disaster management and in managing cyclones, floods and agriculture-related activities." Undistracted by his references to natural disasters, the *Times* editors titled this report "India to Launch Spy Satellite on April 20."[122]

Uncountable changes have taken place since Galileo offered the doge a nine-power spyglass. He could not have foreseen what this monitoring device would turn into. He could not have conceptualized the planetary reach of the telescope's orbiting cousins. But knowing the value of early access to information, he might have been pleased to learn that his name would be attached to the European Union's own emergent global-navigation satellite system. While being interoperable with GPS (as well as with Russia's equivalent system, GLONASS), Galileo will circumvent what was once US military control of information essential to all. As the agency that oversees the system states, "With Galileo, users now have a new, reliable alternative that, unlike these other programmes, remains under civilian control."

Control, but not exclusive use. In 2016 the author of a report on the security aspects of the European Union's space capabilities said that while Galileo and Copernicus, the EU's Earth-observation satellite system, aid in such essentials as coordinating aerial transport and tracking changes in the atmosphere, "we should not be afraid to say that they can also serve the Common Security and Defence Policy."[123]

Aside from facing the threats ceaselessly devised by our fellow Earthlings, all these eyes in the sky are vulnerable to a naturally occurring adversary: space weather. Unbeknownst to nineteenth-century electric telegraph operators and everyone else then living on Earth, the Sun is a giant ball of magnetic plasma that occasionally flares, ejecting blobs of charged particles across interplanetary space. In 1859 the biggest plasma pie in the past five hundred years hit Earth, mysteriously disrupting the world's newborn telegraph systems. The blast was so intense that it merited a name — the Carrington Event, after the English solar astronomer Richard Carrington, who was the first to observe it. Today, with hundreds of military and communications satellites orbiting Earth, and widespread grids feeding our electrically hungry civilization, we are more susceptible than ever to such a burst. In response, power companies are hardening their electronics at major switching stations, and the European Space Agency, Natural Resources Canada, and America's National Oceanic and Atmospheric Administration now have teams whose sole job is to monitor and predict space weather. These predictions will make it possible to switch satellites to safe mode in advance of a solar storm, thereby protecting their electric circuitry from an onslaught of charged particles.[124]

In a now-famous 1961 speech, outgoing president Dwight D. Eisenhower described historical episodes of wartime production — say, the ramping up of optical glassmaking during World War I — as the part-time, temporary making of swords by the usual makers of plowshares, in contrast with the full-time making of armaments that had become standard practice by his time in office. The novelist John Dos Passos had already memorably warned America about the military-financial complex, with a pointed reference to the wealth of J. P. Morgan: "Wars and panics on the stock exchange, machine-gunfire and arson, bankruptcies, warloans . . . good growing weather for the House of Morgan."[125] Now Eisenhower warned America about the military-industrial complex: the underbelly of necessary cooperation among political, scientific, defense, and productive forces. Not the first to issue such warnings but certainly the highest-profile person to do so, he referred to "unwarranted influence, whether sought or unsought" and to the "prospect of domination of the nation's scholars by Federal employment, project allocations, and the power of money." Wanting to have it both ways, Eisenhower also declared that America's armaments must be "mighty, ready for instant action." He worried that America's citizenry might not keep itself well informed enough to guarantee "the proper meshing of the huge industrial and military machinery of defense with our peaceful methods and goals."[126]

Take this meshed military-industrial machinery, add the race for ever-higher high ground, factor in the skyrocketing profit margins invoked by Dos Passos, and you've birthed the military-space-industrial complex: aerospace. Not many commentators have summed it up better than the fictional Madison Avenue creative director Don Draper of the AMC hit television series *Mad Men,* voicing a view that would have been current in late 1962:

> Every scientist, engineer, and general is trying to figure out a way to put a man on the Moon or blow up Moscow — whichever one costs more. We have to explain to them how we can help them spend that money. . . . [Congressmen] are the customer[s]. They want aerospace in their districts. Let them know that we can help them bring those contracts home.[127]

Seven and a half centuries have now passed since Roger Bacon informed the pope that enemy armies could be seen at a distance with the aid of "transparent bodies." Bacon's carefully shaped, suitably arranged refracting bodies have been replaced by a staggering portfolio of detectors, ranging from night-vision goggles to space telescopes. Seeing has become situational awareness and now encompasses a vast swath of wavelengths, well beyond the merely visual. Distances are now measured in light-years rather than stadia. Yet a few armed zealots can now cause more havoc and destruction than entire armies once did, and the future of weaponry may pivot not on how many guided missiles live in your silo, but on how many cyber scientists work in your lab. One factor that hasn't changed is money. Another is the existence and creation of enemies.

THE ULTIMATE HIGH GROUND

5

UNSEEN, UNDETECTED, UNSPOKEN

Invisibility captivates the astrophysicist and the warfighter. Both engage in surveillance. With the aid of a telescope, astrophysicists, in pursuit of knowledge, probe the otherwise nonvisible cosmos at ever greater depths and ever greater distances. Warfighters, in pursuit of defense or dominance, probe the enemy's hidden systems while seeking their own invisibility, gaining control while staying out of harm's way. Besides the pursuit of knowledge, defense, and dominance, there's the pursuit of secrecy, specifically the secrecy of information — yet another side of invisibility.[1]

For most of human history, we understood the world through our five senses. Sight, smell, taste, touch, and hearing gave us encyclopedic amounts of data. There was no particular reason to think that vast quantities of unseen, unheard, untouched, and generally unsensed objects and phenomena might be crammed into the world as well. Eventually the telescope and the microscope cracked open the door to the invisible, yielding extraordinary revelations: "an incredible number of little animals of divers kinds[,] several thousands in one drop" of Earth's water,[2] rilles in the Moon, spots on the Sun, rings around Saturn.

Even so, during their first few centuries, the microscope and telescope deepened human vision only within that narrow band of the electromagnetic spectrum called visible light, enabling us to see better than before but only seeing the same kind of light we were already accustomed to seeing. Yes, we could now detect dimmer things, smaller things, more distant things. But we hadn't yet grasped that much of the physical universe would require means of detection completely different from what our eyes, ears, and skin can provide.

What separates great scientists from ordinary scientists is not the capacity to answer the right question. It's the capacity to ask the right question in the first place, and not let common sense dictate or constrain their thinking. Fact is, there's nothing common about what you never knew existed. The formidable English physicist Isaac Newton, for instance, questioned the fundamentals of light and color. Everyone assumed that color was an intrinsic property of, say, the raindrops in a rainbow or the crystal pendants of a chandelier. Who in their right mind would have thought that ordinary light — white light — was composed of colors at all?

Newton, however, was smart enough to make no assumptions. By directing a ray of sunlight through a glass prism, which caused the visible spectrum to emerge from it, and then reversing the sequence, sending the spectrum back through the prism, whereupon white light emerged, he convincingly demonstrated that white light is indeed composed of multiple colors. Although each color in the spectrum shades gradually into its neighbor, Newton, a proponent of cosmic orderliness and the mystically significant number 7,[3] declared there were not six colors, as most of us today might list, but seven, slotting indigo between blue and violet to round out the set.

As early as the summer of 1672, decades before publishing his great work *Opticks: or, A Treatise of the Reflexions, Refractions, Inflexions and Colours of Light,* Newton sent a letter to the Royal Society with a list of questions about light and color that could be properly answered only through experiments. Two of his earliest queries were "Whether rays, which are endued with particular degrees of refrangibility, when they are by any means separated, have particular colours constantly belonging to them . . . ?" and "Whether a due mixture of rays, indued with all variety of colours, produces Light perfectly like that of the Sun, and which hath all the same properties . . . ?"[4] His prism experiment would answer yes to both.

Might Newton also have wondered, just once, even for a moment, whether there might exist some other, adjacent bands of light that our eyes could not see? He had noticed that red, on one end of the visible spectrum, and violet, on the other end, both just faded away to darkness.[5] He had raised the further possibility of there being "other original Properties of the Rays of Light, besides those already described."[6] Perhaps most important, he was comfortable with the idea of hidden attributes. Yet *Opticks* offers no clear evidence that he ventured there. In any case, a century would come and go before anyone conceived an answer to that unstated query.

Turned out, there were multiple answers. One came early in 1800, when the English astronomer William Herschel — the man who had discovered the planet Uranus two decades earlier — explored the relation between sunlight, color, and heat.

As Newton had so often done, Herschel began by placing a prism in the path of a sunbeam, but took it a step further. To determine whether each color had a different temperature, he placed thermometers in the various regions of the rainbow cast by the prism. And, like any good scientist conducting a well-designed experiment, he placed a control thermometer outside the color range — adjacent to the red side of the spectrum — to measure the ambient air temperature, unaltered by the warmth of the sunbeam. Herschel did indeed discover that different colors register different temperatures, but that turned out to be the second most interesting result of his experiment. The control thermometer, sitting in darkness, registered an even higher temperature than any of the thermometers placed within the rainbow. Only invisible rays could have caused that warming.

Sir William had discovered "infra" red light, the band just "below" red. His finding was the astronomy equivalent of geologists discovering the colossal Nubian Aquifer beneath the sands of the eastern Sahara. Behold his account of it:

By several experiments . . . it appears that the maximum of illumination has little more than half the heat of the full red rays; and from other experiments, I likewise conclude, that the full red falls still short of the maximum of heat; which perhaps lies even a little beyond visible refraction. In this case, radiant heat will at least partly, if not chiefly, consist, if I may be permitted the expression, of invisible light; that is to say, of rays coming from the sun, that have such a

The following year, 1801, Johann Wilhelm Ritter, a German scientist whose main interest was the intersection of electricity and chemistry, picked up where Herschel left off. Philosophically attracted to the concept of polarity in nature, Ritter assumed that infrared must have a companion just off the other side of the visible spectrum. Rather than use thermometers to demonstrate its presence, he used silver chloride, a substance known to decompose and darken at different rates when exposed to different colors of light. Ritter's experiment, like Herschel's, was both simple and smart: he placed a small mound of silver chloride in each visible color as well as in the unlit area alongside the violet, then awaited the results. As expected, the pile in the unlit patch darkened even more than the pile in the violet patch. What's more violet than violet? Ultraviolet.

Detecting without seeing was now a scientific reality.

Skywatching didn't change overnight, though. The first telescope capable of detecting wavelengths outside the slim visible portion of the electromagnetic spectrum wasn't built for another 130 years, well after the German physicist Heinrich Hertz had shown that the only real difference among the different kinds of light is the amount of energy they carry. And that would ultimately include it all: radio waves, microwaves, infrared, ROY G BIV, ultraviolet, X-rays, and gamma rays. In other words, he figured out that there is such a thing as an electromagnetic spectrum — a symphony of vibrating waves, each with a unique wavelength, frequency, and energy. To the astrophysicist, it's all energy, it's all radiation, it's all light.

Sometimes light behaves like particles, which we call photons. Sometimes — in fact, most times in our daily lives — light behaves like waves. Whether light should be conceptualized as waves or particles is an old disagreement: Democritus argued with Aristotle about it, Newton argued with Huygens, and quantum physics says it's both. Hence we're stuck with the phrase "wave–particle duality," even though our brains have a hard time wrapping themselves around the concept. Unfortunately, the word "wavicle" never caught on.

For now, think of light — electromagnetic radiation — as made up of waves made of particles. The word "wavelength" obviously applies to waves. It's the simple measure of length from crest to crest or trough to trough. Gamma-ray wavelengths are shorter than the diameter of an atom; on the far end of the radio band, wavelengths can be longer than the diameter of Earth.[8] The shorter the wavelength, the higher the energy and, broadly speaking, the greater the danger to life as we know it. And whether we're exploiting the electromagnetic spectrum for saintly or nefarious reasons, the shorter the wavelength, the higher the density of information that can be carried by the light beam.

Without technological assistance, garden-variety humans see only the tiniest fraction of the full electromagnetic spectrum, ranging from violet light — with a wavelength of about four hundred nanometers — to red light, with a wavelength not quite twice as long, about seven hundred nanometers. When you consider that the bands of the electromagnetic spectrum we've measured so far span more than a dozen powers of ten in wavelength, our span of barely a single power of two is just plain lame. Crucially for us, the peak of the Sun's energy output lies smack in the middle of the visible part of the spectrum. Since we're daytime creatures, it's evolutionarily sensible that the detection capacity of our eyes peaks in the same place.

Infrared and ultraviolet are invisible to us, but that doesn't mean they're insensible. We experience them through our skin, not our eyes. We sense the Sun's infrared light in real time as heat on our skin, but we sense its ultraviolet light only after our skin has been darkened and perhaps damaged by excessive exposure, otherwise known as sunburn.

Earth itself radiates infrared, as does everything whose molecules are in motion, be it animate or inanimate. In other words, anything and everything with a temperature above absolute zero. Dusty galactic clouds, where stars form deep within, emit infrared. Your kitten, your canary, and your houseplants, dead or alive, all emit infrared. Some species of snakes have small pits on their heads that pick up infrared rays from tasty warm-blooded prey, readily revealed at night against the rapidly cooling surroundings. And alas for the hotel industry and tourists the world over, the antennae of bed bugs have infrared sensors that alert the bugs to a nearby source of warm blood. As for ultraviolet, flying insects — including gnats, moths, mosquitoes, and butterflies — as well as birds, bats, rats, and cats see it quite well.

Just because an object emits infrared doesn't mean it can readily be seen by an infrared detector. You still have to single out your target from any competing sources of infrared light, either surrounding your target or surrounding you. Anything warmer than its surroundings shows up brighter. But if the target is about the same temperature, you'll lose it in the infrared "noise." Skywatchers improve their capacity to detect their chosen infrared target by deeply cooling their apparatus with liquid nitrogen (77 kelvins) or, for the coolest cases, liquid helium (4 kelvins). These tamp down the thermal noise of the detector itself, permitting the celestial object to shine more distinctly in the data. As you might suspect, the needs of the military aviator are the exact opposite. If targeted by a heat-seeking missile, the plane or helicopter will typically deploy infrared countermeasures such as swirling hot flares, which contribute infrared noise to what the warhead "sees" and thus render the engine's hot exhaust indistinguishable from the countermeasures themselves.

Infrared and ultraviolet merely hint at all the light energy we humans cannot see. Further along on the long-wavelength, low-energy end of the electromagnetic spectrum are radio waves (experimentally demonstrated in the 1880s)[9] and microwaves (named as the small-end subset of radio waves in 1964–65, hence the diminutive prefix "micro"); further along in the other direction, on the short-wavelength, high-energy end are X-rays, discovered in 1895, and gamma rays, in 1900. Though we've assigned labels to the various bands, the electromagnetic spectrum is a continuum. Civilization is layered along this continuum. Hundreds of AM, FM, and XM stations are beaming radio waves through your body right now, the phone part of your smartphone is communicating in microwaves with a cell phone tower, and the map features of your smartphone are talking to GPS satellites overhead via microwaves too. You're probably receiving visible light from a nearby lamp and, if its bulb is incandescent, infrared light as well. Meanwhile, across the universe, an ancient, persistent, pervasive sea of microwave radiation forms the cosmic microwave background, a legacy of the Big Bang.

Most celestial goings-on emit light in multiple wavelengths simultaneously. For example, the explosion of a massive star — a supernova — is a cosmically commonplace (though locally rare) and seriously high-energy event that, in addition to visible light, blasts out prodigious quantities of X-rays. Sometimes the explosion is accompanied by a burst of gamma rays or a flash of ultraviolet. When it takes place in our own galaxy, it may emit so much light in visible wavelengths that it remains visible for several weeks without the aid of a telescope, as was true of the supernova spectaculars hosted by the Milky Way in 1572 and 1604. Long after the explosive gases cool, the shock waves dissipate, and the visible light fades, a supernova remnant radiates infrared and radio waves.

The flip side of visibility is detection. When it comes to the pursuit of prey or, conversely, the avoidance of enemies, detection is key to both conquest and survival. Whether you're the victim or the aggressor, never is it more advantageous not to see something than to see it. Either way, but especially if you're the likely victim, you would prefer not only to see the aggressor but also to remain unseen yourself.

Camouflage (a word of French origin, whose earlier meanings ranged from smoke and suffocating underground explosions to costume disguises and criminal sneakiness),[10] the art of remaining unseen, is not uncommon among creatures big and small. Think of the kaleidoscopic changes of the cuttlefish or octopus, the twiglike insect known as the walking stick, or, before the melting induced by climate change, the polar bear's snowy fur against the whiteness of the Arctic snowpack. Camouflage can be about either keeping yourself from getting eaten or closing in on your own dinner.

There's also the distinction, proposed early in the twentieth century by an American artist named Abbott Thayer, between two very different forms of visual camouflage: blending versus dazzling. Nature has chosen the option of blending for both the widespread walking stick and the threatened polar bear. Critters that live in woodsy habitats might blend in by being green on green or speckly brown on speckly brown, while others might dazzle and confuse observers with vivid stripes, prominent spots, or other garish markings that have the effect of breaking up the outlines of their bodies and making them more difficult to track while in motion. In all cases, the goal is to vanish.

Invaders and warfighters love camouflage and stealth — the nearest they can get to invisibility — and they've been attempting it for millennia. In the fifth century BC, the military theorist Sun Tzu advised:

All warfare is based on deception. . . . Hence, when able to attack, we must seem unable; when using our forces, we must seem inactive; when we are near, we must make the enemy believe we are far away; when far away, we must make him believe we are near.[11]

Ten centuries later, Flavius Vegetius Renatus, a prominent Roman court official and author of a military handbook, described traditional camouflage for the scouting craft that accompanied large warships for the purpose of making surprise attacks, intercepting enemy convoys, and monitoring the approaching enemy:

So that the scouting vessels will not be betrayed by brightness, the sails are dyed Venetian blue, similar to the colour of the sea, and the tackle is coloured with the wax that ships are generally coated with. Also, the sailors and marines wear Venetian blue coloured clothing so that not only at night, but also in the daytime, they more easily remain unseen while scouting.[12]

Though the sea continually changes color, at a distance the ships' blue coloration could — under optimal conditions — merge with that of the water. Only at close range would the difference between their Venetian blue and the varied blues, browns, greens, and grays of the sea be readily perceptible. But once the difference had registered, there wouldn't be enough time to organize an attack against the scouts. Distance buys time and advantage, which was precisely Galileo's point when he sought support from the doge of Venice in 1609. A few other proposals for maritime camouflage were more imaginative than a coat of Venetian blue. One early twentieth-century alternative, never implemented, involved swathing ships in billowing white covers meant to simulate clouds.[13]

Cladding troops and vehicles with branches and leaves to simulate forest foliage is another time-honored type of camouflage, whether in the guerilla warfare of twentieth-century Vietnam or in medieval Scotland (recall the dire prophecy in Shakespeare's *Macbeth:* "Fear not, till Birnam wood / Do come to Dunsinane"). But not until World War I, when artists began to paint unrolled canvas to look like roads and observation posts to look like tree trunks, did the word "camouflage" officially enter the English language. Soon the practice of painting whole warships in a dazzle pattern (also called disruptive patterning or, much snazzier, razzle dazzle) was adopted on both sides of the Atlantic. The deciding factor seems to have been the sinking of almost a thousand British ships by Germany's U-boats during the first nine months of 1917, causing a British painter of seascapes who was serving as a naval officer to propose that "since it was impossible to paint a ship so that she could not be seen by a submarine, the extreme opposite was the answer — in other words, to paint her in such a way . . . as to break up her form and thus confuse a submarine officer as to the course on which she was heading."[14]

Creating confusion seemed a better solution than trying to attain invisibility. Suddenly artists became facilitators of military goals, as warriors picked up some of the disintegrationist, scientistic visual strategies of the vanguard movements of Cubism, Futurism, and Vorticism. Picasso and Braque, the fathers of Cubism, were delighted to see what they regarded as their aesthetic invention being applied to ships and weaponry; walking down a boulevard in Paris one evening and seeing a convoy of zigzag-painted heavy guns heading for the front, Picasso reportedly exclaimed, "We invented that!" Franklin D. Roosevelt himself, assistant secretary of the

US Navy during World War I, is said to have shouted, after being shown a bedazzled test ship, "How the hell do you expect me to estimate the course of a God-damn thing all painted up like that?" In the end, though, standardized dazzle camouflage apparently did not live up to its promise. Attacks proceeded at similar rates on ships with and without such a coating. Nevertheless, despite plenty of evidence to the contrary, magical thinking about the efficacy of disruptive camouflage persisted through World War II and beyond.[15]

Several options for disappearing from the visible part of the spectrum have a long wartime pedigree. The simplest is to exploit the darkness of night. Another is to blind the enemy. Set a huge bonfire, and enemy forces who look at it will be unable to see anything other than flames and therefore be unable to target you with any precision. In recent decades, both lasers and smokescreens have been used to blind the enemy: throw a white-phosphorus grenade at your target, and you'll get an instant smokescreen that will also scorch whoever is in the vicinity while it masks your own maneuvers and your own infrared radiation. Blinding also happens to us in the universe, when the light of a host star swamps the much fainter reflected light of its exoplanets. This was a big problem until a couple of decades ago, when space scientists began using a special occulting disk in their telescope's optics to block out the offending starlight, thus achieving the opposite of what a telescope was invented to accomplish.

Another, very different approach to disappearance is transparency — embodied in the clear glass window. Flies, moths, birds, and visitors from outer space who don't know about windows must be baffled by the interposition of something visually imperceptible but impenetrable between themselves and the scenery.

But let's say you want to move around at will, unobtrusively, rather than be locked in place like a window, yet also be functionally invisible. Nowadays you could coat yourself in a foam, fiber, or powder that reflects no light. An adversary would be prevented from illuminating you, although you would still block the scenery behind your position, and a clever hunter would be able to detect you by the exact absence of your human form rather than its presence. You could also cloak yourself in scales or mirrors that redirect the light that hits you, sending little or none of it back to the source — similar to the design principle of stealth aircraft, which redirect incident radar on their fuselage back out in many different directions. Another, quite recent option is a fabric made of tiny light-transmitting beads that can transpose an image from behind you to in front of you. To an observer, it's as though you're not there at all, the same effect as donning a *Star Trek* cloaking device. Other possibilities: If you're an architect designing an offensively gargantuan skyscraper, you could cover it in LEDs that project the surrounding scenery minus its presence. If you're a spy monitoring the doorway down the street, you might want to vanish into thin air, magician-style, via a clever sequence of lenses or mirrors placed between yourself and the doorway.[16]

Achieving invisibility through temporary camouflage is an intuitive, imaginative tactic, with limited reliability. Achieving invisibility through stealth is a scientific tactic, grounded in an understanding of the physical laws of reflection and refraction as well as on centuries' worth of discoveries about the many forms of light energy to which our senses have no access.

By the end of the nineteenth century, we could no longer delude ourselves into believing that the universe communicates with us only through the narrow band of light available to the human retina. With the discovery of multiple bands of light, it became unthinkable to design a defense strategy solely around visible light or to explain the cosmos solely on the basis of observations done in visible light — like composing a symphony from only a single octave's span of notes. A new term was needed — "astrophysics," as distinct from "astronomy" — to clarify the difference between identifying the presence and position of celestial bodies and the more convoluted process of ascertaining their components, their mass, their paths, and their history. Light would become an encyclopedia. The effort to detect things too dim to register on the human eye would become the longest-running show in astrophysics.

All this required new technology and new techniques. The astrophysicist sought detectors capable of capturing every wavelength; the warfighter sought offensive systems capable of exploiting those wavelengths and defensive systems capable of eluding them. The radio band felt like a good bet to both sides. For the military, pre-nukes, it became nearly indispensable; for space scientists, it offered new avenues to new information. Working together, they helped shape the course of World War II.

Although the existence of radio waves was demonstrated as early as the mid-1880s, it took decades of competing theories from physicists and mathematicians, plus mounting experimental evidence, before scientists and engineers could work with them, control them, and exploit them. The first order of business was to understand their behavior: how some radio waves manage to travel intact around the curved surface of Earth and how the upper atmosphere — the ionosphere — affects their journey through space; the causes of radio noise, better known as static; the best shape and material for the antenna; whether the direction of transmission matters; whether the Sun and other celestial neighbors reflect or emit radio waves. And so forth.

By 1919, the biggest transmission question had been answered: radio waves travel not because of being diffracted by Earth's curved surface but because of being reflected by Earth's ionosphere — a several-hundred-mile-thick series of layers in the upper atmosphere that seethe with charged particles (ions) produced by the Sun's high-energy light knocking electrons off our atmosphere's resident atoms and molecules. By 1937, the rest of the answers about transmission were mostly in place. Having approached the question differently, different researchers came up with different pieces of the total answer — and inadvertently advanced such varied endeavors as meteorology and mathematical theory. As one historian of science describes it, "They started from what they wanted to know, and found what they did not expect to learn." While chasing a solution to a practical engineering problem, he writes, the US Navy ended up contributing to pure science.[17]

Through the late 1930s, much theoretical

and practical work focused on sending and receiving radio signals. Not until those twin problems were understood and mastered could the twin problems of detection and its avoidance even be addressed. But something else supremely important happened on the radio front in the 1930s — another practical project that resulted in another serendipitous contribution to science. In fact, it birthed a whole new branch of astrophysics.

The object we know as a telephone started out as a device for relaying radio waves. Nowadays our mobile phones relay microwaves. Back in the medieval era of telephone communication, AT&T — the American Telephone and Telegraph Company — was a giant government-approved monopoly whose motto was "One System, One Policy, Universal Service." AT&T's first long-distance call within the United States took place in 1885 between New York and Philadelphia. Transatlantic calling service via two-way radio (also called radiotelephone) began in 1927, but the only place you could call that year was London. Transpacific calling began in 1934, to Tokyo. One big problem with long-distance service, aside from the price tag, was, as AT&T itself describes the situation, "Telephone service via available radio technology was far from ideal: it was subject to fading and interference, and had strictly limited capacity."[18] Another, double-edged problem was that few channels were available in the low-frequency, long-wavelength part of the radio spectrum, while the higher-frequency, short-wavelength part of it — the part that could carry much more information — was still unfamiliar territory, scientifically and technologically. Not until it was mastered could there be live FM stereo broadcasts from the Metropolitan Opera, which began in the 1970s.

But let's not get ahead of ourselves.

In 1928 AT&T's three-year-old R & D facility, Bell Telephone Laboratories, hired a young physicist named Karl Jansky to study Earth-based radio sources that might account for all the hissing and fading — the noise, the static — in terrestrial radio communications. After constructing a novel rotating antenna, tuned to capture a radio wavelength of 14.6 meters (frequency: 20.5 MHz), Jansky spent several years waiting for signals to drift by his receiver, studying the signal patterns, and scrupulously interpreting the results. In 1932 he published his preliminary findings.

Jansky's tone was modest and careful, his claims limited, his attention to fact honorable. In his 1932 paper, concerning "the direction of arrival and intensity of static on short waves," he cites three identifiable types of static: one from local thunderstorms, one from distant thunderstorms, and an unidentifiable third, "a steady hiss type static of unknown origin" that seemed to be "associated with the sun." In his 1933 paper, after a year of examining only that third type of static, Jansky stated that its origin lay far, far beyond the Sun. It must be somewhere "fixed in space," he concluded, in a location either "very near the point where the line drawn from the sun through the center of the huge galaxy of stars and nebulae of which the sun is a member would strike the celestial sphere."[19] In short, approximately the heart of the Milky Way.[20]

Every twenty-three hours and fifty-six minutes, Earth completes one rotation relative to the stars. Every twenty-three hours and fifty-six minutes, the center of the Milky Way returns to the same angle and same elevation on the sky when viewed from Earth. Every twenty-three hours and fifty-six minutes, Jansky's fixed point in space hissed past his merry-go-round. Hence the inescapable conclusion that his fixed point in space was the center of the Milky Way. If the source had been our Sun, the interval between hisses would have been twenty-four hours, not four minutes less.

That was the birth of radio astronomy, though the end of Jansky's career as a radio astronomer. Rather than agreeing to let him build the hundred-foot dish he proposed as a follow-up, Bell Labs — which now had answers to its practical questions and wasn't about to start funding basic research — assigned Jansky to other tasks.

Fortunately, a young radio engineer from Illinois, Grote Reber — who briefly fell victim to bad timing, having begun a job search just as the Great Depression was deepening — decided to forge ahead and build his own radio telescope in his own backyard. In 1938 Reber confirmed Jansky's discovery, then spent the next five years making low-res maps of the radio sky all on his lonesome. Half a century later, Reber published a reader-friendly article titled "A Play Entitled the Beginning of Radio Astronomy," in which (talk about timing!) he points out that Jansky

was observing near the bottom of a low solar activity minimum. The ionospheric hole at 20.5 MHz was open from zenith to horizon day and night. A few years earlier or later, the observations would have been confused by ionospheric effects, particularly during the day. Jansky is an example of the right man at the right place doing the right thing at the right time.[21]

Every band of light requires its own detection hardware. No single telescope can focus light of all bands. If you're gathering X-rays, whose wavelengths are very short, your reflector will have to be supersmooth lest it distort the rays. But if you're gathering radio waves, your reflector could be made of polished chicken wire that you've bent with your hands, because the irregularities in the wire would be smaller than the wavelength of the radio waves you're trying to detect. The surface smoothness of the mirror simply needs to be commensurate with the scale of the wavelength you want to measure. And don't forget about resolution: if you want a decent level of detail, your reflector's diameter must be much wider than the wavelengths you want to detect.

The detectors built by Jansky and Reber were the first effective radio telescopes — and the earliest invisible-light success stories. Glass mirrors were out of the question, because radio waves would pass right through them. The reflectors would have to be made of metal.

Jansky's hundred-foot-long contraption looked a little like the sprinkler system on a modern corporate farm. The antenna was a series of tall, rectangular metal frames, secured with wooden cross-supports and mounted on the front wheels and axles of junked Model-T Fords. Hooked up to a small motor, the whole thing rolled around a turntable, completing a full 360 every twenty minutes. Inside a nearby shed was a receiver equipped with an automatic temperature recorder that had been rejiggered to record the strength of the radio signals.[22]

On the other hand, Reber's telescope was a single nine-meter-wide dish, progenitor of generations of radio telescopes that rely on the dish — often parabolic, like a tilted half eggshell — to collect the incoming radio waves and then bounce them up to a receiver. In other words, the dish is an antenna that works like a mirror. What Reber achieved was detection; his apparatus wasn't big enough to achieve good resolution. But in the early 1940s, merely detecting an invisible cosmic phenomenon was a huge step forward.

Predictably, dish antennas soon got bigger and better. Mark I, the planet's first really big radio telescope — a single, steerable, solid-steel dish 76 meters wide — saw first light in the summer of 1957 and is still on call at the Jodrell Bank Observatory in northwest England. More recent radio telescopes are not just big; they're colossal. Built into a large natural sinkhole near the north-central coast of Puerto Rico is the 305-meter, non-steerable dish of the Arecibo Observatory. Completed in 1963, this spectacular construction — damaged but far from destroyed by Category 5 Hurricane Maria in September 2017[23] — was under the supervision of the US Department of Defense until 1969.

Initial funding for Arecibo traces to an anti-ballistic-missile program, Project Defender, supported by the Advanced Research Projects Agency. A precursor to the Strategic Defense Initiative, Project Defender addressed US worries that decoys would successfully prevent defensive action against intercontinental ballistic missiles. The Arecibo radio telescope held hope that the radar signature from an actual warhead passing through Earth's ionosphere would differ enough from the radar signature of a decoy that the deadly missile could be identified and shot down. Oh, and the telescope could do astrophysics on the side.

The shape of Arecibo's curved dish is a segment of a true sphere rather than of a traditional paraboloid. Since the dish itself is stationary, an innovative movable detector — positioned high above the dish — serves to "point" the telescope toward different areas of the sky. The optics of a spherical surface uniquely permit this trick. In addition, Arecibo's huge size assured it would detect extremely weak radio signals emitted by objects in deep space as well as by radio-busy layers of Earth's own atmosphere, such as the ionosphere. Plus, the telescope not only detects radio signals; it can transmit them as well. These transmitted signals, beamed out to space in radar mode and then bounced back to Earth when they hit something reflective, can map the shapes and track the orbits of planets, asteroids, and comets.

In 1974, the Arecibo telescope was the first to transmit, on purpose, a radio message to aliens — specifically, to a large and crowded cluster of stars in our Milky Way galaxy presumed to be orbited by planets that might host intelligent life. Another of the observatory's many highlights was its role in the 1993 Nobel Prize in Physics, which went to Russell A. Hulse and Joseph H. Taylor Jr. for their 1974 discovery of a binary pulsar suitable for testing Einstein's general theory of relativity.

For almost fifty years, Arecibo held the title of world's largest single-dish radio telescope. In 2016 that distinction passed to an even more spectacular construction: FAST, the Five-hundred-meter Aperture Spherical Telescope. Set into a huge limestone depression in a thinly populated, mountainous region of southwestern China, FAST's dish is so large that, as the chief scientist at China's National Astronomical Observatories put it, "if you fill it with wine, every one of the world's seven billion people could get a share of about five bottles." As with Arecibo, its shape is a section of a sphere, but that's just an engineering detail. Because of its size, FAST can observe with much greater sensitivity than Arecibo can.[24] At 500 meters in diameter, it enjoys nearly three times the collecting area of the 305-meter Arecibo telescope. Nothing in the world comes close. If something out there falls just below the detection limit of Arecibo, and FAST is pointed in that direction, it will easily extract the signal from the din of cosmic noise. So there's a good chance that the first humans who will ever talk to aliens via radio waves will be Chinese astrophysicists. No nation, after all, has exclusive access to the universe.

But when detail, more than dimness, is what skygazers seek, they turn instead to arrays of smaller dishes, spread across many kilometers of landscape. By pointing all the separate dishes at the same spot on the sky and cleverly combining their signals, these arrays — known as interferometers — achieve the equivalent resolution of one lone dish of unachievably wide diameter, equal to the extent of the array itself. "Supersize me" was the unwritten motto for radio interferometers long before the fast-food industry adopted the slogan, and they form a jumbo class unto themselves. Among their ranks, sprinkled around the world, are the Very Long Baseline Array (ten 25-meter dishes spanning five thousand miles from Hawaii to the Virgin Islands), the Giant Metrewave Radio Telescope (thirty lightweight mesh dishes, each 45 meters across, spanning sixteen miles of arid plains east of Mumbai, India), and the Atacama Large Millimeter/submillimeter Array (sixty-six dishes — some 12-meter, some 7-meter — clustered at an altitude of more than sixteen thousand feet in the driest region of the Chilean Andes).

Not too far in the future, these enormous interferometers will be dwarfed by the Square Kilometre Array's thousands of dishes augmented by battalions of fixed "aperture array" telescopes spread across wasteland — some looking, from a cloud's-eye view, like gargantuan coins with sharply notched edges, others like miniature Eiffel Towers. Installed in a spiral configuration across southern Africa and western Australia, the SKA will have its headquarters at Jodrell Bank.

Even with the finest detectors, there are limitations and irritations. While an extremely low frequency radio wave can be thousands of miles long, the largest individual radio telescope dishes are only several hundred meters across, and interferometer arrays cannot detect light whose wavelength is longer than the width of the array's broadest dish. So, ultra low and extremely low frequency (ULF and ELF) radio waves pass across and through Earth undetected by the kinds of radio telescopes that astrophysicists know and love. Plus, various bands of detectable radio waves get degraded by terrestrial communication towers and other trappings of modern civilization. Then there's the problem of turbulence in the ionosphere, whose various levels propagate radio transmissions as well as interfere with them and whose impact changes according to the time of day and the frequency of the wave.

The ionosphere has figured prominently in the modern pursuits of both warmakers and space scientists. The Third Reich's V-2 rockets — the world's first ballistic missiles — had to pass through it unscathed before falling out of the sky onto their targets. Of equal military significance is the role of the ionosphere and its investigators in the history of radar, an acronym for radio detection and ranging.[25]

Detection, of course, is simply about determining and/or confirming the presence of something. Ranging is about calculating its distance and direction. The idea is straightforward: transmit radio waves toward a distant object — an asteroid, the Moon, a bomber, a submarine — and see if any radio waves bounce back to you. If they do, the time delay as well as the intensity, frequency, and shape of the waves can tell you something about the object's shape as well as how far away it is, in which direction it's moving, and how fast. Nowadays, asteroids are the main cosmic targets for radar studies, allowing the interested party to map the size and shape of the rock and to establish precise orbital parameters lest we discover that one of them is headed toward Earth.

The irrepressible Serbian-American inventor Nikola Tesla raised the basic idea of radar as early as 1900 and formally incorporated it in a 1905 US patent application. A lower-profile inventor named Christian Hülsmeyer, picking up on findings by his own country-man Heinrich Hertz, applied for a similar German patent in 1903–1904.[26] And Gug-lielmo Marconi, the electrical engineer and entrepreneur whose name is inseparable from the early years of radio communication, discussed the idea in 1922 in New York, in an address to fellow engineers:

In some of my tests, I have noticed the effects of reflection and deflection of [radio] waves by metallic objects miles away.

It seems to me that it should be possible to design apparatus by means of which a ship could radiate or project a divergent beam of these rays in any desired direction, which rays, if coming across a metallic object, such as another steamer or ship, would be reflected back to a receiver screened from the local transmitter on the sending ship, and thereby immediately reveal the presence and bearing of the other ship in fog or thick weather [and] to give warning . . . , even should these ships be unprovided with any kind of radio.[27]

"Give warning" is a phrase brimming with military potential. In the early months of World War II, radar was already being deployed for that purpose in much of the world:

Europe east and west, North America, Japan. It was used in South Africa; it was used in the Aleutian Islands. "World War II was the first electronic war, and radar was its prime agent," writes historian Andrew Butrica. "Despite its scientific origins, radar made its mark and was baptized during World War II as an integral and necessary instrument of offensive and defensive warfare."[28]

In *A Radar History of World War II,* on the first page of his preface, physicist Louis Brown proposes that "science and war . . . are unquestionably the two most dissimilar manifestations of the behaviors that distinguish man from beast." But dissimilarity does not preclude marriage:

[W]ar is almost as unique to man as is science. Other than ourselves, only ants organize their violence so that it can be called war. . . . Moreover, from the dawn of civilization science and war have been inseparable companions, locked in a partnership that neither desires and that neither is capable of dissolving.[29]

On both the Allied side and the Axis side, that brand of co-dependence produced military radar. "No weapon," writes Brown, "was ever designed with such intimate collaboration between inventor and warrior."[30] The collaboration, however, was neither automatic nor stress-free. Before their work was officially embraced, not only did radar researchers and advocates face political and institutional roadblocks at several junctures, but their work was also, at first, occasionally sidelined by advocates of other rudimentary technologies competing for primacy: acoustic location and infrared detection. Added to this mix were turf skirmishes between armies and navies and the spotty scientific literacy of key decision makers.[31]

During the first third of the twentieth century, scientists across the Northern Hemisphere were developing components and materials that would eventually enable not only radar but television as well. Foremost among them were cathode-ray vacuum tubes (fused silica was the solution) and insulation for high-frequency cables (polyethylene was the solution). Much of their effort took place at large companies such as DuPont, General Electric, and IG Farbenindustrie AG and was initially spurred by the civilian radio boom rather than by military demands. By the 1930s in Britain, France, Germany, Japan, the Soviet Union, and the United States, the "brothers-in-arms" at military laboratories, electronics corporations, universities, and research institutes were all investigating possibilities for effective radiolocation. Radar was "in the air" as Adolf Hitler became Führer and Germany rearmed.[32]

The direction and pace of radar work varied considerably from country to country. While Britain, for instance, initially focused on defense,[33] Germany focused on offense. While the British government actively sought out scientists and new approaches to weaponry, the German government didn't act until engineers sought out officials and staged demonstrations for them. While Britain put a lot of energy into the development of organizational capability, Germany emphasized the development of sophisticated radar technology and the maintenance of secrecy. Indeed, the secrecy was so extreme that the Kriegsmarine (literally, "War Navy") initially objected to even showing its technology to the Luftwaffe ("Air Weapon"), let alone sharing it, and resisted placing radar officers or even radar instruction manuals on its ships.[34]

At the start of the war, Germany already had three main advanced radar designs, though few of each were in operation. The Kriegsmarine's Seetakt, a surface-search radar for use on warships and in coastal defense, focused on accurate ranging; the Luftwaffe's Freya was a longer-wavelength, mobile air-warning radar for land use that could register targets at greater distances than Seetakt could; and Würzburg, a highly accurate targeting radar, was especially useful for anti-aircraft guns. As World War II escalated, manufacturers came up with variations small and large.[35]

Britain had determined from military exercises during the early 1930s that the nation would be defenseless against an air assault from the modern, all-metal bombers that were rolling off the line in Germany. Some members of the British government quickly grasped radar's strategic possibilities and were willing to commit major resources and personnel to immediate military-radar research and implementation — a commitment not exposed to public discussion in either the Commons or the press. As of July 1935, British radar technology could detect a plane forty miles away; by March 1936, that number had risen to seventy-five. As of late 1937, three early-warning radar stations were operational; by September 1939, a network of twenty, called Chain Home, had been installed along Britain's coastline. A year later, on September 15, 1940, at the height of the Battle of Britain, Chain Home operators helped down so many German planes that the Luftwaffe soon abandoned the large-scale daytime sortie in favor of the nighttime blitzkrieg, plus occasional daytime attacks on specific targets. Germany's plan to invade Britain had to be dropped.

Although the Germans and Americans had superior equipment at the start of the war, the British had done advance threat assessment, chosen a defensive system that could be quickly built, partly reorganized their military forces around the precept of homeland security through radar, and mobilized as well as trained more radar operators — including hundreds of women — than all the other radar nations combined. Swift, concise communication was key. As Brown writes, Britain "had the wisdom to realize that intelligence gained by radar was worthless unless promptly interpreted and acted upon."[36]

But of course, equipment is hardly irrelevant. Underlying the technological side of Chain Home's contribution was a technique, developed in the mid-1920s by American scientists, for measuring the height of the reflective portions of the ionosphere by sending up several-millisecond pulses of radio waves and clocking the duration of the return trip. The UK's Radio Research Board, acting at the behest of the Committee for the Scientific Survey of Air Defence from early 1935 through the end of the war, adapted the technique for purposes of protecting the homeland.[37] Among Chain Home's numerous challenges were distinguishing friendly from enemy aircraft, detecting aircraft that were flying low and close to the coast, providing accurate altitude readings for incoming aircraft, and coming up with accurate counts of enemy aircraft. Chain Home alone couldn't do the job. It needed partners: radio direction-finding sets, good radio telephones, and all those civilian radar operators.[38]

However, a radar ground installation, operating at wavelengths of a meter and a half and transmitting information in short, clear code words to a fighter pilot whose plane was equipped with a radio telephone, could still not provide enough input to enable a British pilot to destroy a German factory, bomb a U-boat, or shoot down a German bomber headed for London in the dark of night. In addition to input from the ground, that pilot would need — on board — a powerful, lightweight, high-frequency device that could serve as a type of searchlight and detect a target in darkness or in fog. This new device could not rely on the same kind of lower-frequency radar that had proved so useful when looking upward from ground level, because, when looking downward from the air, radio energy reflected from Earth itself would swamp the fainter radio echoes bouncing off the enemy craft. Moreover, it had to be portable. Solution: microwave radar, produced by a so-called resonant cavity magnetron. A British version — brought to the United States in a supersecret mission in September 1940 — was described by President Franklin D. Roosevelt as "the most important cargo ever brought to American shores" and by A. P. Rowe, superintendent of Britain's Telecommunications Research Establishment, as "the turning point of the war."

Those statements, it turns out, are only half true. Not only had extensive work been done on microwave radar during the 1930s, but magnetrons of other sorts already existed. The cavity magnetron had been patented by Russians in the 1920s and was already known to the Germans. By the end of the 1930s Japan, too, had them. It's just that the British didn't know about these devices, and the Germans were being ordered to shelve theirs and to concentrate on longer-wavelength radar.[39]

And so, British scientists independently reinvented the invention, and soon Americans sought to improve it. By the spring of 1941, less than a year after the unnecessarily secret mission, the newly created MIT Radiation Laboratory in Boston had produced a three-centimeter version of the resonant cavity magnetron. The Cambridge-based company Raytheon soon began manufacturing most of the magnetrons used in the war effort by both the United States and Great Britain.[40] In fact, the now-indispensable microwave oven can be traced to a Raytheon engineer, Percy Spencer, who found that a candy bar in his pocket had melted because of the microwaves emitted by an active magnetron he'd been standing near.

Concurrently, the US Navy and the US Army Signal Corps were working on longer-wavelength radar, and on December 7, 1941, one of the new mobile radar units in the Army's Aircraft Warning System detected Japanese planes approaching Pearl Harbor almost an hour before the attack. The warning was ignored, and the source of the radar echoes was misinterpreted as B-17s, friendly bombers scheduled to arrive from California that very day.[41]

Unknown knowns also affected the course of the radar war. At certain points in the conflict, one side seemed not to know that the other had effective radar. One telling example of this was the swift Japanese evacuation of the Aleutian island of Kiska in the summer of 1943 during an American naval blockade — an evacuation carried out in heavy fog and made possible by a determined Japanese admiral relying on Japan's new microwave radar, of which the United States was unaware.[42]

Whatever its failures and limitations, radar in a range of incarnations did play a big part in both the Allied and the Axis campaigns. On the Allied side, a popular claim was that the bomb had ended the war but radar had won it — by locating and helping to destroy enemy bombers in darkness, enabling aircraft to "bomb blind," maximizing the accuracy with which anti-aircraft artillery could be aimed, enabling a plane to map the ground or water surface over which it was flying, and, of course, reducing the navigational difficulties posed by the presence of fog and the absence of light. Early in the war, however, accurate aim in blind bombing was still unachievable. So, the precise, selective destruction of German industrial targets such as factories was out of the question. The only obvious alternative was to bomb larger areas. "Translated into practical terms," writes Louis Brown, stressing what the limits of technology can force you to do, "this meant that the targets would have to be of city size." Hence, the air war against Germany turned into the destruction of its cities rather than, as initially strategized, the elimination of its synthetic oil production.[43]

As soon as writers were permitted to discuss radar publicly in detail, a note of hyperbole occasionally crept into Allied accounts: "perhaps the war's most fabulous and zealously guarded secret"; "Warfare today would be more or less impotent without this modern electronic genie"; "the great drama of radar, the war's most powerful 'secret weapon' until the atomic bomb was devised."[44] In early 1946 the British radar pioneer Robert Watson-Watt spoke of "that secret weapon which prevented the cutting of our life-line, which would have resulted had the defeat of the U-boat not been assured."[45] Winston Churchill's assessment was more nuanced: "it was the operational efficiency rather than novelty of equipment that was the British achievement."[46] But whether a great drama or simply an achievement, radar changed war by making the invisible visible. Decades later, Louis Brown would assert that the "introduction of radar, a completely new way to see, in the Second World War altered the basis of warfare more profoundly than any of the inventions that had marked the industrialization of combat."[47]

Yet for planetary astrophysicists, radar also affords a way to track potentially hazardous asteroids that could render humans extinct — the ultimate defensive application of this technology, an agent not of warfare but of survival.

With hostilities ended, rafts of journalists, politicians, warriors, and citizens applauded the proven military benefits of radio waves. They also honored the scientists and engineers who had made those benefits possible. Many scientists launched or resumed their radar research into the ionosphere, and military planners began to think about improved radar countermeasures in the context of new kinds of long-distance battle threats. The stage was now set for even greater and more intricate cooperation between the practitioners of science, the advocates of war, and the seekers of profit.

During and immediately after the war, there

was already widespread sharing and borrowing between scientists and warmakers.[48] Initially, radar scientists supplied the armed forces with basic techniques, while the armed forces, often working with major corporations and universities, undertook large-scale science and technology programs to adapt those techniques for use in militarily beneficial technologies. After the war, radar astronomers further developed the techniques, while private industry attracted many scientists whose skills were no longer required for war work. Former adversaries became allies, and vice versa. The Iron Curtain descended, and Cold War projects multiplied. Postwar research in the radio band swiftly ramped up as astronomers outfitted their observatories with wartime radar surplus, often bought at fire-sale prices or simply rescued from being thrown down a mineshaft. The Jodrell Bank Observatory was furnished just that way.

Early in 1946 radar astronomers at a US Army Signal Corps facility in New Jersey succeeded in bouncing radio waves off the surface of the Moon. Within a month, Hungarian physicists did the same. British researchers found a correlation between their visual sightings of meteors plunging through Earth's atmosphere and the radar echoes that registered on their equipment during the meteors' fiery journey. Through close analysis of paths and velocities, researchers in Britain and Canada determined that detectable meteors are inhabitants of our solar system, not invaders from beyond. Several groups in several countries obtained radar echoes from Venus.[49] Researchers from former enemy nations resumed the normal scientific practice of collaboration — a noteworthy (though later) example being Bernard Lovell, director of Jodrell Bank, and the very same German radio astronomer who in May 1943 had investigated and reported on the blind-bombing radar equipment aboard two downed British bombers.[50]

Ionospheric research contributed to progress on secure, point-to-point, long-distance communication, a goal high on the military's wish list — then and now. In the United States, significant funding and activity came from the Central Radio Propagation Laboratory division of the National Bureau of Standards (nowadays called NIST, the National Institute of Standards and Technology) and from military organizations such as the Air Force Cambridge Research Center, the Army Signal Corps, and the Office of Naval Research. Corporations big and small — ITT (International Telephone and Telegraph), RCA (Radio Corporation of America), the Collins Radio Company of Cedar Rapids, Iowa — were also part of the push. In this milieu, astronomers at Stanford University, the Naval Research Laboratory, Jodrell Bank, and elsewhere explored possibilities for radio-wave communication between Earth and the Moon, including the idea of bouncing signals off the lunar surface. By 1951, several groups of investigators had achieved long-distance wireless voice transmission via the Moon, which they used as a passive relay — a naturally occurring, cost-free, pre-Sputnik satellite.[51]

In the meantime, scientists, generals, futurists, political leaders, and university-based military contractors here, there, and everywhere — from Arthur C. Clarke to Josef Stalin to Project RAND — were thinking hard about rockets.

It was old news that this invention, capable of piercing the ionosphere, could serve equally well as a conduit to space and an agent of terrestrial devastation. Already in the fall of 1931, five years after Robert Goddard demonstrated his first liquid-fuel rocket, a high-school dropout turned MIT engineering graduate named David Lasser, first president of the American Interplanetary Society, could confidently declare to an audience at the American Museum of Natural History in New York: "The perfection of the rocket in my opinion will give to future warfare the horror unknown in previous conflicts and will make possible destruction of nations, in a cool, passionless, and scientific fashion."[52]

Often camouflaged, barely trackable, and inaudible at its target due to its terrifyingly swift supersonic speeds, Germany's V-2 rocket had proved second to none in how technology can deliver terror, and so the United States and the Soviet Union both scrambled to seize small armies of German V-2 rocketeers and shiploads of V-2 rocket parts even before World War II had run its course.[53] Both sides set themselves the task of making a more deadly version of the V-2: a long-range, high-speed missile with a nuclear warhead at its tip, rather than a traditional explosive. Yet at the same time, both sides understood the value of a V-2 pointed beyond Earth's atmosphere toward outer space. Even Wernher von Braun, the life force behind the V-2 rocket, famously quipped in 1944, after the first direct V-2 hit in London, "The rocket worked perfectly except for landing on the wrong planet."[54]

Echoing what Germany had tried to achieve during the war, the United States urged astrophysicists and ionospheric scientists to devise scientific instruments suitable for piggybacking on the first round of twenty-five US-assembled V-2s, which were to be tested in 1946 at White Sands Proving Ground in New Mexico.[55] Members of the V-2 Rocket Panel, charged with shepherding this effort, included the Navy Research Laboratory, the

Army Signal Corps, the Applied Physics Laboratory, the National Advisory Committee for Aeronautics (NACA, the wartime forerunner of NASA), General Electric, Princeton, Harvard, and the University of Michigan. Among the instruments were spectrographs, a shielded Geiger counter, a new type of photographic emulsion, temperature sensors, telemetry systems, and a microwave-band radio transmitter that would propagate its signals through the rocket exhaust. At first, military observers at the V-2 Panel's early meetings assumed it would be necessary to clarify the sorts of data they sought, but soon recognized the almost complete congruence between what they wanted and what the scientists had already pondered. The agendas resonated.

A fall 1946 editorial in *Army Ordnance Magazine* portrays the endeavor in upbeat terms, as a journey toward knowledge: "To accomplish research objectives, the 'war head' of the V-2, with its explosive filling, becomes a 'peace head' filled with scientific paraphernalia for exploring the upper atmosphere and evaluating the performance of the . . . rocket."[56] But whenever the War Department supplies the funding, part the curtains and you'll see the needs of conflict masquerading as the needs of science.

But let's get back to the invisibility of radio

waves and the persistent military goal of stealth.

Earth is one of the noisiest radio sources in the cosmic sky. We broadcast our existence loud and clear. For any aliens who might be skywatching in our direction with a radio telescope, we practice the antithesis of stealth. Terrestrial planets, of which Earth is our best-known example, don't naturally emit radio waves in copious amounts. But think of all our activities that generate radio waves: your mobile phone, your remote car-door opener, the radar guns that identify you as a candidate for a speeding ticket, broadcast television, your wi-fi and that of all your neighbors, the Deep Space Network that communicates with space probes, and of course radio stations themselves. Our planet blazes with radio waves — the aliens' best evidence that we have plenty of technology.

Regarding potential surveillance closer to home, Earthlings are more circumspect. We pay attention to defense. Whenever there's a new threat, we try to come up with a new countermeasure. Radar king Robert Watson-Watt characterized this continual back-and-forth as "the never-ending series of counter-counter-countermeasures in the agelong contest between projectile and armor."[57]

One useful radar countermeasure developed during World War II was what the Americans called chaff, the British called

326

Window, and the Germans called Düppel. The US Secretary of the Navy described it as "a unique method of arranging aluminum foil strips of varying lengths into a package which when released in great numbers by our attacking aircraft had practically the same effect on enemy radar directors as a smoke screen would have upon optical directors."[58]

Chaff was — and is — a decoy. To a radar-equipped plane or guided missile, it looks like a target. In the 1940s its key attraction was its capacity to reflect the radar beamed in its direction — to mimic the radar echoes that would be created by an airplane caught in that same beam. Its requirements were not complicated: it just had to be highly reflective, not subject to clumping, and of a length appropriate to the wavelength of the radar. You'd spray the sky with floating strips, and the enemy's radar tracker would be overcome with confusion, unable to tell the difference between target and chaff. If you didn't know the wavelength of the enemy's radar-targeting system, you could spray chaff of varying lengths and count on some of it succeeding. If you did know the wavelength, you could spray only chaff of a suitable length, thereby intensifying its reflectivity and maximizing the chances of its masquerading as the target, especially if the radar beam was wide and therefore likelier to intercept more of the chaff.

The first person on the Allied side to officially propose chaff as a viable countermeasure was the Welsh physicist Joan Curran, the sole woman scientist at Britain's Telecommunications Research Establishment. Telefunken had already tested Germany's own version two years earlier, in 1940. In retrospect, the concept itself seems fairly self-evident — though, again, as with the resonant cavity magnetron, the decision makers initially resisted authorizing its use for fear that it would soon increase their own side's vulnerability. In this case, the concern was that, once used, the strips could easily be observed, understood, and copied by the other side. Nevertheless, it was finally deployed in 1943, and by war's end, three-fourths of US aluminum foil production went toward the making of chaff.[59]

Chaff wasn't the only World War II radar countermeasure. Other attempts included jamming, blinding, obfuscation (for instance, changing the pulse rate of one's own radio navigation system), noise generation, coating U-boat snorkels with rubber, and radar spoofing, which included jiggering one's technology so that it returned a disproportionately strong echo, causing the other side's radar operators to think a large number of planes were heading their way. Whatever anyone could dream up was fair game until something better came along or until the enemy either became too familiar with a given technique or temporarily forgot about its existence. There was also an electronic instrument called a search receiver, which, when fitted with a directional antenna, could locate an enemy's radar station at greater distances and with greater efficiency than radar itself could achieve.[60]

Then there were the counter-countermeasures. One of these, invented on the German side, was based on the differences in the motion of a bomber plane and a cloud of chaff. In obedience to the Doppler effect, the bomber's high speed caused a shift in the wavelength of the signal reflected from the bomber's surface, whereas the almost weightless ribbons of chaff simply drifted under the influence of the wind. As a result, at least sometimes, the Germans were able to distinguish plane from foil and to direct their flak against the plane.[61]

Chaff is a countermeasure of interest to the astrophysicist because of its reliance on albedo — reflectivity — an attribute almost indispensable to the study of celestial objects in a variety of electromagnetic wavelengths. Biologists and geologists and chemists and physicists do not typically devote themselves to the detection of light; astrophysicists do. The military, too, has an ongoing interest in albedo. Minimizing it is a prime goal in innovative stealth and national-security solutions, except that the military thinks in terms of radar cross-section rather than albedo.

The albedo of an object is the average percentage of light that it reflects compared with the amount of light that hits it. What doesn't get reflected gets absorbed. The lower the albedo, the more difficult it is to detect the object. Earth's moon is shockingly dark, with an albedo of 0.12 — about the same as the sidewalls of your car tires. Meaning that overall, taking both dark areas and bright areas into account, it reflects 12 percent of the light that hits it and absorbs all the rest. Cloud-shrouded Venus, our nearest planetary neighbor, has an albedo of 0.75, rendering it a fine bright object in twilight skies, where it routinely gets mistaken for a hovering UFO. Saturn's moon Enceladus, which is mostly covered in freshly deposited, pristine water ice, has an eye-popping albedo of 0.99. An object that appears bright to your detectors is not necessarily nearby. It could be far but have a highly reflective surface, or it could be nearby but have a surface that's only moderately reflective. So the albedo alone, though containing crucial information, provides only partial data about your target.

The entire industry of stealth is about getting the albedo of an object as close to zero as possible. You want your aircraft to have the radar cross-section of a bumblebee, so that it can disappear from your enemy's radar and thus prevent a coherent signal from being reflected back to them. If you succeed, they won't know whether their signal was absorbed or just kept sailing through space unimpeded. You can also put radio-wave detectors on your plane so that you know when you're being "painted" with a radio signal. You'll then know you've been found, and since you're aware that your adversary might be sending a surface-to-air missile to get you, you can take evasive measures.

But there's another, better option: you can turn the entire surface of your aircraft into a series of facets, at assorted but specific angles, so that radar bounces off it every which way except back toward you, making your plane almost invisible to radio waves and, as the Air Force phrases it, "restoring the element of surprise." Voilà, you have now designed the F-117A stealth fighter, a "low-observable," more or less triangular one-seater aircraft coated with a black, radar-absorbent substance for extra stealthiness. This plane manages simultaneously to resemble an enormous origami crane and an airborne tank. It's not fabulous on the aerodynamic front, but at least for a while — since surprise is perishable — it put the USAF back in the driver's seat regarding time and place of attack.[62]

Developed during the 1970s and early

1980s at the vividly storied, once-secret Nevada salt flat known as Area 51, the F-117A flew hundreds of close-in strikes and bombing missions in Iraq during Operation Desert Storm in 1990–91 and Operation Iraqi Freedom in 2003. Its scientific parent was a monograph written in 1962 by a Soviet theoretical physicist/engineer who established a firm mathematical foundation for calculating "the diffraction of electromagnetic waves by metal bodies of complex shape" — more specifically, "reflecting bodies with abrupt surface discontinuities or with sharp edges (strip, disk, finite cylinder or cone, etc.)." The monograph was translated in 1971 for the US Air Force and studied closely soon thereafter by a radar specialist at Lockheed Aircraft's secretive, cutting-edge Skunk Works unit, which had earlier produced the U-2 spy plane.[63]

Although scientists already understood that certain surface characteristics could enable an aircraft to evade easy detection by radar, the mathematics necessary to a workable physical theory of diffraction did not yet exist. That was the contribution of Pyotr Ufimtsev, author of the 1962 monograph. After spending its first Cold War decade in obscurity, his monograph became "the Rosetta Stone breakthrough for stealth technology," giving rise not only to Lockheed's F-117A stealth fighter but also, later, to Northrop's sleek B-2 stealth bomber, which uses continuously curved surfaces rather than facets for its fuselage. The difference arises from the simple matter of differences in calculating power at the time of their design, between computers of the 1970s and those of the 1980s, which were a hundred times more powerful.[64] If Batman flew a stealth bomber, the B-2 would be his Batplane.

For more than half a century, warmakers have exploited the fact that most military detection takes place under circumstances beyond the reach or domain of visible light. Astrophysics has long dedicated itself to detecting phenomena in every single wavelength of light — a pastime that takes advantage of every possible advance in science and technology to accomplish its task. As of September 2015, you can add gravitational waves to this observational arsenal. Discovered by the LIGO collaboration (the Laser Interferometer Gravitational-Wave Observatory), these signals are ripples in the fabric of space and time, caused by the exotic doings of gravity rather than by light. Even so, gravitational waves from across the universe are so weak by the time they reach Earth that many years will probably pass, possibly even centuries or millennia, before gravitational astrophysics leads to innovative military tactics.

Nowadays the majority of astrophysical revelations derive from detectors designed for invisible parts of the spectrum: from several-hundred-mile-long, extremely low-frequency radio waves on the low-energy end to quadrillionth-centimeter-long, extremely high-frequency gamma rays on the high-energy end. Want to see a gigantic stream of stars 76,000 light-years from Earth and several million times fainter than the dimmest stars detected with the unaided human eye? See it through NASA's infrared Spitzer Space Telescope. How about a sudden flare of gamma rays emitted by a galaxy 7.6 billion light years away, far more ancient than Earth itself? See it with VERITAS, the Very Energetic Radiation Imaging Telescope Array System in Arizona, and confirm it with NASA's Fermi Gamma-ray Space Telescope. And what of a galaxy almost 10 billion light-years from Earth with a mass 400 trillion times that of our Sun? Use data from ESA's XMM-Newton and NASA's Chandra X-ray Observatory to determine the mass.

Today astrophysicists see a universe immeasurably more complex than the one conceptualized by Newton or Herschel. Some things, such as stellar nurseries, glow brilliantly in infrared but are almost completely dark in the visible range. So, too, is the cosmic microwave background. Yet in spite of all the mind-blowing discoveries made in invisible wavelengths since the end of World War II, visible-light detectors still yield surprises. In 2016 astrophysicists using the Hubble Space Telescope announced that they had found the most distant galaxy ever seen, gleaming 13.4 billion light-years from Earth. Its stars would have been made solely of hydrogen, helium, and a tad of lithium, because no other atoms yet existed — no carbon, no nitrogen, no iron, no silicon, certainly no silver or gold.

Each band of light presents its own detection challenges. Earth's atmosphere is transparent to the visible part of the spectrum, which is why we can see the Sun, but it is largely opaque to ultraviolet. Clouds are opaque to visible light but almost transparent to infrared. Brick walls are opaque to our eyes, but to microwaves those walls are transparent, which is why we can talk on our cell phones while indoors. Humans are transparent to radio waves. Glass is transparent to visible light. You may say a brick wall is opaque, but an astrophysicist will ask, Opaque in what wavelength? The astrophysicist will also ask, What's the transmission curve — what fraction of the light of a given wavelength gets through a given medium without being absorbed?

Take microwaves, living out their rather low-energy lives at the longer-wavelength end of the electromagnetic spectrum, ranging from a millimeter up to about thirty centimeters. Only about half the microwave light from objects beyond Earth's atmosphere makes it through to terrestrial telescopes. What happens to the other half? It's absorbed by atmospheric water vapor. That's why microwave astrophysicists locate their Earth-based telescopes in deserts or, even better, in a high-altitude desert, above most of the cloud cover. One place on our planet where both aridity and altitude serve the astrophysicist is the Atacama Desert, a high plateau in the Andes mountains of northern Chile. With its few millimeters of annual rainfall (until climate change brought flash floods and hot-pink flowers in 2015), Atacama is the driest desert on Earth, and it's so high that most of the clouds, and therefore most of the water, lie below it. Not surprisingly, the world's most powerful earthbound microwave telescope, ALMA, the Atacama Large Millimeter/submillimeter Array, has its home there.

When you chart the transmission curve for microwaves through Earth's atmosphere, you find a sudden, transparent window between the wavelengths of eighteen and twenty-one centimeters. At either end of this little band, radio astronomers can detect distinct emissions from the universe's ubiquitous hydrogen atoms (H) and their partners in water, hydroxyl molecules (OH). This band has been dubbed the water hole — a phrase more commonly invoked for places on Earth where wild creatures congregate to drink and wallow. We suspect that any aliens who know of our existence and want to communicate with us might also know the absorptive effects of water on various wavelengths. So if they're clever, they might exploit the dip and try to reach out to us via the microwave frequencies that get through the water hole.[65]

How about a less pacific result of the astrophysical discovery that water absorbs microwaves? How difficult would it be to design a non-lethal weapon that targets the water content of the human body? Three-fifths of our average body mass is water. Such a weapon could operate on the same principle as the microwave oven.

Ask, and it shall be given. Raytheon's Active Denial/Silent Guardian System is America's version. Like peaceable ALMA, it operates at millimeter wavelengths, which are a little shorter than those in a standard microwave oven. This limits the depth of their penetration into the human body. You don't actually want to cook people with nonlethal weaponry. Let's say your mayor thinks there could be property damage during next Saturday's climate protest. He may want to get proactive in the war against domestic terrorists like your Aunt Melissa. The army can send one of its trucks equipped with a millimeter-wave generator to a street corner near the crowd, and when the truck beams its electromagnetic radiation at the center of a crowd of protesters, their skin will feel like it's beginning to fry, even if they're wearing clothes. To avoid pain, the protesters will willingly and rapidly disperse.[66]

There are other small-scale, ostensibly nonlethal weapons, security measures, and crowd-control gizmos that utilize other non-visible wavelengths, notably infrared, and tend to occupy the MOUT (Military Operations on Urban Terrain) portion of the use-of-force spectrum: surface-to-air missiles, airport security systems that disrupt the guidance system of any missile aimed at a plane, weaponized lasers, non-nuclear electromagnetic-pulse generators, pulsed-energy projectiles, PHaSRs. There are battle aids like night vision scopes and goggles capable of image intensification. And of course there are profoundly lethal electro-magnetic weapons — armaments capable of massive devastation. The knowledge that underpins these activities and instruments is what interests the astrophysicist; the instruments themselves are what interest both the destroyers and the defenders.

Whether you're a fighter or an astrophysicist, you can't do much without hard information. Fighters use information in real time, whereas we astrophysicists want our information saved for later — sometimes even years later. Because we analyze at leisure what our observatories have detected in passing, preservation is a huge concern. Galileo could only draw what he saw. Photography was the big breakthrough of the nineteenth century, producing a record of what would otherwise be unprovable. Come the twentieth century, there were multiple breakthroughs. Special-purpose emulsions, the baking of film, spectral filters, photomultiplier tubes, CCDs and their pixels — jointly they yielded a vast archive of information awaiting the engagement, or re-engagement, of ingenious analysts.

Envision a rectangular digital image, a picture. Now envision the smallest possible section of it. That's a picture element, a "pixel." This represents the fundamental unit of detection for charge-coupled devices, or CCDs, which began to transform image-making in the 1970s and had swept away all other approaches by the 1990s. While still in graduate school, I was eyewitness to this revolution, and its impact on my field cannot be overstated.

When the CCD is exposed to light, whether from a nearby street scene or a faraway galaxy, each of its pixels stores some number of electrons, depending on the intensity of the light hitting each of the tiny locations on the CCD's light-responsive computer chip.

The more intense the light, the more electrons get stored — although if the light is too bright, you will saturate the detector and the excess electrons will spill over into neighboring pixels, contaminating their data. Double the exposure, and you get double the number of electrons. The electrons that congregate in each pixel are then collected from the chip, tabulated, and turned into a single electronic tile in the mosaic that constitutes the complete image. The more pixels, the more resolution available to you. Nowadays you can easily download a street scene from Wikimedia Commons that measures 2592 columns × 1944 rows, which translates into a grid of more than 5,000,000 pixels — a crisply detailed photo. But that's nothing: if you're not worried about overtaxing your computer, you can download an image of the Orion Nebula from the HubbleSite Gallery that's 18,000 × 18,000 — a grid of 324,000,000 pixels, packed to the gills with detail.

There's also the issue of "quantum efficiency." In the most efficient detector possible, one photon would give you one electron. Reality isn't quite so cooperative, although CCDs massively outperform film. For every hundred photons of light that landed on the silver halide crystals in Eastman Kodak's now-obsolete astrophotographic emulsion IIIaJ, only about three triggered the necessary chemical reaction to produce an image. That was 3 percent quantum efficiency. What's the quantum efficiency of a CCD today? Some astronomical CCDs are more than 60 percent efficient across a wide band of visible wavelengths. That's a factor-of-twenty improvement in detection power. Other CCDs top out at 90 percent quantum efficiency in selected wavelengths. They also pick up near-infrared and near-ultraviolet. In addition, the CCD can be used with any lens. All these benefits mean that astrophysicists can acquire information from far deeper in space, and from many more regions, than ever before.

Noise can be a problem, though. When a telescope targets something dim, it might not collect enough light to trip the detection threshold. On the other hand, some of what seems to be light might just be noise. Every telescope, every detector, has inherent noise. A CCD, too, has noise — its own warmth is enough to kick some electrons into the pixels — and so the best CCDs and cameras are now chilled during use. In the old days, astrophysicists would have been using photographic plates to record what our telescopes detected, and we would have needed long exposures to get our images. Knowing there were still dimmer things we weren't detecting, we would have yearned for bigger telescopes to collect more light. We would have needed money, engineers, another dome, another mountaintop.

In the early days of CCD technology, chips were small, with few pixels. Some were manufactured in university or industrial laboratories specifically to serve the astrophysicist. But as the CCD became commoditized, especially because of demand for digital cameras, the price, quality, and pace of improvement grew rapidly. The CCD transformed astrophysics, giving new life to small telescopes and endowing large ones with previously unimaginable powers of detection. Some researchers made entire careers of redoing earlier brilliant work whose authors had approximated and speculated about what could be lurking beyond the available data. In the era of the CCD, astrophysicists can tackle the same problems but with greater success. We can push past the earlier limits on data and speculate at yet another level.

Anyone who can't afford to depend on serendipity would say you have to identify your target or goal in advance. Which leads us to the military potential of the CCD.

Knowing what you're looking for is integral to ISR: intelligence, surveillance, reconnaissance. The advent of the CCD did wonders for America's ISR, just as it did wonders for America's astrophysicists. After all, astrophotography and photoreconnaissance differ only in their choice of target, their distance from the target, and the direction of their gaze. In December 1976 the KH-11 KENNAN — one of the KEYHOLE series — became the first spy satellite equipped with CCD technology.[67]

The change was transformative. No longer would the National Reconnaissance Office have to wait days for a spy satellite's parachute-equipped, heat-shielded film canisters to be grabbed in mid-air during a rendezvous with an airplane or, worse, dropped in the ocean and collected by (preferably) US ships, then processed, and finally delivered to the right person's desk.[68] Now the images captured by a KH-11 — for instance, of a Soviet aircraft carrier under construction at a shipyard on the Black Sea — could be almost instantaneously transmitted via a data-relay satellite to a ground station near Washington, DC.

The earliest spy satellites, developed under the CORONA program, were set up to search; their cameras focused on broad coverage. KEYHOLE and GAMBIT satellites, next in line, captured a closer look at specific targets already identified by their CORONA predecessors. HEXAGON satellites further sharpened the resolution of individual targets and improved the search capability. Most carried both a main camera for broad imaging of otherwise inaccessible areas and a mapping camera to assist in war planning. As HEXAGON's maker, Lockheed Martin, described its role in a press release, the country "depended on these search and surveillance satellites to understand the capabilities, intentions, and advancements of those who opposed the U.S. during the Cold War. Together they became America's essential eyes in space."[69]

The camera on the final CORONA spy satellite, launched in 1960 and retroactively renamed the KH-1, could detect objects as small as eight meters wide. A mere six years later, the KH-8 GAMBIT's camera could refine this to fifteen centimeters. A decade later the KH-11 KENNAN, the first to have a CCD, offered much broader coverage, greater recording capacity, and a considerably longer lifetime, but at the cost of lower resolution: two meters. The so-called Advanced KH-11, however, offered both infrared capability and high resolution.

Not surprisingly, there's also a long list of spy satellites launched during the Cold War by the Soviet Union and a short list launched by China. Equally unsurprising is that, although the US programs have usually retained their classified status for decades, there have also been periodic leaks, unintentional disclosures, and episodes of quasi-involuntary declassification. In 1981 a respected aeronautics publication showed a leaked KH-11 image of a Soviet bomber; in 1984 an American naval analyst leaked the KH-11 image of a Soviet aircraft carrier to a respected military publication. KH-11 itself, along with its progeny and cousins, remains classified.[70]

Today there are no more film canisters suspended from parachutes. Rochester, New York — home of Eastman Kodak — is sunk in joblessness, and high-res CCDs are the global standard. There likely now exists a continually updated optical, infrared, and radar image bank of every square foot of every conflict zone and potential conflict zone on the planet. One oft-reproduced Advanced KH-11 image from the 1990s shows a pharmaceutical plant in Sudan said to have been connected with the making of chemical weapons. Another shows a mountain camp in Afghanistan described as an al-Qaeda training facility. More recent satellites — reconnaissance, geospatial, commercial, communications, weather — have imaged and re-imaged such militarily significant targets as Osama bin Laden's compound in Abbottabad, Pakistan. They have detected the sudden appearance of numerous armored vehicles at a military base in Aleppo, Syria, and recorded increased activity just prior to a rocket launch at the Sohae Satellite Launching Station in North Korea.

But spy satellites monitoring conflict zones aren't the only source of such images. Uncountable numbers of commercial satellite images can now be bought by whoever wishes to pay. As William E. Burrows puts it,

> The intelligence establishment itself regularly supplements its own systems' "take" with commercial satellite imagery, and the use of civilian spacecraft for routine intelligence collection and potential war-fighting is increasing because it's cheaper than maneuvering their classified counterparts and processing the avalanche of digital data that keeps coming down in near real time. . . . If the intelligence establishment can in effect use a credit card to buy excellent commercial imagery, so can tyrants and terrorists.[71]

Yes, but so can humanitarian aid agencies and environmental groups.

Will satellite images, whatever their source, never be misused and always make us safer? Probably not. But is it good to have records of the extent of deforestation in the Amazon between 1975 and 2012, and to have been alerted to the breakup of the largest ice shelf in the Arctic in 2003? Probably so. There's now an organization called International Charter: Space and Major Disasters, which supplies free satellite imagery to emergency responders across the world so that they can act more quickly and effectively. Like GPS, those eyes in the sky are dual use.

Q: What do you get when you cross a spy satellite with a ballistic missile, and then launch the result into interplanetary space? **A:** NASA's Deep Impact mission to comet Tempel 1, the first time an intentional collision, rather than a mere flyby, was a mission's main agenda.

On July 3, 2005, after traversing more than 400 million kilometers in less than six months, the Deep Impact spacecraft released an eight-hundred-pound hunk of mass — its "smart" impactor — that smashed into Tempel 1 the following day with the explosive energy of five tons of TNT. It excavated a deep crater, purposefully kicking up loads of dust that could be observed and recorded by the orbiting spacecraft's camera and infrared spectrometer as well as by numerous telescopes around the world. We can now definitively say that Tempel 1 has water ice on its surface, a "very fluffy structure that is weaker than a bank of powder snow," and an abundance of carbon-containing molecules. Those molecules tell us that a comet not unlike Tempel 1 could, in passing, have deposited organic material on Earth during our planet's first billion-plus years of existence, when it was being regularly bombarded from space by all manner of rocks, including comets.[72]

Obviously the impactor had to hit its target — a very dark (0.06 albedo) blob of comet-matter less than four miles in diameter — or the mission would have been for naught, just as a fighting force's artillery has to hit its targets or lose the battle. All concerned parties were in motion: Earth as a launch platform, the spacecraft, the impactor, and the comet. The impactor was fitted with a telescope, a mediumres multi-spectral CCD camera, target sensors, a battery to sustain it during its final day of life, and a dose of hydrazine fuel for brief bouts of propulsion to adjust course. This ballistic projectile had to be released from the spacecraft at a time and angle that would guarantee its subsequent close approach to the comet. Plus, the ultimate collision had to occur on the comet's sunlit side so that the resulting dust could be seen.

Rather than relying on the usual time-consuming practice of ground-based navigation — transmission of data down to Earth, human analysis and execution of commands, relaying of commands back up to the spacecraft — the mission used an onboard system called AutoNav to orchestrate the actual collision. Activated two hours before that final moment, AutoNav took four images per minute so that it could stay current with the position and velocity of both the comet and the impactor. Being smart about keeping the impactor on course, it initiated three targeting maneuvers: at ninety minutes, thirty-five minutes, and twelve minutes before impact.[73] The mission was a success — not because of luck, but because astrophysicists as well as warfighters know how to use multi-spectral data to deploy a ballistic projectile to hit a moving target. We are independent. We are interdependent. We are allies.

6
DETECTION STORIES

Each band of wavelengths in the electromagnetic spectrum is a window to a different component of cosmic reality. As the tally of detectable wavelengths grew, so too did the tally of exploitable collaborations between astrophysics and the military. Some of these were widely known in their day. Others were secret. Still others were accidental alliances that could not have been scheduled, planned, or predicted.

I.

Our first story is about Jodrell Bank — a few muddy acres of fields in Cheshire, England, twenty-odd miles south of Manchester, that at the end of World War II were being overseen by a botanist at the University of Manchester but were shortly turned into the site of a major observatory. The area's suitability as a site for the world's first large steerable radio telescope lay in its low population and especially in its lack of public electricity lines.

As Bernard Lovell wrote in his account of the logistical, financial, and political nightmares connected with bringing the observatory's Mark I radio dish into existence, "Electrical gadgets used in and around houses often spark and radiate more energy into a radio telescope than an entire extragalactic nebula." What made the Mark I steerable was repurposed wartime hardware: two bearing assemblies that had borne the big rotating guns on two British battleships during World War I but could be bought for a song in 1950 from the Admiralty's Gunnery Establishment.[1]

On the night of October 4, 1957, a couple of months after the Mark I became operational — though just barely, as the project was steeped in debt — the Soviet Union launched Sputnik 1. Suddenly the huge radio dish, capable of receiving as well as transmitting signals, and designed for research into cosmic rays, meteors, and the Moon, became the only instrument on Earth capable of radar-tracking the core stage of the intercontinental ballistic missile, the R-7 rocket, that had launched the satellite and had itself achieved Earth orbit. During twilight, a sky-watcher observing in the deepening darkness might manage to see the gleam of the satellite as it passed overhead, high above and still in sunshine. A ham radio operator could easily pick up the satellite's radio beeps on a frequency of 20.005 megahertz. But only the

Mark I could detect the radar echoes bouncing off the rocket.

For the sake of England's prestige and the whole world's benefit, there was no question of refusing to take on the task. Intensive work began on October 7; initial intimations of success came on the 11th; unmistakable triumph occurred on the 12th. Here is Lovell's account of the 12th:

> Just before midnight there was suddenly an unforgettable sight on the cathode ray tube as a large fluctuating echo, moving in range, revealed to us what no man had yet seen — the radar track of the launching rocket of an earth satellite, entering our telescope beam as it swept across England a hundred miles high over the Lake District, moving out over the North Sea at a speed of 5 miles per second. We were transfixed with excitement. A reporter who claimed to have had a view of the inside of the laboratory where we were, wrote that I had leapt into the air with joy.[2]

Soon the Mark I (which Lovell calls the bowl, and which was later renamed the Lovell Telescope), along with Jodrell Bank's newer telescopes, proved indispensable in verifying the telemetry of the earliest Soviet and American space probes. The observatory's cooperation with verification requests from both the US and the USSR during the space race loomed large in attracting desperately needed funds and thereby sustaining its own science agenda.[3] An unsavory bargain? No, realpolitik.

On the first day of 1958, Lovell received a telegram from Moscow, saying, "Every success in your work. Best thanks for satellite operations."[4] Soon there would be more satellite operations. The Soviet Union's subsequent requests involved their pioneering Luna (Lunik) and Venera probes of the Moon and Venus. Confronted with professed international skepticism that it had actually launched Luna 1 on January 2, 1959 — and disappointed that Jodrell Bank hadn't managed to locate it (the spacecraft missed the Moon by more than two diameters) — the Soviet Union sent Jodrell Bank a telex an hour after the Luna 1 launch with transmission frequencies and exact coordinates for its next Moon probe, Luna 2. The Soviets wanted Jodrell Bank to independently verify what they predicted would be a successful lunar landing.

This time the Mark I succeeded in its appointed task, in part because its antenna was already set up to capture the transmission band being used by Luna. By local midnight on September 12, 1959, the Brits were receiving Luna 2's signals on two frequencies. Clearly the rocket was on the right course.

Predicted time of impact was 10:01 the following day. At 10:02 they began to worry, but twenty-three seconds later, the signals stopped. Human-created hardware had made it to the Moon. Some high-profile US politicians persisted in their public skepticism, but the facts were the facts. Less than a month later, and precisely one year after Sputnik 1, Luna 3 reached and photographed the far side of the Moon, a first, while the following month a US Pioneer spacecraft (P-3) designed to orbit that same body exploded on the launch pad. One unnamed American even commented that "it was only necessary for an announcement to be made of American intentions for the Russians to do it first."[5]

The United States, too, had sought Jodrell Bank's support soon after Sputnik's launch, in what was meant to be a superconfidential arrangement, initiated in the spring of 1958 by a US Air Force colonel who had crossed the Atlantic with no announced reason except a desire to meet Lovell. As soon as the pair reached Lovell's office, the colonel asked that the windows be shut and the doors locked, whereupon "the real conversation then began in a scarcely audible near-whisper." The US Army had launched America's first Earth satellite in January 1958; now the Air Force wanted to launch America's first spacecraft to the Moon in August 1958 and to have Jodrell Bank track its journey. Discussion was out of the question; an immediate decision was required. Tracking equipment and technicians would be sent over from Los Angeles before the launch. Everything must remain completely secret.

Except that when the trailer that held all the equipment arrived, it displayed in giant letters the following ID: "Jodrell Bank, U.S. Air Force, Project Able." So much for secrecy.

The *Manchester Guardian* broke the story in July. But the Pioneer 1 launch went ahead anyway, at 8:42 AM on October 11, and the Mark I picked up its signals at 8:52. This was NASA's first-ever launch. Alas, early on October 13 Pioneer 1 fell back toward Earth and burned up upon re-entering the atmosphere, unable to reach the Moon because it never quite attained escape velocity and its launch angle was off by a few degrees.[6] No matter. Occasional failures were inevitable, and officials stopped fretting over secrecy.

The next NASA–Jodrell Bank collaboration, Pioneer 5, was the opposite of a failure. On March 11, 1960, twelve minutes after the spacecraft launched from Cape Canaveral, the Mark I started tracking it. This time the radio telescope — "the only instrument which had any hope of transmitting with enough strength to the probe over distances of tens of millions of miles" — would not merely track the craft but also command it and receive scientific data from its onboard experiments:

> At 1.25 p.m. when Pioneer was 5,000 miles from earth a touch on a button in the trailer at Jodrell transmitted a signal to the probe which fused the explosive bolts holding the payload to the carrier rocket. Immediately the nature of the received signals changed and we knew that Pioneer V was free, on course and transmitting as planned. For the rest of the day Pioneer responded to the commands of the telescope and when it sank below our horizon on that evening it was already 70,000 miles from earth. The next evening it was beyond the moon.[7]

For nearly four months, the radio telescope stayed in touch with the spacecraft. The last communication took place on June 26, 1960, at a distance of thirty-six million kilometers from Earth. In the deep vacuum of interplanetary space, with nothing to force a decay in its trajectory, Pioneer 5 continues to orbit the Sun every 312 days.

II.

Whereas radio waves have yielded all manner of benefits, both near at hand and far away, gamma rays are not generally regarded as beneficial. Quite the opposite.

Occupying the high-energy end of the electromagnetic spectrum, gamma rays were discovered as a by-product of radioactivity in 1900. By the 1950s, gamma rays from space were considered a possibility, but were not actually detected until 1961 by a short-lived, new kind of detector aboard NASA's Explorer XI satellite.

Like X-rays, gamma rays are hard to detect, because they pass right through ordinary lenses and mirrors and thus can't be focused the way radio waves and visible light can. What works for radio waves, microwaves, infrared, visible, and ultraviolet wavelengths doesn't work for X-rays or gamma rays. Detectors in these bands require inventive designs. Plus, film registers only visible and UV light; to register signals from objects emitting in other bands, new recording methods were needed.

Explorer XI's detector was a device called a scintillator, which is as distantly related to a telescope as a whale is to a spider. A scintillator is a tiny block of energy-sensitive material (cesium iodide, for instance) that pumps out tiny flashes of light — charged particles — each time a gamma ray barrels through it. Amplify the flashes with photomultiplier tubes, and you've got yourself a detection device. By measuring the energy of all those charged particles, you can tell what kind of radiation created them. During Explorer XI's four months of tumbling through space, its detector gathered data for twenty-three days and snared a whopping twenty-two certifiable gamma rays.

While "gamma rays" are what we call the shortest wavelengths (and highest energy) of the electromagnetic spectrum, their swath of light is huge. But they're not the only superhigh-energy stuff in the universe. So-called cosmic rays, which actually consist of particles, are competitively energetic. Hardly any of Earth's daily dose of gamma rays that originate in deep space reach our planet's surface. Atmospheric ozone — the three-atom version of the oxygen molecule — shields us nicely, though not entirely, from them, as well as from solar or anybody else's UV and spaceborne X-rays. To detect gamma rays reliably requires specialized satellites in orbit above our atmosphere.

As you might suspect, high-energy phenomena breed high-energy light. Try to imagine the simultaneous detonation of all the nuclear bombs ever made, including those exploded during war or in preparation for war, together with those disassembled in the name of peace. Imagine a star a hundred times as massive as the Sun, collapsing in on itself at the hour of its death. Imagine a sprawling galaxy, formed during the first billion years of our universe's lifetime, and the colossal black hole that lurks at its galactic center, entombing the substance of many billions of long-dead stars and continually swallowing everything within reach. Or imagine the remnant of an exploded giant star — a remnant so dense that a thimbleful would weigh a hundred million tons — spinning in faraway space at tens of thousands of times a second as it crashes into a neighbor. These fierce, violent configurations of matter, these superhigh-energy events, have superhigh-energy consequences. One of those consequences is a sudden, brief, often beamed burst of gamma rays: an explosion of astronomical proportions. A single burst can out-radiate an entire galaxy — as though the energy output of a hundred billion Suns were concentrated into a few moments of overwhelming brilliance. Wildly dramatic . . . and deadly, if you're in the neighborhood.[8]

On average once a day, a gamma-ray burst occurs somewhere in the distant universe. The relatively weaker ones last less than a second; the rarer, highly energetic ones last as much as a few minutes. The source of all that energy is a mélange of gravitational, rotational, magnetic, and thermonuclear phenomena. The object releasing the energy might be a supernova, a kilonova, a hypernova, a blazar, or a quasar. Could also be stuff just before it falls into a black hole, or a nuclear explosion down here on Earth. Repeat: a nuclear explosion down here on Earth. Human ingenuity has conceived, invented, and deployed an equivalent of one of nature's least friendly phenomena.

We still don't have the full story on how cosmic gamma-ray bursts are generated. But before astrophysicists even knew cosmic gamma rays existed, both scientists and politicians knew that a terrestrial version would occur if and when a thermonuclear fusion bomb exploded.[9] Whether the detonation was a test or an actual attack would make no difference, nor would it matter whether it took place in the middle of a desert, in the middle of Manhattan, or on the Moon. Nevertheless, when the twentieth century's second all-out multinational war ended, the design of annihilation-class weaponry proceeded apace, causing as much fear and mistrust among the designers as among the bystanders. Einstein himself, acutely aware of the world's newfound capacity for annihilation, said in a 1949 interview in *Liberal Judaism,* "I know not with what weapons World War III will be fought, but World War IV will be fought with sticks and stones."[10]

Faced with all that progress in destructive capacity and increasingly aware of the longer-term effects of radioactive fallout, the foreign ministers of the Soviet Union, the United States, and the United Kingdom signed the Limited Nuclear Test Ban Treaty in early August 1963. In this case, "Limited" left the door wide open for underground testing. Two months later, after the US Senate had ratified the treaty, President Kennedy signed it. On October 10 it attained the force of law.

Any political skeptic will tell you that abiding by a treaty is an entirely separate matter from signing a treaty. It now became necessary (and possible) to monitor from space any telltale signs of an impermissible detonation on Earth. To do so, the US would send up a few satellites carrying state-of-the-art gamma-ray detectors. These satellites were not telescopes. They were simply orbiting detectors, unable to pinpoint the exact spot of a thermonuclear explosion. But a set of them, each recording the exact arrival time of the gamma rays, would make it possible to triangulate the location. Furthermore, if the orbits were high enough, the satellites would escape the electromagnetic noise created by the Van Allen radiation belt — a region of space that one NASA writer memorably described as "two donuts of seething radiation" enveloping our planet.[11]

The ink was barely dry on the treaty when, on October 16, 1963, in the spirit of the military mantra "Trust but Verify," the United States launched its first pair of Vela Hotel satellites, spaced 180 degrees apart, into a very high orbit — a hundred thousand kilometers, far beyond Earth's atmosphere and well clear of the Van Allen belt. Their mandate was straightforward: to detect any gamma-ray emissions produced by any explosion of any nuclear bomb. Their detector was a scintillator.

The second pair of Vela Hotels was launched in July 1964, the third in July 1965, the sixth and last pair in April 1970. Work had begun in 1959 under President Eisenhower as a project of the Advanced Research Projects Agency, with help from the Atomic Energy Commission's laboratories at Los Alamos. With each new pair, the sensitivity of the detectors and the precision of the timers improved. The Vela Hotels were a durable product, most lasting a decade or more beyond their planned life.[12]

As they circled Earth every four and a quarter days, one or another of the Vela satellites periodically registered hits from high-energy solar particles — nothing to worry about, nothing catastrophic. By contrast, if and when a thermonuclear weapon was exploded, it would register on all satellites in sightline of the event, showing up as an intense gamma-ray burst less than a millionth of a second in duration, followed by a leveling and then a fade-out, followed by hours or days of afterglow. Eventually, on July 2, 1967, the Velas did register a powerful gamma-ray event. The weird thing was, it didn't fit the profile of a nuclear explosion. The recording shows a soaring initial peak lasting less than an eighth of a second, followed almost immediately by a second, somewhat lower peak lasting two seconds.[13] Not a nuclear explosion. Also not a solar flare or a supernova, since none had been observed that day.

A couple of assiduous young astrophysicists at Los Alamos, Ray Klebesadel and Roy Olson, were the first to figure out what it was not and what it might be. But being scientists and also being attached to one of the country's two classified national laboratories dedicated to developing nuclear weapons, they held off in hopes of gathering better evidence — which they got from upgraded pairs of Vela satellites, equipped with better instruments, that were launched in 1969 and 1970. After processing vast quantities of "noisy" Vela data, they and a colleague, Ian Strong, identified sixteen gamma-ray bursts between July 1969 and July 1972 that fulfilled their careful criteria (being recorded by at least two Vela craft within an interval of no more than four seconds). Those sixteen bursts reinforced the investigators' July 1967 finding. In 1973, they published — which, in practice, means declassified — the results. It's a typical scientific article for the *Astrophysical Journal,* calm and circumspect. The closest the authors come to saying they've identified something new and big in the universe is the understated assertion that "[i]nverse-square law considerations thereby place the sources at a distance of at least 10 orbit diameters"[14] — three billion kilometers minimum.

It's not as though gamma-ray research and gamma-ray detectors didn't exist prior to that article. But the findings of Klebesadel, Strong, and Olson stimulated a groundswell of new effort. The military's interest in detecting extremely high-energy explosions ended up exploding what had previously been a low-profile branch of astrophysics. Space-based detectors came online, superseding ground-based detectors made from recycled World War II matériel.[15]

Incidentally, gamma rays and a myriad of subatomic particles are generated by the collision of superhigh-energy cosmic rays with Earth's atmosphere. Within this cascade lurks striking evidence of time dilation, a feature of Einstein's theory of relativity. Cosmic-ray particles move through space at upward of 99.5 percent the speed of light. When they slam into the top of Earth's atmosphere, they break down into many subproducts, each with less and less energy per particle, forming an avalanche of elementary particles that descend toward Earth's surface. Among the subproducts is a shower of gamma rays, which swiftly transform into electrons and their anti-matter counterparts, positrons.

Also in the mix you'll find muons, which are the high-energy, heavy version of the electron. They're not particularly stable. After a half-millionth of a second, on average, they decay into other, less energetic particles, one of which will always be an electron. Compared with the life expectancies of many other subatomic particles, a half-millionth of a second is an eternity. But because the particle shower moves so fast relative to us and our detectors on Earth's surface, the muons experience the passage of time more slowly than we do. Enter the bizarre world of Einstein's special theory of relativity. This branch of physics doesn't care who or what you are, whether you're an animal, vegetable, or mineral. If you travel fast, several weird things happen. One is that your inner time clock will appear to tick more slowly, as seen by all those who observe you. Your time "dilates." And muons in a cosmic-ray cascade offer one of the most striking tests of this phenomenon. Because they travel at such high speeds, we see them living ten times longer before they decay — and, as a result, reaching much deeper into Earth's atmosphere — than they "should."

If you don't happen to be going as fast as a muon, you'll still experience a little time dilation. Spend six months on the International Space Station, which is traveling five miles per second around Earth (a mere 0.0027 percent the speed of light), and you will have aged 0.005 seconds less than everybody else on Earth.

III

Since everything warmer than absolute zero radiates heat, detecting infrared at a distance means, in principle, detecting everything. Period. As a result, every astrophysicist as well as every general, counterrevolutionary, spy, cop, and drone that needs to identify an otherwise invisible target could search for it in infrared. But what the warfighter must also do is distinguish what is from what is not a threat. Simply detecting a patch of oddly intense infrared isn't enough. Surveillants need to know the "heat signature" of their target so they can isolate and differentiate it from the manifold other infrared sources that crowd the theater of operations.

Out there in the cosmos, the cooler residents — those with temperatures below about 1000 kelvins (700 degrees Celsius), which includes planets, failed stars, cosmic dust, and assorted clouds in galaxies, especially those about to give birth to star systems — emit more infrared than any other band of light. Anything hotter than that also begins to glow in the visible part of the spectrum, rendering it plainly visible to anybody looking, initially appearing "red hot" but then, as its temperature rises further, "white hot" and finally "blue hot." So if you want to see cool objects, best use an infrared telescope.

Also, infrared light escapes clouds of gas and dust much more readily than does visible light, even when the visible light is highly luminous. This is where an all-sky survey and the US Air Force's Infrared Celestial Backgrounds program enter the picture.

To distinguish the infrared signature of a missile in the sky headed for the presidential palace from the infrared signature of a cosmic object, the general needs a sky map. The general provides the funding. The astrophysicist provides the map. Whereas the discovery of gamma-ray bursts was a serendipitous by-product of the normal work of military surveillance, comprehensive infrared sky maps were the intentional result of a military initiative meant to furnish surveillance with a necessary tool. As an Air Force fact sheet explains it:

Ballistic missile defense is an important mission with the need to develop technologies for detecting and tracking theater and strategic ballistic missiles from launch to intercept. . . . Effective tracking of cold-body and dim targets in the IR spectral region requires the IR signature of the target to be distinguished from the background against which the target is observed. The issue is that the background can mask or mimic the target. Therefore, [a] key technical goal is to measure and model the full range of backgrounds, particularly challenging backgrounds, in order to design IR sensor systems which will maximize the visibility of the target signature.[16]

This problem is not unique to infrared. Any measurement of anything by any means risks getting confused with background noise. We're all familiar with literal noise. An intimate conversation between you and a loved one can occur without confusion in a quiet room, whereas at a crowded cocktail party you will need to speak well above a whisper to be heard and understood — to be detected. "Noise" includes any unwanted signal that contaminates the target of measurement.

When the earliest infrared investigations took place, most scientists presumed radiant heat and visible light to be two different things, although in 1835 André-Marie Ampère published a note proposing that they were both the result of "vibratory motion."[17] Early IR detectors were upgraded versions of the thermometer, suited to modest achievements such as measuring the heat signature of a cow a quarter mile away. But in the summer of 1878, when infrared wasn't yet called infrared — its discoverer, William Herschel, had used the term "calorific rays" — an unnamed commentator in *Scientific American* described Thomas Alva Edison's ambitious proposal for a sky survey of invisible sources of heat. To carry out the survey, Edison's own heat-sensitive astronomical invention, the tasimeter, would be attached to a large telescope in order to "explore those parts of the heavens which appear blank":

> Hitherto science has given no hint of the possibility of exploring the vast and mysterious beyond, from which no visible ray of light has ever been detected, or is ever likely to be detected, by the most far-reaching and sensitive of optic aids. But now there comes a promise of an extension of positive knowledge to fields of space so remote that light is tired out and lost before it can traverse the intervening distance. . . . If at any point in such blank space the tasimeter indicates an accession of temperature, and does this invariably, the legitimate inference will be that the instrument is in range with a stellar body, either non-luminous or so distant as to be beyond the reach of vision assisted by the telescope. . . . Possibly too it may bring within human ken a vast multitude of nearer bodies — burnt out suns or feebly reflecting planets — now unknown because not luminous.[18]

Fast-forward to the USA during the Cold War. Ballistic missiles are installed far and wide. Broad sky surveys have already been done, but nothing big at infrared wavelengths.

In 1963 the Air Force creates an Infrared Physics Branch within its research laboratories, and the Infrared Celestial Backgrounds program begins.

Not every researcher turned directly to the Air Force for funding. In 1965, for instance, two enterprising astrophysicists at Caltech embarked on an infrared sky survey tuned to a wavelength of 2.2 millionths of a meter. That's 2.2 microns in astrophysical parlance, where one micron equals about one-twentieth the width of a human hair. At this band, Earth's atmosphere happens to be 80 percent transparent. Some war-surplus hardware, a ground-based telescope, a homemade five-foot reflector, and NASA funding enabled the investigators to produce a catalogue of the brightest 5,600 objects in the Northern Hemisphere skies, many of them never seen by visible-light telescopes and a number of them stunningly gigantic and distant. Under the title *Two-Micron Sky Survey*, NASA published their work in 1969.[19]

The next major infrared survey, the *AFCRL Infrared Sky Survey*, did have the Air Force's imprimatur. This was a true military–astrophysical collaboration, conducted in the late 1960s and early 1970s under the auspices of the Air Force Cambridge Research Laboratories at Hanscom Air Force Base in Massachusetts and sponsored in part by the Advanced Research Projects Agency. This time the telescope was built by Hughes Aircraft. The survey itself, implemented via rockets prepared and launched by the US Naval Ordnance Missile Test Facility at White Sands, observed the sky through three longer-wavelength bands than in the prior survey — four, eleven, and twenty microns — and resulted in a catalogue of 3,200 objects covering almost 90 percent of the sky. Therein lies just one of the many advantages of an orbiting telescope: it can access the entire sky, both Northern and Southern Hemispheres. A notable feature of both this survey and its predecessor is that the published report was not classified, so all scientists, no matter what they may have been investigating, would have open access to the data.[20] The same held true for a comprehensive follow-up study, published in 2003: the Two Micron All Sky Survey (2MASS), covering 99.998 percent of the sky and providing IR brightness and co-ordinate data on 471 million objects.[21]

Stephan Price, co-author of the *AFCRL Infrared Sky Survey,* writes that for most of his half-century career in infrared astronomy he was supported by the Air Force, primarily through the AFCRL — and was glad of it, because he not only found himself in a position to do " 'cutting-edge' research that was personally highly rewarding" but also found "the related practical Air Force space surveillance problems both interesting and challenging." His detailed history of the close postwar partnership between astrophysics and the military brims with references to corporations, universities, branches of the Department of Defense, distinguished investigators, significant discoveries, and the intricate braid of military needs and astrophysical quests. Price also chronicles the continual bureaucratic reshuffling and renaming within military research structures, as well as the effects of the Mansfield Amendment to the FY1970 Military Procurement Authorization Act, which mandated that the Department of Defense could not use its funds "to carry out any research project or study unless such project or study has a direct and apparent relationship to a specific military function." Passed in late 1969 during the Vietnam War "in the context of the general public disenchantment with both science and the military at the time," the amendment, intended to trigger increased scrutiny, briefly led to reductions in personnel and reorganization of responsibilities.[22] The FY1971 authorization act turned the amendment's intentions upside down. Now funding decisions would be based on "the opinion of the Secretary of Defense." The secretary, a member of the president's cabinet, would be free to opine on whether projects had a "potential relationship to a military function or operation." The words "direct," "apparent," and "specific" were gone from the legislation.[23]

Whatever the true effects of the Mansfield Amendment, Price's account shows the enduring strength and breadth of military support for both basic space science and utilitarian space science. Few projects could be of greater direct use to both the military and the astrophysicists than an infrared sky map, though of course military support flowed to other infrared projects as well, both before, during, and after the amendment.

Martin Harwit, former director of the National Air and Space Museum and an IR astronomer himself, writes that the development of infrared detectors — the instruments without which IR astronomy could not exist — "was largely guided not by astronomers, but by military needs, such as 'night vision' enabling warm objects to be discerned in the dark."[24] Science historian Ronald E. Doel agrees, referring specifically to US research on planetary atmosphere, the region on Earth through which every ballistic missile must pass:

[T]hese programs introduced astronomers to military agencies eager to fund astronomical research. . . . [M]ilitary patronage helped maintain the viability of American observatories in the lean years of 1946 and 1947. However, military contract funding also encouraged researchers to design proposals with short-term solutions. Those projects that did not achieve these promised ends faced heightened risk of disruption or discontinuance, regardless of their scientific merit; this bound researchers more closely with military missions as the cold war deepened.[25]

But it was the all-sky survey that provided data on the largest possible scale. The infrared sky map characterized the enduring cosmic backdrop against which a real-time incoming threat, whether a ballistic missile or an asteroid, must be distinguished. In war or in peace, this intelligence was, and remains, vital to national security.

In fact, sky surveys have yielded military value not merely because of their infrared info. Take the Sloan Digital Sky Survey, an unprecedentedly ambitious wide-area survey designed to gather ultraviolet, visible light, and infrared brightness readings for hundreds of millions of stars and galaxies, and spectra for millions more. To achieve its ends, SDSS uses a single-purpose telescope at the Apache Point Observatory in New Mexico and a unique calibrated-measurement system and data pipeline invented just for these observations. Begun in the 1990s (and, as of 2018, deep into its fourth survey) and funded in part by the Alfred P. Sloan Foundation, this gargantuan undertaking by hundreds of investigators and dozens of institutions around the world has outstripped all previous ground-based sky surveys in accuracy, scale, and value to astrophysicists.

From its inception, SDSS's central tasks were daunting: the management and analysis of the prodigious quantities of raw data obtained by the telescope. Innovative software to the rescue. So clever, efficient, and effective were the algorithms to turn the light of cosmic objects into analyzable data that SDSS research papers on the analytics of astrophysical data streams appeared not in astrophysics circles but at the 2012 International Conference on High Performance Computing, Networking, Storage and Analysis and the 2014 IEEE High Performance Extreme Computing Conference. The US Department of Defense ultimately took note and, in a reversal of the more usual direction of requests for assistance, asked one of the sky survey's project leaders, Alexander S. Szalay — Johns Hopkins professor of astronomy and computer science as well as director of the Institute for Data Intensive Engineering and Science — to brief a key branch of the Pentagon on how SDSS processed and analyzed stupendous data flows obtained from images and spectra.[26] An instance of astrophysical inventiveness, spurred by the quest for more comprehensive knowledge of the universe and subsequently enlisted in the service of national security.

IV.

The high-energy profile of X-rays, like gamma rays, demands a telescope of very different concept and design from the ones that focus and detect visible light. X-rays are among the several bands of light that do not reach Earth's surface. Our atmospheric layer of ozone simply and completely absorbs them. In the absence of observing platforms above Earth's atmosphere, X-ray phenomena in the universe go unnoticed.

Enter the Italian-born American astrophysicist Riccardo Giacconi. Beginning in the late 1950s, he applied himself to the task of perfecting such a telescope. "Until the space age came about and we could put instruments on satellites and rockets," he said later in life, "we couldn't find out what was out there. So by looking in X-rays, you are seeing aspects of nature which we did not even suspect existed but which are very important in the formation, evolution, and dynamics of the structures in the universe."[27] Giacconi is credited, in fact, with fathering the field of cosmic X-ray astronomy.

In 1959, as a young scientist, Giacconi joined American Science and Engineering (AS&E), a company formed the previous year by a group of investigators from the Massachusetts Institute of Technology. In its early days, the company specialized in making scientific instruments for NASA. But while Giacconi and his team did space science — for example, photographing the Sun in X-rays and discovering the first stellar X-ray source[28] — AS&E began to branch out into medical and security technology. Today, the home page of the company's website highlights its assistance to military and law-enforcement personnel who face challenges at borders, at ports, and in conflict zones. AS&E systems facilitate such procedures as cargo screening, threat detection for military personnel, bomb detection, and drug interdiction. Security is the emphasis. X-rays are the enabler.

The hijacking of commercial airplanes, often American, presented an especially high-profile security challenge during the late 1960s and early 1970s. Cuba, which was subject to a Kennedy-era Cold War trade embargo rendering it off-limits to airline traffic from the United States, was a frequent destination for political dissidents until early 1969, when congressional hearings revealed that, after arriving in Cuba, hijackers were subjected to lengthy interrogation followed by hard labor. Soon the aims of air pirates expanded to include extortion of ransom money, political blackmail, and terrorist revenge. Worldwide in 1969 alone, there were eighty-six hijackings, an average of more than one every four days. American carriers were the most common target.[29]

Clearly the aviation industry needed to screen passengers and their luggage for weapons and explosive devices. AS&E, which had already developed X-ray telescopes for NASA and a parcel X-ray system for the US Postal Service,[30] was able to provide a machine to do precisely that. By late 1972, passenger screening stations had been set up in most US airports. Hijackings plummeted. In early 1973 a Nevada senator introduced legislation to require that, before boarding an aircraft, "all passengers and all property intended to be carried in the aircraft cabin in air transportation be screened by weapon-detecting procedures or facilities before boarding." It became law in 1974.[31] Thenceforth all carry-ons would be scanned at all airports. AS&E scanners were everywhere.

During this period and continuing for several decades, Giacconi would be the principal investigator on four NASA X-ray telescopes, starting with the first one ever, Uhuru, launched in 1970, and continuing through the flagship observatory Chandra, launched in 1999. For pioneering the discovery of highly energetic phenomena in the universe, including black holes dining on stars that have orbited too close, and indeed for birthing an entire subfield of astrophysics, he would share the 2002 Nobel Prize in Physics and receive the 2003 President's National Medal of Science.

At that time, I was serving on the twelve-member National Science Foundation committee tasked with recommending the National Medal of Science recipients to the president. The awards ceremony, to which the committee is of course invited, is held annually at the White House; the ceremony for the 2003 winners took place in March 2005. That's when I met Riccardo for the first time, as we passed together through the visitors' foyer, a semi-detached security area adjoining the East Wing of the White House. Queuing to be scanned, screened, and scrutinized, we placed our belongings on the conveyor belt of an X-ray machine. Its maker? American Science and Engineering: AS&E.

V.

If you like the universe at all, you've probably seen many of the Hubble Space Telescope's gorgeous images of galaxies and nebulae. What you might not have come across is the fact that Hubble is basically a photoreconnaissance satellite whose cameras point upward at the heavens rather than downward at Earth.

During the late 1980s and early 1990s, Eric J. Chaisson served as a senior scientist and director of educational programs at Hubble's "scientific nerve center," the Space

Telescope Science Institute in Baltimore. Just below his original preface to his 1994 book *The Hubble Wars* is a note that reads like a legal disclaimer:

> No part of this book divulges sensitive military-intelligence material not previously having entered the public domain. I have been scrupulous about neither identifying reconnaissance assets unknown to the public nor disclosing the specific capabilities of any known yet classified project.[32]

Right off the bat, the reader is alerted to the largely unspoken and unseen military aspects of this signal achievement of human ingenuity, an instrument that most people around the world know only as a gateway to the glories of the cosmos. But once Chaisson reveals the connections between the Hubble Space Telescope and a certain spy satellite in the top-secret KEYHOLE series, it becomes clear why he added his disclaimer. A couple of decades later he wouldn't have needed to be quite so cautious, since that particular KEYHOLE was declassified in 2011 and put on view for one whole day at the Smithsonian's Air and Space Museum.

When a military program is secret or top secret, mentioning its existence or its codename is verboten. "KEYHOLE," according to a 1964 security memo, was the name given to "the product obtained from U.S. reconnaissance operations from satellites."[33] Capsules of exposed film dropped back to Earth were the main but not the only product; SIGINT (signals intelligence), based on the monitoring and interception of radar and electronic communications, was the other product. The name KEYHOLE is now also given to the overall camera system used by the reconnaissance satellites or, more generally, the satellites themselves. To add another layer of obfuscation, the early KEYHOLEs were part of the CORONA program, overseen by the CIA.

The KEYHOLE that Chaisson obliquely referenced was, it may now be said openly, the jumbo, sixty-foot-long KH-9 HEXAGON. The National Reconnaissance Office — whose very existence was classified for thirty years, until 1992 — launched twenty of them between 1971 and 1986. Repeatedly the KH-9 has been described as being either the size of or larger than a school bus. The same comment has often been made about the Hubble, although the Hubble is a little shorter and less massive than the KH-9. Even before its 2011 declassification, writers would comment every now and again that the KH-9 (a.k.a. Big Bird) looked like a twin of the Hubble.

Not a coincidence. Both could fit equally well lengthwise into the now-retired space shuttle's cargo bay or atop a heavy-lift, Titan-class rocket. Both were fitted with long, narrow solar arrays angled away from their bodies. The biggest differences between the two were that the Hubble focuses at infinity and takes prolonged exposures of extremely dim and distant objects, while the KH-9 focused mostly between one hundred and two hundred miles down on Earth's surface and took quick exposures. When Hubble points at Earth (which it does only occasionally, to help calibrate the telescope's cameras), it registers only smudges and blurs because it cannot focus that closely. When the KH-9's precision mapping camera and twin panoramic rotating cameras pointed at Earth (mostly at the Soviet Union), they registered features such as missile silos, shipyards, airfields, rocket test facilities, submarine bases, even an ICBM under construction, with a resolution of two feet and a horizontal range of more than four hundred miles. In addition, Hubble carries no fuel, whereas the KH-9 had plenty of fuel so that it could change course and make multiple passes over sites of interest.[34]

Early in its post-launch life, Hubble exhibited a bad case of the jitters — bad enough that its capacity to do the long, steady exposures required by scientific research would be seriously undercut if a cure couldn't be found (it was). Orbiting Earth once every ninety-six minutes, Hubble shuddered each time it entered or exited orbital night (total darkness) after spending forty-eight minutes in orbital day (blazing sunlight). Thirty times every Earth day, Hubble pitched, oscillated, and wobbled as unobstructed heat from the reappearing Sun increased the temperature in some parts of the telescope by more than a hundred degrees Celsius within ninety seconds. Then, as the Sun sank from view a mere forty-five minutes later, the telescope would cool rapidly. The result? Hubble couldn't track a target for more than ten minutes at a stretch.

But the main culprit wasn't the telescope itself, which is swathed in heat-reflecting Mylar. Blame the huge pair of solar arrays. Projecting well away from the vehicle and attached only at their center, framed in stainless steel rods and not positioned at the vehicle's center of mass, they bent and flapped too freely. Hubble's built-in compensatory measure — a hull that moves in the direction opposite to the array's displacement — couldn't fully overcome the unfortunate tendency of stainless steel to warp when assaulted by a sudden change in temperature. As soon as sunlight hit the array, the exposed side of the rods shot up to about 50 degrees Celsius, while the side that remained in shadow stayed at about–80 degrees. Each array, as Chaisson describes it, turned into a giant banana, a forty-foot longbow.[35]

As the go-to guy for the media, the White House, and anyone else who wanted substantive scientific information on the status of Hubble, Chaisson presumably knew how to be evasive yet accurate and how to keep control of what got said in public. Soon after the jitters were recognized, but before their cause was diagnosed, he was asked to appear at a closed meeting with several dozen officials involved in military intelligence work. The meeting was labeled "SECRET" and would be held at a secure location. Here's how Chaisson describes what happened:

> [W]hen I discussed the jitter enigma we were experiencing with *Hubble,* I was astonished to see so many nodding heads. Right then and there, midway through the briefing, a rage came over me. I felt like shouting, "Damn it, why didn't you tell us!" For, apparently, these people — some of whom were *Keyhole* controllers — had years ago first noticed specifically this problem. . . .
>
> Later that evening . . . I was stopped by a serious-looking person sporting short hair, gray suit, ID leash around his neck, and absolutely radiating that woods-are-lovely-dark-and-deep demeanor. He told me the name of someone to contact at Lockheed who, he said, might be able to help us. At

which point the intelligence operative did an about-face and marched away.

Having passed along that name to the right person and plumbed other obvious channels, Chaisson began to grasp the closeness of the connections between Hubble and the series of twenty "Hubble-class" vehicles of whose existence he was aware. But since those vehicles, the KH-9 HEXAGONs, remained classified until 2011, Chaisson would have been unaware in the 1990s of salient facts about them. As an Air Force intelligence officer saw fit to inform him one day at the Naval Academy, the Hubble was a KEYHOLE-class satellite, not the other way around.

In any case, Chaisson came to realize that neither ill will nor turf wars was the reason the military hadn't proactively shared its lessons learned:

The jitter problem had been known for several years prior to *Hubble*'s launch, but reconnaissance analysts were not bothered by it, largely because they had never needed to expose their surveillance cameras for more than a fraction of a second. Whether peeking, for example, at work in progress at the Krasnoyarsk radar site, or sensing how many infrared-emitting people inhabit a specific tent outside Tripoli, spying spacecraft need not take long exposures.

They can quickly gather their data whether the spacecraft is stable or oscillating. [As a result,] they probably did not pass along knowledge of it simply because it did not affect their landscape. [T]he industrial contractors had compartmentalized their sensitive intelligence work so thoroughly that there was little or no cross-fertilization — and the civilian world was the loser.[36]

By the way, the Air Force wasn't the only branch of the military whose projects overlapped with Hubble. Another was the National Reconnaissance Office, the agency in charge of America's spy satellites. More unitary in its mission than the much larger USAF, the NRO proactively aided one of NASA's foremost future civilian eyes in the sky — an instrument superior to Hubble.

Every ten years, the National Academy of Sciences facilitates a committee of US astrophysicists to prioritize spending on projects for the upcoming decade — a process that establishes consensus in the field and precludes public arguments about whose pet project should receive federal support. Astrophysicists who raise their own money, personally or institutionally, can spend it however they wish, but when it's time to allocate federal or other shared funds, we follow the priorities of the report. Similar committees in earlier decades gave top ranking to the Very

Large Array in New Mexico, the Hubble Telescope in orbit, and, most recently, the Atacama Large Millimeter/submillimeter Array in Chile.

In 2010 I served on one of these committees. Our final report, *New Worlds, New Horizons,*[37] pegged the spaceborne Wide Field Infrared Survey Telescope (WFIRST) as a number one priority. It promised to revolutionize infrared observations of the universe, whether of nearby exoplanets or of distant galaxies, and, in NASA's words, "to settle essential questions" about dark energy. To make WFIRST happen meant we needed to drum up federal funding for the telescope's costly new mirror and detectors.

Enter the NRO.

Fortunately for the future of astrophysics, the agency happened to have on hand two surplus, freshly declassified, Hubble-size but better than Hubble-class telescope mirrors and, in 2011, offered to donate them to NASA — stripped of their military-grade detectors.[38] NASA, grateful for the gift, could now cross one big budget item off its fund-getting list. Why were these awesome mirrors now available? Because the NRO had begun to use even better ones.

*Un*fortunately for the future of astrophysics, the White House's FY2019 budget request completely eliminates funding for WFIRST, on the grounds that "developing another large space telescope immediately after completing the $8.8-billion James Webb Space Telescope is not a priority for the administration."[39] Let's put that in context. For many years, NASA's budget — covering all ten NASA centers, the astronaut program, the International Space Station, and all space probes and spaceborne telescopes, including Hubble — has been less than one-half of one percent of the federal budget. A year and a half's FY2019 proposed funding for the Department of Defense roughly equals the entire run of NASA funding across the agency's sixty-year history. How much is the universe worth to the president? How much is national security worth to Congress? How much is knowledge of our place in the cosmos worth to the electorate?

As for Hubble itself, once the several post-launch problems were remedied via both emergency and preplanned servicing missions, the telescope was able to begin its working life as a distinguished detective and impresario. On its roster of revelations are the age of the universe; exoplanets; supermassive black holes lurking at the heart of brilliant galaxies; embryonic planetary systems swathed in previously impenetrable disks of gas and dust surrounding young stars; a patch of sky chosen for how devoid of interesting galaxies it was when viewed with ordinary ground-based telescopes but which, after a ten-day exposure by Hubble's camera, showed itself to be populated by thousands of distant galaxies dispersed to the edge of the universe. Hubble was adored not only by scientists but by civilians, who in 2004 took ownership of it. When NASA proposed to cancel the telescope's final servicing mission, the outcry from the general public was greater than that from the scientists. Congress relented, and the mission was reinstated. Hubble's successor, the infrared-tuned James Webb Space Telescope, has, as far as we know, no military doppelgängers — yet.

7
MAKING WAR, SEEKING PEACE

Space is a physics battleground. Gigantic magnetic fields loop through the frigid emptiness. Bursts of plasma erupt from the surfaces of suns. Black holes flay and swallow every object that wanders near. Cosmic rays, gamma rays, and X-rays devastate any speck of living matter in their path. The infancy and youth of every planet consists of a ceaseless hail of rocks. Every day, millions of gigantic stars across the universe blow their metal-rich guts to smithereens, sending shockwaves and radiation across the light-years. Whole galaxies, each containing hundreds of billions of stars, collide and merge, just as will happen with our own Milky Way, doomed to meet and greet the Andromeda galaxy several billion years from now. Here in our solar system, a hundred-meter-wide asteroid sails into Earth every millennium or so at speeds upward of fifty thousand miles an hour, generating a destructive impact equal to 2,500 atomic bombs.

Some members of the human species have wanted to augment all that naturally occurring cosmic mayhem with some space apocalypses of their own doing. Barely had World War II ended when they embraced an even more devastating near-term scenario: the visiting of intentional nuclear disaster across the entire surface of Earth. Thus began a military shopping spree that continues to this day. By now the wish list is quite long.

Space war could take two main forms: direct physical attacks or cyber sabotage. Indeed, today's Air Force Space Command speaks of "space and cyberspace" in the same breath. Cyberwar wouldn't require a physical weapon, only a focused disruption. The seventeen hundred–plus operational satellites that circle Earth are the most obvious potential target. Nearly half of these are American, of which one-fifth are military, supporting contemporary technologies of warfare.[1] As for the remaining satellites, the daily life of nearly every person in the world, but especially in the United States, depends on more than one of them, knowingly or unknowingly, directly or indirectly. Disable enough satellites — by whatever means — and people suddenly can't use their credit cards. They have to reacquaint themselves with paper roadmaps and quickly unlearn their expectations of a reliable power grid and minute-by-minute updates on the weather.[2]

Think of cyberwar against space assets as weaponless sabotage — though "weaponless" can be hard to define, since almost anything, from a hand to a fork to a truck to a plane, can be and has been used as a weapon. Cyberwar's potential reach is broader than that of all but the most unthinkable weapons.

"Space capabilities have proven to be significant force multipliers when integrated into military operations," state the Joint Chiefs of Staff, who stress space situational awareness, strategic deterrence, cyber support, and weaponless cyber interventions rather than space-to-space, air-to-space, or ground-to-space physical destruction — the kind of destruction that would result in huge new batches of space debris.[3] Obviously, if shards and chunks of exploded satellites threaten anything and everything in their path, they're as likely to disrupt one's own space assets as the enemy's. No technologically advanced country welcomes the prospect of being thrown back to the days of the wax candle, the water well, and the electric telegraph, and so, compared with the other options, limited cyberattacks and non-nuclear space-to-ground destruction start to look positively reasonable.

At the same time, the military knows its purpose, and that purpose does not end with awareness and deterrence. The commander of Air Force Space Command is clear about the mandate: "Our job is to prepare for conflict. We hope this preparation will deter potential adversaries and that conflicts will not extend into space or cyberspace, but our job is to be ready when and if that day comes."[4]

In modern times, who are these potential adversaries? Notably China, China, Russia, Russia, and China. Even the most cursory Web search yields extensive evidence of America's alarm about the speed and scope of China's stunning achievements and ambitions in space. The Department of Defense's 2016 annual report to Congress concerning the Chinese military says that China "has built a vast ground infrastructure supporting spacecraft and space launch vehicle (SLV) manufacturing, launch, C2 [command and control], and data downlink" and that it "continues to develop a variety of counter-space capabilities designed to limit or to prevent the use of space-based assets by . . . adversaries during a crisis or conflict."[5] China's own military scorecard of 2015 reiterates the nation's "strategic concept of active defense," including "adherence to the doctrine that 'We will not attack unless we are attacked, but we will surely counterattack if attacked.' "

China also voices alarm about the scope of its adversaries' space achievements and ambitions: "Outer space has become a commanding height in international strategic competition. Countries concerned are developing their space forces and instruments, and the first signs of weaponization of outer space have appeared." Responding to the perceived hostile conditions in language not very different from that of its adversaries, China vows to "keep abreast of the dynamics of outer space, deal with security threats and challenges in that domain, and secure its space assets to serve its national economic and social development, and maintain outer space security."[6] The rhetoric resonates with America's own ambitions in space, although the long-lived US theme of "space superiority" is absent.[7] As for space security, in the summer of 2016 China took a great leap forward in that direction when it launched the world's first quantum satellite, which offers the promise of eventual hack-proof communications for everything from your pet food purchase to the military's surveillance operations.

Say a country has stashed a few ballistic missiles, missile interceptors, and high-energy lasers around the globe to discourage attacks on its own satellites. Now, if it chooses, it can also readily attack another country's satellites, however unwise such a move would be. If that same country adds some surveillance/reconnaissance platforms and satellite communication jammers to the mix, it will have what the United States calls counterspace: defensive as well as offensive measures meant to enable military "agility" and "resilience capacity" for the purpose of ensuring "space superiority."[8] Existence of and access to these technologies enables actions that would be impossible without them, just as access to a military-style semi-automatic assault rifle enables actions that a knife does not.

When most people hear the phrase "space war," they're not thinking weaponless cyber sabotage. They're thinking actual, powerful weapons causing colossal explosions hundreds or thousands of miles above Earth's surface.[9] While possession of an arsenal is not synonymous with war, it can prove either a prelude to war or war's strongest deterrent. A stockpile of bombs, missiles, and lasers is a stockpile of bombs, missiles, and lasers, whether acquired in the name of deterrence, protection, or attack. It can be deployed in both offensive and defensive actions. The difference is not inherent in the weapons themselves.

One de facto category of space weapon has nothing to do with intentional deployment: space junk. It's already up there, the inevitable but inadvertent result of smashups, explosions, rocket launches, space-walk maneuvers, the ordinary dumping of trash, and the inevitable demise of assorted spacecraft. From a distance, it looks like a cloud of dandruff ringing our planet, mostly in low Earth orbit, because that's where most satellites are found. But space junk populates all of nearby space, extending six Earth radii out to the zone of geosynchronous satellites.[10] Besides a few notable mementoes of the late 1950s, such as the final stage of the launch rocket for the USSR's first Sputnik and the entirety of America's first Vanguard, hundreds of thousands of unguided bits of flotsam and jetsam orbit Earth amid our working satellites. Included in the debris are a couple of cameras, a dropped wrench, a glove, multiple bags of garbage, and blobs of unspent rocket fuel. All harmless until they plow into the belly of a satellite or space station at an impact speed as much as ten times faster than a rifle bullet. At that speed, even a paint chip causes real problems. As with our planet's so-called Great Pacific Garbage Patch — a continent-sized region in the Pacific Ocean that's infused with suspended plastic fragments, dumped cargo, fishing nets, and chemical sludge — space garbage is a growing risk, unpredictable and largely unmanageable, at least until some technical breakthroughs give humans more control. Currently the Space Surveillance Network tracks twenty thousand pieces of debris that are grapefruit-sized or larger; half a million more are smaller than a grapefruit but larger than a cherry; millions more are smaller still.

Given the many capabilities, threats, and challenges created by the human presence in space, rational people have long mobilized against warfighting beyond the clouds. The most concrete results of these efforts are a handful of international treaties and resolutions, some of them voluntary. Legal instruments and voluntary agreements, of course, present very low hurdles to anybody willing to prosecute a fight by every means available. All-out war in space is still a hypothetical. But it's certainly on many drawing boards, and many of the weapons that would be used to wage such a war — whether under the banner of deterrence, denial, or destruction — already exist in some form or are in well-funded development.

In past military parlance — notably in *On War,* by the early-eighteenth-century Prussian general Carl von Clausewitz — war was referred to as an art, firmly rooted in strategy and heavily dependent on the wise planning and psychological astuteness of a commander. The presumption was that warriors were fierce and strong and their weapons deadly. But the nature of the weapons themselves was often subordinated to questions of how and when they would be used. Sun Tzu, the oft-quoted sixth-century BC Chinese general, barely mentions weapons in his *Art of War,* singling out only the use of fire.[11] By the beginning of the second millennium AD, detailed writings about weaponry, such as gunpowder-propelled fire arrows and rockets for military use rather than fireworks displays, were appearing in both East and West.

Common early weapons such as arrows, axes, clubs, swords, scythes, and spears were grasped in the hand and meant for close combat. The *Iliad,* epic chronicle of the final year of the thirteenth-century BC Trojan War, offers a cornucopia of grisly details of death by spear:

> and the bronze spear-point plunged in his
>   brow, then penetrated bone;
> darkness covered his eyes,
> and he fell like a tower in the mighty
>   combat.
>
> . . .
>
> he struck him down through the right
>   buttock; straight through
> into the bladder under the bone the
>   spear-point passed;
> he dropped to his knees screaming, and
>   death embraced him.

as well as by rock:

> But the son of Tydeus took in his hand
> a boulder, a great feat, which two men
>   could not lift,

. . . with this he struck Aeneas on the hip
  joint . . .
And the jagged stone crushed his hip
  socket, snapped the tendons on both
  sides,
and forced the skin away.[12]

With the emergence of fortifications and
ship-to-ship or ship-to-shore engagement,
there arose a need for new kinds of weapons
that would be effective at greater distances.
Artillery superseded the arrow and the spear.
The challenges of distance reputedly also
gave rise to practices such as catapulting
beehives or diseased corpses over a city wall,
or even leaving obvious caches of not-
obviously-poisonous honey to be consumed
by an enemy's advancing troops — early versions of bio-warfare. Such innovations in
weaponry sometimes proved as important as
a commander's clever strategizing.

By the late fifteenth century, Leonardo da
Vinci would conclude that the best way to
recommend himself to a potential patron, Lu-
dovico Sforza, the Duke of Milan, was to
foreground his manifold skills in designing
the machinery of battle, from portable bridges
to "big guns, mortars, and light ordnance of
fine and useful forms . . . and other machines
of marvellous efficacy and not in common
use," adding only as an afterthought, "I can
carry out sculpture in marble, bronze, or clay,

and also I can do in painting whatever may be done, as well as any other, be he whom he may."[13] Helping to equip armies promised to be a surer way for Leonardo to support himself than painting portraits and religious murals — a pattern that has held across the ages.

Commentators often invoke Clausewitz's famous dictum concerning the nature of war — that it is "a mere continuation of policy by other means"[14] — but war and weapons can also be considered as problems in physics. Virtually all weapons ever devised are means of moving energy from here to there. "Here" is the device on one side of the conflict; "there," some distance away, is the enemy or the enemy's property. The device can be a boomerang, a bullet, a catapult, a cannonball, a harpoon, a trident, a grenade, a ballistic missile, a bomb, a laser. The energy can be in the form of kinetic mass, fissionable material, explosives, incendiary chemicals, light. The physical agenda — omitting all considerations of politics, law, religion, commerce, history, hatreds, honor, and the like — is to deliver that energy to a preselected location, where it can kill people and break things.

Nonbiological, nonexplosive modern weapons are of two main types: those that propel a certain amount of mass at high speed against a target (kinetic-energy weapons), and those that send destructive energy — chemical, nuclear, or electromagnetic — against a target (directed-energy weapons). A barrage of bullets fired at a line of soldiers a hundred yards away is a kinetic-energy weapon, while a downward-pointing laser aimed by an orbiting satellite at the main generator of a city's water-purification plant would be a directed-energy weapon. Or how about examples from science fiction? Both *Star Trek*'s photon torpedo and *Star Wars*' proton torpedo are kinetic-energy weapons that carry explosive warheads, while *Star Trek*'s ship-mounted and handheld phasers are directed-energy weapons. *Star Wars*' classic personal weapon, the light saber, cleverly combines the directed-energy weapon's futuristic capacity to annihilate at a distance with the ancient practice of hand-to-hand combat.

Through most of history, weapons tended to rely on kinetic energy, whether of a twenty-pound rock, a thirty-two-pound cast-iron sphere, or a twenty-gram lead bullet. Tomorrow's hypervelocity tungsten rods — tall, slim, massive, fairly radar-proof — are a fearsome, though largely fictional, space-age kinetic weapon that would be discharged from a satellite. The destructive potential of these "Rods from God" would be supplied by their gravity-accelerated descent from Earth orbit to Earth's surface.

Sometimes a weapon combines kinetic

energy with another kind of energy. The kinetic energy of a white-phosphorus-filled exploding grenade hurled into a crowd will bruise anyone it hits, but the highly incendiary chemical energy contained within it is what will do the real damage. The kinetic energy of a long-range or intercontinental ballistic missile (ICBM) that falls on a distant city will destroy a building simply by virtue of its mass and speed, but if the ICBM carries a nuclear warhead or two, as most have done since the 1960s, it can destroy entire cities.

Until the space age, weapons largely depended on mass and speed to do damage to their targets. Only when space became a potential domain of warfare, and lasers became the ultimate concentration of light, did nearly massless energy — cheap to launch and inherently able to move at the fastest speed in the universe — become the dream weapon for the highest-altitude battle zones.

Lasers rank high on the wish list for a space arsenal. The laser is the quintessential directed-energy weapon: a needle-thin but intensely strong beam of light that can be precisely aimed at a narrow target. While civilian versions are sufficiently mild to be used as lecture pointers, more powerful versions are used for eye surgery, cosmetic hair removal, and printing. Military versions are meant to range from highly damaging to lethal. There are also peaceable, scientific lasers. One such is perched on NASA's Curiosity rover, which has been trundling across the surface of Mars since 2012. Whenever Curiosity encounters an interesting rock formation, its Chem-Cam laser instrument aims a series of brief million-watt pulses at it. While the laser vaporizes the target area, a camera picks up the flash and determines the chemical composition of the target, the ultimate goal being to assess the Red Planet's habitability.

The word "laser" is an acronym, derived from the phrase "light amplification by stimulated emission of radiation." And the laser has a sibling, the maser, with an "m" standing in for "microwave."[15] Neither occurs naturally here on Earth. Both result from the interaction of photons with specific atoms whose electrons both absorb and emit exactly the same kinds of photons. The task at hand is to generate as many of these identical photons as possible and send them out through a hole. In a laser, the photons of a single frequency accumulate in a customized cavity and are emitted in resonance with one another, with all the crests and troughs of the light waves aligned. Physicists call that state "coherence," and it is singularly responsible for the focused intensity of a laser beam.

Literature — specifically, H. G. Wells's 1898

novel *War of the Worlds,* the inspiration for multiple movies and other spin-offs — was the birthplace of the laser, in the form of a death-dealing beam called the Heat-Ray, which a Martian invasionary force that lands near London "pitilessly flourishes" against anything and anyone in its path. The armor-clad globlike invaders — hundred-foot-tall "boilers on stilts . . . striding along like men" — wield an "invisible sword of heat" emitted by a "camera-like generator." One soldier who has witnessed the effects of the Heat-Ray at close range says of Britain's twelve-pound guns that have been set up to repel the invaders, "It's bows and arrows against the lightning, anyhow."[16]

During the decades that followed publication of *War of the Worlds,* especially those following the uneasy conclusion of World War I, the Heat-Ray morphed into the more general "death ray" and became a recurring warrior fantasy/nightmare. In 1924, during his antiwar days on the backbench, Winston Churchill warned in a widely reprinted magazine article that among the weapons in a coming war would be "electrical rays which could paralyze the engines of motor cars, could claw down aeroplanes from the sky and conceivably be made destructive of human life or vision."[17] Britain's upper-echelon defense planners were not immune to the seductive potential of the fantasy, and Churchill, after becoming a vocal advocate of defensive technologies, urged its implementation. A. P. Rowe, assistant to the director of the Air Ministry's Directorate of Scientific Research and himself soon to be director of the Telecommunications Research Establishment, described the British military's response:

> For many years the "death ray" had been a hardy annual among optimistic inventors. The usual claim was that by means of a ray emanating from a secret device (known to us in the Air Ministry as a Black Box) the inventor had killed rabbits at short distances and if only he were given time and money, particularly money, he would produce a bigger and better ray which would destroy any object, such as an aircraft, on to which the ray was directed. Inventors . . . invariably wanted some of the taxpayers' money before there could be any discussion of their ideas. The Ministry solved the problem by offering £1,000 to any owner of a Black Box who could demonstrate the killing of a sheep at 100 yards, the secret to remain with its owner.
>
> The mortality rate of sheep was not affected by this offer.[18]

Interest in a death ray persisted nonetheless. With radio-wave transmissions and the electrification of the world's cities proceeding apace, Henry Wimperis, Rowe's superior at the Directorate of Scientific Research, wrote that he was "confident that one of the coming things will be the transmission by radiation of large amounts of electric energy along clearly directed channels. If this is correct the use of such transmissions for the purpose of war is inevitable." The assumption was that radio waves would provide the energy, and so Wimperis asked radio researcher Robert Watson-Watt in mid-January 1935 to come up with an answer to what Rowe describes as "the problem [of] whether it was possible to concentrate in an electromagnetic beam sufficient energy to melt the metal structure of an aircraft or incapacitate the crew." Avoiding any reference to planes or people, Watson-Watt handed the problem to an underling, Arnold Wilkins, asking him in a scribbled note to "calculate the amount of radio-frequency power which should be radiated to raise the temperature of eight pints of water from 98°F to 105°F at a distance of 5 km and at a height of 1 km." As eight pints is the amount of blood in an average adult human male, Wilkins wasn't fooled by the obfuscation:

It seemed clear to me that the note concerned the production of fever heat in an airman's blood by a death-ray and I supposed that Watson-Watt's opinion had been sought about the possibility of producing such a ray.

My calculation showed, as expected, that a huge power would have to be generated at any radio frequency to produce a fever in the pilot of an aircraft. . . . it was clear that no radio death ray was possible.

But Wilkins suggested another idea to Watson-Watt: exploiting an earlier discovery by a couple of radio engineers that metal airplanes invariably interfered with radio communication. That interference, he said, amounted to an announcement of their presence, even when they couldn't be seen. Watson-Watt rushed off a memo to Wimperis, omitting the fact that Wilkins had supplied the information.[19] Thus was born the concept of radar.

Was the death ray dead? No, only the radio death ray. As Rowe opined in 1948, "The idea of a death ray however was not absurd and something of the kind may come within a hundred years."[20]

And so it has.

While astrophysical lasers are a rarity, astrophysical masers are semi-common. You can find one variety deep within colossal gas clouds scattered across spiral galaxies. In dense, bright, star-forming regions within those clouds, countless electrons that belong to molecules of hydroxyl (OH) or water (H_2O) or ammonia (NH_3) are primed to emit resonant photons.

Picture a large cavity within a blob of gas. Now picture the light from nearby stars bathing the region. The photons get absorbed by selected molecules. What happens next is part of the weird world of quantum mechanics. The same bath of photons that the molecules absorbed stimulates those same molecules to emit photons of the same wavelength — the same energy — mostly in the microwave part of the spectrum. Excite a gas with microwaves; the molecules of gas emit microwaves; the microwaves cause the molecules to emit more microwaves. Next, the microwave energy punches through the cloud, creating a powerful, concentrated beam that funnels out in one direction. Behold an astrophysical maser, aimed wherever the opening in the cloud happens to face.

Unlike their astrophysical relatives, human-made lasers must be pointed with exactitude. Bad aim brings disaster. A common ground-based, military-grade thirty-kilowatt laser (with its six million times the power of an ordinary laser pointer) can punch a hole in a truck engine or in the fuel tank of a booster rocket sitting on a launch pad.[21] Space-based lasers, once perfected, would be dominant and lethal. Deployment presents challenges, however. While in motion around Earth, the laser must generate and direct colossal power to a specific target, which is also in motion. What's more, the laser beam must be delivered swiftly, unattenuated by clouds and unmolested by atmospherics.

Coming up with the initial power is the first step, and there are many possible sources. A common chemical laser, for instance, depends on the conversion of stored chemical energy into intense infrared energy — harnessing the energy of molecules that are zealously engaged in a chemical reaction and then channeling that energy into a beam of light. A less benign option would be a small nuclear bomb or nuclear reactor investigated in the 1970s and 1980s for use in weapons such as the space-based X-ray lasers of the failed US missile-defense program Project Excalibur. Once the energy has been produced, you need a cavity of some kind that will hold and, ideally, further stimulate the already hyped-up molecules. A recent development is the fiber laser, in which extremely long, hair-thin, light-transmitting glass or plastic fibers, saturated with rare-earth elements, are the operative technology. Notable attributes of this kind of laser include its potential to offer huge amounts of power and to be bent into a compact shape. Furthermore, it's stable at high temperatures, and the inherent light-wave-guiding properties of a fiber produce an extremely precise, intense beam.

Once you've got the power, the medium, and the beam, you face the challenge of aiming. For that, you'll need an optical apparatus capable of high resolution, so that the target area can be clearly seen. Does that sound like a telescope? It should, because it is. Today, the largest optics (mirrors rather than lenses) are the ones developed for telescopes.

Early in the twenty-first century, a comprehensive report from the RAND Corporation, a think tank devoted mainly to US military policy, proposed that the optics required for a space-based laser weapon designed to destroy terrestrial targets could become available and affordable once the optics required for the "next-generation space telescope" had been mastered. Today's real-life next-generation space telescope, the seven-ton James Webb Space Telescope, has a 6.5-meter gold-coated mirror made up of eighteen separate hexagonal segments constructed from pure beryllium, a metal that's both strong and light. But all that high design doesn't come close to the report's estimate of what would be needed for a space-to-Earth laser: a mirror measuring more than ten meters across, millions of watts of available power, and the capacity to withstand overwhelming heat.

After all is said and done, according to RAND's researchers, much more progress would have to happen before a space-based laser becomes "feasible," let alone "reasonable."[22] Many other analysts arrived at the same position, including, a few years later, the authors of a report from the American Academy of Arts and Sciences, who concluded that the technology for a "usable" laser of this sort "does not currently exist and will not for the foreseeable future."[23] Even the people who themselves develop directed-energy technologies or oversee the acquisition of these technologies for the Department of Defense now cite time frames such as the 2020s for the first flight test of an aircraft-mounted 60–150-kilowatt laser.[24] In other words, space-based, long-distance, million-watt laser weapons remain the stuff of fantasy.

Aware of this mismatch between needs and expectations, RAND offered an interim solution: laser weapons aimed at space could be located on high, dry mountaintops down here on Earth. Yes, but . . . Those mountaintops happen to be the very places where we astrophysicists like to put our telescopes. Shouldn't be a problem, said the researchers — just put the weapon and the science in the same location, since both groups use laser beams to monitor and correct for atmospheric turbulence in their observations. The scientists "might even welcome the laser if its large optics could also be used to increase observing time when not needed for weapon operations, maintenance, or training," said RAND.[25] Given that astro-folk tend to be a peaceable crew, "accept" might be a more suitable word than "welcome."

Since its founding in 1948, by the way, the RAND Corporation has supported a fair bit of "thinking about the unthinkable."[26] To this and other ends, it has consistently hired impressive individuals, ranging from Daniel Ellsberg, who made public the Pentagon Papers, to Donald Rumsfeld, who served as a RAND trustee for a quarter century before becoming George W. Bush's secretary of defense. RAND's very first report, commissioned by Major General Curtis LeMay and published in 1946, bears the exciting sci-fi title *Preliminary Design of an Experimental World-Circling Spaceship* — i.e., a satellite — and its very first client was the Air Force.[27] In 1958 a RAND author, Robert W. Buchheim, produced a classified space tutorial, *The Space Handbook,* for the enlightenment of the House Select Committee on Astronautics and Space Exploration. The following year, Buchheim and the RAND staff updated and declassified the handbook for publication by Random House.[28] And by the beginning of the twenty-first century, RAND had generated hundreds of policy papers on space science, space exploration, and space warfare.

■ ■ ■ ■

Speaking on national radio and television in late March 1983, halfway through his first term of office, President Ronald Reagan drew an alarming picture of the defunding and decay of US military forces and military technology during the 1970s in contrast to the USSR's concurrent military buildup. According to Reagan, the imbalance was overwhelming and terrifying:

> For twenty years the Soviet Union has been accumulating enormous military might. They didn't stop when their forces exceeded all requirements of a legitimate defensive capability. And they haven't stopped now. During the past decade and a half, the Soviets have built up a massive arsenal of new strategic nuclear weapons — weapons that can strike directly at the United States.

With US military spending overdue for a major boost, the president was publicly announcing a new initiative "to counter the awesome Soviet missile threat with measures that are defensive." He first posed a rhetorical question: "What if free people could live secure in the knowledge . . . that we could intercept and destroy strategic ballistic missiles before they reached our own soil or that of our allies?" He then called for "the scien-

413

tific community in our country, those who gave us nuclear weapons, to turn their great talents now to the cause of mankind and world peace, to give us the means of rendering these nuclear weapons impotent and obsolete."[29]

In its report on the speech, the *New York Times* relayed statements from White House officials who said that "the new program might involve lasers, microwave devices, particle beams and projectile beams" — all of which, though still "in a very early stage of development," would be capable "in theory" of being "directed from satellites, airplanes or land-based installations to shoot down missiles in the air."[30]

The administration's vision of missile defense was formally called the Strategic Defense Initiative, or SDI, but given the release of the third film in the *Star Wars* trilogy two months after Reagan's speech, a colloquial rename was irresistible, and SDI was dubbed Star Wars. Its primary goal was to destroy a nuclear-tipped ICBM in the course of its swift journey in our direction, preferably soon after launch. It was to be a matter of interception: knocking out the other side's missile well before it reached our side and knocked us out.

That's quite a technical challenge, often and accurately described as a bullet hitting a bullet. Richard L. Garwin, a renowned physicist whose military work has ranged from the hydrogen bomb to spy satellites, warned that if the missile carried a hundred targeted bomblets rather than just one big bomb, an interception during the terminal phase (the release of all the bomblets) would fail. A successful interception could be done earlier, he added — up to about four minutes into the boost phase — if and only if the interceptor issued from somewhere nearby.[31] Burton Richter, recipient of the 1976 Nobel Prize for Physics, declared, "The intercept-in-space, hit-to-kill system now under development is the most technically challenging of the possible alternatives. . . . The proposed system is not ready to graduate from development to deployment, and probably never will be."[32] Journalists echoed the skepticism. In a piece about one SDI concept, Brilliant Pebbles — tens of thousands of small, smart orbiting rockets that would hurl ten-pound projectiles at incoming enemy ICBMs, fatally puncturing them — John M. Broder, then at the *Los Angeles Times,* wrote, "Any space-based anti-missile system confronts a difficult technological task — to identify ballistic missiles in the very early stages of flight, to pick out the rocket body from its large plume of fire, to track and then home in on the target before the weapons-carrying 'bus' separates and releases the nuclear warheads."[33]

Nevertheless, beginning in the 1950s, while

415

Ronald Reagan was still a Hollywood actor, both the USA and the USSR conducted considerable R & D on missile defense, pausing briefly in acquiescence to the 1972 Anti-Ballistic Missile Treaty but picking up again within a few years. During the decade that followed Reagan's announcement, Congress put $30 billion into the Strategic Defense Initiative Organization. In 1993, describing how well that money was spent, the organization's director, General James A. Abrahamson, said it had yielded "major hardware assembly and field experiments necessary to prove available technologies can be integrated together to operate as an effective defensive system in a hostile and reactive environment" and had caused "a sea change in our negotiations with the former Soviet Union and, by informed and authoritative accounts, the end of the Cold War." Indeed, claimed he, the evaporation of the need for Cold War spending had "more than repaid the $30-billion investment of the past decade in just a couple of years."[34]

Meanwhile, large quantities of sober analysis, concerned commentary, damning survey results, and carefully worded petitions — much of it issuing from Nobel laureates and other unassailable experts such as Richard Garwin and Carl Sagan, who felt obliged to demolish Star Wars' precepts, politics, and prospects — were rapidly accumulating. Possibly swayed by the scale of the opposition, Congress periodically reduced SDI funding and attached strings to it. The Pentagon responded by classifying information on the costs of SDI.[35]

Problem is, when more than 90 percent of the 450 physicists, engineers, and mathematicians in the National Academy of Sciences who answer a 1986 Cornell University questionnaire on SDI say that the technology would be unable to effectively defend the US population against a Soviet missile strike; and when 1,400 "scientists and engineers currently or formerly at government and industrial laboratories" send a letter to Congress declaring their "serious concerns" about SDI and their belief that its stated goal "is not feasible in the foreseeable future" and "represents a significant escalation of the arms race"; and when more than 3,800 senior faculty members of physics, computer science, and other "hard science" departments at "leading" US universities, including almost 60 percent of all faculty in America's "top twenty" physics departments, sign a pledge to reject funding from the Strategic Defense Initiative Organization — that's when it becomes hard to tout the achievements of SDI unless you're the guy in charge. Even some scientists who were doing the actual research told Senate staffers in 1986 that "there had been *no major breakthroughs*" that

417

would make comprehensive deployment possible by the late 1990s.[36]

In November 1987, three weeks before a Reagan–Gorbachev summit, a group called Spacewatch organized a debate titled "Is the Strategic Defense Initiative in the National Interest," in which Sagan and Garwin spoke against the emergent, non-Reaganesque version of SDI, while General Abrahamson and Richard Perle, then an assistant secretary of defense, spoke in its favor. The former pair offered physics and logic; the latter offered primarily politics and fear plus a couple of superficially reasonable statements, such as the validity of attempting a partial defense if a comprehensive defense proved unachievable.

In his opening remarks, Sagan points out that there are almost 60,000 nuclear weapons in the world, more than a third of them "designed to go from the homeland of one nation to the homeland of another." Since the world has only 2,300 cities of a hundred thousand people or more, there is obviously a "grotesque disproportionality between the power of the nuclear arsenals of the United States and the Soviet Union and any conceivable use." Having established the degree of nuclear peril, Sagan goes on to say that because Reagan's promise of population defense would be so hard to achieve, SDI's champions have given in to the "temptation to shift the ground, to invent more modest objectives." So,

> SDI is fine if it is perfect — that is, if no significant number of Soviet warheads leaks through the shield. The most optimistic numbers you can hear from technically competent advocates of Star Wars is 70, 80, or maybe even 90 percent of incoming Soviet warheads destroyed. Well, take the more optimistic number: If 90 percent are destroyed, 10 percent get through. Ten percent of, say, ten thousand warheads is one thousand warheads. One thousand warheads is much more than is needed to obliterate the United States. The shield is leaky.[37]

Well, Star Wars has survived the decades, even though major components have been abandoned, postponed, reframed, or scaled back. Both Brilliant Pebbles and a space-based, nuclear-pumped X-ray laser called Project Excalibur were canceled in the early 1990s. By then, some commentators maintain, SDI had effectively served a central Cold War purpose: the further weakening of the Soviet economy. That space-based missile defense was largely unachievable was beside the point; if the Soviets could be convinced it was achievable, they would pour money they barely had into trying to make it happen. As

419

Gorbachev's military advisor told Soviet studies specialist Dimitri Simes in 1990, "while SDI was unlikely to achieve its stated goal of serving as an impenetrable barrier against nuclear attack, it was nevertheless a full-scale military-technical offensive planned simultaneously to overcome Moscow militarily and ruin the USSR financially."[38]

Within a few months after 9/11, the Strategic Defense Initiative became the Missile Defense Agency and was exempted from the Pentagon's standard procurement and oversight procedures. February 2010 marked a modest milestone: the first lethal boost-phase intercept, with both weapon and target in motion (a megawatt-level laser, mounted on a plane, destroyed a ballistic missile at close range within two minutes after the missile's launch). By 2017, Star Wars had barely made it past the concept stage. As the world witnessed a flurry of North Korean missile successes, accompanied by a volley of insults exchanged between the US and North Korean heads of state, a land-based US missile-defense installation broke ground in South Korea.[39]

So, for the time being, the armed forces will have to do without orbiting laser weapons.

Asteroids — large rocks and even larger rock piles held together by gravity — may present other military possibilities. Some are the size of cars, others the size of houses, still others the size of stadiums. The largest are the size of mountains. From time to time, these cosmic missiles have hit and obliterated entire regions of Earth. Many more have come close but not hit us.[40]

A good solution to the danger would be to identify the trajectories of such objects and destroy or deflect any that threaten to hit us. The first step is to find them and their co-metary brethren, together classified as near-Earth objects, or NEOs. For good reason, space organizations around the world have prioritized finding and tracking NEOs. Celestial mechanics decrees that any NEO whose path crosses Earth's orbit will strike Earth sometime within the next hundred million years or so. Size matters. By now, the NEO catalogue lists more than sixteen thousand, about a thousand of which are larger than a kilometer across — large enough to disrupt civilization. Yet size isn't the whole story. What about NEOs that might strike in the next thousand years, the next century, or the next decade? A threatening subset of NEOs known as PHAs, short for potentially hazardous asteroids, harbor a high probability of swooping within about twenty times the Earth–Moon distance during the next century.[41]

But maybe there's an asteroid with Earth's name on it that isn't yet known. If found and tracked, it must then either be destroyed or forced into another path — and that part of the challenge persists.

In the 1990s, when NASA was about to begin surveying the skies for near-Earth asteroids that measured more than a kilometer across and that might be heading our way, the most efficient available technology for deflection would have been a series of multimegaton nuclear explosions. At that time, Carl Sagan and his colleague Steven Ostro of NASA's Jet Propulsion Laboratory saw grave dangers inherent in both the survey, known as Spaceguard, and the deflection technology:

> If we can perturb an asteroid out of impact trajectory, it follows that we can also transform one on a benign trajectory into an Earth-impactor. . . . With a Spaceguard-like inventory of such asteroids and launch-ready deflection system of nuclear-armed missiles, it might take only a few years to identify a suitably large asteroid, alter its orbit with a series of nuclear explosions . . . and send it crashing into Earth. [Thus] a few nuclear weapons could by themselves threaten the global civilization.

The two scientists were also concerned that "in the real world and in light of well-established human frailty and fallibility," knowledge could too easily be turned from a tool of protection into a tool of destruction:

> Given twentieth-century history and present global politics, it is hard to imagine guarantees against eventual misuse of an asteroid deflection system commensurate with the dangers such a system poses. Those who argue that it would be prudent to prevent catastrophic impacts with annual probabilities of 10^{-5} will surely recognize the prudence of preventing more probable catastrophes of comparable magnitude from misuse of a potentially apocalyptic technology.[42]

A decade later, focusing not on the prevention but rather on the creation of catastrophe, RAND researchers studied the feasibility, relative costs, and psychology of turning asteroids into weapons. They concluded that since "much cheaper, more responsive weapons of mass destruction are readily available, this one is likely to remain safely in the realm of science fiction."[43]

If deploying an actual asteroid against an enemy is infeasible, we still face the problem of rendering an incoming one harmless to ourselves. Two 1998 blockbuster movies, *Deep Impact* and *Armageddon,* solved that problem by nuking their NEOs.[44] In the domain of nonfiction, too, a few investigators have been assessing the possibility of having a spacecraft set off a nuclear explosion deep within the unwelcome space rock (thereby yielding scads of unwelcome orbital debris).[45] But nukes could be used for deflection, not merely destruction. For instance, you could deploy one to create a proximal explosion near one side of an asteroid, causing a recoil by the asteroid and thus forcing a change in its orbit.

Misuse of a potentially apocalyptic technology was the specter raised by Sagan and Ostro — a perspective that matters to us all, whether or not asteroids are the subject. But even a technological glitch or a minor accident could lead to an apocalyptic event. Safety mechanisms have prevented most such outcomes, but no real-life mechanisms are foolproof, infallible, or always applicable. The Three Mile Island, Chernobyl, and Fukushima nuclear disasters made that clear.

In July 1961, the nuclear reactor on a Soviet submarine stationed not far from a NATO base in the North Atlantic developed a disastrous leak in its cooling system. The engineering crew managed to rig a substitute coolant arrangement and prevent a nuclear explosion, but within three weeks those engineers had all died of ionizing radiation. Today *Pravda* is on record saying that if the reactor had in fact exploded, it could have triggered World War III.[46]

In September 1980, during routine maintenance on a nuclear-tipped Titan II missile far underground in rural Arkansas, a socket that was inadvertently dropped from a wrench punctured the missile. This damage caused a fuel leak, leading to the collapse and explosion of the entire apparatus. It could also have led to the inadvertent detonation of the nine-megaton nuclear warhead perched atop the missile, resulting in the destruction of practically everything and everyone from Little Rock to New York City. But by pure dumb luck, the nose cone that housed the warhead blew apart during the explosion, separating the warhead from its source of electricity. Without electricity, its detonators wouldn't work. That was just one US accident, and there have been thousands.[47]

Besides glitches and accidents involving deadly technology, there's the fact that benign technology can be adapted to serve apocalyptic ends. Dual-use space hardware, which dominates the world of space assets, can serve either military or civilian purposes — and dual use can slip into misuse. Even a kindly little weather satellite can be reprogrammed, repurposed, and deployed as a platform to support weapons of mass destruction.

On the other hand, a solar sail cannot redirect a threatening asteroid on short notice, nor can an app for identifying new asteroids solve the problem of deflection. Mapping the paths of space rocks is not the same as moving them. You'd still need a way to eliminate the danger, whether by destruction or deflection, plus sufficient lead time to make it happen. Are there any non-scary proposals for planetary defense? Yes. Current consensus favors the gravitational tractor. Park a massive space probe beside the offending NEO. Although their mutual gravity gently urges them toward each other, station-keeping retro-rockets on the probe preserve the gap. The space probe slowly draws the asteroid out of its deadly path, yielding no debris at all.[48]

Enough hypotheticals. What's in the real-life, readily available space arsenal? Surely there are X-ray lasers, high-power microwave beams, hypervelocity metal rods, miniature autonomous attack spaceplanes with self-adapting warheads, space-based smart munitions, nuclear electromagnetic pulse warheads designed to be detonated at high altitudes, co-orbital ASATs, weaponized microsatellites, orbiting battle stations? Nope.

Definitions of space weapons vary. Here are two: (1) "terrestrially-based devices specifically designed and flight-tested to physically attack, impair or destroy objects in space, or space-based devices designed and flight-tested to attack, impair or destroy objects in space or on earth";[49] and (2) "attack munitions that are themselves orbital objects or that are intended to destroy space objects."[50] What's actually available? Kinetic or explosive interceptor missiles and tactical, modest-kilowatt lasers. That's it. They can each target satellites, terrestrial installations, and long-range missiles, though that last one is still a challenge. When their task is to kill or otherwise neutralize a satellite, they're called ASATs — "A" for "anti-"; "SAT" for "satellite."

Satellites would seem to be prime targets. They're central to modern life, especially the GPS constellation, and they're not easy to camouflage, so their orbits are obvious to all who look. Furthermore, geostationary satellites, such as those used for nearly all space-based communications, are always present at the same altitude above the same places on our planet, hence "geo-" + "stationary." At 36,000 kilometers up, they're the highest-orbiting class of satellites but present an especially easy target for evildoers. For societies and militaries that depend heavily on global positioning, communications, surveillance, navigation, early warning, and weather satellites — and no society or military is more dependent on them than America's — an attack on our space assets would be terrifying. Precisely because of the world's growing dependence on satellites, ASATs work best as a threat. To visit intentional doom upon someone else's satellite spells retaliatory doom for one's own, although in fact the United States would suffer more from a successful hit than would its adversaries.

ASATs, like so many other elements of contemporary military thinking and technology, have taproots in the Cold War, when threat inflation held sway at the Pentagon.[51] In tandem with their earliest work on satellites, both the USA and the USSR actively pursued antisatellite weapons. By 1962 the United States had produced interceptors fitted with nuclear warheads; by 1968 the Soviet Union had carried out the first successful test of a non-nuclear, kinetic-kill interceptor. During the next few decades, while continually voicing anxiety over each other's ASAT tests, the two sides kept designing and sometimes building ASATs based on land, sea, and air as well as an antisatellite orbital station or two. Although many of these weapons were eventually abandoned or mothballed for political and self-protective reasons — a mutually soothing policy that some political scientists call "contingent restraint" — the planning and execution continued in a whirl of simultaneous escalation and de-escalation, confrontation and quasi-cooperation, anxieties and pullbacks.

Finally, in August 1983, five months after

Reagan's Star Wars speech — as certain members of Congress were working up legislation aimed at achieving a joint US–Soviet moratorium — Yuri Andropov, general secretary of the Soviet Communist Party, met with a bipartisan delegation of nine US senators in Moscow and committed the Soviet Union to a moratorium on the deployment of any new ASAT systems in space, even for testing purposes. The United States did not follow suit. In October 1985, an American ASAT — a small missile launched from an F-15 fighter jet — took out an aged American scientific satellite, spreading debris throughout low Earth orbit.

By the way, the Soviet Union regarded America's space shuttle as a possible ASAT. National security specialist Joan Johnson-Freese suggests they feared its robotic arm, which might "pluck satellites out of the sky." UK military space specialist Matthew Mowthorpe proposes instead that the Soviets feared its possible cargo of nuclear missiles.[52]

One of the more imaginative ASAT designs was the Kinetic Energy Antisatellite (KE-ASAT) interceptor, birthed in the United States. It would not only smash into and destroy its target satellite but also envelop the resulting space debris in a giant Teflon sheet. Notwithstanding the unlikelihood of this environmentally tidy outcome, KE-ASAT has not completely disappeared from America's portfolio.[53] Nor has the ASAT as a category disappeared from the global arsenal. The United States, under the banner of safeguarding its extensive space assets, continues to invest far more heavily than any other country in antisatellite research and development. Other countries, too, are pursuing the ASAT: Russia tested one in November 2015, Israel and India are working on theirs, and North Korea keeps demonstrating how fiercely it wants one and how close it is to getting what it wants. The most striking demonstrations of ASAT power in recent years, however, were carried out on their very own satellites by China (2007) and the United States (2008).

Let's say you want to harm a satellite or, at a minimum, be recognized as capable of doing so. Harm covers a lot of tactical territory, ranging from temporary disruption to obliteration. Moving lethal levels of energy across the distances inherent in space warfare is still a stretch. Even if you could do it, tomahawking an enemy satellite out of the sky would be as expensive and dangerous to your own and your allies' space assets as it would be to those of your enemies. Deploying a missile from an airborne platform would be easier and less expensive but would make no less of a mess in space. Simple disruption seems the way to go.

If you've got a strong enough laser, you could focus it at the satellite's circuit box or transmission antenna, which would disable the satellite completely and cheaply with no muss or fuss. Or how about swamping the satellite's sensors with a laser brighter than whatever the satellite might be trying to monitor or record — an act of high-tech vandalism called dazzling. If your laser is energetic enough, you could even melt, evaporate, or fracture parts of the satellite's sensor, partially blinding it. You could also consider having your own spacecraft sidle up to the enemy satellite and spray-paint its optics or physically break its antenna. A cheaper, easier approach to disruption — at least until quantum satellites take over — would be to interfere with satellite communications either cybernetically or electronically. A powerful Earth-based transmitter tuned to the right frequencies can compete with the signal that an enemy receiver needs to receive. That transmitter could drown out the enemy's real signal with meaningless noise, otherwise known as jamming, or mimic the real signal with a fake one of similar power, otherwise known as spoofing. In these cases, there's no need to destroy anything as you turn the transmitter into a useless hunk of junk.[54] Strictly speaking, your jammer or spoofer would not be classified as a space weapon, nor would the interventions of a hacker. Plus, most of the measures we've mentioned could be accomplished more readily and cheaply from land, sea, or air than from an orbiting platform. In the end, of course, carrying out any of these attacks risks retaliation in kind.

What, then, is the status of the arsenal for a space war? "Modern warfare can be fought on so many delightfully different levels," says the creepy Baron Ver Dorco in the sci-fi classic *Babel-17*.[55] Yes, there's a cornucopia of imaginary choices: directed mass, directed energy, chemical, biological, electronic, nuclear, cyber, terrestrial, submarine, aerial, orbital, parasitical, face-to-face, close-range, remote-controlled, robotic, boost-phase, mid-course, targeted, carpeted, smart, dumb. But the space weaponry piece of this picture has little to do with actual warfare. It's mostly about threat and deterrence. It's about potential, power, perceived and projected superiority. Nonetheless, warfighters and national security planners everywhere will not stop trying to actualize the imaginary.

Given the many human-made threats to human life and property, the General Assembly of the United Nations has (among its many other frustrating endeavors) struggled since the dawn of the space age to establish rules of the road for achieving "freedom of scientific investigation [and] international cooperation in the exploration and use of outer space" and to keep outer space free of weapons in order to "avert a grave danger for international peace and security."[56] Within a few decades, the United Nations also began to tangle with the mounting problems of space debris and global space security.[57]

Some might say such efforts are naive: that whoever owns space assets should take charge of protecting them, that the militarization of space in the interests of protecting one's assets is inevitable, that one unbalanced person in power can undo any space agreements the community of nations has adopted. Others might reply that everybody's space assets would be far more vulnerable if there were no international agreements or resolutions in place, if there were no collective effort to preserve what we each separately have. As James Clay Moltz, a specialist in matters of conflict, nuclear and otherwise, points out, "unilateral military approaches to space security can go only so far."[58] Fear of retaliation and the costs of escalation constrain most unilateralists.

Diplomacy is one of the few paths forward. However arbitrary it may seem to name a starting date and place for the intricacies of space diplomacy, let's go with October 4, 1954, at a planning meeting of the International Council of Scientific Unions in Rome. There they conceived and planned for the first ever (and only ever) International Geophysical Year. Oddly, it was to span a year and a half, from July 1957 through December 1958. IGY represented a Cold War thawing of frozen scientific interchange concerning oceanography, seismology, glaciology, meteorology, solar activity, and related topics. Sixty-seven countries collaborated on IGY, including the United States and the Soviet Union.

At that meeting, the US representatives proposed that satellites equipped with observation instruments be launched during IGY. Historian Walter A. McDougall writes that, shortly after the end of World War II, observation/reconnaissance satellites topped US space thinkers' wish lists but that such satellites "could not have been more delicate from the standpoints of international law, diplomacy, and strategy." Only a scientific satellite could neatly embody the principle of freedom of space — the Americans called it Open Skies — and so the IGY proposal, made in an international context, amounted to a fortuitous match between need and opportunity. Though the Soviet representatives contributed no comments regarding the US proposal, the committee as a whole unanimously welcomed it. Their approval "pulled back the hammer on the starter's gun in the satellite race."[59]

Both US and Soviet scientists had already been developing artificial satellites and a suitable launch rocket or rocket package for almost a decade.[60] In early October 1945, one month after the formal end of World War II, the US Navy had established the Committee for Evaluating the Feasibility of Space Rocketry. Also formed in 1945, and also charged with producing a feasibility study, was the Army–Navy joint Guided Missile Committee. Two top-secret operations, Overcast and Paperclip, had transported hundreds of tons of equipment, vast quantities of technical documentation, and scores of newly laundered Nazi rocket scientists and engineers to the United States, including Wernher von Braun and Arthur Rudolph. By 1946, new consultative frameworks such as Douglas Aircraft Company's Project RAND and President Truman's Air Policy Commission were hard at work (recall that RAND's first report presented a preliminary design for a satellite). That year, a Yale astrophysicist named Lyman Spitzer produced a report for RAND called "Astronomical Advantages of an Extra-Terrestrial Observatory." Free of atmospheric attenuation, such an instrument would be better able to detect visible light from the universe than could any ground-based telescope and also able to detect bands of light almost entirely blocked by the atmosphere, such as ultraviolet and infrared. The Hubble Space Telescope is Spitzer's legacy.[61]

Soon the newly independent US Air Force, which until 1947 had been the Army Air Force, began to compete internally with the

Army and Navy for military primacy in space R & D.[62] The Air Force and RAND focused on the feasibility of the satellites themselves, while the Army and Navy focused on the missiles — the rockets — that would boost the satellites into orbit. Different factions had different priorities. By early 1949, the satellite's potential for prestige and reconnaissance clearly outweighed its potential as a weapons platform. The idea that a satellite could be an excellent meteorological tool also emerged. By the time Harry Truman left the Oval Office, in January 1953, the groundwork had been laid for a US space program that would be politically and militarily advantageous but not simply a pipeline for new generations of weaponry.[63]

Cold War rocket research in the Soviet Union started off as a means to a different end: the delivery of a nuclear bomb to the continental United States. Stalin — for whom "the danger was not the atomic bomb as such, but the American monopoly of the bomb" — fast-tracked work on a Soviet bomb within days after Hiroshima, devoting seven times as much funding to it from 1947 through 1949 as was allocated in the same period to developing a rocket capable of carrying that bomb to any target on Earth's surface.[64] The first Soviet nuclear test, an atomic bomb with a plutonium core (similar to the A-bomb that the United States dropped on Nagasaki), took place in August 1949. The first Soviet test of an H-bomb took place four years later, with an explosive yield almost twenty times that of the A-bomb. Now the issue of delivery moved to the forefront.

Despite Stalin's early lack of interest in his country's missile program, Soviet progress was swift and substantial. The Soviet "trophy brigades" that plundered Germany's V-2 work sites in the spring and early summer of 1945 initially regarded the huge rocket as "nothing more than a glorified artillery projectile." Yet by 1947, Soviet missile designers, under the supervision of the indefatigable Sergei Korolev and aided by captured German rocket scientists, had not only mastered the construction of the V-2 themselves but had convinced the nascent missile industry to develop an ICBM with a range of almost two thousand miles — ten times farther than that of the V-2. Within a few years, Korolev's first deputy had proposed that the goal be at least twice that. By late 1953, the missile designers were being told to develop an ICBM capable of carrying a six-ton payload. Such a huge capacity was twice the mass Korolev and his team had been expecting. But that unexpected challenge had an upside for the USSR: any rocket powerful enough to carry a heavy bomb would also be able to lift a satellite into Earth orbit.[65]

While some members of the military-industrial sector focused on the Soviet bomb and others focused on the Soviet missile, a few took up the mantle of their countryman Konstantin Tsiolkovsky, who decades earlier had thought about multistage rockets as an efficient way to launch a satellite. Well-placed Tsiolkovsky followers, along with legions of civilian space enthusiasts and popularizers, dreamed of a Soviet entry into space. Foremost among them was an aeronautical engineer named Mikhail Tikhonravov, a comrade of Sergei Korolev's who worked at the think tank NII-4, the Soviet counterpart to RAND, and who was a key part of the USSR's space program from Sputnik to Gagarin. In 1951 Tikhonravov created a small satellite research team at NII-4; in the fall of 1953, half a year after Stalin's death, NII-4 expanded the team into a full-scale secret project, Research into the Problems of Creating an Artificial Satellite of the Earth, codenamed Theme No. 72. The problems ranged from putting a satellite in orbit to using a satellite as a bombing platform.[66]

So, by the end of 1953 — less than a year after the termination of both Stalin's and Truman's time at the helm — the two Cold War adversaries had established their space agenda as well as their space personnel. The following year, the IGY resolution forced them to deliver. Competition crystallized.

IGY's satellite project soon metamorphosed from an idealistic, supranational collaboration of truth-seeking scientists (to whatever extent it had been so) into a fight for alpha status between American Imperialism and the Red Menace.

Meanwhile, East Asia, though caught up in damaging, convoluted confrontations involving both superpowers, was also mobilizing for space. Rocket scientists in Japan used the upcoming IGY as the rationale for developing homegrown rockets for atmospheric research. Mao's China was on the brink of welcoming back the man who would jumpstart its space program: Qian Xuesen, a Chinese-born professor of aeronautics at Caltech and MIT, a founding member of the Jet Propulsion Laboratory, and a member of the elite Scientific Advisory Group set up during World War II to advise the US military on the possibilities for wartime air power. So valuable was Qian to America's rocket research that in April 1945, despite his not being a US citizen, he was given the title "expert consultant" and the temporary rank of colonel in the US Air Force so that he could deploy to Germany and interrogate the V-2 scientists, including Wernher von Braun, who had just handed themselves over to the Americans. Yet in 1950, during the heyday of America's Red Scare, Qian was accused (without evidence) of being a member of the

Communist Party and was soon robbed of his security clearance and professional opportunities. In 1955, minus his papers and belongings, he was deported to China. As a former undersecretary of the Navy famously said, his deportation was "the stupidest thing this country ever did."[67] Within a few years, Qian Xuesen had become his country's Sergei Korolev.

Vilification of Communism by the West — and of imperialism by the East — was standard practice by the time Qian returned to his country of birth. The House Committee on Un-American Activities had poisoned US politics. Winston Churchill had introduced the term "iron curtain" to the "free world" in 1946, after having tried out "iron fence." Truman had delivered his Truman Doctrine speech to Congress in March 1947, calling for the United States henceforth to support "free peoples" and oppose "totalitarian regimes" anywhere and everywhere, at a 1947 price tag of $400 million. Peacetime militarization, too, was on the rise. "Security" had become a prime focus of policy. The National Security Act of 1947 completely overhauled the structure of the US military, establishing the Department of Defense to replace the three separate military services and creating the National Security Council and the Central Intelligence Agency. The secretary of war disappeared, his cabinet post taken over by the secretary of defense. The North Atlantic Treaty, which gave birth to NATO, was signed in April 1949; its purpose, according to the American consensus, was "to create not merely a balance of power, but a preponderance of power."[68]

Alarmed at the possibility of "capitalist encirclement," the USSR responded forcefully to the flurry of "free world" rhetoric and military/economic institution-building in the West. In 1946 the Soviet Union refused to join the newly formed, US-dominated World Bank and International Monetary Fund, dashing American hopes that an influx of dollars would induce a Soviet retreat from Eastern Europe. In 1947, having failed to gain traction during planning sessions for the Marshall Plan, the Soviet foreign minister began work on the Molotov Plan for the Eastern Bloc. In late June 1948 the Soviet Union imposed what turned into a yearlong blockade of all surface routes from Allied-occupied western Germany to the Western-occupied sectors of Berlin — a blockade made possible because the entire city of Berlin is located far inside what was then the Soviet occupation zone. During Stalin's last couple of years in power, as his hopes, fears, demands, and missteps in Germany, Korea, China, Japan, and much of Eastern Europe gave rise to increasingly unpalatable conditions, he began a second round of political purges — reviving the 1936–38 campaign that had thrown Sergei Korolev into a series of labor camps and penitentiaries for many years and sent hundreds of thousands of other Russians, notable and ordinary, military and literary, to their death.[69]

One top-secret document that exemplifies the fraught politics and charged political language of the early Cold War and encapsulates America's foreign policy during much of the second half of the twentieth century is NSC 68, "A Report to the National Security Council, by the Executive Secretary, on United States Objectives and Programs for National Security," dated April 14, 1950.[70]

In NSC 68, the United States has a lofty "fundamental purpose": "to assure the integrity and vitality of our free society, which is founded upon the dignity and worth of the individual," while the Soviet Union has an insidious "fundamental design": "to retain and solidify their absolute power, [which] calls for the complete subversion or forcible destruction of the machinery of government and structure of society in the countries of the non-Soviet world." Once the usual paeans to freedom have been reiterated in a suitable number of paragraphs, hardly a page omits mention of the pressing need to increase US military power as a counterweight to the USSR's program for "world domination."

The Soviet Union's increased "atomic capability" and pursuit of militarization combine to "back up infiltration with intimidation." On the one hand, the report acknowledges, "Resort to war is not only a last resort for a free society, but it is also an act which cannot definitively end the fundamental conflict in the realm of ideas"; on the other hand, "Only if we had overwhelming atomic superiority and obtained command of the air might the U.S.S.R. be deterred from deploying its atomic weapons as we progressed toward the attainment of our objectives."[71]

What to do? Business as usual won't work, isolationism won't work, negotiation on its own won't really work (though we must appear willing to participate), and outright war would be unpalatable to the population. The only sensible course of action is a "rapid build-up of political, economic, and military strength in the free world." The United States will need a "military shield" to shelter all nonmilitary initiatives.[72]

Which is exactly what happened: US military spending soon tripled, from 5 percent of gross domestic product in 1950 to more than 14 percent of a growing economy in 1953 — a tripling concurrent with the Korean War, often called the first hot conflict of the Cold War. For its part, from 1951 to 1952 the Soviet Union almost doubled the size of the Red Army and increased military spending by 50 percent.[73]

On July 29, 1955, nine months after the Rome resolution, the National Science Foundation and the National Academy of Sciences issued a joint press release declaring the United States' intention to construct "a small, unmanned, earth-circling satellite vehicle to be used for basic scientific observations during the forthcoming International Geophysical Year." The target of those observations would be "extra-terrestrial radiations and geophysical phenomena." On the same day, echoing the NSF/NAS press release, Eisenhower's press secretary announced the president's approval of plans for "the launching of small unmanned earth-circling satellites" so that "scientists throughout the world [could] make sustained observations in the regions beyond the earth's atmosphere."[74] Soon afterward, the Navy's satellite proposal, Project Vanguard, beat out those of the Army and the Air Force — even though the Army, which had rocket scientist Wernher von Braun on its payroll, may well have had a better shot at succeeding. Within a year, Vanguard was incurring cost overruns and technical problems. Various Eisenhower administration officials, however, now convinced of the importance of being the first nation to place a satellite in orbit, fought its termination.[75]

Once the Americans had declared their intent, the Soviets visibly quickened their pace. As Bernard Lovell, director of Britain's Jodrell Bank radio-telescope observatory and witness to the first months of the space race, later wrote, "At that stage no one could accuse the Soviet Union of lacking in frankness about its space programme."[76] On July 30, 1955, just one day after the NSF/NAS announcement, the Soviet Union came out with a similar announcement. On September 25, Sergei Korolev gave an unprecedented public lecture at which he proclaimed that his country's goals should rightly be "that the first artificial satellite of the Earth be Soviet, created by the Soviet people[, and] to have Soviet rockets and rocket ships be the first to fly in the limitless expanse of the universe!"[77]

The year 1957 was mentioned far and wide as the target date. At the end of January 1956, the USSR's Council of Ministers decreed that the country would launch an artificial satellite sometime in 1957. In late 1956, the CIA warned the president that the Soviet Union "would orbit a satellite any time after early 1957" and alerted the National Security Council in early 1957 that the Soviets had tested a "one to five kiloton atomic weapon affixed to a missile" — a missile powerful enough to launch a satellite.[78] The early October launch of Sputnik erased all remaining doubts.

Some analysts maintain that the Eisenhower administration, intentionally or inadvertently, either allowed the Soviet Union to go first or was hugely relieved when it did so, because the historic flight of the first world-circling satellite effectively resolved the fraught issue of "freedom of space": whether flights through the airspace above the territory of another country violated that country's sovereignty. Insistence on "vertical sovereignty" and prohibitions against overflights would mean that a country deemed military satellite reconnaissance illegal. But now, having launched the first satellite, the "Soviets had unwittingly placed themselves in a position where they could hardly argue the illegality of the trespass of their own Sputnik."[79] Thenceforth, in principle, anyone could go anywhere in space.

Whatever machinations and mishaps may or may not have taken place behind the scenes, Sputnik was a world-changing, banner-headline event.

The first phase of its success was the perfection of an extremely powerful intercontinental ballistic missile, the R-7 rocket. On August 21, 1957, the rocket flew four thousand miles to the remote northeastern Kamchatka Peninsula.[80] Once the USSR had mastered the means of transport — pointedly announced by the Soviet news agency, TASS, as a military achievement[81] — it could focus more intensively on the new payload: a satellite rather than a bomb.

At 22:28:34 Moscow time on October 4, 1957, very early in IGY and three years to the day after the Rome resolution, the Soviet Union launched the first Sputnik, a glossy, silvery, radio-beeping, 184-pound sphere the size of a beach ball. The front page of *Pravda* called it "A Great Victory in the Global Competition with Capitalism."[82] A month later, the Soviets launched a second Sputnik, six times as heavy as the first. The score was now USSR 2, USA 0.

Three days later, President Eisenhower announced the appointment of America's first science advisor, a post that could almost have dated from the days of President Lincoln. In 1863, with much to distract him, Lincoln nonetheless signed into existence the National Academy of Sciences, an association of independent scientists whose task, then and still, was to provide informed advice to the executive and congressional branches of the government. America's agenda in space demanded a more sharply defined role for science. Eisenhower, who had served as supreme commander of the Allied Expeditionary Forces during World War II, recognized that no modern country could be militarily preeminent without also being scientifically eminent.

On the last day of January 1958, the US Army successfully launched its Explorer 1 satellite. A couple of months later, the US Navy's Vanguard 1 reached orbit as well.[83] Americans who had been paying attention "as one American rocket after another turned into a greasy fireball down in Florida" could now hold their heads a little higher. But multiple failures followed in the wake of the first successes. The Cold War was in high fever. America's pundits, policymakers, and professors were suddenly, justifiably frantic. Space journalist William E. Burrows caustically frames the situation: "Not only were the Reds militarily muscular and infinitely devious, but it turned out that they were apparently superbly educated as well, particularly in science and engineering." Every Soviet student took five years each of physics and math, it was reported, while a mere quarter of "their lackluster American counterparts" took one lone course in physics.[84]

Time for some government mobilization. Early in 1958, mere months after Sputnik, brand-new standing committees on space and aeronautics were rushed into existence in both houses of Congress, with Lyndon Johnson, then Senate majority leader, as chair of the upper chamber's committee.[85] In February, the Advanced Research Projects Agency (ARPA, soon to become DARPA, with a "D" for "Defense") was established, serving as

America's de facto national space agency until the beginning of October, when the National Aeronautics and Space Administration (NASA) opened for business.[86] At the end of July, the National Aeronautics and Space Act of 1958 was passed. In mid-August the National Security Council issued its secret "Preliminary U.S. Policy on Outer Space," which unequivocally stated that any use of outer space, "whatever the purpose it is intended to serve, may have some degree of military or other non-peaceful application."[87] In September the National Defense Education Act, emphasizing support for math and science, was signed into law. Not since the influential 1945 report to the president, *Science, The Endless Frontier,* had the link between science education and national security, and the necessity of government support for science, been so explicitly formulated. Finally, in October 1958, came the official launch not only of NASA but of America's first Pioneer space probe.

Science, The Endless Frontier, written by the director of the wartime Office of Scientific Research and Development, had flatly declared that science "is a proper concern of government" and that it was essential to create a civilian-controlled organization funded by Congress, committed to freedom of inquiry, and empowered to "initiate military research which will supplement and

449

strengthen that carried on directly under the control of the Army and Navy."[88] The result was the creation of the National Science Foundation in 1950. Thinking along similar lines specifically about space, Eisenhower was convinced that "the highest priority should go of course to space research with a military application." But, he added, "because national morale, and to some extent national prestige, could be affected by the results of peaceful space research, this should likewise be pushed." The way to push it, urged his vice president, Richard Nixon, would be to set up a separate agency.[89] That separate agency was NASA, the most peaceable version of a space agency that a nuclear superpower could be expected to create.

At the same time, the Eisenhower administration funded investigations into new antimissile technologies under the umbrella of a new venture, Project Defender. One Defender proposal was known by the name of a sweet little Disney deer: BAMBI, the acronym for Ballistic Missile Boost Intercepts. Envisioned as hundreds of space-based battle stations that would use infrared to track enemy missile exhaust and then release a rocket-propelled weapon, BAMBI was anything but sweet. To help disable the ascending enemy missile, the weapon would release a huge rotating wire net studded with steel pellets. Although canceled in 1963, BAMBI presaged

Star Wars' Brilliant Pebbles two decades later. Another Project Defender aspirant was an orbiting battle station with several thousand nuclear weapons.[90]

While all this space activity was happening in Washington, on the other side of the planet the Soviet Union was no less busy, launching its third Sputnik in May 1958 and preparing its Luna satellites. One of them would orbit the Moon, another would land on the Moon, a third would photograph the Moon — all during 1959. By then, some US space scientists had ceded space preeminence to the Soviets, as evidenced in a formerly top-secret report on lunar research submitted to the Air Force Special Weapons Center in 1959. The report suggests that US scientific committees shouldn't concern themselves with potential lunar contamination by visiting astronauts, "since the first moonfall is very likely to be by a Soviet vehicle."[91]

The space racers had good reason to fear each other. Bombardment satellites were on both sides' agendas, as were plans to detonate a nuclear bomb on the Moon. Four decades later, the primary author of that top-secret 1959 US report on lunar research said in an interview:

It was clear the main aim of the proposed detonation was a PR exercise and a show

451

of one-upmanship. The Air Force wanted a mushroom cloud so large it would be visible on earth. . . . The US was lagging behind in the space race.

The explosion would obviously be best on the dark side of the moon and the theory was that if the bomb exploded on the edge of the moon, the mushroom cloud would be illuminated by the sun. . . .

Thankfully, the thinking changed. I am horrified that such a gesture to sway public opinion was ever considered.[92]

In any case, some sectors of public opinion were already heading in a more constructive direction. On November 14, 1957, within an overall disarmament resolution, Res. 1148, the UN General Assembly invoked outer space for the first time, calling for the "joint study of an inspection system designed to ensure that the sending of objects through outer space shall be exclusively for peaceful and scientific purposes." On the same day, the General Assembly declared its general alarm in Res. 1149 — "Collective Action to Inform and Enlighten the Peoples of the World as to the Dangers of the Armaments Race, and Particularly as to the Destructive Effects of Modern Weapons" — which called for a global publicity campaign to help alert the entire populace of the world that "the armaments race, owing to advances of nuclear science and other modern forms of technology, creates means whereby unprecedented devastation might be inflicted upon the entire world." Thirteen months later, in December 1958, the General Assembly proposed and adopted its first resolution devoted specifically to space, Res. 1348 — "Question on the Peaceful Use of Outer Space" — followed by more resolutions in 1959, 1961, 1962, 1963, 1965, and 1966.[93]

From the viewpoint of a US diplomat deeply involved in disarmament and space policy at the time, the two-year lead-up to the 1963 UN resolution was the breakthrough, setting the terms under which peace might be preserved in space.[94] Although a resolution is less potent than a treaty, it may have been the more achievable option, given the prevailing political climate: hot on the heels of the terrifying superpower standoff known as the Cuban Missile Crisis, in the fall of 1962, and the Senate's 80–19 approval in September 1963 of the groundbreaking Limited Test Ban Treaty, the very existence of which could be substantially chalked up to the near-catastrophe over Cuba.[95]

At long last, in October 1967, during the presidency of Lyndon Baines Johnson, the United Nations' pioneering Outer Space Treaty (full title: Treaty on Principles Governing the Activities of States in the Exploration and Use of Outer Space, Including the Moon and Other Celestial Bodies) became international law. Space, it declared, would be "the province of all mankind." Only peaceable and scientific activities would be acceptable — no weapons testing, no fortifications, no military maneuvers — although the "use of military personnel for scientific research or for any other peaceful purposes" was to be permitted. The keyword here is "peaceful." Like "defense," it's a slippery concept.[96]

In their day, presidents Eisenhower, Kennedy, and Johnson wanted to differentiate the "non-aggressive militarization of space" from the "weaponization of space." Their collective time in office was strongly shaped by America's official identification of the Soviet Union as "the primary threat to the security, free institutions, and fundamental values of the United States" and by America's identification of itself as "leader of the free world."[97] They wanted to distinguish between the right to use weapons under certain circumstances and the sustained condition of weaponization; between extensive military preparedness and militarism; between passive military satellites and active space weapons; between stabilizing deterrence and the proactive pursuit of dominance coupled with a readiness to destroy.[98]

Juxtaposed with disarmament, such distinctions might seem overly subtle, maybe even duplicitous. But they're not irrelevant. An unarmed US Air Force reconnaissance satellite, mindlessly collecting data as it circles Earth, is undoubtedly military but also, strictly in itself, nonaggressive. Its mission is information, not destruction. Whereas a US Air Force Space Command satellite circling nearby, if fitted with a constellation of interceptor missiles, could have lethal consequences for wide swaths of life and civilization from one moment to the next.

During the lead-up to the Outer Space Treaty, all three presidents did what presidents often do: they operated simultaneously on multiple, often opposing tracks and on both sides of almost every fence. As they gave something to this faction, then placated another faction, engaging in brinksmanship here, avoidance there, and accommodation in between, they maneuvered along a quasimiddle path through a dense forest of contradictions amid hundreds of simultaneous competing conflicts large and small. They pursued weapons along with diplomacy; they vociferously proclaimed the benefits of international cooperation along with the necessity for anti-Soviet mobilization; they tried, as one military historian put it, to "convince the world of America's noble intentions while also ensuring that the United States maintained the ability to fight for the peaceful use of space." In the words of Sinclair Lewis,

from his 1935 novel *It Can't Happen Here,*
"every statesman and clergyman praised
Peace and brightly asserted that the only way
to get Peace was to get ready for War."[99] All
the while, close at hand, loomed the specter
of nuclear confrontation between the superpowers. Only international cooperation on
disarmament could banish it.

Dwight D. Eisenhower — military but not
militaristic, a five-star general who had served
as Supreme Allied Commander in Europe
from December 1943 through May 1945 —
wanted both peace and preparedness.

Pre-Sputnik, Eisenhower dismissed huge
expenditures on weaponry as "just negative
stuff adding nothing to the earning capability
of the country." He maintained that he
wanted to "get the Federal Government out
of any unnecessary activity" and to stop the
"hysterical approach . . . to cur[ing] every ill"
through infusions of cash.[100] In the spring of
1953, three short months into his first term,
he told the American Society of Newspaper
Editors:

> Every gun that is made, every warship
> launched, every rocket fired signifies, in the
> final sense, a theft from those who hunger
> and are not fed, those who are not
> clothed. . . . We pay for a single fighter
> plane with a half million bushels of wheat.

We pay for a single destroyer with new homes that could have housed more than 8000 people. . . . This is not a way of life at all, in any true sense. Under the cloud of threatening war, it is humanity hanging from a cross of iron.[101]

At the beginning of his second term, in his State of the Union address of 1957, Eisenhower described the nation's military as "the most powerful in our peacetime history" and "a major deterrent to war," warning that it could "punish heavily any enemy who undertakes to attack us." But he also proclaimed the nation's quest for peace:

A sound and safeguarded agreement for open skies, unarmed aerial sentinels, and reduced armament would provide a valuable contribution toward a durable peace in the years ahead. And we have been persistent in our effort to reach such an agreement. We are willing to enter any reliable agreement which would reverse the trend toward ever more devastating nuclear weapons; reciprocally provide against the possibility of surprise attack; mutually control the outer space missile and satellite development; and make feasible a lower level of armaments and armed forces and an easier burden of military expenditures.[102]

President Eisenhower's preferred agenda

may have been peace, but the National Security Council's was preparedness. In early 1955, in answer to the NSC's concerns, forty-two prominent scientists, engineers, corporate CEOs, and university presidents — the Science Advisory Committee's Technological Capabilities Panel — produced the top-secret document known as the Killian report, "Meeting the Threat of Surprise Attack." They examined the likely timetable for increases in both US and Soviet capabilities in multimegaton weapons, jet bombers, and intercontinental ballistic missiles, the latter two being the delivery systems for the former. In addition, they recommended a slew of measures to fortify America's arsenal. Foremost among them was the immediate funding of an intercontinental ballistic missile with a range of 5,500 nautical miles, which would put Moscow within striking range of launch locations in the continental United States. Other measures urged by the panel: convince Canada to give the United States "authority for instant use of atomic warheads" over Canadian territory; extend radar coverage hundreds of miles northward and seaward from US continental boundaries; develop "interceptor aircraft" to conduct air-to-air combat at very high altitudes, including the launch of guided missiles; design an "artificial satellite transmission system" for stronger and safer communication of critical strategic-warning data; and develop advanced technology for intelligence acquisition. That last-listed task was soon turned over to Lockheed Aircraft's aerospace futurists, Skunk Works, who swiftly devised the U-2, a stratospheric photoreconnaissance plane that would fly beyond the range of contemporary anti-aircraft measures.[103]

By late 1957, many of the panel's recommendations were being implemented. In a radio and television address on science and national security, broadcast one month after the triumph of Sputnik, Eisenhower assured his listeners that the United States "has today, and has had for some years, enough power in its strategic retaliatory forces to bring near annihilation to the war-making capabilities of any other country." He spoke about the US arsenal in some detail and painted a picture of a militaristic Soviet Union, but then switched to his preferred theme, now framed in terms of space: "What the world needs today even more than a giant leap into outer space, is a giant step toward peace." In January of the following year, he sent a letter to the Soviet premier Nikolai Bulganin (Nikita Khrushchev's predecessor) proposing that "we agree that outer space be used only for peaceful purposes"; another détente-seeking letter, sent in February, suggested "wholly eliminating the newest types of weapons which use outer space for human destruction."[104]

The dual-track approach of regularly calling for peace while intensively preparing for war was firmly established by the time Congress passed the cornerstone of America's space program, the National Aeronautics and Space Act of 1958. Its opening paragraph, Sec. 102(a), declares that "activities in space should be devoted to peaceful purposes for the benefit of all mankind." Sec. 102(b) then expends many more words declaring space activities to be the provenance of a civilian agency . . . except when they are not:

[A]ctivities peculiar to or primarily associated with the development of weapons systems, military operations, or the defense of the United States (including the research and development necessary to make effective provision for the defense of the United States) shall be the responsibility of, and shall be directed by, the Department of Defense.[105]

Note that "the development of weapons systems" is listed first — separately from "defense of the United States." While giving a diplomatic nod to "peaceful purposes," Congress leaves no doubt that the militarization and weaponization of space are inevitable.

President Eisenhower maneuvered through

the war/peace/diplomacy minefield in the usual way: back and forth, periodically invoking the benefits of peace as well as the drawbacks and high costs of the arms race. All the while, he presided over a swift post-Sputnik expansion of the nation's intercontinental and intermediate-range ballistic-missile programs as well as initiatives dedicated to military and dual-use space hardware. Reconnaissance satellites remained high on the agenda.[106]

Besides BAMBI and Project Defender's other costly, top-secret initiatives, Eisenhower-era military space efforts included the WS-117L ("WS" for "weapons system") reconnaissance satellite; the MIDAS (Missile Defense Alarm System) missile-detection satellite, intended to provide a thirty-minute advance warning of an incoming Soviet ICBM attack rather than the fifteen-minute warning available through ground-based early-warning systems; the SAMOS (Satellite and Missile Observation System) photographic and electromagnetic data-collection satellite; and Project Corona, a joint effort of the CIA and the Air Force that superseded both the WS-117L and SAMOS and that largely took over reconnaissance tasks previously performed by the CIA's U-2 spy planes. Among the dual-use systems — valuable to both military and nonmilitary users — were TIROS (Television

461

Infrared Observation Satellite), a weather satellite jointly developed by NASA and the Army Materiel Command, and the Navy's Transit navigation satellites. Both began as strictly military programs but were eventually made available to civilians, as would later happen with the NAVSTAR Global Positioning System (GPS), whose value to the US economy will soon be upward of $100 billion a year.

Other militarily important initiatives from the Eisenhower years included first-generation communications satellites: Echo, an inflatable Mylar sphere that could serve as a passive relay for radio signals; SCORE, which could transmit a prerecorded message; and Courier, which could both store and transmit data. Almost all communication satellites circle Earth in a geostationary orbit (GEO), synchronized with Earth's rotation. Just like Earth, they take exactly one day to complete one orbit, so they appear to hover over a selected spot. The idea that GEO would be a good place for communication satellites to live and work was first explored in 1945 by the futurist Arthur C. Clarke, in a detailed article that posed the question, "Can rocket stations give world-wide radio coverage?" Yes, they can. And yes, they do.[107]

Ways to intercept enemy satellites and ballistic missiles also began in earnest under Eisenhower, as did the idea of putting constellations of bombardment satellites into orbit. The Air Force, hoping to displace NASA as the leading edge of America's space exploits, independently designed a crewed, reusable surveillance and bombing spaceplane called Dyna-Soar, which — if it hadn't been canceled before being flown even once — would have been launched on a rocket, aerodynamically glided around Earth in the mid-stratosphere at an altitude of sixty miles, and landed like an airplane. Suborbitally the Air Force teamed up with NASA to research the effects of hypersonic speeds, extremely high altitudes, and atmospheric re-entry on their joint airplane, the X-15.[108]

Much of the actual and potential militarization of space took place because of Senator Lyndon B. Johnson, a Texas Democrat, who had seized on space as a promising issue for his party.[109] The Republican president may have wanted to downplay the importance of Sputnik and not explicitly race the Reds, but that soon became impossible. A couple of weeks after the launch of the second Sputnik, Johnson, acting as chair of the Preparedness Investigating Subcommittee of the powerful Senate Armed Services Committee, began several months of hearings on what would be needed for America to dominate space. These hearings, according to one of the drafters of the National Aeronautics and Space Act of 1958, "were conducted in an emergency atmosphere of deep concern with the status of U.S. national defense." Senator Johnson was ascendant. The day before Eisenhower's State of the Union speech in January 1958, Johnson told the Senate Democratic Caucus:

Control of space means control of the world. . . . [I]f out in space, there is the ultimate position — from which total control of the earth may be exercised — then our national goal and the goal of all free men must be to win and hold that position.[110]

The Democrats crushed the Republicans in the November 1958 congressional election. Almost immediately, Johnson began jockeying to preempt Eisenhower's foreign-policy team. He also convinced President Eisenhower to let him address the UN General Assembly in mid-November 1958 in support of a US draft resolution calling for a Committee on the Peaceful Uses of Outer Space.[111] The resolution passed. In his speech, Senator Johnson — for whom, on a Wednesday in January, US dominance had been the only tolerable agenda — was ready to declare on a Monday in November that cooperation was the only true path forward:

Today outer space is free. It is unscarred by conflict. No nation holds a concession there. It must remain this way.

We of the United States do not acknowledge that there are landlords of outer space who can presume to bargain with the nations of the earth on the price of access to this new domain. . . . We know the gains of cooperation. We know the losses of failure to cooperate. If we fail now to apply the lessons we have learned or even if we delay their application, we know that the advances into space may only mean adding a new dimension to warfare. . . . Men who have worked together to reach the stars are not likely to descend together into the depths of war and desolation.[112]

Persuasive rhetoric. But by now it's clear that "[w]hat passed for attempts at cooperation consisted mostly of fig leaves meant to embarrass or set back the other side's progress," writes James Clay Moltz of the Department of National Security Affairs at the Naval Postgraduate School.[113] To Everett Dolman of the School of Advanced Air and Space Studies at the US Air Force's Air University, cooperation was a fiction: "expansion into near-Earth space came not as the accommodating effort of many nations joined as one, but rather as an integral component of an overall strategy applied by wary superstates attempting to ensure their political survival."[114] Khrushchev himself saw cooperation as an avenue to be pursued from a

465

position of strength: "We felt we needed time to test, perfect, produce, and install [an effective weapon]. Once we . . . provided for the defense of our country, *then* we could begin space cooperation with the United States."[115]

In the fall of 1960, during his last few months in office, President Eisenhower proposed to the UN General Assembly that there be a targeted ban, subject to verification, on the orbiting or stationing of weapons of mass destruction in space. It would be a small step toward cooperation, to keep outer space from becoming "another focus for the arms race — and thus an area of dangerous and sterile competition."[116] As with its predecessors, it remained a proposal, a glimpse of the possible — a little like recent proposals to reduce global greenhouse-gas emissions or to guarantee "universal" health care in America. Implementation lay well in the future. Soon both President Kennedy and Chairman Khrushchev would take up the issue, each in his own way.

In the 1960 election, John F. Kennedy ran in part on closing the fictive "missile gap" with the Soviet Union and on the need to beat the Soviets in space. After all, Khrushchev had said he was turning out ICBMs "like sausages."[117] Kennedy's victory over Richard Nixon meant that money might be lavished on both a high-profile civilian race to the

466

Moon and closed-door military work. The names of Eisenhower-era space-reconnaissance satellites vanished under a shroud of codenames. Also, unlike Eisenhower's public position that the two strands of space activity must be kept separate, the Kennedy administration's (and the Air Force's) position was that both strands were part of a single mission: to preserve space as a domain of non-aggression.[118] Kennedy's budget for military space spending in 1963 was $1.5 billion — almost triple what Eisenhower's had been in 1960. Meanwhile, NASA's budget skyrocketed sixfold, from $400 million in 1960 to more than $2.5 billion in 1963.[119]

Congress and the American people had been prepped for the civilian portion of those billions during Kennedy's speech to a joint session of Congress on May 25, 1961, in which he proposed landing on the Moon as "a great new American enterprise." Six weeks earlier, Yuri Gagarin had become the first human to orbit Earth, while America had not yet perfected an astronaut-ready rocket that wouldn't explode on launch. Intent on shoring up America's faltering prestige through a stunning commitment to space projects, Kennedy sought to make America the guarantor and facilitator of world peace. In space, he maintained, US nonaggressive militarization would be able to neutralize Soviet aggression: "Our arms do not prepare for war — they are efforts to discourage and resist the adventures of others that could end in war." The classic distinction between defensive and offensive weapons, between the good intentions of the good guys and the bad intentions of the bad guys.

Kennedy began his speech with a crusader's vow to defend freedom — "Our strength as well as our convictions have imposed upon this nation the role of leader in freedom's cause" — and concluded with a biblical call for an end to war:

> [W]e will make clear America's enduring concern is for both peace *and* freedom[,] that we are anxious to live in harmony with the Russian people — that we seek no conquests, no satellites, no riches — that we seek only the day when "nation shall not lift up sword against nation, neither shall they learn war any more."[120]

The Soviet Union had heard that claim before, when Stalin was still alive and Molotov was his foreign minister. After more than a decade of repetition, the assertion that the expansionist United States had no desire to conquer surely rang hollow to its expansionist adversary in the East.

In early June, less than two weeks after Kennedy's speech to Congress, a team of authors from NASA and the Department of Defense issued a report titled *The National Space Program,* which, despite its classified status, scrubbed virtually all mention of military applications. No references to ASATs or ballistic missile defense programs occur in these pages, even though work on them had already been pursued for half a decade. Outdoing the Soviet Union in space science and technology was the primary agenda. America's post-Sputnik space failures must be swallowed by strings of successes. Prestige would be the prize.[121]

Speaking sixteen months later in the open-air stadium at Rice University in Houston, Texas, Kennedy rhapsodized about science, space, and leadership. Along the way he also mentioned that the year's space budget exceeded those of the previous eight years combined. But could America afford it? Sure. The $5.4 billion budget, Kennedy deftly pointed out, was less than America's annual expenditure on cigars and cigarettes. Noting the many American space successes since the beginning of his term of office, he invoked the interconnectedness of US leadership with his earlier themes of peace and freedom:

[T]his generation does not intend to founder in the backwash of the coming age of space. We mean to be a part of it — we mean to lead it. For the eyes of the world now look into space, to the moon and to the planets beyond, and we have vowed that we shall not see it governed by a hostile flag of conquest, but by a banner of freedom and peace. We have vowed that we shall not see space filled with weapons of mass destruction, but with instruments of knowledge and understanding.

Yet the vows of this Nation can only be fulfilled if we in this Nation are first, and, therefore, we intend to be first.[122]

By the time Kennedy spoke at Rice, civilian scrutiny of weapons programs and overall military spending had mounted. Add to that the shock of several US debacles, and arms control and denuclearization began to look increasingly attractive — almost as attractive (or imperative) as being first.

Like Eisenhower, Kennedy sent up several trial balloons on arms control and cooperative endeavors. Just a few months after he took office, his State Department produced a document titled "Draft Proposals for US–USSR Cooperation," which paints scientific cooperation between the superpowers as both fiscally and strategically sensible and as a path toward working cooperatively in other important fields. One proposal urged "early cooperation in fields (e.g. meteorological activities that might eventually lead to weather control or manned exploration of the moon) in which unchecked competition may ultimately be dangerous as well as wasteful." Half a year later, as the US ballistic missile program was revving up, the Kennedy administration created the Arms Control and Disarmament Agency, an initiative loathed by much of the military. As one general commented, "the U.S. is attempting the exercise of trying to dress and undress at the same time."[123]

By mid-1962, Kennedy's stated space agenda was to permit militarization but to forbid weapons of mass destruction. Khrushchev's was to forbid all weapons. Subsidiary issues added to the complexity: Should the ban on nuclear weapons in space be a separate agreement or part of a general disarmament treaty? What about inspection? What about advance notification regarding all space launches? Everyone in the US administration wanted to preserve some militarization of space for purposes of reconnaissance, communications, navigation, and weather monitoring. Some officials wanted a swift ban on nuclear weapons in space; others had grave doubts about any ban.

But multiplying fiascos, fears, and losses were beginning to force America's hand. Among them were the worsening prospects of the Vietnam War; the attempted invasion of Cuba at the Bay of Pigs in April 1961; two atmospheric US nuclear tests in July 1962 that interfered with radio transmissions, disabled several satellites, and contaminated four Midwestern states with radioactive iodine; and a plan by the Atomic Energy Commission to detonate up to six hydrogen bombs at the mouth of a valley on Alaska's seacoast in order to create an instant artificial harbor.[124] Add the humiliation of a Soviet citizen becoming the first human to orbit Earth. Add, too, the growing awareness that space debris and nuclear radiation posed enormous dangers to astronaut flights and orbiting satellites.

Soon after the Cuban Missile Crisis pushed the USA and USSR alarmingly close to nuclear war in the fall of 1962,[125] the Soviet Union increased its retaliatory arsenal. Secretary of Defense Robert McNamara glimpsed the possibility of "a more stable balance of terror."[126]

Diplomatic work to address at least the testing of nuclear weapons — work that had been creeping along since the end of World War II — now rocketed ahead. On August 5, 1963, the Limited Test Ban Treaty was signed in Moscow by the United States, the United Kingdom, and the Soviet Union. What the signatories agreed to was "to prohibit, to prevent, and not to carry out any nuclear weapon test explosion, or any other nuclear explosion . . . in the atmosphere; beyond its limits, including outer space; or under water."[127] Note the absence of a reference to explosions underground. Note also the phrase "any other nuclear explosion": under the treaty, deadly explosions in space were verboten, but not the wherewithal to create those explosions.

A few weeks later, on September 19, the Soviet foreign minister told the UN General Assembly that "the placing into orbit of objects with nuclear weapons on board" must be banned and that his government was ready to sign an accord with the United States. The following day, Kennedy replied that, yes, the time to reach such an arrangement had arrived. On October 17, 1963, the General Assembly adopted Resolution 1884 — sometimes called "Stationing Weapons of Mass Destruction in Outer Space" but officially named "Question of General and Complete Disarmament." The resolution beseeched all nations to "refrain from placing in orbit around the earth any objects carrying nuclear weapons or any other kinds of weapons of mass destruction, installing such weapons on celestial bodies, or stationing such weapons in outer space in any other manner" — language that was carried forward into the Outer Space Treaty of 1967.[128]

Did the prolonged effort to ban deadly weapons from space mean that the United States stopped all R & D on such weapons during JFK's presidency? No — in part because the Soviet Union had an aggressive space weapons program of its own, designed by Sergei Korolev: the FOBS, or Fractional Orbital Bombardment System, a long-range ballistic missile armed with a nuclear warhead. A FOBS would spend part of its time in a low polar orbit — the shortest path from Russia to America — undetected by US early-warning radar networks. It would then brake and discharge its warhead over the continental United States.[129] As for the Kennedy-era space weapons program, both the Air Force and the Army designed nuclear-armed long-range ballistic missiles outfitted as satellite interceptors. They were to be ground-based, launched from Earth into space but not capable of entering Earth orbit. To some people, this distinction, like the one between offensive and defensive weapons, is elusive and artificial. But to the government of the United States, which repeatedly claimed the right to counteract aggression as well as the obligation to preserve space as a sanctuary, the distinction was fundamental.[130]

Lyndon Johnson did not need a crash course in space mavenry when he assumed the presidency on November 22, 1963, following Kennedy's assassination. During the Eisenhower presidency, he had chaired the Senate's

Satellite and Missile Programs Subcommittee. During his own vice presidency, he had chaired the National Aeronautics and Space Council, as well as various other space and military committees. Johnson's entire position might be summarized in ten of his own words: "we cannot be first on earth and second in space."[131] And his clout might be attested by the choice in 1961 of his home state of Texas for NASA's Manned Spaceflight Center (now the Johnson Space Center) as the home of America's astronaut corps and Mission Control.

Post-Sputnik, the idea that technological achievement is a straight shot to prestige and primacy among nations became a mantra. But prestige can also result from sharing the benefits of mastery, either through collaboration with equals or assistance to those in need. Johnson subscribed to both forms of sharing, and his concept of technological achievement included not just the mastery of space but also the practical applications of science in the service of civilization. He wanted cleaner air, more access to potable water, and fewer pesticides. While President Eisenhower had used Atoms for Peace[132] and Project Plowshare[133] to paper over the nuclear nightmare and his administration's growing nuclear arsenal, President Johnson signed the Pesticide Control Act and launched Water for Peace.[134]

As for nuclear weapons and various other weapons of mass destruction, Johnson, like Kennedy, held that a ground-based weapons system was not a space weapon, even if the active life of the target and its ultimate destruction were to play out in the theater of space. The point was, our side required the means to defend against the other side's space weapons, and we would mount that defense with weapons that did not live idly in orbit. That's how you weaponize space without weaponizing space. By observing that guideline, the United States could maintain that it — unlike the dangerous other side, purportedly bent on "world domination"[135] — was honoring and preserving the serenity and sanctity of space. Once enshrined in the 1967 Outer Space Treaty, this distinction between ground-based weapons and space-based weapons, however strained, kept the world marginally safer for a couple of decades, in ways similar to the 1963 Limited Test Ban Treaty or the 1972 Anti-Ballistic Missile Treaty.

Dating back to his days as a senator, Johnson wanted American military might to secure freedom the world over. While chair of the Senate's Special Committee on Space and Astronautics during the lead-up to the passage of the National Aeronautics and Space Act of 1958, he oversaw a committee report that proclaimed:

> We have no intent to plant flags of conquest upon the planets or lay extensive claims to the stars. We do propose that space shall never become the route of march for tyrants and totalitarians and, as we have dedicated our resources in the past to maintain the freedom of the seas and security of the skies, so shall we dedicate our capacity to maintain the neutrality of space.[136]

Johnson's opponent in the 1964 presidential election, Senator Barry Goldwater — a brigadier general in the Air Force Reserve and a man comfortable with the use of nuclear weapons — was not a fan of neutrality. In his view, space research should be directed by the military, "with national security and control of the access to space as primary goals." What America needed ASAP were antimissile missiles, laser weapons (the laser itself had been invented by American scientists at Bell Labs just four years earlier), and a manned space station in near-Earth orbit. Daily surveillance of nearby space would be crucial. According to Goldwater, America had to "move beyond just sailing into space." The idea that the United States could collaborate with the Soviet Union was "too ludicrous for comment." Goldwater opposed disarmament. He voted against bills and treaties aimed at fostering peace. He thought that low-yield nuclear weapons should be used to defoliate key areas of South Vietnam and wanted senior military commanders to be preauthorized to use nuclear weapons in an emergency. Sounds extreme, but Goldwater was not an outlier. The president of the Aerospace Corporation demanded to know, "Why do we place an evil cast on military activities in space?" A senior air-power advocate and *Reader's Digest* editor called the US space program too peaceable and "the wrong race with Russia."[137]

Goldwater lost, overwhelmingly. Once elected, Johnson — an arm-twisting, New Deal kind of Democrat and as anti-Communist as his predecessors — presided over the passage of the 1964 Civil Rights Act, the 1964 Food Stamp Act, the 1965 Voting Rights Act, and the 1968 Fair Housing Act, all part of his "War on Poverty." On his watch, Medicare and Medicaid were introduced, the federal minimum wage was increased, thirty-five national parks were established, and the National Endowment for the Arts and the Corporation for Public Broadcasting were created. Also on his watch, the country would witness the escalation of the Vietnam War, the ostracizing of Cuba by nearly every nation in Latin America, eruptions in America's cities and on America's campuses, and, shortly before the end of his term in office, the successful launch and return of the Apollo 8 spacecraft and its three-astronaut crew, who became the first humans ever to orbit the Moon — the first humans ever to leave Earth for another destination. Johnson halved the official poverty rate but also sent a million and a half Americans to Vietnam and convinced Congress to give him a wide berth to do whatever he deemed necessary in Southeast Asia. He further strengthened America's thriving space program, and he pushed America farther down the path of substantive arms control, both on Earth and in outer space — not merely bans on testing nuclear weapons but prohibitions against using them at all.[138]

Under Johnson, the Air Force competed with the Army and the Navy to become the space power within the Pentagon, a fight for primacy that dated back to its formal split from the Army in 1947.[139] Beginning in 1958, the Air Force faced an additional, civilian competitor: NASA.[140] But putting a man on the Moon became NASA's mandate, not the USAF's — even though half the astronaut corps came from the Air Force — and Johnson was strongly committed to it, despite opposition from many quarters.[141]

Soon NASA's outlays far surpassed the Pentagon's space budget. In an attempt to swing the pendulum back, the military unsuccessfully stepped up its efforts to dominate

America's crewed space program, efforts doomed to fail because the president's secretary of defense, Robert McNamara, was on the lookout for Pentagon programs that could be cut, not expanded. The race to reach the Moon had become the core of Johnsonian space policy, and NASA's share of total space spending soared to 74 percent in 1965, with much of the remainder going to military reconnaissance satellites.[142] But the largesse couldn't last. For 1967, NASA requested five and a half billion dollars but received five. As James Clay Moltz writes:

> [The] boom years for space could not go on forever. In domestic politics, Johnson's Great Society programs in particular had caused social spending to soar. Even in peacetime, the simultaneous drains on the federal budget of space programs and expansive new social programs were bound to cause eventual fiscal strain. With the growing cost of the Vietnam War[,] cuts had to be made somewhere. . . . [D]ual pressures from liberals favoring more social spending and conservatives favoring more military spending began to catch NASA in a scissors.[143]

In keeping with Johnson's (and Kennedy's) vision of a single, broad, unified space program, the Vietnam War presented both the need and the opportunity for cooperation between NASA and the Pentagon. The title alone of a 1964 joint NASA/USAF document, highlighted in *US Presidents and the Militarization of Space 1946–1967* by military historian Sean Kalic, makes clear the military benefits of open borders between military space and civilian space: "Summary of Suggestions by NASA Headquarters Personnel as to Ideas That May Have Application to the War in Southeast Asia." Among the possibilities proposed by the co-authors were satellites that could identify "instantaneous cloud cover, synchronize altitude communication, and locate downed pilots" and research into "supersensitive seismic sensors, lightweight power supplies, and infrared technology."[144] By the mid-1960s, infrared detection devices such as tank-mounted searchlights and the earliest handheld thermal imagers appeared in the American war against Communism in Vietnam, Laos, and Cambodia.

Off the battlefield, new varieties of danger to space assets were gaining visibility. Any and every spacecraft, military or civilian, US or Soviet, faced the possibility of collision with orbital debris. Thanks to the 1962 series of missile-launched American nuclear tests collectively called Operation Fishbowl, we learned that the brief electromagnetic pulse of a high-altitude nuclear test could temporarily paralyze unshielded communications and reconnaissance satellites, while the detonation's longer-lasting radiation would suffuse the upper atmosphere and make human spaceflight even more problematic than it was already was. Down on the ground, even an unfueled rocket sitting on a launchpad could become a site of disaster: during an Apollo 1 launch rehearsal on January 27, 1967, three astronauts asphyxiated within seconds from an electrical spark gone wild, because the pure oxygen atmosphere in their sealed capsule spread fire instantly, igniting the interior nylon netting, the Velcro straps, and the polyurethane foam insulation.[145]

No thinking person imagines that being president of the United States, especially in wartime, is a job for the faint of heart. Johnson's presidency, rife with tumult, was marked by a widely questioned war and social upheaval as well as pioneering safety-net programs and technological triumphs. Like his predecessors and successors, he walked both sides of the space street. Carrying on the postwar practice of US "technological anticommunism," he promoted what Walter A. McDougall called the "benign hypocrisy [of] cooperation in science and competition in engineering."[146] Johnson's support of cutting-edge military reconnaissance and his public acknowledgment of successful American ASAT tests served as accident insurance for peace initiatives. For him, national security demanded the presence of the military in space but the absence of orbiting weapons.

Just days after his sudden accession to the presidency, Johnson told the space industry that the United States would uphold its commitment to the "peaceful purpose of space for the good of all mankind." The United States would have to appear to the world as the champion of peace so as to nudge the Soviet Union away from its work on bombardment satellites. One way to look peaceable would be to open previously military programs to general public use — the civilian spin-off argument. Two such programs were Transit, the US Navy's satellite navigation system, and Nimbus, a system of NASA satellites that recorded and photographed cloud cover, atmospheric chemistry, ozone, and sea ice. Another way to look peaceable would be to keep supporting the kinds of cooperative scientific ventures that had started during International Geophysical Year and had continued under Kennedy — not the Moon landing program, which Johnson wanted to remain a unilateral first, but less high-profile programs such as satellite studies of Earth's magnetic field, satellite relay of communications, and, as discussed by Soviet and US scientific delegations from December 1964 through September 1966, a joint civilian mission to space.[147] A third way would be to press directly and conspicuously for a UN treaty on arms control in outer space.

By this point, contends Moltz, political leaders "had begun to recognize that space was now too valuable to be used for war."[148] In December 1963 the UN General Assembly adopted a declaration of principles affirming that "the exploration and use of outer space should be carried on for the betterment of mankind and for the benefit of States irrespective of their degree of economic or scientific development" and asserting the shared desire "to contribute to broad international co-operation in the scientific as well as in the legal aspects of exploration and use of outer space for peaceful purposes."

The time was ripe for a full-fledged treaty. The pieces were in place.

And so, on the afternoon of January 27, 1967, in the East Room of the White House, US Secretary of State Dean Rusk, followed by UN Ambassador Arthur J. Goldberg representing the United States, UN Ambassador Sir Patrick Dean representing Great Britain, and finally UN Ambassador Anatoly F. Dobrynin representing the Soviet Union, put their signatures to the Treaty on Principles Governing the Activities of States in the Exploration and Use of Outer Space, including the Moon and Other Celestial Bodies. Representatives of fifty-seven other nations then added their signatures, and President Johnson said a few words:

> We have never succeeded in freeing our planet from the implements of war. But if we cannot yet achieve this goal here on earth, we can at least keep the virus from spreading.
>
> We can keep the ugly and wasteful weapons of mass destruction from contaminating space. And that is exactly what this treaty does.
>
> This treaty means that the moon and our sister planets will serve only the purposes of peace and not of war.
>
> It means that orbiting man-made satellites will remain free of nuclear weapons.
>
> It means that astronaut and cosmonaut will meet someday on the surface of the moon as brothers and not as warriors for competing nationalities or ideologies.[149]

Today, five UN space treaties are in force, along with hundreds of resolutions, initiatives, conventions, reports, declarations of principles, and arms-control treaties that include language directly addressing space. Among the arms-control treaties are the 1963 Limited Test Ban Treaty (Treaty Banning Nuclear Weapon Tests in the Atmosphere, in Outer Space, and Under Water) and the 1972 Anti-Ballistic Missile Treaty (Treaty Between the United States of America and the Union of Soviet Socialist Republics on the Limitation of Anti-Ballistic Missile Systems). While it was still in force, the ABM Treaty banned the development, testing, and deployment of widely distributed ballistic missile systems on land, in the seas, in the atmosphere, and in space. It also banned interference with the signatories' "national technical means of verification," which — though unspecified — meant satellite photography, aircraft over-flights, and electronic and seismic monitoring. Many nations regarded the provisions as indirectly including a ban on space-based anti-satellite weapons.[150] Upon the Bush administration's withdrawal from the treaty in 2002, that ban evaporated.

Among the resolutions, PAROS — Prevention of an Arms Race in Outer Space — is of considerable interest. Its several-decade history highlights America's preference for stonewalling on disarmament. First presented in December 1981 as a plea for measures and guarantees that would deepen those delineated in the 1967 Outer Space Treaty, PAROS was reintroduced in essentially the same form every year thereafter until finally, in January 1994, it came to a vote. No country opposed it; the United States alone abstained. Year after year, the General Assembly voted on this resolution. In 1996 and 1997 the United States was joined in abstention by numerous allies, but beginning in 1999 it was joined consistently, either in abstention or in outright opposition, only by Israel, with the occasional addition of Micronesia, the Marshall Islands, or Haiti.[151]

Also starting in 1981 — and accelerated by US disinformation regarding the Strategic Defense Initiative, coupled with genuine as well as artificially inflated Soviet alarm at SDI and the space shuttle — the Soviet Union asked the UN General Assembly to place on its agenda a Soviet draft treaty that sought to prohibit stationing weapons "of any kind" in outer space. By the summer of 1983, no longer a draft, it bore the title Treaty on the Prohibition of the Use of Force in Outer Space and from Space Against the Earth. In 1984, the USSR proposed further clarifications.

If just the 1983 provisions had been promulgated as a treaty, its signatories would have been agreeing not to test, place in orbit, or station on celestial bodies "any space-based weapons for the destruction of objects on the Earth, in the atmosphere or in outer space"; not to use already orbiting or already stationed space objects "as means to destroy any targets on the Earth, in the atmosphere or in outer space"; not to "destroy, damage, disturb the normal functioning or change the flight trajectory of space objects of other States"; not to test or create new ASATs and to destroy any ASATs they might already have; and not to test or use manned spacecraft (read: the US space shuttle) for military or ASAT purposes.[152]

In 2008, Russia and China, working through the UN's Conference on Disarmament, presented their draft of a successor to PAROS called the PPWT: Treaty on Prevention of the Placement of Weapons in Outer Space and of the Threat or Use of Force Against Outer Space Objects. Its prospects can be gauged in part by paragraph 34 in a 2015 Russian–Chinese letter describing the lack of useful input from the United States: "instead of constructive proposals on the contents of the draft PPWT, we once again see the appalling attempts of the United States of America to impose on the international community its politicized assessment of the space programmes of certain States." The United States, they say, is "avoiding having to shoulder any additional international legal obligations as far as outer space is concerned, including in order to ensure that outer space remains free of weapons of any kind" and is doing nothing to "facilitat[e] progress towards a mutually acceptable resolution of issues involving the security of space activities."[153] Rough language from diplomats.

Other concerned parties made other proposals. A senior German diplomat, Detlev

Wolter, drew up detailed guidelines in 2006–2007 for a Treaty on Common Security in Outer Space. Regarding the need for "an explicit prohibition of active and destructive military uses in outer space," Wolter suggested that signatories "commit themselves to refrain from any deployment or use of any object in space or on Earth, that was designed or modified specifically for the purpose to inflict permanent physical damage on any other object through the projection of mass or energy respectively." Adding to the explicitness, he also suggested that ballistic missile defense and anti-satellite systems be forbidden in outer space, the sole exception to be a UN system whose purpose would be "implementing and enforcing a non-proliferation regime and . . . protecting against unauthorized and accidental missile launches."[154]

Pending ratification of anything that resembles current attempts to consecrate space as a permanent haven of peaceable activity for the common good of all humanity, the centerpiece and foundation of international space law remains the Outer Space Treaty of 1967, "the high-water mark of [the United Nation's] sponsorship of the demilitarization of outer space."[155] To date, it has been signed and ratified by 105 countries, including not only the United States and Russia but also China, India, Pakistan, Israel, the United

Kingdom, Canada, and every country in Western Europe that isn't a microstate, and has been signed by an additional twenty-five — a total of 130 of the United Nations' 193 member states as of early 2017.[156]

Ideally, ratification should signal that each of these countries recognizes and honors "the common interest of all mankind in the progress of the exploration and use of outer space for peaceful purposes." But, as with other peace-seeking agreements, it is honored almost as much in the breach as in the observance. Have a look at several of its provisions:

Article II
Outer space, including the moon and other celestial bodies, is not subject to national appropriation by claim of sovereignty, by means of use or occupation, or by any other means.

Article III
States Parties to the Treaty shall carry on activities in the use and exploration of outer space, including the moon and other celestial bodies, in accordance with international law, including the Charter of the United Nations, in the interest of maintaining international peace and security and promoting international cooperation and understanding.

Article IV

States Parties to the Treaty undertake not to place in orbit around the earth any objects carrying nuclear weapons or any other kinds of weapons of mass destruction, install such weapons on celestial bodies, or station such weapons in outer space in any other manner.

An ordinary person would interpret these clauses as saying that nobody may own or control any part of space, violate the peace there, undermine international cooperation there, or introduce weapons capable of catastrophic damage there.

Do not assume governments and militaries share that interpretation. The US Air Force, for example, embraces more open-ended guidelines for observance of the Outer Space Treaty (if and when the treaty is mentioned in its documents, which is not always the case).[157] Take the Air Force doctrine document *Space Operations* (2006), which states that although "the space legal regime imposes a few significant constraints, the bulk of this regime provides a great deal of flexibility for military operations in space." The document then turns Article III of the treaty, regarding the role of international law in the preservation of peace, on its head:

The right of self-defense, as recognized in

491

the United Nations Charter and more fundamentally in customary international law, applies in outer space. Also, law of war precepts such as necessity, distinction, and proportionality will apply to any military activity in outer space.

As for Article IV, regarding weapons, the Air Force clarifies that "the placement of weapons other than weapons of mass destruction in outer space is permissible . . . , as is the transit of nuclear weapons, such as ICBMs, through space." Finally, with regard to the treaty's overall goal of upholding peace, it asserts:

> The majority of nations have traditionally held that the "peaceful purposes" language does not prohibit military activities in outer space; such activities have taken place throughout the space age without significant international protest. The phrase, rather, has been interpreted to require that activities in space be non-aggressive . . . to refrain from the threat or use of force except in accordance with the law, such as in self-defense.[158]

At the heart of US military objectives is an expansive definition of self-defense — active, not passive. To embrace that definition means the potential use of weapons, and potential use opens the door to an actual program of weaponization.

In 2002 the RAND Corporation's Project Air Force division published *Space Weapons Earth Wars.* Intended as a tutorial for anyone concerned with national security, it focuses on facts, options, feasibilities, tactics, costs, positives, negatives, and possible scenarios.[159] Everything you need to know about warmaking from space.

The first sentence informs us that since at least the late 1950s, "the Air Force has espoused the full use of the medium of space for national security." Hardly a shock that an ambitious air force, anywhere in the world, would view "the medium of space" as part of its workplace and would seek full use rather than limited use. Also hardly a shock that national security is the stated goal and rationale. Who among the sane does not wish that their country and its inhabitants be kept safe and well?

But national security is a malleable concept. From the US Air Force's point of view, many national-security uses of space are not open to debate, at least when it's the United States doing the using. Those uses include communications, environmental monitoring, position location, weather, and warning, as well as intelligence, surveillance, reconnaissance, and weapons guidance. Several of those activities may sound like benign defensive measures meant to protect the state and its citizens. But in fact, all of them possess potential offensive military value as well.

Early on, the report proposes that the most urgent reason to develop space weapons may be "the possibility that other nations will decide to acquire them." That's like saying we'd better not only keep up with the Joneses but also acquire an Olympic-size backyard pool before they do. The report also contends that the military's periodic statements on doctrine and strategic planning "give space weapons an air of inevitability."[160]

So, picture this: Two adversaries possess weapons capable of immense devastation. Neither side will hesitate to use those weapons to their fullest. Both await a reason to do so — such as an aggressive move by the other side. The likely scenario is inescapable escalation toward a renewed state of mutual assured destruction (MAD), in the course of which the line between defensive and offensive weaponry will be blurry at best. As the political essayist I. F. Stone wrote back in the 1950s, when US–Soviet nuclear war felt almost imminent, "Both sides in every war always claim to be *aggressed.*"[161]

Uncompromisingly hardheaded, *Space Weapons Earth Wars* simply presents a panoply of available or conceivable options for the waging of space battles. Its brief historical overview takes the militarization of space as axiomatic:

> The use of space in conflict and the use of weapons against space systems are both historical fact and current reality. From its beginning, man's use of space has included conflict, wars cold and hot: finding targets, warning of threats, relaying commands, aiding navigation, and forecasting weather.[162]

If militarization is the basic condition, if all past and current activity in space has a military and therefore potentially aggressive character, then outright weaponization of space is simply a logical extension of an already existing condition. By that line of reasoning, attempts to demilitarize space are naive, and attempts to exclude weapons from it are quixotic. But do everybody's space weapons have equal access? The United States habitually treats the possession of advanced weaponry by non-allies as an intolerable threat.[163] Iran and North Korea are today's examples.

The RAND authors take note of accelerating international concern during the 1960s that space be preserved solely for peaceful purposes but have almost nothing to say about the wisdom or likelihood of abiding by the resulting treaties. "[O]ne of the uses of space weapons of current interest to the

United States is explicitly illegal," states their report, with studied evenhandedness. Not that explicitly expressed reservations are entirely absent. On grounds of clear self-interest, for instance, it opposes the use of a deadly space weapon against any high-altitude target carrying biological or chemical agents, because much of the resulting hazardous debris would be broadly dispersed by wind and its harmful effects broadly distributed. "As a deterrent," warns the report, "the possibility of poisoning one's own homeland should give a rational actor more reason not to employ such weapons.[164] An optimist might contend that nobody except a rational actor would ever be permitted to make such decisions, but anybody who has watched the 1964 movie masterpiece *Dr. Strangelove* or witnessed the rampant irrationality of the 2016 US presidential campaign is unlikely to agree.

Joan Johnson-Freese, analyzing what would induce countries without space assets to acquire them, argues that the combined existing, imminent, and projected US space capabilities would make it effectively not only "unassailable and able to impose its will virtually without question" but also capable, if it chose, to undertake "highly intrusive forms of coercion . . . without the burdens of occupation." After all, space capabilities can readily be repurposed. The same rocket can launch a nuclear-tipped missile or a crop-monitoring satellite; the same adaptive optics can eliminate a star's twinkle or a laser weapon's dispersion. Other countries, alarmed by (but also dependent on) America's abundant space assets, would understandably want to develop their own space assets to help preserve their sovereignty and autonomy. Europe, Japan, France, Israel, China, and others have already gone that route.[165]

Keep in mind that a single nation, the United States, far outspends all others on military space systems. We are the big buyer. We are the patron. We're also the biggest exporter.[166] And our own definitions of self-defense lean toward preemption. But beyond that, says the RAND report, "it is conceivable that the United States could decide to acquire space weapons in advance of a specific, compelling threat" — that is, in a deliberate, unilateral, preventive move to maintain technological superiority.

Besides, not all decisions to acquire would be deliberate. They could arise incidentally from the development of commercial, reusable space systems — not yet widespread when *Space Weapons Earth Wars* was being written but now proceeding rapidly at corporations like Blue Origin and SpaceX and at the space agencies of France and Germany. To quote the report again, "Any capability to deliver and retrieve large quantities of material economically to and from space could be adapted to emplace and deliver conventional weapons from space."[167] If you can go back and forth cheaply and easily, you can take almost anything along for the ride in either direction. Laser weapons or multivitamins — doesn't really matter.

Once the problem of transportation has been solved, the question of the advantages and limitations of space assets arises. One imagined advantage is ready access to targets via overflights, because (ever since Sputnik) political constraints on overflights apply only to lower-altitude aircraft.[168] Another advantage is global reach, which holds true only if the assets are sufficiently numerous and strategically deployed. Two others are the potential for swift response, compared with the time needed to mobilize terrestrial forces, and the adversary's difficulties in defending against space weapons. One big limitation is the stable, observable, predictable orbit of each satellite: to assemble a constellation of already orbiting satellites would mean having to change some established orbits, a formidable task requiring more fuel for thrust than a satellite typically carries in orbit. Other limitations include the $10,000-per-pound cost of placing any object in space and the need to have plenty of satellites in orbit (preferably hardened against attack) so that at least one will lie within timely reach of any likely target. Still other limitations, in theory if not in practice, are the treaty provisions that prohibit putting weapons of mass destruction in space bases and that assign liability for damages under certain circumstances. Finally, the paucity of time between decision and deployment can be good or bad or both.[169]

But aside from the disputable morality of various space weapons, and aside from the budgetary constraints and the optimal tactics for guaranteeing preemption and preeminence, there's a more basic question. Even a middle-school kid who's been told about Hiroshima and Nagasaki and the firebombing of Tokyo, about defoliants and napalm, about bunker buster bombs and sarin gas attacks, might ask: What kinds of space weapons are the people in power willing to use today? Is there a line they will never cross?

Statements to the effect that the United States must commit itself to space supremacy and embrace the full spectrum of space weapons — or, indeed, weapons in general — have never been in short supply. "Full spectrum dominance" is the main theme of the Joint Chiefs of Staff's *Joint Vision 2020.*[170] At times, US rhetorical insistence on space superiority has sounded more like an insistence on space exclusivity. As one Air Force officer wrote during the George W. Bush administration, "the United States does not explicitly support other nations' rights to operate militarily in space, reserving this right for itself."[171]

Militarization is often exponential. Intensified, it breeds weaponization. Each further step on the continuum escalates the danger: from simply operating in space, to operating militarily, to operating aggressively, to operating lethally. Beyond limited lethality lie extermination and annihilation, the predictable outcomes of waging war in space using the full spectrum of available and conceivable weapons.

For more than half a century, the nuclear bomb has stood at the far end of the full spectrum, whether launched from a submarine or a silo, dropped from a plane, or delivered via suborbital missile. It is the emblematic weapon of annihilation. The scale of damage in Hiroshima and Nagasaki is well known. By contemporary standards, both cities represent mild demonstrations of what nuclear weapons can deliver. But even smallish nuclear explosions can cause serious collateral damage. On July 9, 1962, for example, in a nuclear test known as Starfish Prime, the United States exploded a 1.4-megaton hydrogen bomb four hundred kilometers up in the air, disabling half a dozen orbiting satellites and stoking the already disruptive population of electrons in the Van Allen radiation belt, which encircles Earth many thousands of kilometers above our planet's surface. Nevertheless, tests and plans continued. U Thant, who was then secretary-general of the United Nations, described the US program to test nukes in space as "a manifestation of a very dangerous psychosis."[172]

That psychosis ruled the body politic for a couple of decades. From the time President Truman learned about America's atomic bombs, in April 1945, he expected to use them. Though initially built to devastate Germany, which was now on the verge of surrender, the bombs would instead be deployed to shorten the war with Japan. As Cold War historian Walter LaFeber put it, "Roosevelt had built this bomb to be used. Truman was going to carry out Roosevelt's policy. Billions of dollars had been put into the bomb project. Truman was not going to waste that money."[173]

Three years later, in 1948, during the US airlift of supplies to West Berlin following Stalin's blockade of surface traffic, Truman told his secretary of defense and secretary of state that while he "prayed the bomb would not have to be used," nobody should think he wouldn't order a bombing "if it became necessary." By the early 1950s, America had four hundred nuclear bombs and a fleet of intercontinental B-29 bombers that could be refueled in midflight.[174]

Nor was this nuclear psychosis an exclusively American illness. On July 24, 1945, at the trilateral negotiations on reparations, postwar reconstruction, and borders that took place in Potsdam, Germany — just two weeks before the United States dropped "Little Boy" and "Fat Man" on Hiroshima and Nagasaki — President Truman casually (he thought) dropped mention of the fact that the United States now possessed a horrible new weapon. Stalin's response appeared so nonchalant that Truman thought he might not have understood the comment. Far from it. Stalin immediately ordered the USSR's existing atomic project to accelerate its work. Whole forests were soon cleared to make way for laboratories. Electricity was redirected from civilian areas. "Just hours after the atomic age began," writes LaFeber, "its arms race was escalating."[175] Soon US policymakers and generals were speaking out and strategizing in favor of maintaining a nuclear arsenal, not simply as a deterrent but as a commitment to act.

Hundreds of detonations followed forthwith. August 1949, Soviet, with an explosive yield equivalent to 22 kilotons of TNT. November 1952, US, 10.4 megatons. August 1953, Soviet, 400 kilotons. March 1954, US, 15 megatons. November 1955, Soviet, 1.6 megatons. November 1957, British, 1.8

megatons. February 1960, French, 70 kilotons. October 1961, Soviet, 50 megatons. October 1964, Chinese, 22 kilotons.[176] The kiloton blasts were typically A-bombs, exploiting the fissionable properties of uranium and plutonium — elements named, by the way, for the planets Uranus and Pluto. The more deadly megaton blasts were H-bombs, which derive their energy from the thermonuclear fusion of hydrogen to helium. Same as the Sun's been doing in its core for the past five billion years.

Seeing this proliferation of nuclear testing, the conservative public-policy group American Security Council, proponents of a full-spectrum US stance, demanded that "atomic test-ban negotiations at Geneva be discontinued and that underground nuclear tests be resumed immediately."[177] Some military strategists agreed. Nuclear weapons small enough to be transported on ballistic missiles were already available. The good news is that by the early 1960s, few in the US military saw the installation of nuclear bombs on a satellite as a reasonable option. After 1967 the option was dead.

Recalling his frame of mind just after World War II, the merciless general Curtis LeMay, commander of the singularly deadly fire-bombing of Tokyo and subsequently the first head of the US Air Force's Strategic Air

Command (SAC), told interviewers:

> SAC was the only force we had that could react quickly to a nuclear attack. It did not make much sense to me to be in a position of not being able to act because I had no weapons. . . . [W]hen I first came back from Germany [1948], there wasn't any doubt in my mind that if we had to go to a full scale war, we would use nuclear weapons. . . . We didn't consider any unit really combat-ready unless it had a nuclear capability [because] we were planning on a nuclear war.[178]

In 1953 the executive secretary of the National Security Council, James S. Lay Jr., issued a top-secret presidential directive, NSC 162/2, that warned of the Soviet Union's intent to dominate the world, its growing nuclear stockpile, its suspicious peace gestures, and the possibility that it might soon be able to deal "a crippling blow to our industrial base and our continued ability to prosecute a war." This multifaceted "Soviet threat" had led many US allies to regard negotiation as "the only hope of ending the present tension, fear and frustration." Lay, on the other hand, downplays the usefulness of negotiation and stresses the need for deterrence through military might, as in paragraph 34, which states that the "risk of

Soviet aggression will be minimized by maintaining a strong security posture, with emphasis on adequate offensive retaliatory strength and defense strength. This must be based on massive atomic capability." Other factors would contribute to the posture, such as bases, a continental defense system, deployed forces, an effective intelligence system, superior scientific research, and "the determined spirit of the U.S. people." But atomic capability headed the list. One cold sentence in paragraph 39b(1) sums up the US position: "In the event of hostilities, the United States will consider nuclear weapons to be as available for use as other munitions."[179]

The following year, General Bernard Montgomery, who had served as Deputy Supreme Allied Commander in Europe, told a London audience, "I want to make it absolutely clear that we . . . are basing all our operational planning on using atomic and thermonuclear weapons in our own defense."[180]

In 1956, the Strategic Air Command compiled a target list for a conceivable war three years later. The goal was "systematic destruction." Eight hundred pages long, top-secret, and titled "Atomic Weapons Requirements Study for 1959," the list included 179 targets in Moscow, 145 in Leningrad, and 91 in East Berlin. Airfields, factories, infrastructure, government buildings, and agricultural equipment were prime targets; any unlucky humans who happened to be on the premises would become part of the target. In addition, one target in each city is simply listed as "Population."[181]

In a 1957 document labeled MC 14/2 and often referred to as "Massive Retaliation," NATO plainly states that "[i]n case of general war, therefore, NATO defense depends upon an immediate exploitation of our nuclear capability, whether or not the Soviets employ nuclear weapons" and declares its commitment to a ready-to-counterattack version of deterrence:

> Our chief objective is to prevent war by creating an effective deterrent to aggression. The principal elements of the deterrent are adequate nuclear and other ready forces and the manifest determination to retaliate against any aggressor with all the forces at our disposal, including nuclear weapons, which the defense of NATO would require.[182]

Couldn't be clearer. During the heyday of nuclear buildup, there was (and, given the "fire and fury" rhetoric of President Trump, probably still is) no line the United States will not cross in the name of security, real or imagined. There have always been and will always be powerful people in key positions who are willing to use any and every weapon against an enemy.

Any country that expects to fight a nuclear war should reflect on possible outcomes: mere survival, all-out victory, or something in between. By the mid-1960s, some US and Soviet military thinkers assumed that all-out nuclear war was in fact survivable, even though the traditional goal of victory was unachievable. Historians Richard Dean Burns and Joseph M. Siracusa describe the mindset:

> A nominal military victory, to many political leaders, no longer seemed possible; rather, with sufficient forces to survive a first nuclear strike, a badly mauled state could still launch its own devastating response, which should give an aggressor crucial pause before starting a nuclear conflict.[183]

War may be politics by other means, but politics drops out of the equation when countries and whole civilizations cease to exist. Yet with only occasional slackening, the design, manufacture, and stockpiling of nuclear armaments continued apace for decades and may again be on the upswing.

For diplomats, a cessation of testing these armaments seemed a feasible entry point into overall arms reduction. Late in the Eisenhower years the United States initiated a three-year moratorium on nuclear testing. It didn't last. By the end of 1961, the United States had resumed testing, and the Soviet Union had detonated its fifty-megaton Tsar Bomba in the air thirteen thousand feet above the Barents Sea, generating human history's biggest explosion ever. Tsar Bomba unleashed fifteen hundred times the energy of "Little Boy" and "Fat Man" combined, an explosion so powerful that its blast wave circled Earth three times. As intended, the Soviet Union had proved itself supremely capable and supremely dangerous.

Still, the question of survival remained sufficiently vital to force American presidents and Soviet leaders to sit in adjoining armchairs every once in a while and try to work something out. In the meantime, American and European citizens by the millions, Catholic bishops, former Cold Warriors, and even staunch anti-Communists began to press for an end to the arms race. By the fall of 1986, Gorbachev told his aides, "[O]ur goal is to prevent the next round of [the] arms race. . . . [T]he leitmotif here is the liquidation of nuclear weapons, and the political approach prevails here, not the arithmetical one." By spring 1987, he and his foreign minister, Eduard Shevardnadze, shocked the Cold Warriors by agreeing to an earlier American proposal known as the zero option (devised by Richard Perle in 1981 and never intended to be acceptable).[184] It would cancel US positioning of hundreds of intermediate-range missiles in Western Europe if the Soviet Union would destroy its own arsenal of more than a thousand of the same. Now the Soviets also proactively offered to cut back on short-range missiles. And so, on December 8, 1987, in Washington, DC, Gorbachev and Reagan sat at the same table and signed the Intermediate Nuclear Forces Treaty.[185] Some saw this as Soviet capitulation, others as a modest victory for humanity.

In our own day, the issue of nuclear victory has been recast as a duel around the "right" of first use. Both sides of this duel contend that their approach offers the stronger deterrent and therefore the stronger promise of peace. And remember: given the ever-mounting population of satellites, almost anything that anybody says about nukes in general applies to nukes in, through, and from space.

The United States has long refused to relinquish the option of being first to use nuclear weapons in a conflict. "Success," states a recent US Air Force doctrine document, *Nuclear Operations,* "depends upon air, space, and cyberspace superiority. They provide freedom *to* attack as well as freedom *from* attack [italics added]. This is as true for nuclear missions as it is for any other form of attack." The Obama administration's *Nuclear Posture Review Report,* after noting that the "massive nuclear arsenal we inherited from the Cold War era of bipolar military confrontation is poorly suited to address the challenges posed by suicidal terrorists and unfriendly regimes seeking nuclear weapons," takes a position quite compatible with that of *Nuclear Operations:* "This does not mean that our nuclear deterrent has become irrelevant. Indeed, as long as nuclear weapons exist, the United States will sustain safe, secure, and effective nuclear forces."[186] Obama's successor, not in an official document but in a pre-Christmas tweet shortly before the start of his presidency, declared, "The United States must greatly strengthen and expand its nuclear capability until such time as the world comes to its senses regarding nukes," and followed that tweet the very next day with a boast on MSNBC: "Let it be an arms race. We will outmatch them at every pass and outlast them all." After one month in the White House, he said in an interview with Reuters:

I am the first one that would like to see everybody — nobody have nukes, but we're never going to fall behind any country even if it's a friendly country, we're never going to fall behind on nuclear power.

It would be wonderful, a dream would be

that no country would have nukes, but if countries are going to have nukes, we're going to be at the top of the pack.[187]

Official and off-the-cuff US policy may still embrace the option of first use, but many former officials now decry that option. During the throes of the 2016 election campaign, two of them made their case in the *New York Times.* One is a former vice chair of the Joint Chiefs of Staff and former commander of the US Strategic Command (USSTRAT-COMM), the other a former USSTRAT-COMM officer and former senior fellow at the Brookings Institution:

> [N]uclear weapons today no longer serve any purpose beyond deterring the first use of such weapons by our adversaries. . . . [B]eyond reducing [the] dangers, ruling out first use would also bring myriad benefits. To start, it would reduce the risk of a first strike against us during global crises. Leaders of other countries would be calmed by the knowledge that the United States viewed its own weapons as deterrents to nuclear warfare, not as tools of aggression.[188]

Writing for a blog rather than for the "paper of record" a few days after the end of the 2016 Republican National Convention, an oft-cited analyst in the fields of nuclear disarmament and outer-space security, Michael Krepon, was more blunt:

> The United States is not going to use nuclear weapons first in a conflict. Allies who believe otherwise are attached to a fiction and a psychological crutch. . . . Nuclear deterrence works best in the abstract. It relies on ambiguity and uncertainty. The belief system built around nuclear deterrence implodes once the first mushroom cloud appears. Since one nuclear detonation is very likely to lead to the next, prospects for escalation control depend on No First Use.[189]

Of course, the discussion doesn't stop at the US border, and a pledge of no first use is not synonymous with disarmament, even if every country were to make that pledge. NATO still firmly subscribes to keeping a stockpile of nuclear weapons: "Deterrence, based on an appropriate mix of nuclear and conventional capabilities, remains a core element of our overall strategy. . . . As long as nuclear weapons exist, NATO will remain a nuclear alliance." Today the global nuclear situation is a patchwork of warring fears, goals, and commitments: Brazil's constitution commits the country to only peaceful uses of nuclear energy; Iran has nullified its capacity to quickly produce nuclear weapons;

whereas Pakistan, traditionally at odds with its giant neighbor India, has the world's fastest-growing stockpile of such weapons. In July 2017 North Korea test-launched two intercontinental ballistic missiles, alarming not only its near neighbors but also the distant United States. A North Korean ICBM capable of delivering a nuclear bomb to San Francisco is now a near-term possibility. Also in July, at the UN General Assembly in New York City, 122 countries agreed on the language of a treaty to ban nuclear weapons forevermore. Not one of the world's nine nuclear-armed nations participated in drafting it. The treaty prohibits all signatories from developing, testing, producing, possessing, transferring, deploying, stationing, using, or even threatening to use "nuclear weapons or other nuclear explosive devices." Once fifty countries have signed and ratified the treaty, it will become law.[190]

Writing for the *Bulletin of the Atomic Scientists* three weeks after Krepon's blog post appeared, Ramesh Thakur, a former UN assistant secretary-general, characterized the global situation as precarious, especially in Asia:

[U]nlike the superpower tête-à-tête of the last century, the second nuclear age features a multiplicity of nuclear powers with crisscrossing ties of cooperation and conflict, fragile command-and-control systems, critical cyber vulnerabilities, threat perceptions occurring among three or more nuclear-armed states simultaneously. . . . This is a situation that needs all the de-escalation measures it can get.[191]

The *Bulletin of the Atomic Scientists,* by the way, is the home of the Doomsday Clock. The cover of its June 1947 issue featured a schematic clock set at seven minutes to midnight, indicating "the urgency of nuclear danger." Since then, the clock's hands have been moved twenty-one times in accordance with "whether events push humanity closer to or further from nuclear apocalypse." Seventeen minutes to midnight, in 1991, was the safest the world has been. Today, climate change and other potential threats influence the clock. In January 2015 the hands were moved to three minutes to midnight, the most dire setting since the height of the Cold War. They remained at three minutes to midnight until January 26, 2017, when they were moved half a minute closer. One year later — because "major nuclear actors are on the cusp of a new arms race," because of the "momentum . . . of nations' investments in their nuclear arsenals," because of "reckless language in the nuclear realm heat[ing] up already dangerous situations" — they were moved half a minute closer still.[192]

■ ■ ■ ■

Reshaping a military for potential nuclear holocaust was a foreground political project of an earlier era. Reshaping an air force for the potential deception, disruption, denial, degradation, and destruction of unfriendly space assets is a foreground military project of the present one. But let's be clear, the current project incorporates the earlier one. Ever since the triumph of the intercontinental ballistic missile, it's been one long unified endeavor, in which nuclear weapons have never disappeared from the catalogue of options and satellites have become ever more indispensable to the practice of war.

Simply because it exists, a lavish and partly nuclear arsenal purports to serve as a powerful deterrent. The bigger, deadlier, more diverse, more agile, and more numerous the weapons, the stronger the deterrence. Scare people enough, and they'll cower and retreat. Not that people don't disagree on the dynamics of deterrence and the best means of achieving (or ruining) it. But the proponents of biggest-arsenal-as-strongest-deterrent have generally held the bullhorn. As the US Air Force's doctrine document *Nuclear Operations* puts it, "Although nuclear forces are not the only factor in the deterrence equation, our nuclear capability underpins all other elements of deterrence."

Small problem: deterrence based simply on armaments may not succeed with every adversary. Some leaders may be undeterrable. Some may be smitten with fantasies of supremacy. No matter how lethal and savage the weapons, at least a few people envision circumstances in which they would be ready to use them. In fact, being seen as ready and willing to use those weapons might serve as an important component of deterrence.

Most analysts concede that an all-out space war is a remote possibility. Most of the weapons to conduct it don't yet exist, and the dangers to one's own territory, population, and satellites are colossal. Nevertheless, argues a former science and technology specialist at the Council on Foreign Relations, putting a few weaponized vehicles into low Earth orbit would be an exquisitely effective deterrent:

Space-based weapons would . . . provid[e] the ability to exercise gunboat-style diplomatic pressure anywhere on the globe, continuously and instantly. Currently grossly underappreciated, but no doubt soon to be realized, is the fact that space weapons will afford countries an omnipresent ability to influence the politics of other states by the mere possibility of force application.[193]

What about diplomacy not of the gunboat variety? Very iffy, even though indispensable. Diplomatic negotiations are never easy and are rarely a guarantee of lasting peace. Existing international agreements on outer space can be ignored. Diplomats can have too little room to maneuver and too few carrots to offer. Adversaries can be driven by fervor, enmity, or denial to the point of extreme inflexibility. Countries retain the right to renounce a treaty if and when its provisions come into conflict with the safeguarding of its national security.[194] There's also the option of just not ratifying a treaty, even if everyone else in the world has done so.

On the other hand, while the exercise of national sovereignty may seem relatively straightforward down here on the surface of our planet, it's almost unachievable in space. Space cannot (yet) be physically controlled by the military the way other battlespaces can. So nations don't have much choice. When it comes to space, they must resort to diplomacy.[195]

Another level of the diplomacy problem is that the Outer Space Treaty doesn't cover every possible weapon or intervention. That's one of the reasons so many countries and analysts have pushed for the broadening and clarification of its mandates and for a comprehensive definition of "space weapon." The orbiting of conventional weapons (kinetic-energy or directed-energy) is not explicitly forbidden by the Outer Space Treaty. Electronic interference, including interference with ground-based control of spacecraft, is not forbidden. Ballistic missiles that merely pass through space but don't achieve orbit are not forbidden. Antisatellite weapons launched from land, sea, or air are not forbidden. Space-based ballistic missile defense systems are not forbidden. The testing of antisatellite weapons in space is not forbidden. Orbital weapons based on cutting-edge physical principles (quantum entanglement, particle beams) would be permissible as long as they couldn't be characterized as weapons of mass destruction. Parasitic microsatellites and space mines would not be forbidden. Vehicles that would descend from orbit to attack terrestrial targets would not be forbidden. Short-term "pop-up maneuvers" in space would be permissible. On top of all this, verification and inspection remain thorny issues.[196]

All these omissions and limitations cause one to wonder how conducive to personal, national, or global security the Outer Space Treaty can be. Consider the 1996 Comprehensive Nuclear Test-Ban Treaty. The CTBT was an international diplomatic effort to extend and strengthen the watershed 1963 Limited Test-Ban Treaty. What the CTBT obligates its signatories to do is "not to carry out," "to prohibit and prevent," and "to refrain from causing, encouraging, or in any way participating in the carrying out" of "any nuclear weapon test explosion or any other nuclear explosion." It sounds comprehensive, especially given the phrase "any other nuclear explosion." But this is a ban on testing, not on weapons. It focuses on monitoring, on-site inspection, and verification, and there are several giant loopholes. One is simply the term "explosion." Preparations for an explosion are perfectly acceptable, just not an actual explosion, and experiments don't necessarily include explosions. Another loophole is that, under the oft-invoked Article 51 of the UN Charter, the use of weapons for purposes of self-defense is perfectly reasonable ("the inherent right of individual or collective self-defence if an armed attack occurs"), so the use of "defensive" nuclear weapons is not explicitly prohibited by the CTBT. A third loophole is that countries are perfectly free to keep a stockpile of nuclear weapons, and many do.[197] In any case, before the CTBT can become a tenet of international law, forty-four specified countries must not only sign but also ratify it. Of those forty-four, three have not even signed it, and another five — including the United States and China — have signed but not ratified it.[198]

By comparison, the Outer Space Treaty has

wide support: a hundred-plus states have ratified it or confirmed support through an equivalent process, and twenty-plus have signed but not ratified it. Might this be in part because weapons of mass destruction continually circling Earth a thousand miles above our planet's residents are — and are seen as — so much more dangerous to everybody and everything than such weapons sitting on Earth's surface?

The political options going forward are not numerous. We can have continued diplomatic efforts, unceasing military escalation, unceasing public protests, actual violence, or a state of permanent, paralyzing fear — mostly the same options envisioned in NSC 68 seventy years ago.

Diplomacy is a state-sanctioned, elaborate version of talking. It's how one approaches adversaries, rivals, neighbors, rogues, and bullies. As East Asia historian Bruce Cumings has said, "It's not something you do among friends."[199] If every person thought, felt, believed, and wanted exactly the same things, if individual, corporate, and national interests were always completely congruent, if priorities never shifted, if there were no such thing as a conflict of interest or a thirst for power, then diplomacy would be unnecessary. But when conflict looms, survival demands decades-long, on-again-off-again efforts at diplomacy, even if successes are sporadic and outcomes amount to little more than a few brief ceasefires or slightly fewer weapons in Earth orbit.

In 1935, best-selling author Sinclair Lewis wrote the dystopian novel *It Can't Happen Here,* which depicts the US presidential election of 1936 — an election that transforms the nation into a repressive, heavily militarized dictatorship. Most of Lewis's characters are bigoted, small-minded, small-town Real Americans. Very early in the novel the reader meets Brigadier General Herbert Y. Edgeways, U.S.A. (ret.), who is delivering after-dinner remarks at a Rotary Club fête in the town of Fort Beulah, Vermont, on a spring evening half a year before the election. His topic is "Peace through Defense."

"For the first time in all history," proclaims General Edgeways, "a great nation must go on arming itself more and more, not for conquest — not for jealousy — not for war — but for *peace!*" However, urged by the other after-dinner speaker, an indefatigable Daughter of the American Revolution, to " 'fess up!" and admit that "a war might be a good thing," the general answers,

I better confess that while I do abhor war, yet there are worse things. Ah, my friends, far worse! A state of so-called peace, in which labor organizations are riddled, as by plague germs, with insane notions out of anarchistic Red Russia! A state in which college professors, newspapermen, and notorious authors are secretly promulgating these same seditious attacks on the grand old Constitution! A state in which, as a result of being fed with these mental drugs, the People are flabby, cowardly, grasping, and lacking in the fierce pride of the warrior! No, such a state is far worse than war at its most monstrous!

. . . What I'd really like us to do would be to come out and tell the whole world: "Now you boys never mind about the moral side of this. We have power, and power is its own excuse!"[200]

Back in 1935, power was less powerful than it is today. It had fewer weapons at its disposal. It could do less damage. Today we have both megalomaniacs and megaweapons. We have elected officials who invoke the pro-armaments motto "Peace through strength" and talk about "not taking any cards off the table."

Fortunately, we also have spacefarers from many nations living, talking, and investigating biology and chemistry and medical research and astrophysics cheek by jowl on the International Space Station for months at a time — a test case for peace through cohabitation and collaboration. The space station is their little world. Unlike their mobile contemporaries down here on Earth, they can't up and leave at a moment's notice. And when spacefarers look at Earth, the separate countries of the schoolroom globe are nowhere to be found. All they see are blue seas, green and tan landmasses, and the white of cloud tops and glaciers: one world, indivisible, humanity's only home thus far.

Some of us may be waiting for the chance to colonize Mars. Not happening tomorrow. In the meantime, maybe we could try pretending we're astronauts — because in fact, considered in terms of the galaxy, not to mention the universe, we are.

8

SPACE POWER

Power is the capacity to achieve a specified outcome. Its sources, trappings, abuses, and allure can be detected everywhere. Sometimes power itself is the goal. "Power is not a means, it is an end," the mastermind of the Inner Party in George Orwell's dystopian novel *1984* tells his prisoner, a former skilled falsifier at the Ministry of Truth who has begun to exhibit a dangerous allegiance to facts. "The object of persecution is persecution. The object of torture is torture. The object of power is power."[1]

Whatever else power may be, it is not mysterious. Space, on the other hand, is. Its vastness challenges measurement and comprehension. Most of what drives it remains a puzzle, yet it contains everything we can ever hope to verify. It is unaffected by humankind. Ceaselessly in motion, space is the ultimate stage, presenting continual cycles of creation and destruction.

Space power is about having the knowledge,

the material capability, and the will to take strong, daring actions far beyond the limits of Earth's atmosphere. When politicians talk about space power, they're referencing nations that belong to the small but influential spacefarers club. When warfighters talk about it, they're referencing the means to deter, defend, and destroy and also, if warranted, to deny adversaries access to space for their own military or even civil purposes.

Space power enables communication, intimidation, surveillance, dominance, threat assessment, and, yes, scientific research in ways and at distances never before possible. It's the prime agent of remote control and instant action. The space update of Mao Zedong's aphorism "political power grows out of the barrel of a gun" might be "political power grows out of the high ground of space." Just as an eighteen-pound iron cannonball emboldened a dozen or so of General Washington's revolutionary troops to attack a line of British soldiers from a clump of trees a mile away, so has the satellite enabled a fighter drone operated by an American pilot sitting in a shed near Las Vegas to attack insurgents in the mountains of Afghanistan.

The path to multi-spectrum space power is long, difficult, and costly. First comes science, then investigation, engineering, multiple failures, and eventual mastery. Finally the aspirants arrive at control and, if sought, occupation and exploitation. Contemporary China understands this sequence and has been willing to pay the tab. Today China boasts four spacecraft launch facilities, compared with three each in Russia and the United States, and for some years it has been chalking up about the same number of successful launches as the other two space powers. In the fall of 2016 China put its second space station into orbit and plans to assemble a large, multi-module station in orbit by the early 2020s. It also expects to land on the far side of the Moon and launch a probe to orbit and sample Mars.

"Nations aspiring to global leadership in the 21st century must be space-faring," says the 2002 report of the Commission on the Future of the United States Aerospace Industry (on which I served).[2] China obviously agrees with this assertion. In 2013 President Xi Jinping declared that "the space dream is an important component of realizing the Chinese people's mighty dream of national rejuvenation." Two ideas coalesce in Xi's declaration: first, space power is key, not incidental, to overall power, and second, restoration of China's former greatness is key to the greatness of its future and therefore its power.

In that spirit, China released a series of twenty-first-century white papers announcing its intentions and accomplishments, so that the rest of the world could know how regularly the former turns into the latter. Before launching into its stunning portrayal of both progress and upcoming tasks, the 2006 white paper on space adopts a determined tone: "China has set the strategic goal of building itself into a well-off society in an all-round way, ranking it among the countries with the best innovative capabilities in the first 20 years of the 21st century." The 2011 white paper on space again includes a roster of achievements and agenda items but states up front that "[t]he Chinese government makes the space industry an important part of the nation's overall development strategy, and adheres to exploration and utilization of outer space for peaceful purposes" and that, having created the right conditions for rapid development of its space industry, "China [now] ranks among the world's leading countries in certain major areas of space technology." The 2015 white paper on military strategy characterizes space and cyberspace as "new commanding heights in strategic competition among all parties," so that "[l]ong-range, precise, smart, stealthy and unmanned weapons and equipment are becoming increasingly sophisticated" and "[t]he form of war is accelerating its evolution to informationization." The December 2016 white paper on space declares: "To explore the vast cosmos, develop the space industry and build China into a space power is a dream we pursue unremittingly." All these public statements stress that space — without which there would be no cyberspace — offers a coveted path toward comprehensive national power, a path China has been following at top speed in recent years.[3]

Back in the USA in the 1960s, President Kennedy sent America along the same arc, at a similar speed and with resonant goals. In September 1962, after vowing that his nation would neither "founder in the backwash" of the space age nor tolerate space being "governed by a hostile flag of conquest," he told the throngs at the Rice University stadium in Houston,

> [S]pace science, like nuclear science and all technology, has no conscience of its own. Whether it will become a force for good or ill depends on man, and only if the United States occupies a position of preeminence can we help decide whether this new ocean will be a sea of peace or a new terrifying theater of war.[4]

A couple of months later, Kennedy met with a few top NASA officials at the White House to tell them in less soaring language why an American lunar landing had to happen without delay. To achieve American "preeminence in space," he emphasized, going to the Moon had to become NASA's "top priority project":

> This is, whether we like it or not, in a sense a race. If we get second to the Moon it's nice but it's like being second any time. . . .
>
> I'm not that interested in space. I think it's good, I think we ought to know about it, we're ready to spend reasonable amounts of money. But we're talking about these fantastic expenditures which wreck our budget and all these other domestic programs and the only justification for it in my opinion to do it in this time or fashion is because we hope to beat them [the Soviet Union] and demonstrate that starting behind, as we did by a couple of years, by God, we passed them.[5]

Kennedy's demand couldn't have been clearer: the United States must become nothing less than the supreme space power. And yes, the expenditures were fantastic.

Less than seven years and $16 billion of Project Apollo spending later, Americans set their boots down in the Moon's dusty regolith just as the US troop presence in Vietnam was hitting its all-time high of more than half a million. In 1965 and 1966, while NASA's costly Moon-shot infrastructure was being built, total NASA spending exceeded 4 percent of total federal spending. Following the US lunar triumphs of 1969 through 1972, NASA's share of federal spending has hovered around (and mostly below) 1 percent.[6]

Plenty of US government expenditures on space don't show up in NASA budgets, however. As you might suspect, the Department of Defense also has a space budget. In 2012 it was $27.5 billion — half again as much as NASA's budget; in 2015 it was $23.6 billion — a third more. Far and away the world's most lavish military spender overall, the United States is also the most lavish spender on military space. In 2008 America shelled out almost ten times as much as the rest of the world combined. And that doesn't include spending on dual-use technology, which includes anything that can, with equal aplomb, carry out a nonmilitary task today and a military task tomorrow.[7] By 2016, America's military-space spending had fallen to twice that of all other countries combined. American spending dominance may level off further as other countries establish a stronger space presence, a recalibration likely to reverberate through international politics and investment.

For half a century, US space rhetoric echoed the assertive tone of Kennedy's public declarations. Military and quasi-military policy documents dating to the opening decade of the twenty-first century emphasized space (reflexively paired with cyberspace, like Adam and Eve or rice and beans)[8] as the battlefield of the future, a domain to capture, control, and exploit. By contrast, policy papers on robotic and human exploration of deep space have extolled the pursuit of science. Usually they invoke the need to collaborate rather than dominate — again, the idiom of science rather than conflict. But since the earliest days of the space race, when the word "race" was an unambiguous call to scientific arms, those in power have been adamant that science plays second fiddle to military capability.

Take the 2006 Air Force doctrine document titled *Space Operations.* Twenty-first-century warfare, it states, must be "space-enabled warfare":

> Today, control of the ultimate high ground is critical for space superiority and assures the force-multiplying capabilities of space power. Tomorrow, space superiority may enable instant engagement anywhere in the world. . . . [T]he Air Force views space power as a key ingredient for achieving battlespace superiority. . . . Space power should be integrated throughout joint operations as both an enabler and a force multiplier.

As recognized enablers, nonmilitary organizations would be subject to conscription: "Today, many civil, commercial, and foreign organizations contribute space capabilities to military operations [that] often must be requested on an unplanned basis. For example, the military may request NASA to redirect focus from a scientific mission to support a military operation."[9] So, neither NASA, Google, Intelsat, ExxonMobil, the Harvard–Smithsonian Center for Astrophysics, nor the European Global Navigation Satellite System Agency sits beyond the reach of the Department of Defense.

Counterspace Operations, a slightly earlier Air Force doctrine document, strikes a more combative note. It details the US military's commitment to disable or destroy anybody else's asset that does, can, or might interfere with any activities conducted in space by the United States or its allies. To protect and preserve America's space superiority, along with the option to go on the offensive when doing so, is a recurrent theme. Precluding a potential adversary from "exploiting space to their advantage" is essential. Deterrence is paramount. But if deterrence isn't enough, the military has available a portfolio of options, from camouflage and dispersal of space assets to satellite repositioning and "suppression of adversary counterspace capabilities." In other words, an offensive attack.

Attacking on the basis of mild indications of probability suggests a broad definition of self-defense. Taking the offensive to achieve this self-defense involves what the Air Force calls the Five D's: deception, disruption, denial, degradation, and destruction, carried out by technology ranging from radio frequency jammers and malicious codes to remotely piloted aircraft, missiles, antisatellite weapons, and lasers. Special-ops forces, too, may join the fray. Counterspace demands a broad reach.[10]

Leadership was President Kennedy's prime theme. Like assertiveness, it has had a long thematic life. Through at least the first decade of the present millennium, it was almost axiomatic that a world in which the United States did not lead would be intolerable:

- "To achieve national security objectives and compete successfully internationally, the U.S. must maintain technological leadership in space," states the 2001 report of the Commission to Assess United States National Security Space Management and Organization.
- "Other nations, against whom we compete for jobs in the global economy, are also intent on exploring space. If not us, someone else will lead in the exploration, utilization and, ultimately, the

commercialization of space, as we sit idly by," states the 2004 report of the President's Commission on Implementation of United States Space Exploration Policy (another commission on which I served).

- "The United States is the leading space power. . . . Therefore, the failure of the United States to remain in the forefront of space technologies would have both military and commercial implications. Advances in the military or civilian sectors will overlap, intersect, and reinforce each other. Consequently, the development in the United States of a dynamic and innovative private-sector space industry will be indispensable to future U.S. space leadership," states the 2009 report of the Independent Working Group on Missile Defense, the Space Relationship, and the Twenty-First Century, a project of the Institute for Foreign Policy Analysis.
- "For over 50 years, our space community has been a catalyst for innovation and a hallmark of U.S. technological leadership. Our space capabilities underpin global commerce and scientific advancements and bolster our national security strengths and those of our allies and partners," states the 2010 National Security Strategy of the

Obama administration.

- "U.S. national security is . . . increasingly predicated on active U.S. leadership of alliance and coalition efforts in peacetime, crisis, and conflict. . . . U.S. leadership in space can help the United States and our partners address the challenges posed by a space domain that is increasingly congested, contested, and competitive," states the unclassified summary of the 2011 National Security Space Strategy.

Meanwhile, the rest of the world has not stood idle. That phrase "congested, contested, and competitive" from 2011? In certain quarters, presumptions of US primacy are now being undercut by recognition of vulnerability, or at least multiplicity. Even back in 2006, *Space Operations* recognized three levels of advantage — space parity, space superiority, and space supremacy — and conceded that supremacy "may sometimes be an unrealistic objective because sources of space power include commercial and third party space capabilities, and it is difficult to completely deny an adversary's access to these capabilities." Quite true — unless you're willing to turn space into a permanent battlefield.

Power doesn't disappear as a goal just because there's competition. In 2014 a deputy assistant secretary of defense for space policy stated: "Space remains vital to our national security. . . . It's a key to U.S. power projection, providing a strong deterrent to our potential adversaries and a source of confidence to our allies." But that same assistant secretary, in the same testimony to a Senate Armed Services subcommittee, also stated that the strategic environment of space is evolving in ways that challenge the United States. Space, he conceded, is now a frontier open to all.[11]

By now, that last idea is a self-evident truth. It's clear that America's current asymmetric advantage in space is not impervious to erosion. A multilateral space regime has shot well beyond the launch phase. US power is under threat. And it's becoming ever clearer that the heavy US military, commercial, and civilian dependence on space assets creates its own heavy problems.

Here's how General John E. Hyten, commander of US Air Force Space Command, portrayed the situation in the summer of 2016:

> Despite world interest in avoiding militarization of space, potential adversaries have identified the use of space as an advantage for U.S. military forces, and are actively fielding systems to deny our use of space in a conflict. This is not without precedence.

Through the centuries, nations formed armies, navies and air forces to defend the right to use the global commons of land, sea and air. Securing our right to use space is simply an extension of an age old principle to guarantee use of global commons.

Space as a global commons is vital to commerce and is an essential element of Joint Warfare and global stability. Space is no longer a sanctuary where the United States or our allies and partners operate with impunity. Although Air Force Space Command (AFSPC) has a long history of providing space capabilities vital to the defense of our nation, the training and skills that sustained our space operations for the last several decades are not the same skills we need to fight through threats and win in today's contested, degraded and operationally-limited (CDO) environment.

Gone are the repeated references to US leadership, here replaced by "our competitive advantage in space." Gone, too, are the days when the job of US space forces was "simply [to] provide space services." Now the focus is on "improving combat capability against ever-increasing threats and complex scenarios."[12]

A year later — and a day after being sworn in to her new office — the secretary of the Air Force, Heather Wilson, took a similarly combat-focused stance at a Senate Strategic Forces subcommittee hearing. Flanked by three top Air Force space officials, she testified that space is no longer simply "an enabler and force enhancer for U.S. military operations." Now, it is "a warfighting domain just like air, land, and sea."[13] Space war: just another option.

Like General Hyten, European Union defense officials invoke stability as a vital goal. But they don't introduce it in the same sentence as warfare. Nor do power projection, operating with impunity, or space superiority figure in their discussions. One recent EU document lists stability alongside inviolability of borders, human rights and fundamental freedoms, rule of law, media freedom, and fair democratic elections. Another maintains that the EU's reliance on soft power has long been a point of pride for Europeans, even though they see that soft power isn't enough to address evolving realities. As for the global commons, the European Union sees it as primarily a civil, not a military, domain: "[O]ur security and prosperity increasingly rely on the protection of networks, critical infrastructure and energy security, on preventing and addressing proliferation crises, as well as on secure access to the global commons (cyber, airspace, maritime, space) on which our modern societies depend in order to thrive."[14]

The European Commission construes the lengthening list of space actors mostly in economic terms: Europe now faces "tougher global competition," "high dependence on non-European critical components and technologies," "a global value chain that increasingly attracts new companies and entrepreneurs." On the political front, Europe is far less concerned about contestation of leadership — America's usual worry — than about cooperation, the upholding of international standards, and sustainable access to space for all who reach for it.[15]

In short, European postwar interest in space — and science in general — has been predominantly nonmilitary. CERN, the European Organization for Nuclear Research, was created in 1954 to investigate the fundamental structure of the universe. The 1962 document that created the European Launcher Development Organisation — tasked with building a rocket that would end European dependence on US launch technology — stipulated that the organization limit itself to peaceful applications. The European Space Research Organisation, created in 1964 to develop European satellites, "should," argued one of its early proponents, "have no other purpose than research and should therefore be independent of any kind of military organisation and free from any official secrets act." The mandate of the European Space Agency, created in 1975 as the successor to the earlier organizations, has always been to "promote, for exclusively peaceful purposes, cooperation among European states in space research and technology." Plans for Galileo, the European global-positioning system, were initiated in 1999 so that the Continent would no longer need to rely on America's military-controlled GPS.[16] Getting out from under US dominance and control was a central motivation.

Contemporary Western Europe, having emerged from two devastating twentieth-century wars on its own soil within three decades, prefers conflict resolution to military confrontation. Between 2006 and 2008, the aggregate military spending of the almost thirty member states of the European Union hovered around €200 billion, about half that of the United States and the second highest globally. In 2010 it slipped to little more than a third of the US amount. But in response to the following decade's waves of desperate refugees from Africa and the Middle East, and multiple terrorist attacks in European cities, Europe's policymakers began to press for more military spending. Even amid these new pressures, however, as of 2012 Asia had displaced Europe as the number two military spender.[17]

Important fact: At present there is no such

thing as a standing, full-spectrum European Armed Forces, let alone a European Space Command. There's the European Defence Agency, which describes itself as a catalyst — promoting collaborations, launching new initiatives, introducing better defense capabilities. There's the Common Security and Defence Policy, which stresses peacekeeping, conflict prevention, and crisis management. There's the Organization for Security and Co-operation in Europe, with its three-sided view of security: political/military, economic/environmental, and human. There are the individual militaries and gendarmeries of the European Union's member states, and the United Nations' frequent multinational efforts at peacekeeping. None of these is a permanent warfighting machine.

Even less war-oriented are the EU's space-related agencies: the twenty-plus-member European Space Agency, the European Organisation for the Exploitation of Meteorological Satellites, the European Global Navigation Satellite System Agency, and the European Union Satellite Centre.

Other than its multinational, rapid-response Battlegroups, intended for crisis management rather than all-out war, the European Union's closest approach to a conventional military force is NATO, the North Atlantic Treaty Organization. Though the territory it defends now stretches from the western shores of

Canada and the United States to the eastern border of Turkey, far indeed from the North Atlantic, its name persists. Article 5 of its founding treaty, signed in 1949, declares that an attack against one member is an attack against all. This is the principle of collective defense, invoked for the first time in NATO's history following the terrorist attacks of 9/11. NATO has also intervened in conflicts between and within nonmember states, such as the Bosnian War, the Kosovo War, and the Libyan Civil War.

But NATO has no explicit space policy and no integrated space command. Should there be another Article 5 attack, the only space assets NATO could muster would be the space assets of its member states. And what Europe could presently contribute consists mainly of remote sensing, commercial communications, and data from weather and scientific satellites, along with Galileo and Galileo's comrade EGNOS, the European Geostationary Navigation Overlay Service, which augments GPS signals over Europe. Surveillance is something Europe could contribute, but any battle-ready spacecraft would have to come from the United States, whose forty-fifth president did not merely express displeasure at the substantial level of US funding for NATO but also, early on, pronounced the organization obsolete.[18]

The nonmartiality of European space policy

matches that of overall European security and defense policy. "Space Strategy for Europe," a European Commission plan adopted in late 2016, stresses that space technologies contribute to economic competitiveness, assistance to refugees, climate monitoring, and sustainable management of natural resources. The role of space in security and defense is presented as a side benefit, useful for addressing the increase in people fleeing their home countries and the consequent demand for increased border controls and maritime surveillance. Other named threats and risks include the proliferation of space debris, the effects of Sun-driven space weather on satellites and ground-level infrastructure, cyber sabotage of space assets, and the increase in space actors and space objects. Here the term "situational awareness" refers to space debris, not battle plans. Overall, Europe embraces a characteristically internationalist approach. Space power must be shared power.

But the focus continually returns to economics. "Space Strategy for Europe" pegs the EU's space sector, including manufacturing and ancillary services, at more than 20 percent of the total global value, and while the EU is now third globally in military spending, it remains second globally in public space spending, with a projected seven-year space budget of €12 billion.[19]

Not that the military dimension is absent.

The "Space Strategy" plan describes Europe's Earth-observation and global-navigation satellite programs as "purely civilian programmes entirely under civilian control." But that isn't the full picture. Services and data generated by those programs can, by request of an individual member state, be made available for emergency services, crisis management, border management, peacekeeping, and police operations. And the EU's other space capabilities

> can provide additional operational capacity for the implementation of the common security and defence policy, notably with regard to precision navigation (Galileo), surveillance (Copernicus), communications (Govsatcom), autonomous access to space (launchers) and situational awareness (SST), and can contribute to European strategic autonomy and non-dependence. Space and defence technologies are also closely interlinked.[20]

Sounds not too different from the US Air Force's declared power to conscript NASA or Google or even a foreign entity's satellite assets and data when needed. The biggest difference might be that in post-conquistador, post-Holocaust Western Europe, "defence" sounds more peaceable than "defense" does in America.

Compared with the EU's Galileo and Copernicus, the European Space Agency may be a less likely conscript for crisis management or peacekeeping. ESA's stated foreground issues are innovation, inspiration, and industry — not migration or terrorism. Science is fundamental to every ESA endeavor. Safety and security are auxiliary, mentioned only in connection with the need for European space scientists to pursue their work unimpeded by threats, whether from orbital debris or cyber sabotage. Glance at the press releases and webpages produced for ESA's triennial policy meeting of space ministers in December 2016, and you won't find the word "military." What you will find, repeated with variations, is simply the wish to pursue economic betterment and scientific enlightenment for all.[21]

A straight comparison between Europe's and America's official visions of space power in the twenty-first century reveals two partners with divergent agendas. But ask European and American astrophysicists about their agendas, and what you'll hear will strongly converge.

Space power — or, as some military commentators prefer to write it, spacepower — moved to the front lines of warfighting in early 1991. For forty-three days between January 17 and February 28, more than eighty thousand tons of bombs were dropped on the formerly thriving, fourth-largest military power in the world: the oil-endowed but heavily indebted nation of Iraq, which half a year earlier had invaded its small, still-thriving, also-oil-endowed neighbor and unyielding creditor Kuwait and had refused to leave despite a string of condemnatory UN resolutions. It was the first major US air war since Vietnam and the first major conflict following what then appeared to be the end of the Cold War. It was "the world's first satellite war" (Arthur C. Clarke), "the coming-out party for space support" (Everett C. Dolman). The US military, which had dropped most of the bombs, repeatedly called it the world's "first space war." Never before had a military force been so dependent on Earth-orbiting satellites for extensive support of its war effort: strategy, tactics, planning, communications, identification of targets, weapons guidance, troop movements, navigation, long-range weather prediction. Satellites reshaped them all, while also providing early warning of Iraqi missile launches and, of course, live TV coverage.[22]

Starting in the late 1980s, components of the military's own advanced space systems had been called into action for mission planning in Libya, minesweeping in the Persian Gulf, communications and weather updates in Panama. But not until the first Gulf War did the huge military potential of such systems become evident. In the words of a US Space Command assessment issued in early 1992, "Space systems supported every aspect of planning, control and execution of the war with Iraq."[23] Or, in the words of a retired Royal Navy rear admiral and a retired Royal Air Force captain, "It was the first real test under war conditions of the $200 billion US space machine."[24]

A tangled web of factors instigated the Gulf War and its consequences. The national boundaries of Iraq had largely been drawn by the League of Nations in the early twentieth century when the Ottoman Empire was dismantled, and Iraq, resentful at having been nearly cut off from the Persian Gulf, had thrice claimed Kuwait as a proper part of its territory and campaigned for its annexation.[25] The United States had, since the late 1970s, treated Saddam Hussein's Iraq as a favored partner against Iran, ignoring his regime's use of chemical weapons and giving Iraq tens of billions of dollars' worth of armaments on credit. Now Iraq's oil revenues were crashing while Kuwait was helping to keep world oil prices low through its own overproduction. America's demand for an uninterrupted supply of low-priced oil was an unspoken but key motivator for its attack on Iraq. The American-led Coalition forces' intensive 1991 bombing of Iraq's telecommunications infrastructure, power plants, water treatment facilities, government ministries, bridges, roads, airfields, ammunition depots, petroleum refineries, food-processing factories, retreating soldiers, and civilians shopping at markets turned out to be the prologue to the later, fuller US destruction of Iraq as a modern nation.

Here, however, our concern is the ascendancy of the satellite as enabler of war — all kinds of war, from assaults waged by the armies, navies, air forces, and cyber forces of nation-states to the scattershot terrorist acts of an individual propagandized through the Internet and in possession of a mobile phone.

GPS — the US Air Force's NAVSTAR Global Positioning System — was a prominent innovation of America's prosecution of the Gulf War. Today's GPS is 24/7, real-time, accurate and precise down to a few meters. With input from widely available augmentation systems, those few meters can become a few centimeters.[26] GPS is now absolutely ho-hum for most users. But back in 1991 it was an extraordinary idea that American soldiers, whose commanders had seen many more jungles than deserts, would be able to navigate the blowing sands of Kuwait, Iraq, and Saudi Arabia with the help of orbital assets that included just sixteen participating satellites of the planned constellation of twenty-four; that could provide longitude, latitude, and elevation data for about nineteen hours a day; and that yielded measurements accurate to fifteen meters, at best, but infinitely better than a paper map could offer. Ground troops could even traverse regions that would challenge seasoned Iraqi navigators.

GPS also assisted pilots. For the initial air strikes on Iraqi radar installations, for instance, Pave Low helicopters equipped with GPS partnered with Apache helicopters equipped with old-style Doppler radar. The Pave Lows led the way and pinpointed the targets, which the Apaches then attacked with Hellfire missiles. Even non-stealthy B-52 bombers equipped with GPS could enter the theater shielded by electronic silence. Early on January 17, 1991, seven B-52G Stratofortresses — described as "flying bomb trucks" by one aviation source — flew nonstop from Louisiana to Iraq, where they launched thirty-five GPS-equipped cruise missiles at key parts of the communications infrastructure, destroying most of their targets. GPS-equipped Air Force F-117A Nighthawks, the first stealthy plane ever used in combat, delivered laser-guided, semi-smart bombs at an average hit rate of 50 percent.[27]

When fully functional, GPS has three components: (1) a minimum of twenty-one orbiting satellites plus at least three spares, all of which continuously signal their ever-changing position and the atomic time at that position; (2) the individual receivers, which automatically calculate their own position based on the signals they pick up from multiple satellites in different positions; and (3) the ground control network of monitoring stations and antennas, which manage the satellites' flight paths and atomic clocks. The satellites occupy six different orbital planes in medium Earth orbit, at an altitude of about twenty thousand kilometers. The signals are sent as radio waves and, as with radar, can be distorted by the electrically charged ionosphere, through which they must pass on their way down to Earth's surface. To establish its position in longitude, latitude, elevation, and time, a receiver must detect signals from at least four different satellites. If you exclude elevation, it needs just three. The extensive sky coverage of today's GPS constellation enables receivers to pick up ample signals for all commercial and military applications.

It would not be unfair to call Gulf War–vintage GPS rudimentary. Six of the sixteen satellites were old R & D units pressed into wartime service; one of the sixteen had malfunctioned two months earlier. Also, as GPS signals made their way down to Earth's surface, they were susceptible to jamming.[28] Then there was the problem of the receivers, specifically that there weren't enough of them. At the start of the war the stockpile was negligible. In 1989 GPS had an encrypted military channel, with an accuracy of about fifteen meters, and an unencrypted civilian channel, with one-sixth the accuracy. But receivers capable of using the military channel were in short supply, causing not only the Pentagon but also the soldiers' family members to order thousands of portable commercial receivers for America's sons on the open market. This, in turn, forced the Pentagon to switch off the encryption, making the now-insecure military channel accessible to all. By the end of the war, some 4,500 commercial and 850 military receivers had been officially deployed to the US forces, plus all those unofficially supplied by loved ones — still far too few to serve the hundreds of naval vessels, the sixteen hundred combat aircraft, the forty thousand tanks, armored vehicles, and heavy artillery, and the half million US troops. The paucity of receivers had grave consequences; as a recent Air Force news story stated, "After the Gulf War, the U.S. Army announced it would install GPS receivers in all armored vehicles to help minimize fratricide, which became a major source of casualties in Desert Storm, most often caused by armored unit commanders lost in the featureless Iraqi desert or out of position during ground attacks."[29]

This was boom time for manufacturers of

GPS receivers. Non-military sales had already doubled year over year as hikers, boaters, pilots, and land surveyors learned of the device, which had moved past the experimental stage only at the very end of 1988. Initially, military sales accounted for only one-fifth of the foremost US manufacturer's total revenues. But as soon as Iraq invaded Kuwait, the Department of Defense ordered some eight thousand lightweight receivers from just that one company — a sale of more than $40 million, an amount that exceeded its total 1989 revenues. Suddenly the factory was running three shifts a day to meet demand.[30]

Despite its limitations, the Gulf War's GPS turned time-honored methods of navigation into an arcane, archaic craft and forever changed the planning and prosecution of war. Early commentators were blown away by its transformative power. Writing for the *New York Times* in 1988, at the dawn of the GPS era, science and war journalist Malcolm Browne evoked the momentousness of the neonate satellite constellation in a piece titled "New Space Beacons Replace the Compass":

To the captain of a clipper ship, a pocket-size gadget linked to artificial beacons in the sky that could infallibly guide a traveler to any point on earth might have seemed as remote from reality as a winged horse. But next year, anyone with a few thousand dollars will be able to buy just such a magic navigator. . . .

For the Defense Department, the completion of the Global Positioning System will be a milestone. Guided by the system, a missile traveling to the opposite side of the world could hit within a few dozen feet of its target, Pentagon officials say. The system could infallibly lead an assault team through trackless jungle to an enemy stronghold, a bomber to a single enemy building or a boat to a gap in a dangerous shoal. . . .

Without landmarks, sextant, star almanac, dividers or the other paraphernalia of conventional navigation, the user of a portable G.P.S. receiver can effortlessly read off the direction and exact distance along a great circle of any of 50 destinations stored in its memory. A glance at the gadget's liquid-crystal display also tells the traveler his own latitude and longitude to within 100 feet, his speed and course over the ground, and the probable time of arrival at his destination.[31]

Browne's exuberant prose harkens back to the accounts of those who waxed poetic at the power and joy of using the first navigational compass, the first spyglass, the first seaworthy chronometer.

Greater momentousness was soon to come, when a GPS precision-guidance package would be integrated into smart bombs and not just be part of the bomber aircraft's navigation system. But in the mid-1990s, when Gulf War–era laser-guided weapons were still the last word in precision — and GPS-guided gizmos such as the Joint Direct Attack Munition (JDAM, a conversion kit for smartening dumb bombs) and the Joint Standoff Weapon (JSOW, a winged, air-to-surface actual weapon) were not yet readily available — military theorist Colin Gray felt it useful to note the obvious:

> Systems that gather and provide information do not themselves fight the enemy. Ultra intelligence in World War II, whose potency as an enabling influence is beyond question, did not in itself sink any submarines or destroy any aircraft, although it did empower tactical combat units to do those things. The NAVSTAR global positioning system (GPS) permits economies of force in mission planning, but NAVSTAR itself can put no weapons on target. . . . It is not always obvious where space power begins and ends when information from satellites augments the potency of naval, air, and terrestrial military operations.

By the late 1990s, the full-scale GPS system was up and running, ready to direct the first generations of JDAM and JSOW plus an updated, GPS-endowed model of the Army

Tactical Missile System. Do these develop-ments weaken Gray's contention that GPS could not put a weapon on a target? No, he might answer, technology is always ancillary to strategy.[32]

Back in 1988, Malcolm Browne's sources had told him that handheld GPS receivers would start being issued to American soldiers in 1992 and that the first to get them would probably be special forces units. While GPS wasn't ready in time for the Coalition's inva-sion of Iraq in January 1991, it was definitely ready by March 2003, when a now mainly US and British invasion force overwhelmed the country a second time. The reasons, ac-cording to President George W. Bush, were "to disarm [it], to free its people and to defend the world from grave danger" and to "remove . . . an outlaw regime that threatens the peace with weapons of mass murder."[33]

In 1995 the system had achieved its full complement of twenty-four satellites; by 2003 there were twenty-eight. In 2000 the practice known as selective availability — the inten-tional degradation of publicly available GPS signals, implemented for reasons of national security — had been terminated. For a while at least, everyone, civilian or military, would always and everywhere have access to the same degree of accuracy. During the 2003 invasion, in fact, precision improved from just over fifteen meters to just over two meters. That improvement came from updating every satellite's navigation package each time the satellite began to rise above the horizon of the battle zone. "Errors that accumulated over time — [such as] ionospheric distortion and relativity effects — were driven to zero for a period of time," enthused one of the space ops commanders who carried out war support long-distance from the 50th Space Wing's operations center at Schriever Air Force Base in Colorado. "We hit them all a half-hour before satellites came into theater, and created this sweet spot over Iraq of less than 4 meters [precision]."[34]

Today GPS comprises thirty-one operational satellites plus a few that are decommissioned but still orbiting, ready for reactivation if needed. The twelve oldest ones were launched between 1997 and 2004; the twelve newest, between 2010 and 2016. Next-generation GPS III is waiting in the wings, the first two satellites to be launched by SpaceX. Now that the practice of degrading the signals available to civilian users has been discontinued, GPS transmits on several different frequencies that require different access codes, some of which are strictly or partially military. As essential a public good as roadways and running water, basic GPS nonetheless remains an Air Force program, funded primarily through the Department of

Defense, with modest input from the Department of Transportation.[35] In itself, that management and funding arrangement shows how militarily indispensable the system has become since its debut in Iraq.

Although the first, GPS is no longer the only global positioning system, and the names for the system itself are multiplying, two recent acronyms being PNT (position/navigation/timing) and GNSS (global navigation satellite systems). China has nearly completed its own version, Beidou, while the European Union's Galileo is well on its way. For some years, however, the most important one has been Russia's GLONASS. As with GPS, its full complement is twenty-four satellites. After becoming fully operational in 1995, the same year as GPS, it languished for several years as some of its satellites ceased operating and were not replaced until President Putin prioritized the system during the first decade of the twenty-first century. Available only to the military until 2007, GLONASS is now a more or less joint operation of the Ministry of Defense and the space agency, Roscosmos. Whereas GPS, as of spring 2017, had thirty-one satellites slotted into six orbital planes, GLONASS had twenty-seven, launched between 2006 and 2014, slotted into three planes just slightly nearer Earth's surface and inclined at a different angle to the equator — an angle that makes GLONASS more effective than GPS at higher latitudes, where so much of Russia is located. Moscow, for instance, sits farther north than the northern-most point of the contiguous United States.[36]

Today, many international smartphones use both GPS and GLONASS for greater coverage. Russia, in fact, has mandated that all its state and security apps use the dual system, although such a mandate could be dropped in response to a deteriorating political climate. But as the head of GLONASS said in 2014, the capability to receive and process signals from both systems increases not only the speed at which coordinates are processed but also their reliability, from 60–70 percent to "practically 100 percent" for ordinary urban conditions. Pointing out how dependency on a single system makes users vulnerable to denial or interruption of service, he argued that no individual country should unilaterally control infrastructure so crucial to every country and every economy:

The operator of the navigation system . . . has the option of either switching off the civilian signal for a specific area or of desensitizing it artificially. . . . This is not even about military conflict, as the threat of turning off the navigation switch can in itself be used to achieve political or economic aims. Therefore it is just a small step from a technological dependence on a narrow satellite navigation field to economic, political and military dependence.[37]

In other words, the deterrent power of threat must be circumvented by the power of independence. Which is an important reason for the global proliferation of separate global positioning systems — though interdependence can have its own deterrent power.

The Defense Advanced Research Projects Agency (dubbed "the Pentagon's brain" by its chronicler Annie Jacobsen) is taking independence one big step further. Having already funded the miniaturization of GPS receivers, DARPA has lately been funding the development of battery-powered, chip-scale atomic clocks that could function even in the absence of a satellite connection. Physicist Robert Lutwak, the program manager for this effort, called ACES (Atomic Clock with Enhanced Stability), notes that precise timing is essential not only for the Department of Defense but also for the infrastructure of everyday civilian necessities such as banking and electrical power distribution. Reducing reliance on signals from satellite-based navigation systems is key to improving resilience.[38]

The European Space Agency, too, has an ACES project, except that in this case the acronym stands for Atomic Clock Ensemble in Space, and its material incarnation is a large, peace-loving payload carried on the International Space Station, not a small object sized to fit in a warfighter's pack or pocket. Its main agenda is scientific inquiry into the fundamental laws of physics, not the facilitation of battle planning or banking.[39] But at the heart of both versions of ACES is the atomic clock, a mechanism that tracks the passage of time by registering the frequency of light emitted by a specific quantum leap — a transition — of electrons within the atoms of a chosen element. Currently, the second is defined as exactly 9,192,631,770 cycles of the light emitted from such a transition within the cesium-133 atom. It's easily replicable in the laboratories that need it, so scientists the world over can be sure their measured second is consistent with everybody else's measured second.

For now, though, the world relies on satellite-dependent global positioning systems, which have already transformed how billions of people and things navigate their appointed or chosen paths. For better or worse, GPS and its relatives have made countless inventions possible, from America's Predator and Reaper warrior drones to Dubai's single-passenger driverless air taxis to the management of crops via precision farming. So, unless you hope to re-create the conjectural voyage of Polynesians to Peru a thousand years ago, you need never again consult a map or compass, let alone an astrolabe or *kamal.* Whether you're in Moscow or Manhattan, rural Rajasthan or central Shanghai, global positioning will direct you to your destination — assuming you don't mind having your whereabouts tracked by eyes in the sky.

Ease of navigation through sand and cloud was only one of the benefits that Gulf War space assets supplied to Western warfighters. Swift communication — both long haul and intra-theater — was another.

General Colin Powell, chair of the Joint Chiefs of Staff during the Gulf War and later George W. Bush's secretary of state, maintained that satellites were the most important factor in building the command, control, and communication network for Operation Desert Shield. From a British military point of view, satellites "brought the Coalition Supreme Commander within a telephone call of the White House, Downing Street and the Elysée Palace."[40]

Not every channel on every communications satellite used during the war could be made available to the Coalition forces, so to maximize capacity, communications satellites were drawn from many sources, including the US Defense Satellite Communications System, the US Navy's Fleet Satellite Communications System, the US Air Force's Satellite Data System, NATO, and the British Skynet system. Also commandeered were several one-of-a-kind US military and experimental satellites as well as leased assets from the likes of AT&T, Bell, Sprint, and especially the International Telecommunications Satellite Organization (Intelsat) and the International Maritime Satellite Organization (Inmarsat). A pioneering constellation called Tracking and Data Relay Satellite System (TDRSS), operated by NASA and intended to support communication with spacecraft in orbit, was also called upon. The satellites were not all tuned to the same frequency. The network was cumbersome and complicated. As the ground forces advanced, hundreds of satellite ground terminals had to move with them. Because of the diversity of hardware and the multiplicity of controlling agencies, the parts didn't always work well as a whole. The system could be overwhelmed when communications conveyed not just words but images, which consume high bandwidth.

Yet despite the many problems, all those satellites at the military's disposal accelerated the pace of command during the hundred-hour ground war. Information could now pass rapidly from battlefield to tactical commanders and onward to the strategists. The benefits of those communications satellites accrued to television networks, too, and thus to the public at large. CNN became the go-to place for war news. Home TV reached the troops, while TV viewers at home could watch the skies of Baghdad erupting.[41]

Besides facilitating communication, satellites also provided early warning of Iraq's nighttime Scud missile launches against US allies Israel and Saudi Arabia. Three Defense Support Program satellites continuously scanned the skies for bright splashes of infrared, indicating the heat of a rocket plume. The DSP satellites relayed their data to US Space Command in Colorado, which would immediately confirm whether a given plume came from a Scud launch and, if so, would predict the impact zone and then rush the analysis (via satellite, of course) to Central Command in Saudi Arabia. By that time, the missile would have been airborne for about five minutes, traveling at thousands of miles per hour toward its target, leaving at most two minutes for Central Command to issue a warning and for its recipients to take action.[42]

Another group of satellites provided crucial data on weather. They warned of increased risk of a sudden sandstorm; determined wind direction, important for predicting the spread of any chemical agents; assessed dense coastal fogs that could drastically reduce visibility; monitored midday conditions in remote deserts that could presage evening thunderstorms, dust storms, or sandstorms; and tracked the dark smoke plumes from any of the hundreds of Kuwaiti oil wells deliberately ignited by the Iraqis. Coalition forces needed to know of any weather that could compromise weapons systems or force commanders to cancel air attacks. Hence the reliance on weather satellites, not only from US military (Defense Meteorological Satellite Program) and nonmilitary sources (National Oceanic and Atmospheric Administration) but also from the European Space Agency, Japan, and Russia. Weather imagery was "so important," states *Conduct of the Persian Gulf War,* the Pentagon's exhaustive final report to Congress, that the Joint Force Air Component Commander in Riyadh, Saudi Arabia, "kept a light table next to his desk to review the latest DMSP data," and the Tactical Air Command Center didn't finalize the daily "air tasking order" until after they'd received the most up-to-date DMSP images. Out in the field, though, DMSP weather reports were harder to access. The US Army dealt with the problem by buying German commercial receivers, which supplied data directly from civilian weather satellites passing above the Middle East, while the US Air Force dealt with it by faxing DMSP images to the field via landline, which slowed down the flow of information.

But again, as with GPS, the weather satellites offered key input that was otherwise unavailable to the warfighter in 1991. Sky conditions during the entire Gulf War were the cloudiest they'd been in more than a decade, and updates on cloudiness affected daily tactical decisions. Clouds could interfere with the laser beams used to illuminate targets, causing laser-guided bombs to lose guidance. Suddenly too cloudy? Scrap today's LGBs. Clouds could also cause a switch in targets, which is what happened because of the changes in cloud cover visible in two satellite images taken on the same January day less than two hours apart. One showed a clearing of cloudy skies over Baghdad; the other showed sunny skies in Al-Basrah beginning to cloud over.[43]

Now for the spy satellites. Another major enabler of the 1991 assault was the mountain of information — maps, photographs, multispectral images — made possible by the many surveillance, photoreconnaissance, and remote-sensing satellites that had been scrutinizing Earth's terrain from afar for decades. Their purposes varied, but their yield was uniformly useful. Millions of dollars' worth of images purchased from commercial remote-sensing satellites helped to track Iraqi troops, select targets, plan amphibious operations, execute aerial bombing campaigns, and establish land access routes for Coalition ground forces entering from the deserts of Saudi Arabia in the final hundred hours of the war. Today nobody has to be informed that Earth observation satellites can, as a 1992 US Space Command assessment phrased it, "show what is hidden from normal view."[44] But back then, a decade before the release of Google Earth, that was still a capability worth celebrating.

During the Gulf War, wide-field images — supplied by commercial remote-sensing satellites, both American Landsat and French SPOT — provided information on the overall lay of the land, substantial changes in terrain compared with previous images, open areas suitable for helicopter drops, and large-scale movement of troops or matériel. Landsat images showed features in both the optical and infrared bands of the spectrum, and they came only in color, with a resolution of ninety-eight feet at best — about the size of a blue whale. SPOT data came in either color or black-and-white and offered better resolution (thirty-three feet for the b & w) for making detailed maps. Arms-control specialists say those levels of resolution are fine for overall planning but insufficient to identify a military ground unit, which requires resolution of about fifteen feet, much less a tank, which requires about three feet.

Narrow-field, higher-resolution "spotlight" images — useful for planning and executing strikes against specific targets and in assessing bomb damage — came from America's KEYHOLE photoreconnaissance satellites, the KH-11 and the Advanced KH-11, both equipped with CCD technology, and from the Lacrosse cloud-penetrating radar-imaging satellite, which in 1991 was the sole spacecraft undeterred by bad weather 24/7. Russian and Japanese satellites also provided multi-spectral high-res images to the Coalition's war effort.

In early August 1990, when US military forces first deployed to the region, their maps of Kuwait, Iraq, and Saudi Arabia were between ten and thirty years old — old enough to make them undependable. Using then-current Landsat data, the Defense Mapping Agency produced an initial batch of updated maps by early September. At that point, Saudi airfields were little more than runways in the sand, so Landsat imagery was converted into the engineering drawings used to build huge modern air bases. Also, whenever and wherever the ground was disturbed — by troop movements, road building, a jeep's path over sand or grass — that disturbance would show up as a change in reflectivity when compared with an earlier image of the same locale. When two scenes are flashed quickly back and forth — as was done with two Landsat images of the Kuwaiti–Saudi border, the first taken in August 1990 and the other in December 1990 — most of what's there remains the same, making it possible to readily identify the site of any changes. This technique, pioneered by astrophysicists using analog photographs in the early twentieth century, was readily adapted to digital imagery in modern times, with a computer instead of a human reviewing the differences between images. It's a simple yet potent means to discover changes from one moment in time to another. Astrophysical discoveries generated by such comparisons include fleeting supernovas in the distant universe; Barnard's Star, the fastest moving star seen in Earth's sky; Pluto, the first object found in what would later be called the Kuiper Belt; salty water oozing down the inner wall of an impact crater on Mars; and the rapid movement of stars at the center of the Milky Way, suggesting the presence of an otherwise undetected supermassive black hole.

A few months after the end of Desert Storm, the French minister of defense declared, "Without allied intelligence in the war, we would have been almost blind." At that point, Europe had not a single military spy satellite, only France's remote-sensing SPOT. America, by contrast, as described by British military space advocates in the winter of 1991, had a "multi-purpose space armada with [a] massive supporting processing and communications chain."[45] No comparison: America had put heavy resources into military space; Europe hadn't.

By the beginning of the next engagement in Iraq, in March 2003, the US space armada was even more massive. Lessons learned from shortfalls and screwups in Desert Storm had catalyzed the creative squads at military space corporations, military laboratories, and military research agencies and had accelerated work already under way. Their varied efforts during the next decade were aimed at better connectivity; greater numbers of smaller, lighter tactical satellites to support commanders in the field; ultraviolet sensors for military surveillance spacecraft; increased numbers and carrying capacity of small communications spacecraft; better GPS, with better ground antennas and direct input to precision-guided weapons; computer networks that could fuse input from multiple sensors; electric propulsion for the transfer of high-mass payloads to geosynchronous orbit without resorting to huge rocket stages; lightweight solar arrays; a standard, all-purpose basic space "bus" to which mission-specific hardware would be attached the way a warhead attaches to a rocket; an experimental satellite to test autonomous navigation systems; infrared tunnel sensors that would be thousands of times more sensitive than piezoelectric sensors; supercomputers; a radiation-hardened, superhigh-speed integrated circuit chip set. All these and more were in the works by the spring of 1991, just months after the end of the first Gulf War.

As *Aviation Week & Space Technology* noted at the time, "The development of advanced military space technology for future defense and intelligence satellites continues to grow despite pressures on the U.S. defense budget." The 1991 budget for the Air Force's new Phillips Laboratory was more than $600 million, while the budget of the Defense Advanced Research Projects Agency, DARPA, was $187 million, set to jump the following year to $290 million. Money was flowing; miltech was soaring. NASA's budget was way larger, yes — almost $14 billion — but dual-use tech was a meaningful share of that. Furthermore, as a 1992 task force on US space policy stressed,

> There are not two space industrial bases, one for defense and one for the civil space program; they both draw from the same well. . . . [T]hey largely use the same industry, require virtual identical technologies, share the human skills, often use common facilities and certainly draw new entrants from the same academic institutions. Preserving the base for one helps the other, and vice versa.[46]

It's no wonder the multitrillion-dollar Iraq War started off in 2003 on a better space-technology footing than had its predecessor — although it might be more correct to say a better space-operations footing.

The intervening dozen years had yielded many technology upgrades. The GPS system now had its full complement of satellites. Most cruise missiles were now GPS-guided. Crucial satellite data could now be transmitted directly to users, bypassing a time-consuming layer of analysts. At least one of the multiple surveillance satellites would now image the battlefield every two to three hours while passing overhead. Even if you had continuous video taken by a strategically positioned geostationary satellite, it couldn't match the resolution of an ordinary surveillance satellite orbiting a hundred times closer to the ground. Plus, there were now many more, and more capacious, satellite ground terminals, with six times the bandwidth as before, and enough lightweight GPS receivers to supply the land forces with at least one per nine-person squad.

But in speaking to *Aviation Week & Space Technology* shortly after the end of Operation Iraqi Freedom — the first twenty-six days of the 2003 invasion — the senior military space officer at the Combined Air Operations Center in Saudi Arabia emphasized the organizational, not simply the technological, improvements: "Our whole intent was to bring an integrated effect to the battlespace. . . . It's not space for space's sake; it's space integrated with everything else to produce effects in the kill chain." Addressing the House Armed Services Committee at about the same time, Deputy Secretary of Defense Paul Wolfowitz agreed:

Our approach to OIF [Operation Iraqi Freedom] reflected the concept of the "battlespace," replacing the concept of the "battlefield."

On previous battlefields, we massed forces and achieved jointness by deconflicting rather than integrating forces, and conducted relatively symmetrical attrition warfare.

In this joint air, land, sea battlespace — which also includes space and the electromagnetic spectrum — we massed information and knowledge, used smaller formations that employed both lethal and nonlethal force in rapid and asymmetric ways, and conducted effects-based operations directed by flexible, dynamic command and control relationships. This synergistic battlespace makes each of our military service members more powerful in the effects they can achieve and confers greater protection from the enemy.[47]

There's more than a little salesmanship in those paragraphs. Bravado aside, the interval between Desert Storm and Iraqi Freedom did raise US commanders' comfort level with space capabilities, which translated into even greater speed of information processing and even more rapid communication at all levels of the campaign. To help carry out Iraqi Freedom's approach to warfighting, thirty-three thousand space-savvy support personnel were dispersed across the joint services and throughout the area of operations at twenty-one US and fifteen foreign sites. The combination of all those people and all that technology yielded unprecedented overall results. As summed up minus the hyperbole by Anthony H. Cordesman, a senior national-security analyst,

[T]his was the first large-scale war in which the United States could fight with 24-hour continuing intelligence satellite and sensor coverage over the battlefield, as well as the first major conflict where it could take advantage of full 24-hour coverage by the global positioning satellite (GPS) system.

The United States and Britain did not have total dominance of space. Iraq had access to satellites for television transmittal[,] had purchased large amounts of satellite photography . . . before the war, and it could make commercial use of the global positioning satellite system.

The coalition had so great a superiority in every area of space, however, that Iraq's capabilities were trivial in comparison. . . . The range of space-based communications and sensor assets, and the vast bandwidth the United States could bring to managing global military operations, allowed it to achieve near-real-time command and control and intelligence collection, processing, and dissemination. At the same time, GPS allowed U.S. and British forces to locate friendly and enemy forces and both target and guide weapons.[48]

Did America's increased space power benefit the twenty-five million Iraqis left leaderless by the fall of Baghdad and the capture of President Saddam Hussein in April 2003? In the end, the second US-led rapid assault on Iraq, followed by year upon year of every kind of violence, ripped the country apart. The victory proclaimed in the spring and summer of 2003 by so many US officials and commentators, and the "government in a box, ready to roll in" touted by one US commanding general, were delusions. Cordesman cautions in the opening paragraph of his 2003 book *The Iraq War: Strategy, Tactics, and Military Lessons,* "History is filled with efforts to make instant judgments about the lessons of war that ultimately proved to be based on false information and assumptions." He cannot bring himself to refer to the war's end or the declared peace without alerting his readers to the falsity, even the hubris, of using those words, and so he encloses them in scare quotes: "the cost of the fighting since the 'end' of the war"; "the 'peace' that has followed." He warns that US operations did not succeed in reducing violence and that the scale of the fighting might well broaden.

Cordesman's warning has come true. America's "shock and awe" campaign gave way to the disintegration of modern Iraq, the renewal of sectarian conflicts, and the emergence of regional jihadi armies. Even today in parts of the country, going to the market for a kilo of rice can be a gamble with death, and going to school an impossibility. In the wake of the war, aside from the role of reconnaissance and remote-sensing satellites in detecting the hidden presence of ancient ruins and the looting of ancient sites, America's formidable space assets did little to improve the lives of the people of Iraq.[49]

As for America's forthcoming wars, they will be waged with even more formidable space assets, fewer ground troops, more autonomous aerial vehicles, more nimble satellites and drones, more remote control, more cyber sabotage, and less blanket physical destruction. Space power and cyber power will produce less messy, less bloody results, which will also be less dangerous to those who impose them. Relying on satellite data to pinpoint the presence of purported or actual enemies such as Osama bin Laden, and to obliterate them and their associates or at least their families and neighbors, will become ever more commonplace. Video-gaming the destruction of a building will become ever easier and ever more removed from the torments of the battlespace.

Meanwhile, even if the costs of warmaking in dollars might diminish, the human costs on all sides will continue to defy restitution. Some of America's formerly cutting-edge space assets will "fade to black," as one space-tech advocate put it, as those of other space powers supplant them. At the same time, continued refinements in cyber sabotage could make calls for physically destructive space weapons less convincing. Analysts who have been withholding their use of the term "space war" until such time as weapons are launched from Earth orbit or until space-to-space attacks begin to occur may have to wait a lot longer. Space-based aggression amid mounting orbital debris will become an ever less sane approach to conflict, while space-based surveillance could become an ever more potent or invasive means of defense. As waged not only by America and its allies but also by jihadists and their social-media propagandists, war will likely remain more space-enabled than space-situated for some time to come. Milspace — the warfighter's segment of space, the site of surveillance and communications platforms, the locus of easy entry and easy control, so unlike the astrophysicist's vast and vastly hostile expanses — will foreseeably remain restricted to near-Earth orbit.[50]

All told, the space economy worldwide — encompassing military, civil, and commercial spending, both government and nongovernment — is pushing past US$330 billion, more than the current GDP of economic powerhouses such as Hong Kong and Singapore and more than fifteen times the annual budget of NASA.

The biggest chunk of the world's space economy is the fast-rising level of commercial activity, which amounted to more than three-quarters of global spending on space in 2016. Included in this category is everything from telecommunications satellites, satellite TV and radio broadcasting, launch services, insurance, the manufacture of spacecraft, the satellite monitoring of long-distance food shipments, high-res Earth-observation imagery, and space tourism to position/navigation/ timing (PNT) support for the delivery of your latest unnecessary Internet purchase or the optimal location for your imminent fishing weekend. In other words, a lot of freewheeling space power and profit looks ready for the taking — or, as the Space Foundation puts it, "the ongoing process of integrating space technology into all aspects of life" has spawned companies that "seek to monetize the growing torrent of information flowing from and through space systems."

Which is not to say that the governmental side of space activity is peripheral. Fifty-plus space agencies now exist around the globe, some of which are in nations handicapped by poverty, debt, inadequate infrastructure, and other ills. Yet in 2016, most countries increased their space budgets.[51]

Today no country can achieve or preserve economic viability and national security while ignoring space as a source of data, a channel of communication, and a domain of potential threats. Connectivity is key to modernity, to participation in the global economy and the global polity. No connectivity means no access, no presence, and no power. As Joan Johnson-Freese writes, "The imperative for connectivity to avoid being roadkill in the globalized economy makes access to the benefits of space a matter of national security that countries will not be denied."[52]

Consider China and India. Together, they're home to more than a third of the planet's population. Neither, despite the tempests that buffet their economies, treats space R & D or S & T as dispensable.

China's space program began in the 1960s, with a first successful satellite launch in 1970. Mao Zedong saw missile and space research as a bulwark against the Cold War superpowers, and so the excesses of the Cultural Revolution did not kneecap the space program. Subsequent leaders have given it far more than token support. China is the third global space power — the chronologically third nation to send its own citizens into space in its own spacecraft, launched by its own rockets from its own space center. Having achieved this status, it has, in the words of one Chinese general, reached a "new commanding height for international strategic competition." The rhetoric and emphasis have been military, though pure science has been gaining ground of late. China's space agenda has reached well beyond mastery and on into inquiry. Between 2016 and 2030 Chinese space-science missions will address two fundamental questions: the origin and evolution of matter, and the relationship between the solar system and human beings. The "strategic goal," as presented in the official publication of the Chinese Academy of Sciences, is "great scientific discoveries and . . . innovative breakthroughs."

Among China's recent breakthroughs was its August 2016 launch of the world's first quantum satellite, Micius (named for an ancient Chinese philosopher), which uses fiber optics to beam signals to stations on Earth in the form of "entangled" photons. Entanglement involves pairs of photons, simultaneously born and subsequently separated. Measuring one of them instantly grants information about the other, no matter the distance between them. Long viewed as a curious laboratory experiment across laboratory-length distances, this frontier method of quantum communication may one day establish a hack-proof global Internet.[53] And Micius is only the beginning of a larger program of satellite physics called QUESS, Quantum Experiments at Space Scale. As of this printing, the Chinese hold the entanglement distance record: photon pairs sent by Micius to two receiving stations on mountaintops in China, twelve hundred kilometers apart.

India, though still afflicted by what its first post-Independence prime minister, Jawaharlal Nehru, called "the problems of hunger, insanitation, and illiteracy" — problems that China has been more successful in alleviating — has had a space agency since 1969. India's emphasis has been more civil than military, with a dash of scientific research: the Indian Space Research Organisation (ISRO) plans, for instance, to study the Sun's coronal mass ejections and their impact on space weather. But as the space program's founder, physicist and industrialist Vikram Sarabhai, put it, "If we are to play a meaningful role nationally, and in the community of nations, we must be second to none in the application of advanced technologies to the real problems of man and society." In 1980 India joined the small community of nations to have used its own carrier rocket to launch a satellite. Soon India began to build the satellites themselves, dedicated mainly to two tasks: telecommunications, bringing its people not only Bollywood productions but also educational and public-health programming; and Earth observation, providing data on flood and drought threats, availability of freshwater, pest infestations, ocean conditions, arable land, mineral resources, and so on. But as we've seen, there's a paper-thin line between Earth observation and military reconnaissance. Any satellite that does the one can do the other, too.

India has also become a provider of launch services, and in mid-February 2017 staged a spectacular deployment of 104 satellites — most of which weighed just ten pounds — from a single rocket within the span of eighteen minutes, displacing Russia as the world's record holder, based on Roscosmos's 2014 launch of thirty-seven satellites from a single rocket. And as night follows day, January 2018 saw India successfully test a long-range ballistic missile capable of delivering a nuclear warhead to a target three thousand (or, according to Chinese sources, five thousand) miles away. An editorial in the *Delhi Defense Review* heralded the achievement, saying it "marks the arrival of India as a missile power."[54]

By way of comparison, consider the space histories of two wealthier nations, Canada and Japan, in which literacy is high, hunger is rare, access to affordable health care is a birthright, and life expectancy exceeds eighty years. Unlike China and India, Canada and Japan are partner states of the International Space Station. Also unlike China and India, their military expenditures, according to the Stockholm International Peace Research Institute, constitute less than 3 percent of government spending, compared with China's 6 percent and India's 9 percent — not to mention the even greater military expenditures by the two traditional space superpowers, the United States (more than 9 percent of government spending, but 36 percent of global military spending) and Russia (15 percent of government spending).[55]

For decades before the Canadian Space Agency was established in 1990, Canada had partnered with other countries on space initiatives, especially with the United States. Among the many early partnerships were an Ontario-built storable communications antenna used during the pioneering US manned flights in 1961–62 and Québec-built landing

582

legs for Apollo 11's lunar module, the vessel that safely landed Neil Armstrong and Buzz Aldrin on the Moon in 1969. Canada was the third country to build its own satellite and the first country to have its own communications satellite in geostationary orbit. More recently, Canada has made high-profile robotics contributions to the US space shuttle and the International Space Station. First came the highly versatile Canadarm, the robotic arm whose main job during the entire three-decade span of the shuttle program was to maneuver space stuff in and around the shuttle's payload bay. Then came the much more complex and versatile Canadarm2, which can move around and latch onto power and data fixtures throughout the International Space Station and which has not only maneuvered hundreds of tons' worth of payload but has also helped assemble most of the space station itself, docked and undocked visiting spacecraft, and given astronauts a foothold on more than a hundred spacewalks. Most recent is Dextre, a precision "robotic handyman" capable of executing delicate tasks on the ISS that Canadarm2 can't do. Eight Canadian astronauts have collectively logged more than eleven thousand hours in space, primarily on the ISS.[56]

All of the above are civil S & T efforts, which is not to say that Canada eschews military space capabilities. Together with the

United States, Canada is responsible for aerospace warning and control through NORAD, the North American Aerospace Defense Command. Its duties include monitoring artificial objects in space as well as detecting, validating, and warning of any attack against North America from air or space. But as late as 1997, Canada had no official military space strategy. And as late as 1998 — years after the American assault against Iraq in Operation Desert Storm had decisively demonstrated the uses of space systems in warfighting — a Canadian colonel would complain politely in print that "our military forces do not seem to be forward-looking in an attempt to make full use of space," that "Canadian space doctrine is virtually non-existent," and that the "lack of direction on the use of space at the operational or strategic level of war is striking."

Not until 2013 did the country's armed forces have their own surveillance satellite, Sapphire, built in Canada and launched in India. This satellite, however, takes no part in warfighting. It serves to guarantee the safety of the world's, not merely Canada's, space assets by monitoring every piece of space junk larger than ten centimeters across. Think of it as an eminently peaceable example of the military's traditional obligation to protect and defend, as well as the opening salvo in Canada's amplified space capabilities. Beginning in 2014 the Canadian Air Forces Space Cadre has provided Joint Operations Command with 24/7 support, which includes missile warnings, launch notifications, GPS status updates, and detection of any electronic interference directed at satellites. Today Canada's space operations are often joint efforts — for instance, its partnership in the US Air Force Space Command's secure communications constellation-in-progress, AEHF (Advanced Extremely High Frequency). An important independent Canadian space capability is the rapid processing of data from maritime radar surveillance, combined with a satellite system enabling automatic identification of ships.[57]

Now take Japan, which in 2016, spending triple what Canada spent as a percentage of GDP, was one of the world's top five space spenders.[58]

While JAXA, the Japan Aerospace Exploration Agency, was established in 2003 as a merger of three pre-existing aerospace organizations, the country's space history spans well over half a century. From the 1950s onward, Japan deployed scores of continually improved suborbital rockets to measure high-altitude phenomena such as electron density and ozone-layer depletion; by 1980, these rockets were powerful enough to serve as launch vehicles for payloads of several hundred kilos. The Institute of Space and Aeronautical Science at the University of Tokyo (absorbed into JAXA in 2003) launched its first satellite in February 1970, making Japan the fourth nation — after the Soviet Union, the United States, and France — to succeed in a satellite launch, with China becoming the fifth less than three months later. The following year Japan launched its first scientific satellite into orbit. Over the following decades, the Japanese orbited dozens of satellites dedicated to Earth observation and local positioning as well as astronomical observation and lunar and planetary exploration. These satellites have investigated tropical rainfall, greenhouse gases, global changes in Earth's climate and water, lunar minerals and topography, Venus's climate, and more. In the fall of 2005 Hayabusa, a self-navigating Japanese spacecraft that had left Earth in the spring of 2003, efficiently propelled by ionized xenon gas — a plasma ion drive — became the first to visit an asteroid, scoop up some of its surface material, and, five years later, return to Earth with the samples intact. Most notably, as its contribution to the International Space Station, Japan operates a large science module, Kibo (Hope), which has its own robotic arms and can simultaneously accommodate as many as ten experiments inside the module and about the same number outside, exposed to the space environment.[59]

Like Canada, Japan delayed its involvement with military space, but for different reasons. The delay was caused by the constraints of Article 9 of the country's constitution, which was drafted in 1946 by American occupation staff and presented to the Japanese as a fait accompli. It states, in part: "Aspiring sincerely to an international peace based on justice and order, the Japanese people forever renounce war as a sovereign right of the nation and the threat or use of force as means of settling international disputes." This approach was reinforced in 1969 by a resolution that the country's space program be restricted to peaceful, nonmilitary, nonnuclear purposes and that it be dedicated to the principles of independence, democracy, openness, and international cooperation.

Openness notwithstanding, space surveillance began in the 1980s, in the form of information-gathering satellites, which were presented as garden-variety technology that would assist the Self-Defense Forces and hence not be a blow to peace. But it was the test-firing of North Korea's first intermediate-range ballistic missile, Taepodong-1, in the airspace above Japan in August 1998 — combined with a series of high-profile Japanese space failures in the 1990s and 2000s — that slammed the door on the idea that Japan's space program could continue to dedicate itself solely to peaceful purposes. Taepodong-1 triggered defensive rearming and created pressure to develop further military cooperation with the United States, Japan's longtime space supplier and arbiter. The succeeding decade of rethinking and institutional rearrangements led to the 2008 Space Basic Law, which was passed by the Japanese parliament after a mere four hours of discussion. Watching China increase and diversify its military space capabilities, highlighted by its 2007 test of an antisatellite kinetic-kill missile, provided splendid motivation to pass that law. National security concerns about the proximity of a nuclear China and a nuclear North Korea moved to the foreground. In 2016 Japan released its fourth Basic Plan for space, the nation's first fully funded national-security-focused space program — a central element of which is an emphasis on space cooperation with the United States.[60]

And what about the largest country on Earth: the twenty-two-million-square-kilometer Union of Soviet Socialist Republics, succeeded by the still-largest, seventeen-million-square-kilometer Russian Federation? From the end of World War II until China, North Korea, Iran, Latin American drug cartels, and Islamist terrorists arose as threats to US national security, Russia remained America's primary adversary and space competitor. As treaty partner on disarmament and, in the post-shuttle era, America's sole means of carrying astronauts and supplies to and from the International Space Station, Russia became America's frenemy. But the gradual increase of US–Russia conflict zones around the world — along with widespread evidence of Russian cyber intervention and Internet-based fabricated news stories, some amounting to state-sponsored propaganda targeting the 2016 US election — has resurrected the political atmosphere of the formerly dead-and-buried Cold War.[61]

For decades, American analysts had presented conflict as binary, carrying forward with varying intensity the dualistic rhetoric of the Kennedy presidency: two superpowers, two mutually exclusive economic and political systems. In this vision, conflict between the superpowers was inherent and inevitable, while cooperation was unlikely or at best untrustworthy. Even so, US–Soviet cooperation in space had early diplomatic successes, such as the Outer Space Treaty of 1967, the bilateral "Cooperation in Space" agreement of 1972, and the joint Apollo–Soyuz Test Project of mid-1975, when manned spacecraft of the two superpowers docked and co-orbited for two days in low Earth orbit.[62] Clearly, space was not simply a sci-tech frontier but a frontier of détente as well.

During the 1980s — as Star Wars boosters and Star Wars denouncers intensified their tactics in the United States — the Soviet Union began to come apart at the seams, assailed by internal economic and political turmoil and by burgeoning non-Communist movements in the satellite states of Eastern Europe. Not a good time for basic research in space science, but a fine time for military applications. In 1986, both sides of the Cold War had some terrible weeks and months. In February, the US space shuttle Challenger disintegrated just seconds after launch, killing seven crew members. In April, a nuclear reactor at the Chernobyl power plant in Ukraine caught fire and released radioactive material into the surroundings, killing scores of workers and causing thyroid cancer in thousands of children because of contaminated milk. As William E. Burrows writes in *This New Ocean,* "The end of the beginning started in 1986. It was marked by two ruptures, widely thought to be technological, but whose causes ran far deeper. . . . Both made the planet wince."[63]

In March 1989 the Soviet Union held an election for the newly created legislature, in which voters had a choice of candidates for the first time since 1917. Many officials of the Communist Party went down to defeat. On the evening of November 9, 1989, at a press conference broadcast live on TV, a tired

East German bureaucrat blundered his way into an unexpected, half-unauthorized announcement that the Party had "decided today (um) to implement a regulation that allows every citizen of the German Democratic Republic (um) to (um) leave the GDR through any of the border crossings." Within hours, East Berliners were dancing atop the hated Wall, smashing it with hammers, and pouring into West Berlin.[64] In early December 1991, eleven of the fifteen former Soviet republics formed the Commonwealth of Independent States. Just weeks later, on the 25th of December, Soviet leader Mikhail Gorbachev, champion of *glasnost* and *perestroika* — "openness" and "restructuring" — resigned his appointed post. The following February in Washington, DC, independent Russia's leader, Boris Yeltsin, and America's president, George H. W. Bush, announced that the Cold War was officially over.

Consider even a smattering of Soviet economic statistics from 1990–91, depicting a Soviet Union in collapse and a newborn Russia beset by hardship. Forty percent of the USSR's huge 1990 grain harvest was left to rot or was devoured by vermin because of the sorry state of food processing and transportation. Factories faced crippling increases in the price of materials: the wholesale price index for industrial production rose 200 percent year over year during the first half of

1991 and kept on rising. The prices for ordinary goods and services increased by 240 percent from December 1990 to December 1991, but the incomes of ordinary people did not keep pace. Exports dropped by 33 percent in 1991. While the state paid its bills by printing more rubles, workers paid some of their bills with vodka. Aspirin, not to mention antibiotics, became rare. Life expectancy dropped.

Given the strife and privation across the USSR during Gorbachev's later years in office, you might assume that the formidable Soviet space program would have collapsed alongside the plummeting gross national product.[65] Not quite. It's complicated.

In 1985, during Gorbachev's first year as general secretary, writes James Clay Moltz in *The Politics of Space Security,* he and his advisors were fed up with "tit-for-tat military competitions," convinced that the achievements of space exploration could "serve as an effective locomotive for the scientific and technological revolution," and intent on breaking the military's stranglehold on the space program. Their first step was to create the outward-looking space agency Glavkosmos, designed to market some of the Soviet Union's impressive space services, hardware, and research to the West, thereby attracting much-needed hard currency, among other benefits. "Soviet space-related technologies now constituted one of the few areas where the USSR was on par with — and in some cases, even ahead of — world leaders," Moltz argues.[66] Among those marketable Soviet technologies were the Proton launch system and the Energiya booster rocket, capable of hoisting a hundred tons into orbit. And lest anyone forget, the list of Soviet space firsts is far longer than that of the United States. By 1985 the Soviet Union had successfully orbited six space stations in fourteen years, while America had orbited one.

In February 1986 the Soviet Union launched its long-lived space station Mir — eighth in its Salyut series but now given a new name, meaning "Peace." Later in the year Ronald Reagan and Mikhail Gorbachev met at a nuclear disarmament summit in Iceland, which might have succeeded but for the unshakable US commitment to Star Wars, the Strategic Defense Initiative. In fact, unbeknownst to Gorbachev at the time of the summit, the Soviet Union's own offensive versions of Star Wars were well under way: an orbiting laser cannon, Skif, and an orbiting missile-armed battle station, Kaskad. Though displeased to learn of Skif, Gorbachev permitted a demonstration launch via Energiya in May 1987 — minus the laser. The giant rocket did its part admirably, but the Skif tumbled into the North Pacific. Funding for Soviet Star Wars evaporated soon thereafter.

Once again, space moved to the frontier of détente. Civilian Soviet space officials presented proposals to the West, including a joint US–Soviet mission to Mars and a Soviet-funded international space station. By April 1987, even before the death of Skif, the future seemed to have brightened sufficiently to permit the US secretary of state and the Soviet foreign minister to sign an agreement on sixteen cooperative space-science projects, including several Mars missions.[67] In early December 1988, mere months after a couple of Soviet space disasters that had been intended to showcase the USSR as a "reliable and still innovative partner," Mikhail Gorbachev addressed the UN General Assembly. Besides announcing that his country would unilaterally reduce its armed forces and armaments, he reiterated the Soviet position that "activities in outer space must rule out the appearance of weapons there" and offered a Soviet radar station to serve as an international space facility under UN control.[68] Potential space conflict cost too much and imperiled far more. The Soviet Union could no longer afford any of it.

In his engaging 1994 memoir, *The Making of a Soviet Scientist,* plasma physicist Roald Sagdeev, director of the Russian Academy of Science's Space Research Institute (Institut Kosmicheskih Issledovany, IKI) from 1973 to 1988 — a man whom Carl Sagan describes in the foreword as having "instituted glasnost before Gorbachev" and as "helping to forestall an acceleration of the nuclear arms race both in space and on Earth" — details the downgrading of Soviet space science. Describing the "military-industrial iceberg" and his sense that the prime national task remained the "building of a huge military machine," he writes, with more than a hint of sarcasm, "In my space career, when I had to deal extensively with the defense industries, rockets in space were provided as a show of philanthropy from the military's enterprises." Secrecy and lies — "small lies for a noble cause" — were customary (not that the USSR invented the tactic) and continued into the Gorbachev era. Sagdeev does not mince words, calling the managers of his nation's space program "the corrupt barons of the Soviet space mafia" and "men from the stone age." Near the end of the memoir, he distinguishes between the agenda of science and the agenda of the military-industrial complex, a phrase he embraces, even using it as the title of one of his chapters:

The difference between the space science community and the space industry community rests on the fact that while industry instinctively prefers contracts that repeat projects and models already in existence, scientists need novelty. Old mundane results have no real scientific value. Our profession by definition requires us to move on to new designs. The difference between space science and technology is, essentially, a philosophical conflict between two ways of life.

... The space industry had developed a special capability to survive in [an] environment [where] everything was kept under secrecy. Now they were afraid to start a new life with glasnost.[69]

The day Gorbachev stepped down, his adversary Yeltsin assumed power in the new Russia. Yeltsin had been demanding that all space programs be suspended for several years. First things first. The economy, already reeling, suffered more blows from aggressive privatization of both resources and industrial capacity. Add to this picture the rise of oligarchs and mobsters and breakaway republics, fights over oil development and oil pipelines, slumping oil prices, and outflows of money to Swiss banks. According to the International Monetary Fund, in 1992 Russia's GDP dropped more than 14 percent, while prices escalated by more than 1,700 percent; in 1993 GDP declined another 9 percent, while inflation proceeded at an annual rate of nearly 900 percent. Not until 1997 did the Russian economy improve.[70] In the interim, the Russian Space Agency (partnering with American firms) began to sell time on its formerly secret, top-of-the-line spy satellites. Civilian flights on MiG-29 fighter jets became available for a few thousand dollars. And in 1993, Sotheby's in New York auctioned off two hundred pieces of the Soviet and Russian space programs, from logbooks and used space suits, to a slotted chess set designed to work in zero gravity, to a recovered burnt Soyuz capsule. That last item sold for $1.7 million. I was there. Not quite a garage sale, but the auction room smelled of victors divvying up the spoils of war — a long-fought Cold War victory. One heavy buyer was anti-Communist Texas billionaire and 1992 independent presidential candidate H. Ross Perot, who later donated his purchases to the National Air and Space Museum in Washington, DC.[71]

By 1996 Russia owed Kazakhstan hundreds of millions of dollars in unpaid rent for use of the main space-launch facility at Baikonur, in what had suddenly, in December 1991, become a separate country. With a 1996 space budget of $700 million, Russia was now next to last in space spending, just above India. While pleading for a pittance equal to one-twelfth of the US civil space budget, the general director of the Russian Space Agency told the Russian parliament in 1996 that almost half the space program's engineers and technicians had left because they couldn't survive on the average monthly salary of a hundred dollars. This had consequences. Russia's early-warning satellite network mistook a Norwegian scientific rocket for an attack by a US sea-launched Trident missile. Russian space reconnaissance went blind for half a year because some of the short-lived spy satellites were failing and couldn't be replaced. Yet another deep-space collaborative mission fell into the ocean, owing to the failure of Russia's Proton rocket.

Then, writes Burrows, "it got worse. Early in 1997, time and the severely shrunken budget began to catch up with the world's only space station, then in its eleventh year in orbit." Mir — by now a "jury-rigged flying tool kit that had gone six years beyond its design life" — almost bit the dust. Saving money had superseded safety. Instead of getting routine maintenance, Mir's parts were being used until they died. In 1998 GLONASS, too, began to falter, with no money to replenish the system with new satellites, as had been planned. Having carried out ten space launches in 1998, Russia's military managed to fund only four launches in each of the next two years. As for space science, said the director of the Institute of Space Research in Moscow, "We were barely functioning."

Corporate behemoths swooped in to offer their version of a rescue: joint ventures. Lockheed linked up with Khrunichev to market the Proton rocket; subsequent mergers and acquisitions pulled in Martin Marietta and the Energiya corporation, resulting in International Launch Services, which by 1995 had garnered 15 percent of the global commercial space-launch market. The Ukrainian design group Yuzhnoye teamed with Boeing, Energiya, and Norway's Kvaerner Maritime Group to form the Sea Launch partnership. Pratt & Whitney in America joined with Energomash in Russia to make and service the RD-180 rocket engine, for which the main customer became the US military. The French company Arianespace, the world's first commercial space-transportation company, joined with the Russian design bureau Starsem to market the Soyuz launcher.

The US government, too, found ways to support Russia and simultaneously serve its own national security interests. One was to funnel money through the International Space Station budget to ensure completion of the mission-crucial Russian service module. The US also paid the Russian Space Agency to fly seven American astronauts on Mir. With the crumbling of the Soviet state, the Reagan-era bugaboo of malign Soviet space stations orbiting above American cities evaporated, and the United States embraced the participation of former Soviet scientists and engineers in American space projects —

anything to keep brilliant but institutionally orphaned Russians engaged in science rather than bomb-making, whether for Russia or for an adversary of the United States.[72]

Vladimir Putin's accession to power on the final day of 1999 brought changes. Gradually the full constellation of twenty-four GLO-NASS satellites was replaced. A new but considerably over-budget space launch facility, Vostochny, was built inside the borders of the Russian Federation. Roscosmos was transformed into a state corporation following a slew of rocket failures and corruption scandals. And following the end of the US space shuttle program in 2011, Soyuz space-craft became — and remain — the only way to transport crew of any nationality, as well as supplies, to and from the International Space Station.

Yet despite the Putin-era improvements, Russia's space program continues to suffer overall. While Russia's deputy prime minister for space and defense personally benefits from sales of Russian-made rocket engines for US launch vehicles, a Roscosmos engineer earns less than three hundred dollars a month. While China's space program undertakes ever larger projects, Russia's becomes ever more precarious. While European and Japanese astronauts conduct manifold scientific experiments in their areas of the International Space Station, Russian cosmonauts lack even a proper laboratory in theirs. Roscosmos now operates within a ten-year (2016–25) budget of about $20 billion — barely more than NASA gets each year. Cutbacks and constraints on space science are the result of ongoing low oil revenues for the Russian government, Western sanctions against supplying dual-use items to Russia, even for obviously scientific projects, and prioritization of military spending.

Collaboration opens a path out of this cul de sac, although sanctions and scarce funds still throw up roadblocks. Among the current partnerships of Russia's Space Research Institute are ExoMars, with the European Space Agency, to undertake robotic exploration of the Martian surface and atmosphere; Venera-D, with NASA and the Jet Propulsion Laboratory, to develop an orbiter and lander to resume Russia's pioneering investigations of Venus that date back to 1961; and the Spektr-RG orbital X-ray observatory, with the German Space Agency, DLR, to perform a wide-angle survey of large-scale structures of the universe, including a hundred thousand galaxy clusters.[73]

Between 2013 and 2016, the annual Russian government space budget dropped by two-thirds, from just under $5 billion to $1.6 billion. Russia provided more than half the resupply missions to the International Space Station and all of the crew transportation in

2015, but the bad news is that two of its launches failed that year and fifteen failed between 2011 and the end of 2016. Fortunately, none were crewed capsules. But space experts are retiring, available labor doesn't always meet earlier standards, and quality control is waning. Evidently, one of the two original space superpowers has been sidelined, at least for now.[74] Instead, mastery of near-Earth and underwater nuclear-powered weapons — superpowerful, invincible, maneuverable missiles capable of unlimited range — has become Russia's trump card, as Putin declared in his 2018 state of the nation address:

> We talked about missiles that are capable of bypassing, avoiding, defensive barriers. We made no secret of our plans. We spoke openly of what we were about to do. We wanted to motivate our counterparts. This was in 2004. . . . Russia is a major nuclear power, but nobody wanted to talk to us seriously. They kept ignoring us. Nobody listened to us. So, listen to us now.[75]

The International Space Station — a massive, solar-powered orbiting habitat and laboratory that today involves fifteen countries, represented by the space agencies of Canada, Europe, Japan, Russia, and the United States, as well as support from commercial ventures — is humanity's quintessential space collaboration of the early twenty-first century. Its current primary purpose is neither astrophysical nor aggressive. Mostly it serves as a technological, physiological, psychological, sociological, and even agricultural trial run for human survival in deep space. Challenges faced by the rigorously trained inhabitants during the weeks and months they pass within its confines help show what is, and is not, possible in the non-terrestrial future of our species. Surely the capacity to live and work off-planet long-term would, for *Homo sapiens,* be the ultimate in space power.

Announced as a goal by Ronald Reagan in 1984, this sprawling vessel has had quite a history. The foremost triumph is that the ISS — all 460 tons of it, gradually assembled in low Earth orbit — actually exists and functions. But of course, struggles have accompanied that triumph.

For instance, the US Department of Defense, still fighting the Cold War and wary of international participation, inserted itself early in the negotiating phase of what had been billed as a civilian endeavor for peaceful purposes. In December 1986, DoD suddenly demanded assurances that the military would be able to do national security work on the station, unvetted by other nations, and that any notion of equal partnership would not "displace either the reality or symbol of U.S. leadership."[76]

Struggles also arose on other fronts. With the sudden end of the Soviet Union, Europe and Russia considered building a joint space station, but the United States decreed that Russia would become a partner in America's space station. China has repeatedly been frozen out of participation. Cost overruns far in excess of the absurdly low early estimates caused components to be canceled and even threatened an early miscarriage of the project as a whole. And, of course, terrestrial politics acts as a permanent irritant. On the other hand, little things like dinnertime on the station — when Russians have been known to exchange their savory canned stews for American ice cream — regularly elicits a positive renewal of international relations.[77]

Back in the 1950s, American talk of a space station was heavily infused with inflammatory language. Conceived as a laboratory and weapons platform, it would be constructed as a strictly national asset. The first nation to build it would assuredly control all of Earth. US space superiority would be the only acceptable state of affairs. That vision was superbly conveyed in a popular 1952–54 eight-part *Collier's* magazine series on the conquest of space, based on several workshops hosted by New York City's Hayden Planetarium that brought together engineers, scientists, artists, futurists, and journalists. It wasn't mainly pie-in-the-sky space enthusiasm, though. Rocket scientist supreme Werner von Braun shaped much of the content. William E. Burrows notes that the series also includes "one of the earliest and most influential examples of boilerplate cold war space rhetoric." Although these dramatically illustrated articles amounted to the public's first glimpse at their spacefaring future, including trips to the Moon, Mars, and beyond, they were also seasoned with fear-mongering:

> A ruthless foe established on a space station could actually subjugate the peoples of the world. Sweeping around the earth in a fixed orbit, like a second moon, this man-made island in the heavens could be used as a platform from which to launch guided missiles. Armed with atomic warheads, radar-controlled projectiles could be aimed at any target on the earth's surface with devastating accuracy.

The initial Soviet concept for a space station was only somewhat less terrifying than the American one. It was to have space for three cosmonauts, remote-sensing capabilities, capsules for sending imagery down to Earth, and a defensive cannon for use against an American attack.[78] Fortunately, both sides eventually set aside their plans for a death-dealing version of a space station.

Enter President Ronald Reagan. Halfway through his January 1984 State of the Union speech, in which the words "free" or "freedom" occur twenty-five times, Reagan told Congress,

> Tonight, I am directing NASA to develop a permanently manned space station and to do it within a decade.
>
> A space station will permit quantum leaps in our research in science, communications, in metals, and in lifesaving medicines which could be manufactured only in space. We want our friends to help us meet these challenges and share in their benefits. NASA will invite other countries to participate so we can strengthen peace, build prosperity, and expand freedom for all who share our goals.[79]

It was to be an American-run project with some input from selected subordinates — not a partnership of equals. Proposed at a time of escalated Cold War tensions, it was at least partly motivated by an ungenerous agenda; as political scientist Michael Sheehan suggests, "Reagan was trying to bankrupt the USSR, not only by re-igniting the arms race with SDI, but by re-launching the space race through the space station."

Soon the hypothetical space station would be named Freedom. Costs would soar, Congress would protest, reports would be written, plans would be revised, likely failure rates of individual components would metastasize, and predicted hours required for maintenance spacewalks would mount. In 1984 the estimated price tag was $8 billion; within a couple of years the National Research Council pegged it at $28 billion. Eventually the cost in billions would reach triple digits.[80]

With vilification raining down on the project, NASA and its contractors promised manifold scientific benefits and tens of thousands of jobs. Following the apparent end of the Cold War, the space station acquired its straightforwardly descriptive current name. The first component launched into space, in November 1998, was the Russian-built service module. The first American-built component went up early the next month. Canadarm2 joined the evolving station in 2001. The European Space Agency's Columbus laboratory linked up in February 2008, followed in March by the first segment of Japan's Kibo laboratory. More than once, Russia was bailed out by the United States so that it could deliver its agreed contributions. Today an elaborate web of intergovernmental agreements, memoranda of understanding, utilization rights, intellectual property rights, barter agreements, subcontracts, liability conventions, and public-relations campaigns more or less govern what takes place on, to, from, and around the International Space Station. Animating all that complexity, opines Joan Johnson-Freese, has been a two-pronged American agenda of nonproliferation and job creation. "If it had not been for domestic jobs and international politics," she wrote in 2007, "there would be no ISS."[81]

Truth be told, a number of people — including notable American scientists — would prefer that there be no ISS. They want space science, not space hardware, and especially not astronauts housekeeping the hardware.

In September 2007 Stephen Weinberg, a Nobel laureate in physics, declared during a workshop at the Space Telescope Science Institute in Baltimore, "The International Space Station is an orbital turkey." And that was just a warm-up. He went on to say,

No important science has come out of it. I could almost say no science has come out of it. And I would go beyond that and say that the whole manned spaceflight program, which is so enormously expensive, has produced nothing of scientific value.

This is at the same time that NASA's budget is increasing, with the increase being driven by what I see on the part of the president and the administrators of NASA as an infantile fixation on putting people into space.[82]

These sentiments, held by many accomplished academics who were in fact deeply influenced in childhood by the manned missions to the Moon, is neither new nor recent. A similar dissatisfaction with the back-burnering of science is evident in an impassioned, frustrated resignation letter from Donald U. Wise, chief scientist and deputy director of the Apollo Lunar Exploration Office, sent to NASA Associate Administrator Homer Newell in August 1969, just one month after Apollo 11 returned safely to Earth:

I came to the Agency because the scientific advisory boards to NASA on which I sat seemed to have little influence on the manned lunar program. After working inside the system to give science a more effective voice, I became convinced that the system was equally refractory to internal scientific advice. . . .

Until such time as the Administrator, together with the Associate Administrators, determines that science is a major function of manned space flight and is to be supported with adequate manpower and funds, any other scientist in my vacated position would also be likely to expend his time futilely.[83]

Some forty years separate those two critiques. Yet during that entire time, science averaged about 25 percent of NASA's budget, despite considerable variance from year to year. Two things are certain: when NASA is healthy and well-funded, so too is the agency's science portfolio; and money not spent on the International Space Station does not automatically flow into the science budget. Given the Cold War underpinnings of NASA's very existence, no astrophysicist should see NASA as our personal science-funding agency. We are the wagging tail on a large geostrategic dog, which makes decisions without direct reference to the desires of astrophysicists. Hegemony drives science because science piggybacks on geopolitics.

Collaboration and its slightly less demanding cousin, cooperation, are fundamentally hard to achieve. If achieved, they can then be stonewalled, unembraced, sabotaged, or used as a bargaining chip, which, for the ISS, could threaten its image, mandate, operability, and life span. For the United States, official manager and "hegemonic partner"[84] of the space station, to collaborate or not to collaborate is a highly politicized question. And usually the questioning involves either Russia or China.

On the morning of April 2, 2014, two weeks after Russia's annexation of Crimea, NASA issued an internal memo stating that the agency was suspending "all NASA contacts with Russian Government representatives." This struck some commentators as a risky move, because Russia could simply cut off US access to the Soyuz shuttle service, which had been the United States' only way of reaching the space station following the shutdown of its own shuttle program. Later that same day came another statement: "NASA and Roscosmos will, however, continue to work together to maintain safe and continuous operation of the International Space Station. NASA is laser focused on a plan to return human spaceflight launches to American soil, and end our reliance on Russia to get into space." Soon the United States and the European Union levied a range of sanctions against Russia, including the supply of certain high-tech US components important to Russian industry.

In mid-May 2014 Russia retaliated. Prior to this freeze, the United States had planned to keep the ISS operating until 2024, four years longer than the previously agreed mission end of 2020. Now Russia's deputy prime minister announced that his country would not only not collaborate on the ISS past 2020 but would also, in certain instances, stop exports of Russian rocket engines to the United States — a potentially crippling move that would ensure Russian space dominance in decades to come.

US Atlas III and Atlas V rockets use Russian-made RD-180 engines. US Antares rockets use Russian-made RD-181 engines. Dozens of important US missions, from deep-space probes such as the Mars Curiosity rover to spy satellites and early-warning satellites — as well as ISS cargo shipments — were boosted to their destinations by Atlas or Antares rockets. US military reliance on Russian rocket engines had become a dependency, a collaboration of sorts: United Launch Alliance, a partnership of Lockheed Martin and Boeing that in 2014 held a near-monopoly on launching US military satellites, manufactures the Atlas V rocket. Under the letter of Russia's new export restrictions, only rockets for military launches would be prohibited. But in practice it would become hard for the United States to import any rocket engines at all.

Congress responded to the 2014 Russian engine ban with an engine ban of its own. Although the ban was lifted in December 2015, certain members of Congress continued to press for its reinstatement. As of early 2016, according to the website NASASpaceflight.com, those who contended that Congress was "placing Russia's economy ahead of U.S. military and national security interests" were decrying those who wanted to guarantee "launch-market redundancies" until a US-made successor to the Atlas V's engine was truly ready for prime time. The pro-ban forces lost. By the end of 2016, Russian sales of rocket engines were back on track. TASS, the Russian news agency, announced that Russia would be delivering nineteen engines to the United States in 2017, pointing out that "the US Congress [had] imposed a ban on the use of these engines after 2019 amid deteriorating relations with Russia but later lifted it when it became clear that the United States would be unable to develop engines of its own in the next three years." All those threats, counter-threats, accusations, and grandstanding served only to temporarily rattle a billion-dollar agreement signed twenty years earlier for delivery of 101 RD-180 engines.

As for US–Russian issues around the ISS, the grandstanding has come to naught there, too. Russia's threat in 2014 to cease collaboration as of 2020 would have meant, in practice, that the United States would have to abandon its own part of the station. Why? Because the Russian part (specifically the service module Zvezda) includes systems used by the entire spacecraft, notably for life-support functions such as cooling, dehumidifying, and separating oxygen from water.

Back in 2014, Russia's straight-talking deputy prime minister had said, "The Russian segment can exist independently from the American one. The U.S. one cannot." By the following year, as space journalist Anatoly Zak writes, "cooler heads in Moscow prevailed." Russia would stick with the ISS collaboration all the way through 2024.[85]

At no point during all these tensions did Russia stop ferrying astronauts to and from the ISS via its Soyuz capsule. Maybe, just maybe, it's because NASA pays Roscosmos $82 million per seat for a round trip in 2018, up from $21 million in 2006.[86]

As the world's most ambitious space power, the People's Republic of China presents the United States of America with unique conundrums. To policymakers who are committed to American global dominance, China is an adversary, a threat — not an ally or potential partner. Collaboration, in their view, is unwise.

China is the twenty-first-century economic powerhouse. For 2016, the World Bank ranked China number one in a nuanced version of gross domestic product (GDP), based not simply on the total market value of all the country's goods and services but on the relative purchasing power of its currency. By that measure, the United States ranked number two. The economic fallout is telling.

For eight years, until it began to dump US Treasury bonds in late 2016, China was the USA's biggest foreign creditor. In 2017 the United States ran a balance-of-trade deficit — far more imports than exports — of $375 billion with China, exceeding the $350 billion racked up in 2016.[87]

As for space power, Joan Johnson-Freese contends that "the potential for China as a peer competitor to the United States is a consistent concern of those who view zero-sum competition as inevitable." Michael Sheehan contends that China "has no reason to accept America's self-appointed hegemonic dominance of space" and has "sought to negate it through a policy of encouraging multipolar modifications to the international space regime." The 2016 annual Department of Defense report on Chinese military capabilities states that although during 2015, "China demonstrated a willingness to tolerate higher levels of tension in the pursuit of its interests, [it] still seeks to avoid direct and explicit conflict with the United States." Longer-term, however, "China's military modernization is producing capabilities that have the potential to reduce core U.S. military technological advantages." One example of that modernization is a new branch, the Strategic Support Force, created by the People's Liberation Army on the final day of 2015. Its focus is space, cyber, and electronic warfare capabilities. Another aspect of the modernization, according to the 2017 version of the report, is China's increasing interest in missions beyond its periphery, including power projection, disaster relief, and the building of foreign bases. China's leaders, states the report, now seek stable relations and want to be seen as pursuing policies of peaceful development. But China is also increasingly assertive and "sees the United States as the dominant regional and global actor with the greatest potential either to support or disrupt China's rise." As for the Strategic Support Force, it's going great guns on counterspace capabilities, the manufacture of spacecraft and space launch vehicles, human spaceflight, and the in-space assembly of its very own space station.[88]

Unlike Europe, the United States has persistently opposed Chinese participation in the International Space Station. In the early days of space station planning, before China was seen as a real or imagined threat to US space dominance, the denunciation of Chinese human-rights violations was an easy way to exclude the People's Republic from America's space sandbox. The year the Berlin Wall fell, 1989, was also the year of the student-led Tiananmen Square protests in Beijing, which snared global headlines, ended with the shooting of both students and workers, and engendered an international outcry. The noisier the professed outrage against "Red China" and the cruelties of Tiananmen Square, the less likely became any invitation to participate in the ISS. Yet during the next decade, as space writer Brian Harvey recounts, "China made several pitches to join the ISS project, dropping heavy hints to visiting journalists and officials of other space programs, especially the Europeans." The US response was "an uncompromising brush-off."

That brush-off stemmed from the Cox Committee in the House of Representatives. Formed in 1998, the committee was driven by US readiness to view China as a malicious global actor. Its mandate was to investigate "any transfers of technology, information, advice, goods, or services that may have contributed" to the improvement of China's weaponry or intelligence capabilities. Its approach had the flavor of 1950s-style Mc Carthyism.

Johnson-Freese calls the Cox Committee's efforts "melodramatic," "sensational," "technically challenged and politically trumped-up." "Seemingly overnight," she writes, communication satellites "were described as threats to U.S. national security," and "as dual-use technology, [they] became subject to the same government controls as military satellites, tanks, or guns, for purposes of sale overseas." No longer was civilian space technology impressive, multipurpose engineering. Now it was potential armaments, subject to AECA, the Arms Export Control Act; USML, the US Munitions List, and its definitions of sensitive US technology; COCOM, the Coordinating Committee on Multilateral Export Controls; the Wassenaar Arrangement on Export Controls for Conventional Weapons and Dual-Use Goods and Technologies; and especially to ITAR, the International Traffic in Arms Regulations, which, when broadly implemented, thwarted the free exchange of scientific and technological ideas in the name of national security. A curious portfolio of political posturing, given that various other countries, aided by US efforts that had begun in the 1960s, were quite capable of producing that technology by the 1990s, and large swaths of relevant engineering advances were already in the public domain.[89]

The Cox Committee's final report opened with accusations against the People's Republic of China: that it had "stolen design information on the United States' most advanced thermonuclear weapons" and had "penetrat[ed] our national weapons laboratories." The highest-profile individual victim was Los Alamos scientist Wen Ho Lee, a Taiwan-born US citizen who in 1999 was labeled a spy, fired without review, and held for 278 days without bail in solitary confinement for downloading restricted data, until a federal judge ordered his release, expressing his "great sadness" at having been "led astray last December by the executive branch of our government . . . who have embarrassed our entire nation . . . by the way this case began and was handled." The government ended up dropping fifty-eight of its fifty-nine charges.[90]

All told, these measures to restrict space exports and freeze out a potential competitor may have created more problems than they solved. US aerospace jobs and global market share took a major, lasting dive. China began to pursue an independently cooperative path in space, including its own space station, which would welcome attachable foreign modules for long-term stays and foreign crew-transport vehicles for short stays. One bizarre piece of the picture is that a few months after the Cox Report came out, Congress voted to make China one of America's "most favored nation" partners in trade — except for aerospace.

Meanwhile, France and Britain began to work with China to develop "ITAR-free" spacecraft, circumventing US restrictions. Russian–Chinese cooperation continued. China's Long March rockets became an attractive, moderate-cost launch option available to other countries. By 2007 China had displaced Europe from third place as a satellite-launching power, behind Russia and the United States. By 2011 it had displaced the United States. James Clay Moltz concludes that China had gone a long way toward "successfully outflank[ing] the U.S. sanctions, although it also attracted unwanted attention to the continued, central role of the military in its space program."[91]

Despite America's obvious inability to keep pace with the fast-growing international space community, Congress was obstinate. In late 2011 both houses agreed to include a provision in the annual appropriations act that explicitly prohibited NASA from engaging in any contract or collaboration with China that would enable Chinese access to advanced Western technologies associated with "national security or economic security." Nor could Chinese officials even set foot in a NASA facility.[92]

So counterproductive, constraining, and ungenerous were these policies of exclusion that US officials, including not only heads of NASA but even presidents, periodically ignored them or at least pushed the envelope. In 2005 the United States gave China tracking data on space debris to help ensure a safe trajectory for a Shenzhou manned mission. In 2006 President George W. Bush proposed to President Hu Jintao that America and China would benefit from significant space cooperation in the unspecified future. That same year, NASA administrator Michael

Griffin met the assistant administrator of the CNSA on American soil and later, despite opposition from some members of Congress, traveled to China — though once there, he was forbidden access to the military-controlled space facilities. Following their summit in 2009, President Obama and President Hu called for "expanding discussions on space science cooperation and starting a dialogue on human space flight and space exploration." In 2010 Charles Bolden, the next NASA administrator, had a better visit to China than his predecessor's. In 2014 satellites were removed from the ITAR list. By 2015 discussions about inviting a Chinese taikonaut — *yuhangyuan,* "traveler of the universe" — aboard the International Space Station were under way. John Logsdon, founder of the Space Policy Institute at George Washington University, points out that the 2011 prohibitions are merely bilateral and that welcoming China into the multilateral ISS would be a politically smart "escape route from current limitations." Johnson-Freese and Sheehan point out that the docking mechanism for China's manned Shenzhou spacecraft is a Russian design, already in use on the International Space Station to accommodate both Russian and US vehicles. Nevertheless, no taikonaut has yet served as an ISS crew member. But as Johnson-Freese suggests, China's own evolving Tiangong space station, which will enjoy the co-operation of the European Space Agency, stands a chance of becoming "the de facto international space station" when America and Russia say goodbye to the ISS.[93]

On January 20, 2017, Donald J. Trump was sworn into office as president of the United States. One week later he issued an executive order permanently barring refugees from Syria and temporarily refusing entry to immigrants from seven Muslim-majority nations, whether the immigrants were first-time applicants or returnees from travel abroad. Thousands of people and scores of lawyers massed at airports in protest, and the order ended up in court.

Experts in many fields used their power and prestige to decry the ban on immigration and travel. Scientists were no exception — after all, the laws of the universe transcend nationality, ethnicity, and genetic heritage. The International Astronomical Union's secretary general urged that "any new or existing limitations to the free circulation of world citizens . . . take into account the necessary mobility of astronomers as well as human rights at large." The Royal Astronomical Society characterized the ban as "hinder[ing] researchers from sharing their work with their peers, a fundamental tenet of scientific endeavour." Almost two hundred American scientific organizations and universities added their names to a letter to the president warning that the ban would "discourage many of the best and brightest international students, scholars, engineers and scientists from studying and working, attending academic and scientific conferences, or seeking to build new businesses in the United States." Even before the new president took office, physicist Richard L. Garwin organized an open letter to Trump that was signed by several dozen specialists in "the physics and technology of nuclear power and of nuclear weapons," including a number of Nobel laureates. The signatories contended that the multilateral Iran Deal — often derided by the president during the election campaign as "the worst deal ever negotiated" — was in fact "a strong bulwark against an Iranian nuclear weapons program." To date, this deal is still in place.

In late January, just days after huge women's marches took place across the world, concerned scientists made a decision to hold a March for Science in Washington, DC, on April 22, 2017. More than three hundred scientific organizations, including the American Astronomical Society and the Planetary Society, signed on as partners. On the designated day, the global roster of "satellite" marches numbered 610. Why march? "People who value science have remained silent for far too long in the face of policies that ignore scientific evidence and endanger both human life and the future of our world," answered the organizers. "We face a possible future where people not only ignore scientific evidence, but seek to eliminate it entirely. Staying silent is a luxury that we can no longer afford."[94]

What *can* we afford?

As of late March the new administration's budget ideas for the remainder of FY2017 included a $52.3 billion (10 percent) increase for the Department of Defense, coupled with large percentage cuts in most other departments, agencies, and programs, from the State Department (29 percent) to the Environmental Protection Agency (31 percent) and the Advanced Research Projects Agency–Energy (50 percent), along with the elimination of funding for indulgences such as the National Endowment for the Arts, the National Endowment for the Humanities, and the Corporation for Public Broadcasting. NASA was to squeak by with a one percent cut. But in early May 2017, Congress had other ideas: a 3 percent increase for NASA Science inside a 2 percent increase for NASA overall. ARPA–E got a 5 percent increase, while the NEA, NEH, and CPB were held flat. For FY2018, funding for the Department of Defense's S & T programs jumped 6 percent, with applied research and advanced technology development leading the way.

Turning designs into operational capabilities was the goal, with directed-energy weapons a popular line item.[95]

Reaction against America's forty-fifth president, his administration, his supporters, and his party has come from many quarters. But with any increase in popular resistance comes an increased possibility of retaliation by those with more obvious, more weaponized power. In a confrontation between the many and the few, military power is never irrelevant. Deployed on orders from any government, military power is not autonomous. It is a tool of policy. And as the military has told us in hundreds of ways in thousands of documents, space power — especially the many kinds of satellites in Earth orbit — is now an indispensable piece of the warfighter's arsenal. As with other forms of power, near-Earth space power can be used for both protection and persecution, against an individual or a group, a building or a bridge, a domestic threat or a foreign militia. It can be used against the citizenry at large. The many varieties of surveillance can be both legitimate and illegitimate, snaring known enemies and unforeseen attacks as readily as random passersby and random trysts. Satellite surveillance of North Korean missile-launch sites provides a crucial source of forewarning for North Korea's neighbors, whereas the satellite tracking of cars in Xinjiang Province, where the

Chinese government has mandated that a positioning system be installed in all vehicles as part of its anti-terror campaign, smells more like government overreach and invasion of privacy.

Space power as embraced and exercised by the military seems worlds away from the power of space as understood by the astrophysicist. Yet as we've seen, they intersect surprisingly often, to mutual advantage. Across history, the roster of nations that wielded the most power on the world stage — military as well as economic — strongly coincides with the roster of nations whose scientists were the most knowledgeable about the doings of the universe.

What propelled America to the Moon was not science or exploration, but fear and competition with the Soviet Union. A contest of worldviews. A battle of political and economic philosophies. Might the subsequent rocket-rise of China, in every way that matters on the world stage — economic, political, technological, military — propel America back into space? Is a formidable spacefaring adversary a more powerful inducement than a peaceable ally? In the 1960s, faced with a Cold War space adversary, the United States placed the boot-prints of twelve astronauts on the dusty lunar surface. Since then, faced with peacetime ISS collaborators, our astronauts have stayed in low Earth orbit, boldly going where hundreds have gone before.

Cosmic discovery is often enabled even when it's not the driver — and even when the show of force is not weapons but technological bravado. All Mercury, Gemini, and Apollo astronauts except one served in the US military. Yet it was NASA, a civilian agency, that sent them into space. Soft power at its finest. Science budgets didn't pay for the Moon landings, but science certainly benefited. The astrophysical history of the Earth–Moon system and the geology of the lunar surface came into sharp focus only after the rocks collected by Apollo astronauts were returned to Earth for analysis. Interplanetary space is the next arena where the soft power of technological bravado, augmented by the lure of unlimited resources, urges us to take flight.

9
A Time to Heal

On July 21, 1969 — the day the *New York Times* banner headline read "MEN WALK ON MOON: *ASTRONAUTS LAND ON PLAIN; COLLECT ROCKS, PLANT FLAG*" — the paper also provided space for reactions from several dozen notable individuals: the Dalai Lama, R. Buckminster Fuller, Jesse Jackson, Charles Lindbergh, Arthur Miller, Pablo Picasso. Some were enthusiastic, some were ambivalent, Picasso was completely uninterested. The admired historian of cities and technology Lewis Mumford was disgusted.

Five years earlier, Mumford had received the Presidential Medal of Freedom. Now he felt impelled to describe the foremost scientific and technical achievements of the modern era — rockets, computers, nuclear bombs — as "direct products of war," hyped as research and development

for military and political ends that would shrivel under rational examination and

628

candid moral appraisal. The moon-landing program is no exception: it is a symbolic act of war, and the slogan the astronauts will carry, proclaiming that it is for the benefit of mankind, is on the same level as the Air Force's monstrous hypocrisy — "Our Profession is Peace."

Mumford also painted America's Moon program as a ravenous beast, maiming or devouring all other human enterprises:

It is no accident that the climactic moon landing coincides with cutbacks in education, the bankruptcy of hospital services, the closing of libraries and museums, and the mounting defilement of the urban and natural environment, to say nothing of many other evidences of gross social failure and human deterioration.

Saying technological triumphs had brought the "moonstruck" human species to the brink of catastrophe, Mumford called out the proponents of space exploration for their duplicity in lavishing support on the "power elite" while making "the scientifically uninformed believe that a better future may await mankind on the sterile moon, or on an even more life-hostile Mars."[1]

Yet many of the world's inhabitants derive conspicuous collateral benefits from scientific and technical advances that started life as military projects. Communications and weather satellites, GPS, medical technologies, and mobile phones help both the farmer in rural India and the surgeon in a Manhattan hospital.

As a form of protection, militarization of space might seem inevitable, even desirable, as a kind of shield for our growing orbital assets. But weaponization arrives close on the heels of militarization. On the other hand, humanity has officially embraced a peaceable space agenda. Drawn up by the UN Committee on the Peaceful Uses of Outer Space, the 1967 Treaty on Principles Governing the Activities of States in the Exploration and Use of Outer Space, Including the Moon and Other Celestial Bodies is ambitious and inspiring. Yet who among us believes that humans will act peacefully in space? Space is not a magical place where somehow, suddenly, everybody is friendly. We remain the same species, with the same primal urges as our tribal ancestors. How about working on the peaceful uses of Earth? Once we figure those out, maybe we'll be able to non-delusionally envision the peaceful uses of space.

One way to assess a society is to examine how it rewards or punishes those who act on primal urges, how it attempts to encourage, channel, or inhibit those urges. But is war primal? That civilization exists at all, that at any given moment most people and most nation-states are not waging war on one another, implies that we are not entirely hapless victims of an opportunistic compulsion awaiting a time to kill. We may also be capable of opportunistically seizing a time to heal.

Being a scientist, when I think of how and where and when healing could take place, I think of knowledge, rational analysis, co-operation. I think of what it would be like to live in a country — let's call it Rationalia — in which all decisions that affect the population as a whole would flow from a single constitutional tenet: "Laws shall be based only on the weight of evidence." Which means that where evidence is inconclusive, there can be no law.

In Rationalia, I contend, space exploration could conceivably serve as the ultimate healer, offering the high road to peace. To talk about sources of peace, you have to ask, What have been the causes, costs, and casualties of war? One is a scarcity of natural resources: oil, freshwater, salt, nitrates, ores, guano, shipworthy timber. Dwindling or interrupted access to each of these commodities has figured in past armed conflict.[2] So-called rare earth metals, such as yttrium, dysprosium, and neodymium — along with others that complete an entire row of the

631

periodic table of elements — could easily join this list.

Tech sectors thrive on rare earth elements. Without them, America's electronics, defense, and green-energy industries would implode. We wouldn't have satellites, smartphones, lightweight laptops, jet engines, missile guidance systems, antimissile defense systems, nuclear-reactor shielding, lasers, catalytic converters, rechargeable batteries for hybrid vehicles, magnets for speakers and headphones, advanced wind turbines, LED lighting systems, MRI scans, or energy-efficient air conditioners. About 90 percent of the world's supply currently comes from China. Other sources, in descending order of productivity, are Australia, Russia, and India. Until 1989 the United States — specifically, the open-pit Mountain Pass Mine in California — was the world's main producer. But after supplying plenty of europium for the red tones in color TVs while leaking radioactive wastewater into the surroundings for a decade or two, the mine stopped operations and eventually declared bankruptcy. China offered a cheaper alternative, forcing the United States to sell off its stockpiles. Now every industrialized country is in thrall to Chinese suppliers, who are acutely aware of the economic and strategic implications of being the dominant supplier of scarce resources with inelastic demand.[3]

But there's a remedy. What's contested on Earth because of scarcity is typically common in space. Selected asteroids contain unlimited quantities of metals and minerals. Comets have unlimited quantities of water. And solar energy is boundless in the empty space between planets. Access to space gives us access to these resources. Even if control of that access rests in the hands of people you'd hate to be in control of anything, the resources themselves will not be scarce — and it's scarcity that breeds conflict.

Asteroids are fragments of planets that didn't stay planetized. They start their lives through accretion. Debris collects in space, and any speck that's slightly bigger than the surrounding specks will have more gravity and attract more debris. Soon you'll have blobs rather than just specks. A big blob gets bigger faster than a small blob. Meanwhile, a lot of energy is getting deposited on what we would now call a protoplanet, as the kinetic energy from other colliding specks and blobs accumulates. For a couple hundred million years during the late childhood of our solar system, a period sensibly called the Late Heavy Bombardment, those collisions were significant and continual. With kinetic energy converting entirely to heat on impact, the deposited energy renders the protoplanet molten. And when you're molten, dense ingredients (such as pure heavy metals) fall toward your middle, and less dense ingredients (such as silicates) rise to your surface. By this process, Nature pre-sifts heavy things from light things, which geologists label with a six-syllable word: differentiation.

All of Earth was once molten. That's why it has an iron-rich core, containing abundant quantities of other metals that are rare on the surface.[4] Rare earth metals are not actually rare. They're simply not found in any significant concentration in Earth's crust, and we have no access to Earth's core, where they lie in abundance. The deepest we've ever drilled is less than one five-hundredth the distance to the center of our planet, and the core extends to half the planet's radius.

Eventually every molten object cools and solidifies. But if a big, fast-moving object then slams into it, you get a shattered, scattered field of pre-sifted space debris. That's how you get entire asteroids made of pure rock and others of pure metal. What matters to the future space miner is that some asteroids came from a protoplanet's shattered differentiated core, and they're packed with rare earth metals, as well as other metals we deem precious, including gold, silver, platinum, iridium, and palladium.

Once you have access to multiple sources of rare earth metals, you no longer have to worry about anybody's unilateral control of the strategic supply. Yes, Space Prospectors No. 1 — a country or a private company — will be the first to start mining the nearest rare-earth-laden asteroid and will therefore control that part of the supply. But so what? Space Prospectors No. 2 will just plan to get to a different asteroid and start mining that one. At which time normal economic and political forces begin to kick in. SP1, the pioneer, will not want to see anyone starting up a mine on a different asteroid. They'll want the rest of us to buy the rare earths they've mined. So they'll price their product at the point where it's cheaper for everyone to buy SP1 metals than to send their own missions to other asteroids. If SP1 goes above that price point, the rest of us will just go out and mine our own asteroids.

Unquestionably, asteroid mining will one day be a trillion-dollar industry, even if the vast increase in supply depresses the high prices at which rare earths are currently traded. As the price of highly useful goods drops, the number of affordable applications tends to grow. In the shorter run, however, since asteroid mining won't start tomorrow — although startups are multiplying, and the Finnish Meteorological Institute, for instance, is proposing a fleet of solar-wind-powered nanosatellites to collect data on the composition of several hundred asteroids — we'll have to come up with other solutions.[5] Maybe someone will invent a smartphone that doesn't need dysprosium. Maybe someone else will finally invent a storage mechanism in lieu of batteries for stockpiling solar energy.

Asteroids aren't the only small celestial bodies that can bring us a little more peace and security. Some comets contain as much water as the entire Indian Ocean, and it's not saltwater; do a bit of filtering, and you get freshwater. The way to snare a comet is to match orbits with it and break off a piece, which should be very easy. Comets are loosely held together, like snowballs made of dry snow. They look for excuses to break apart. Even the gentlest nudge from the tidal forces of a passing planet will do. Once you've grabbed a piece of the comet, you could put it in orbit around the site where the need exists — Earth, the Moon, Mars, wherever — and intermittently go up and grab iceberg-size chunks of it. Of course, you'll have to figure out how to accomplish all that, but you'd be working on engineering problems, not scientific ones. Any clever engineer would delight in being tasked to solve them.

There you have it: one vision of a future avenue to peace and healing. In the centuries-long alliance between warfighters and sky-watchers, the two sides have more often been in sync than at odds. Now astrophysicists and space scientists — heirs of the skywatchers of yore — may hold the power to erase a perennial rationale for war.

But we're not there yet. For millennia, war between nations, regions, religious factions, clans, or generally disagreeing or competing humans seems to have been always on the horizon or under way. Yet despite its ubiquity and persistence, "we (or at least we Americans) have forgotten the meaning of war," wrote the noted historian Tony Judt not long before his death. "In part this is, perhaps, because the impact of war in the twentieth century, though global in reach, was not everywhere the same." In Africa, in Europe, in Latin America, in Asia, in the Middle East, war in the last century "signified occupation, displacement, deprivation, destruction, and mass murder," the loss of family and neighbors, homes and shops, personal safety and national autonomy. For both victors and losers, and both sides in the long strings of civil wars, the memories of horror were similar. The United States, on the other hand,

avoided all that. Americans experienced the twentieth century in a far more positive light. The U.S. was never occupied. It did not lose vast numbers of citizens, or huge swaths of national territory, as a result of occupation or dismemberment. Although humiliated in neocolonial wars (in Vietnam and now in Iraq), it has never suffered the other consequences of defeat. Despite the ambivalence of its most recent undertakings, most Americans still feel that the wars their country has fought were "good wars." The USA was enriched rather than impoverished by its role in the two world wars and by their outcome[, and thus] for many American commentators and policymakers the message of the last century is that war *works*. . . . For Washington, war remains an option — in this case the first option. For the rest of the developed world it has become a last resort.[6]

If an all-out space-enabled war should ever occur, it would bear no resemblance to the world wars portrayed in *All Quiet on the Western Front* or *The Naked and the Dead* or the poems of Siegfried Sassoon and Wilfred Owen. Nor would it be like Vietnam or Iraq or Afghanistan. There would be no muddy, stinking trenches or sweltering, unforgiving deserts; nineteen-year-old boys would not blindly stagger through jungles half a world away; no Marine would see his buddy's head blown half off a yard from where he crouched. True space-age war would be sanitized, emotionless, thorough, and likely brief. Nations would fail in a day.

However often American public figures proclaim their country's prominence or dominance, the work that must be done in this century is inescapably cooperative — a point made by President Barack Obama in a speech to the UN General Assembly eight months after taking office:

> [M]y responsibility is to act in the interest of my nation and my people, and I will never apologize for defending those interests. But it is my deeply held belief that in the year 2009 — more than at any point in human history — the interests of nations and peoples are shared. . . . The technology we harness can light the path to peace, or forever darken it. . . .
>
> In an era when our destiny is shared, power is no longer a zero-sum game. No one nation can or should try to dominate another nation. No world order that elevates one nation or group of people over another will succeed. No balance of power among nations will hold.[7]

Were this understanding — that dominance cannot be the cornerstone of security in an interconnected world — ever to take root, the resulting cooperation would not only help forestall an arms race in outer space but could also help rescue our home planet from some of the upheavals of climate change.

The Paris Agreement — the 2016 United Nations climate accord, accepted by 197 parties as of early 2018[8] — represents the first

639

time that rigorous scientific consensus has shaped the political agenda of the world. People in power have learned that air and water molecules do not carry passports, as the American astrophysicist Carl Sagan was fond of saying. A melting glacier raises the sea level of all the world's coastlines. Greenhouse gases generated in one area of Earth mix swiftly with air currents that carry them to all areas of Earth. Warming air and warming ocean currents do not observe national boundaries or property rights. Neither would the thousands of deadly fragments of wayward orbital debris that an attack on a satellite would produce. No longer can the inhabitants of Earth survive as a collection of tribes, each looking out for only its own members. The world itself has become a tribe.

The same day Obama spoke at the United Nations, the journal *Nature* published grave news about the drastically accelerated melting of ice sheets in Antarctica and Greenland in 2003–2007 compared with that of the preceding decade. This was a finding by British climatologists, who based their determination on fifty million laser readings from a NASA satellite: an instance of international cooperation, in this case between allies. But adversaries, too, sometimes toss a little cooperation in with their confrontations. It's diplomacy's forte.

In July 2015, US–Russian relations pointed

toward the dawn of Cold War 2.0. Inflammatory rhetoric had been ratcheted up in the wake of Russia's annexation of the Crimean Peninsula and Russian military incursions across the Ukrainian border. In response, the United States had led the call for Western sanctions against Russia. Yet all that bad blood did not keep Russia from sending an unmanned cargo ship packed with food, water, oxygen, and equipment to the International Space Station to do what the community of spacefaring nations needed done following the failure of three supply missions within seven months (two US failures and one Russian). Russia deployed its reliable Soyuz-U rocket — not merely because the space station's crew consisted of two Russians and an American, not merely because Russia and America are founding partners of the ISS, but also because of the hefty sums Russia had been getting as sole provider of transport to the ISS.

Yes, it's complicated. And yes, there's no shortage of contradictions. But in the end, off-planet survival among spacefaring comrades can override them all.

One notable twentieth-century result of the countless alliances between astrophysics and the military is the thermonuclear fusion bomb, whose design principles arise in part from the astrophysicist's investigations of the cosmic crucible that occupies the center of every star. A less explosive example, from our own century, is the ChemCam instrument (short for Chemistry and Camera) atop the Curiosity rover, which began trundling across Mars in August 2012. From its skybox position on the rover's mast, ChemCam fires laser pulses at rocks and soil and then uses its spectrometer to analyze the chemical makeup of what got vaporized.

Who or what built ChemCam? The Los Alamos National Laboratory: birthplace of the atom bomb, originator of hundreds of spacecraft instruments designed for use by the military, and home to the Center for Earth and Space Science, a division of the National Security Education Center as well as a hub of support for astrophysics. Los Alamos Lab operates under the auspices of the National Nuclear Security Administration, whose mission is to maintain and protect America's stockpile of nuclear weapons while simultaneously working to undercut the proliferation of such stockpiles elsewhere in the world. And the lab's astrophysicists use the same supercomputer and similar software to calculate the yield from hydrogen fusion within the heart of a star that physicists use to calculate the yield of a hydrogen bomb. You'd have to look far and wide to find a clearer example of dual use.

Say you want to know what takes place during the explosion of a nuclear bomb. If you were to tabulate the many varieties of subatomic particles, and track the ways they interact and transmute into one another under controlled conditions of temperature and pressure — not to mention the particles that get created or destroyed in the process — you'd quickly realize you need more than pencil and paper. You need computers. Powerful computers.

A properly programmed computer can calculate crucial parameters for nuclear bomb design, ignition, and explosive yields, so it can predict what to expect from an experiment. Of course, "experiment" means the actual detonation of a nuclear bomb, either in a test or in warfare. During the Manhattan Project, in the 1940s, Los Alamos used mechanical calculators and early IBM punchcard tabulators to calculate atomic bomb yields. Decade by decade, as computing power increased exponentially, so too did the power to calculate and understand in detail the nuclear happenings in a nuclear explosion. And the needs of Los Alamos fostered the sustained quest to build the fastest computer in the world.

Second-generation computers of the 1960s, furnished with transistors that greatly accelerated their performance, in part made the 1963 Nuclear Test Ban Treaty possible. While later generations of computers didn't stop the arms race, they did offer a viable way to test weapon systems without actually detonating anything. By 1998, the Los Alamos supercomputer Blue Mountain could run 1.6 trillion calculations per second. By 2009, the lab's Roadrunner had increased that speed more than six hundredfold, to the milestone of one quadrillion calculations per second. And by late 2017, its Trinity supercomputer had racked up another factor of fourteen in computing power.[9]

We know that stars generate energy in exactly the same way that hydrogen bombs do. The difference is that the controlled nuclear fusion that happens in the star's core is contained by the weight of the star itself, whereas in warfare the nuclear fusion is positively uncontrolled — the precise objective of a bomb. And that is why astrophysicists have long been associated with Los Alamos National Lab and its supercomputers. Picture scientists working away on opposite sides of a classified wall. On one side, you have researchers engaged in secret projects that are "responsible for enhancing national security through the military application of nuclear science."[10] On the other side, you have researchers trying to figure out how stars in the universe live and die. Each side is accessory to the other's needs, interests, and resources.

If you seek more evidence, search the SAO/

NASA Astrophysics Data System[11] for research published in 2017 whose co-authors are affiliated with Los Alamos National Laboratory. You'll recover 102 papers. On average, that's an astrophysics paper published every 3.6 days. And that's the *un*classified research. Next, peruse the titles of Los Alamos–affiliated papers over the years. Supernovas turn out to be a perennial favorite. Published in the year 2013, for instance, there's "The Los Alamos Supernova Lightcurve Project: Computational Methods." In 2013–14 there's a three-paper sequence: "Finding the First Cosmic Explosions. I. Pair-instability Supernovae," "II. Core-collapse Supernovae," and "III. Pulsational Pair-instability Supernovae." For 2006 you'll find "Modeling Supernova Shocks with Intense Lasers." For earlier years, you'll see titles such as "Testing Astrophysics in the Lab: Simulations with the FLASH Code" (2003) and "Gamma-Ray Bursts: The Most Powerful Cosmic Explosions" (2002).

Born in Cold War fear, the alliance between space and national security remains alive and well in the unstable geopolitical climes of the twenty-first century. And it swings on a double-hinged door.

Some alliances, however, are forced on everybody in all domains on all sides because there's no other choice, as with the swarms of dreck passing overhead in Earth orbit and posing a volitionless threat not only to everything else circling up there but also to our wholly space-dependent way of life down here. Orbital debris is widely recognized as so grave a danger that Bill Maher, in the great American tradition of political satire — the necessary-for-survival alliance of truth, parody, pain, and healing — did a routine about it:

STAR DREK

Human beings are such slobs that, from now on, pigs must declare us the other white meat. Do you know that right now there is so much discarded trash in outer space that three times last month the International Space Station was almost hit by some useless hunk of floating metal — not unlike the International Space Station itself? So really, you've got to give the human race credit: only humans could visit an infinite void and leave it cluttered. Not only have we screwed up our own planet; somehow we have also managed to use up all the space in space.

Now, history shows over and over again that if the citizens of Earth put their minds to it, they can destroy anything. It doesn't matter how remote or pristine, together, yes, we can fuck it up. The age of space exploration is only fifty years old, and we have already managed to turn the final frontier into the New Jersey Meadowlands.[12]

One place you won't find comic relief is a US presidential commission report or a military doctrine document about national security space/ milspace/counterspace. Some of the language in these things might lead a reader to assume that America's military already has at its disposal not merely scores of dedicated satellites, which it does, but also a panoply of fully functional space weapons suitable for various kinds of confrontations, which it does not. The reader might further assume that other countries will shortly have such weapons too and that all sides are ready, willing, and able to deploy them. Not true.

Back in 2009, Major Scott A. Weston, USAF, published a piece in the Air Force's own *Air & Space Power Journal* in which he seeks to separate the factual from the fictional regarding prospects for space war. The major, who envisions a sky filled with hazardous debris under any scenario of overt space conflict, dismantles "the very concept of a space Pearl Harbor." That specter was raised repeatedly in the January 2001 final report of the Commission to Assess United States National Security Space Management and Organization, chaired by Donald Rumsfeld. Two pages into the executive summary, the report asserts that an attack on American space assets during a crisis or conflict is not improbable. "If the U.S. is to avoid a 'Space Pearl Harbor,' it needs to take seriously the possibility of an attack on U.S. space systems."[13] Weston emphatically disagrees:

If a conflict occurs in the next five to 10 years, the long acquisition process for space systems and limited space-launch schedules will confine the main space systems involved to those now fielded. . . .

Many works about space weapons quickly move from what the United States and its adversaries can do now to what they could possibly do soon, principally because few fielded terrestrial weapons can attack space assets and because no declared space-based attack assets exist. We could probably field a few promising technologies rapidly in wartime conditions, but as former defense secretary Donald Rumsfeld commented, "You have to go to war with the army you have, not the army you want." Fielded weapons include only the ones tested and turned over to military forces trained to employ them as an integrated part of battlefield forces. . . .

The United States has just one counter-space weapon — an electronic countercommunication system specifically designed and fielded with the intent of disrupting enemy satellite communications. . . .

After all the hype about space warfare and space weapons, an examination of currently fielded forces capable of direct counter-space operations against satellites clearly shows that few countries can conduct this type of warfare. Most threats envisioned in the US military's space doctrine simply do not exist in an operationally deployed form.[14]

That last contention apparently still holds.

The opening sentence of an eight-page white paper produced by the Office of the Assistant Secretary of Defense for Homeland Defense and Global Security in September 2015 reads: "Today's space architectures, designed and deployed under conditions more reflective of nuclear warfighting deterrence than conventional warfighting sustainability, lack, in general, the robustness that would normally be considered mandatory in such vital warfighting services."[15] Recast into everyday English, this is a complaint that America can't readily wage a space war.

Deep within the National Defense Authorization Act for Fiscal 2017, we discover that Congress's findings as of December 2016 included:

- "The advantages of the United States in national security space are now threatened to an unprecedented degree by growing and serious counterspace capabilities of potential foreign adversaries, and the space advantages of the United States must be protected."

- "The Department of Defense has recognized the threat and has taken initial steps necessary to defend space, however the organization and management may not be strategically postured to fully address this changed domain of operations over the long term."
- "Space elements provide critical capabilities to all of the Armed Forces in the joint fight, however the disparate activities throughout the Department have no single leader that is empowered to make decisions affecting the space forces of the Department."[16]

Again, in everyday English: US dominance in space is a thing of the past, and the future defense of US space assets will require restructuring of the military.

Following the high point of the Apollo program's Moon landings, there's been an enduring chasm between rhetoric and realization, between grandiose mandate and inadequate follow-through — a lot of PR and not much implementation. For more than a decade, US space policy was shaped by the combative tone of the Rumsfeld Commission's final report, which crystallized a view of outer space as a potential battleground. Notwithstanding some twenty occurrences of the words "peace" or "peaceful" in the report, its stance is anything but:

- "The Commissioners believe the U.S. Government should vigorously pursue the capabilities . . . to ensure that the President will have the option to deploy weapons in space to deter threats to and, if necessary, defend against attacks on U.S. interests."
- "In the coming period, the U.S. will conduct operations to, from, in and through space in support of its national interests both on earth and in space."
- "Unlike weapons from aircraft, land forces or ships, space missions initiated from earth or space could be carried out with little transit, information or weather delay. Having this capability would give the U.S. a much stronger deterrent and, in a conflict, an extraordinary military advantage."[17]

This report, followed a few weeks later by the start of Rumsfeld's stint as President George W. Bush's secretary of defense, sounded the alarm bell abroad in somewhat the same way as have the campaign comments, acerbic tweets, and unrestrained threats of nuclear escalation made more

651

recently by President Donald J. Trump.[18] The director of the Arms Control Program at Tsinghua University in Beijing, for instance, noted in 2003, "We have seen some explicit moves in the United States in recent years in preparing for space wars," including directives to the military "to engage in organization, training and equipment for swift, continuous, offensive and defensive space operations" and initiatives for the corporate development of "weapons for offensive space operations." He concluded that "US decision makers prefer war preparation in space rather than peaceful approaches" and "may believe that the US can certainly win a space war."[19]

Nobody can certainly win a space war, just as nobody can certainly win a war fought with nuclear weapons. Do you declare victory after all nukes have reached their targets, and you've got fewer incinerated cities than your enemy does? After almost two decades of the proliferation of both civilian and military space efforts by a number of countries, Rumsfeldian–Trumpian truculence on the part of the United States seems misplaced.[20] As national security specialist Joan Johnson-Freese has written, "If technology could offer the United States a way to 'control' space, then pursuing that course would make sense. But it does not. Politicians do not want to hear that because they want to believe otherwise."[21] Nor do defense corporations want them to believe otherwise. As mandates, "space situational awareness," "freedom of action in space," "maintaining space superiority," and "resilience of space architecture" yield reliable profits.

Eventually, though, in one form or another, reality will intervene: economic, political, environmental, social, physical. When that happens, the United States will almost certainly be forced to adopt a more peaceable persona, simply because it cannot — nor can any other country — achieve the degree of space superiority, let alone space control, regularly envisioned by its military strategists not so very long ago.[22] America in the foreseeable future is unlikely to satisfy such aspirations, and many in the military already acknowledge this.[23] As a result, mastering the intricacies of calm coexistence will probably show up on the agenda well before the fruits of extractive forays to comets and asteroids succeed in quelling some of the salient sources of international tension.

In the meantime, as you'd expect, people who are convinced that militarism does not promote national security or a safer world are not sitting on their hands waiting for a spontaneously generated peace or optimal conditions for a multilateral treaty on space weapons. Brian Weeden, a former Air Force officer with the US Strategic Command's

Joint Space Operations Center, has been pushing for more easily achievable moves — the demilitarization and internationalization of space situational awareness, for instance. The Council of the European Union has come up with a code of space conduct that stresses safety and sustainability. A Canadian–Australian–Chinese–American partnership has been publishing an annual *Space Security Index* since 2004. A raft of civil society organizations are each doing their bit to keep space from becoming another combat zone.[24]

Laudable goals. But at present we're uncomfortably close to open season up there in near-Earth space. The old two-superpower space-scape is long gone. So, too, is the vision of America as the space hegemon. Multiple smaller nations and private companies are becoming spacefarers. New projects and problems keep presenting themselves: potentially profitable mining ventures, lucrative space tourism, an increasingly crowded geostationary Earth orbit for communications satellites, maneuverable satellites that could conceivably be used as attack vehicles, launch services for sale by competing countries, insufficient coverage in the five existing UN space treaties of issues relevant to private ventures, frequent but legally mushy invocation of the "global commons," the reawakened nightmare of nuclear escalation and proliferation, everybody's growing reliance on satellite capabilities. Space law does not enshrine a single firm definition of "space weapon." There are no recognized borders marking territories in space. There's no single entity, governmental or otherwise, that holds the mandate to keep order in space. The potential for both unprecedented conflict and unprecedented cooperation is considerable. Some of those who diagnose the state of national security advise diplomacy first, technology next, and a big dose of proactive prevention. Others point out that true space security is not about foregrounding the interests of particular countries or corporations, but the security and sustainability of outer space for all.[25]

Among the three zones of Earth orbit — low, medium, and geosynchronous — you'll find most space telescopes, Hubble included, circling in the low zone, LEO, between 250 and 400 miles above Earth's surface. At these accessible altitudes, treasured orbital assets are vulnerable to attack by adversaries. But low Earth orbit is hardly the only zone of exploration available to the modern astrophysicist. The nature of the universe also reveals itself to the telescopes and probes we launch into the uncrowded, uncontested regions of deep space. And this is where full-spectrum collaboration abounds.

Modern astrophysics is unlike most other

sciences. The collective objects of astrophysical affection sail far above everyone's head. They do not sit within the borders of one or even several countries — at least not until nations claim ownership of planets. Multiple researchers, scattered across the globe and hailing from historically conflicting nation-states, can study the same object at the same time with similar or complementary tools and telescopes, whether those instruments are based on the ground, circling a few hundred miles above Earth, or orbiting in deep space. Scientists' urge to collaborate transcends religion, culture, and politics, because in space there is no religion, culture, or politics — only the receding boundary of our ignorance and the advancing frontier of our cosmic discovery.

One of our chief tools has been the Hubble Space Telescope, by far the most fertile scientific instrument ever built. Since its launch in 1990, Hubble has yielded more than fifteen thousand research papers, written by collaborators in nearly every country of the world where astrophysicists reside, and those papers have generated three-quarters of a million (and counting) citations in peer-reviewed journals.[26] Today Hubble has many extraordinary cousins, each hosting international collaborators.

What things wondrous and strange have these astrophysicists discovered?

Researchers from Canada, Germany, the Netherlands, the United Kingdom, and the United States have found a colossal wave of hot gas — 200,000 light-years wide, twice the width of the Milky Way, and so torrid it glows copiously in X-rays — that has been barreling through the supermassive Perseus cluster of galaxies for several billion years, caused by gravitational discombobulations from a smaller cluster grazing Perseus as it journeyed through space.

A team of two dozen researchers — from Australia, France, Portugal, Spain, Switzerland, and the United States, led by an astrophysicist from the Harvard–Smithsonian Center for Astrophysics — has identified a promising exoplanetary candidate for alien life: LHS 1140B, a rocky, metal-cored planet a bit bigger than Earth that orbits in the habitable zone of a cool star and quite possibly has retained its atmosphere.

The Laser Interferometer Gravitational-Wave Observatory (LIGO) — a collaboration of more than a thousand scientists from more than a hundred institutions dispersed across eighteen countries — has detected gravitational waves from colliding black holes billions of light-years away.

A huge team from Belgium, France, Morocco, Saudi Arabia, South Africa, Switzerland, the United Kingdom, and the United States, led by an astrophysicist from the

University of Liège in Belgium, has identified a system of seven Earth-sized, probably rocky exoplanets — TRAPPIST-1 — closely orbiting a single star whose surface temperature is less than half that of our Sun. Three of those exoplanets live in the habitable zone.

Various permutations of astrophysicists from Canada, Chile, France, Israel, Italy, Poland, Spain, the United Kingdom, and the United States have been studying the quantum effects of the intense magnetic field surrounding a neutron star; a vast intergalactic void that is helping to propel our galaxy through space by repelling it; an as-yet-unexplained cool region in the cosmic microwave background (imprint from the Big Bang) that may offer the first evidence of the multiverse. They've found a large, dim, relatively nearby spheroidal galaxy, similar in total mass to the Milky Way, that was only recently discovered because 99.99 percent of it consists of dark matter. They've witnessed an interstellar asteroid, the solar system's first visitor from elsewhere in the Milky Way, which plunged past the Sun and onward toward Mars at 300,000 kilometers per hour in the fall of 2017.

Besides making discoveries, astrophysicists have speculated that aliens might use lasers to broadcast obviously purposeful signals of their existence that would be picked up by skywatchers carefully monitoring known and suspected exoplanets. Some of us also speculate that aliens may power their interstellar probes with continuous beams from gigantic star-powered radio transmitters, which might explain the brief, otherwise unexplained flashes of radio waves that have been picked up by Earth's largest radio telescopes and that appear to come from billions of light-years away.

True, some of our mind-altering discoveries and speculations may pique the interest of warfighters and weapons developers. But others may undermine any notion that such a thing as long-term space superiority would ever be possible.

One mind-altering discovery that predates Hubble and all of its spaceborne cousins by decades was the origin of elements in the universe.

Key atoms of our biochemistry and of all life on Earth are traceable to thermonuclear fusion in the hearts of stars. We exist in the universe, and the universe exists within us. This insight, this almost spiritual gift from twentieth-century research to modern civilization, did not arise from a lone, sleepless researcher's eureka moment but rather from a seminal collaboration of four scientists during the 1950s.

The origin and abundance of the chemical elements had been a long-standing mystery in modern astrophysics. Research into radio-activity — the natural transmutation of elements — led to strong suspicions that some kind of natural nuclear process lurked behind it all, perhaps the same nuclear process that liberated sufficient energy to keep the stars shining.

In 1920, with the carnage of the Great War freshly ended, the English astrophysicist Sir Arthur Eddington offered prescient reflections on the source of stellar energy at a meeting of the British Association for the Advancement of Science:

> A star is drawing on some vast reservoir of energy by means unknown to us. This reservoir can scarcely be other than the subatomic energy which, it is known, exists abundantly in all matter; we sometimes dream that man will one day learn how to release it and use it for his service. The store is well-nigh inexhaustible, if only it could be tapped. . . .
>
> If, indeed, the subatomic energy in the stars is being freely used to maintain their great furnaces, it seems to bring a little nearer to fulfillment our dream of controlling this latent power for the well-being of the human race — or for its suicide.[27]

Major advances in quantum physics unfolded in the 1920s and continued through to 1932 with British physicist James Chadwick's discovery of the neutron, a new subatomic particle. Until then, everything known about stellar structure had told us that, in spite of the extreme temperature and pressure in a star's core, elements could not be forged there. But that didn't stop Eddington from engaging in rational speculation or from commenting in his 1926 book *The Internal Constitution of the Stars,* "We do not argue with the critic who urges that the stars are not hot enough for this process; we tell him to go and find a hotter place."[28] Might he have been telling his detractors to go to hell?

In any case, quantum physics as it stood in the 1930s accounted for the basics of how the Sun converts hydrogen into helium, generating energy as a by-product. But the origin of all the heavier elements remained elusive. Nuclear weapons — developed by the Manhattan Project, in which Chadwick participated — would yield answers.

The only way to know how atomic nuclei combine to make heavy nuclei under high temperatures and pressures, such as the state of affairs you'd find within the core of a star, is to study all the ways, all the places, and all the chances that one specified nucleus can slam into another specified nucleus. These so-called collision cross-sections can be theoretically estimated but, ideally, are measured directly in laboratory experiments.

Fresh access to declassified nuclear physics data from World War II and from the flurry of nuclear bomb tests that followed (underground, on the ground, in the water, and in the air) became just the kind of laboratory needed. By the mid-1950s, enough data was available on what subatomic particles and atomic nuclei do when they collide for Margaret and Geoffrey Burbidge, William Fowler, and Fred Hoyle to figure out how and why the life and explosive death of a star makes heavy elements.

In a preview of that work, published early in 1957, Fowler reflects on the value of access to declassified data:

[W]e think that [the element] californium-254 is produced in supernova explosions and that its especially energetic decay with a conveniently observable lifetime makes its presence stand out, but presumably other heavy elements are produced in a similar manner. . . . This highly unclassified result came to light within less than 4 weeks after the publication of the Bikini test results after a lapse of almost 4 years.[29]

Twenty-three nuclear bombs were detonated by the United States at Bikini Atoll in the South Pacific between 1946 and 1958.[30] Displaced people. Radioactive terrain. Incinerated flora and fauna. A steep price to pay for data.

The Burbidge team's research was published in October 1957 — the same month the Soviet Union launched Sputnik, starting gun of the space race. While their paper was neutrally titled "Synthesis of the Elements in Stars" and its tone was unvaryingly objective, their work was supported in part by a joint program of the Office of Naval Research and the US Atomic Energy Commission.[31] As Fowler had written earlier, the californium 254 produced at Bikini contributed significantly to the team's conclusions. And if, ignoring the arcane science, you read the last few pages of Burbidge et al.'s paper, you cannot help but pick up an implicit expectation or hope that Bikini-like tests will continue, in part because of the notable benefits to astrophysics:

> The identification of Cf^{254} in the Bikini test and then in the supernova in IC 4182 first suggested that here was the seat of the r-process production. Whether this finally turns out to be correct will depend both on further work on the Cf^{254} fission half-life and on further studies of supernova light curves.[32]

No endeavor is ceaselessly noble or electrifying. Eventually the question of money intrudes. Space probes, space telescopes, and frontier research hardware do not come cheap. Yet it's clear that the bill for worldwide astrophysics research is many orders of magnitude less than the bill for worldwide war[33] — that other collaboration of nations besides the Olympics and the World Cup. Even when the world isn't actually waging all-out war, we spend trillions preparing for it.

Today, astrophysics around the globe is funded at less than $3 billion a year,[34] while global military spending is nearing $1.7 trillion. With a 2016 world GDP of almost $76 trillion, that amounts to .004 percent for astrophysics and 2.2 percent for the military.[35] One year's worth of that level of military spending could lavishly fund every astrophysicist in the world for half a millennium.

Now for America. Consider the US contribution to World War II. In just a single year, 1943, military spending on the war swallowed 42 percent of America's national income.[36] Direct, upfront spending on American military operations was $75 billion a year. If the United States funded a war today at the same rate, relative to GDP, that $75 billion would turn into almost $7 trillion a year, or $19 billion a day.[37] Two hours' worth of that level of war spending could fund American astrophysics for an entire year.

You've heard the journalists' maxim "follow the money"? What a country funds is what that country prioritizes. By definition. Decades ago, *la dictadura fascista* Benito Mussolini, speaking about the Italian economy, declared that "the state will only take up the sectors related to defense, the existence and security of the homeland."[38] Well, the American economy has been sliding in that direction. It's what General and President Dwight D. Eisenhower lambasted, and it's a dubious route to genuine security. In 2015 the US government allocated $600 billion — 54 percent of its discretionary dollars — to military spending, versus $30 billion, or 3 percent, to science and engineering. In 2016 the United States accounted for a greater share of global military spending — $611 billion of the world's $1.7 trillion — than the next eight countries combined (China, Russia, Saudi Arabia, India, France, UK, Japan, and Germany, in descending order).[39]

Given all those billions flowing through the system, is it possible there's no money available to modernize New York City's century-old subways, to keep New Orleans from drowning again, to build truly affordable housing for the people who collectively make our cities run, to help the Metropolitan Museum of Art reinstate its voluntary admission fee for all visitors, and to expand the search for other habitable planets?

■ ■ ■ ■

The almost-final word goes to the anonymous carrier of a placard at one of the six-hundred-plus Marches for Science that took place around the world on April 22, 2017. "THINK WHILE IT'S STILL LEGAL," urged the placard. And while you're thinking, try to imagine that each of us is a transient assemblage of atoms and molecules; that our planet is one small pebble ambling in orbit through the vacuum of space; that astrophysics, a historical hand-maiden of human conflict, now offers a way to redirect our species' urges to kill into col-laborative urges to explore, to uncover alien civilizations, to link Earth with the rest of the cosmos — genetically, chemically, atomically — and protect our home planet until the Sun's furnace burns itself out five billion years hence.

Try to imagine such things not because they are imaginary, but because they are true.

ACKNOWLEDGMENTS

We are separately and jointly grateful to the innumerable individuals and institutions whose writings, lectures, emails, conversations, fact checks, critiques, queries, responses, and resources enabled us to construct this book.

Without the space-time intersection provided by *Natural History* magazine, we would never have begun our long association. Without the staffs of the Hayden Planetarium, the American Museum of Natural History, and the museum's Research Library — especially Tom Baione, Gwen King, Mai Reitmeyer, Elizabeth Stachow, and Rachel Wysoki — we would have had far more difficulty carrying out our collaboration.

For sharing their worldviews and expertise over the past two decades, whether or not they knew a book was percolating at the time, Neil would like to thank Buzz Aldrin, former NASA astronaut; Wanda M. Austin, former president and CEO, Aerospace Corporation;

Ashton Carter, former US secretary of defense; John W. Douglass, former president and CEO, Aerospace Industries Association; Commander Sue Hegg, US Navy (ret.), formerly of Intelligence and Security Systems, Boeing; General Lester L. Lyles, US Air Force (ret.), former commander, Air Force Materiel Command; Joanne M. Maguire, former executive vice president, Lockheed Martin Space Systems; Arati Prabhakar, former director, Defense Advanced Research Projects Agency; the late Elliot Pulham, former CEO, Space Foundation; William Schneider Jr., former chair, Defense Science Board, US Department of Defense; Michael Shara, astrophysicist, American Museum of Natural History; Robert J. Stevens, retired chair, president, and CEO, Lockheed Martin; Robert Walker, former chair, House Science Committee, Reagan administration; and Heidi Wood, former managing director and senior aerospace and defense analyst, Morgan Stanley.

For their generous assistance, Avis is especially grateful to Vivek Chibber of New York University; Mark Harrison of the University of Warwick; Alexander R. Jones of the Institute for the Study of the Ancient World, New York University; John P. C. Moffett of the East Asian History of Science Library, Needham Research Institute, University of Cambridge; James Clay Moltz of the Naval Postgraduate School; Stephen C. Sambrook of the University of Glasgow; and Louise S. Sherby of the Archives and Special Collections, Hunter College Libraries, City University of New York. For their gracious response to requests for information, translation, confirmation, or elucidation, she thanks Peter Abrahams, Linda J. Bilmes, Michael Buckland, Anita Cochran, Larry J. Curtis, Neal Evans, Gary W. Ewer, Alexander Field, Toby Huff, Mark Johnson, Mary Knight, Walter F. LaFeber, Norton D. Lang, Theresa Levitt, Russ Levrault, John Logsdon, Lu Xiuyuan, Steve Maran, Andy Martin, Surendra Parashar, Kenneth Pomeranz, Charles Post, Jessica Rawson, Stéphan Reebs, M. Eugene Rudd, Michael Scholtes, Anwar Shaikh, Maryline Simler, Steven Soter, Steven Topik, Jason Walkowiak, Micah Walter-Range, and James G. Wilson.

Avis is more than grateful to Neil for presenting her with an immense challenge and having confidence that she will deliver. In addition, for their counsel and encouragement throughout the book's prolonged gestation, she thanks Nan Bauer-Maglin, Josely Carvalho, Nivedita Majumdar, Fran Nesi, Julia Scully, Gerry Wallman, and Shelly Wallman. And on the broadest planes of life itself, she is grateful to Elliot Podwill — minister of culture, cuisine, tourism, and human services in the Lang–Podwill domestic polity.

Throughout our decade and a half of work on *Accessory to War,* our publisher, W. W. Norton, has tolerated our delays and supported our progress. We particularly thank our editor, John Glusman — editor-in-chief at Norton — for his persistent patience, his attentiveness to every word in our manuscript, his flexibility, and his capacity to share our enthusiasm for this project even when our rate of progress could not have justified it. We further thank our agent, Betsy Lerner, who greatly valued where we were coming from, where we were going, and why.

Finally, the authors want to thank each other — for perseverance, for ruthless mutual editing, for the willingness to rationally adjudicate the other's justifications for a word, a phrase, a deletion, or a digression.

NOTES

Abbreviations Used in Titles of Periodicals
Amer. American
Assoc. Association
Brit. British
Bull. Bulletin / Bulletin of the
Int. International
J. Journal / Journal of the
Proc. Proceedings / Proceedings of the
Rev. Review
Transac. Transactions / Transactions of the

1. A Time to Kill

1. Edmund Phelps interviewed by Steve Evans, *Business Daily,* BBC World Service, Dec. 11, 2008.
2. Christopher Bodeen, AP, "China Breaks Ground on Space Launch Center," *US News & World Report,* Sept. 14, 2009.
3. Christiaan Huygens, *The Celestial Worlds Discover'd: Or, Conjectures Concerning the Inhabitants, Plants and Productions of the*

Worlds in the Planets (London: Timothy Childe, 1698), 39–41.

4. See Superfunction "National Defense" within the Office of Management and Budget's Historical Table 3.1, "Outlays by Superfunction and Function: 1940–2021," www.whitehouse.gov/omb/budget/Histor icals (accessed Apr. 3, 2016). Regarding the updating of the table over time, the OMB notes: "To the extent feasible, the data have been adjusted to provide consistency with the 2017 Budget and to provide comparability over time." As of spring 2016, the lowest spending for national defense during the 1970s was $76.7 billion in FY1973; the highest was $116.3 billion in 1979. By FY1983, the defense budget had exceeded $200 billion; by FY1989, it had exceeded $300 billion.

5. Opening phrase of a 1984 campaign ad for Ronald Reagan, at Museum of the Moving Image, "The Living Room Candidate: 1984 Reagan vs. Mondale," www.livingroom candidate.org/commercials/1984/prouder -stronger-better (accessed Mar. 20, 2016).

6. Ronald Reagan, "Inaugural Address, January 20, 1981," American Presidency Project, www.presidency.ucsb.edu/ws/?pid= 43130 (accessed Mar. 20, 2016).

7. For the story of the Pulitzer Prize–winning photograph and its subject, see BBC News, "Picture Power: Vietnam Napalm Attack,"

news.bbc.co.uk/2/hi/4517597.stm (accessed Apr. 5, 2016).

8. As of early 2008, members of Iraq Veterans Against the War (IVAW) began to deliver public testimony in a campaign called Winter Soldier, which culminated in a four-day event in March 2008 near Washington, DC; see www.ivaw.org/wintersoldier; www.ivaw.org/blog/press-releases; www.ivaw.org/blog/press-coverage (accessed Apr. 5, 2016). Re demonstrators attending a mobilization on Feb. 17, 2003, the BBC states: "Between six and 10 million people are thought to have marched in up to 60 countries over the weekend — the largest demonstrations of their kind since the Vietnam War." BBC News, "Millions Join Global Anti-War Protests," Feb. 17, 2003, news.bbc.co.uk/2/hi/europe/2765215.stm (accessed Apr. 5, 2016). Public opinion polls show a steady, and sometimes mounting, opposition to the war; see compilation of polls on the Iraq War from Pew Research, CNN, ABC News/*Washington Post,* and others at "Iraq," PollingReport.com, www.pollingreport.com/iraq.htm (accessed Apr. 5, 2016).

9. Although the US Constitution grants Congress the sole power to declare war, no Congress has done so since 1942. Instead, Congress has agreed to resolutions authorizing the use of military force, controlled

appropriations, and exercised limited oversight. Throughout World War I, World War II, the Korean War, and the Vietnam War, the Democrats held the majority in both the Senate and the House. United States Senate, "Official Declarations of War by Congress" and "Party Division," www .senate.gov/pagelayout/history/h_multi _sections_and_teasers/WarDeclarations byCongress.htm and www.senate.gov/ pagelayout/history/one_item_and_teasers/ partydiv.htm; United States House of Representatives: History, Art & Archives, "Party Divisions of the House of Representatives, 1789–Present," history.house.gov/ Institution/Party-Divisions/ Party-Divisions/ (accessed Oct. 10, 2017).

10. For a poison gas attack, see Dexter Filkins, "The Fight of Their Lives," *New Yorker,* Sept. 29, 2014, 44–45. See also Chris Maume, "It Was Better to Live in Iraq under Saddam," *Independent,* June 12, 2014; Costs of War Project, "Education: Universities in Iraq and the U.S.," Watson Institute for International Studies, Brown University, costsofwar.org/article/ education -universities-iraq-and-us (accessed June 27, 2014); Benjamin Busch, " 'Today Is Better Than Tomorrow': A Marine Returns to a Divided Iraq," *Harper's,* Oct. 2014, 38.

11. FTSE 350 Aerospace & Defence Index. See "Global Defence Outlook" ("Lex"

column), *Financial Times,* Jan. 26, 2007.

12. Commission on the Future of the United States Aerospace Industry, *Anyone, Anything, Anywhere, Anytime,* final report, Nov. 2002, 7–2, 7–4, history.nasa.gov/AeroCommissionFinalReport.pdf (accessed Apr. 3, 2016). Appointed by George W. Bush in 2001, Neil deGrasse Tyson was a member of the presidential commission that produced this report. In his analysis of aerospace consolidation, Andrew Cockburn ("Game On," *Harper's,* Jan. 2015) cites a watershed meeting in 1993 between deputy defense secretary William Perry and a "group of industry titans." At this meeting ("the Last Supper"), Perry warned that budget cuts would necessitate consolidation and force some of them out of business. Cockburn writes, "Perry's warning sparked a feeding frenzy of mergers and takeovers, lubricated by generous subsidies at taxpayer expense in the form of Pentagon reimbursements for 'restructuring costs' " (68).

13. Commission to Assess United States National Security Space Management and Organization, *Report* — Pursuant to Public Law 106-65, Jan. 11, 2001, www.dod.gov/pubs/space20010111.pdf (accessed Apr. 3, 2016). The phrase "Space Pearl Harbor" occurs seven times and echoes another intentionally fear-inducing phrase, common

during the 1950s: "nuclear Pearl Harbor." In the executive summary, the other phrases occur at xvi, viii, and xi.

14. William D. Hartung et al., "Introduction," *Tangled Web 2005: A Profile of the Missile Defense and Space Weapons Lobbies* (New York: World Policy Institute — Arms Trade Resource Center, 2005), www.worldpolicy .org/projects/arms/reports/tangledweb.html (accessed Apr. 12, 2017): "From its inception during the Reagan administration to the present [2005], the current generation of missile defense development has cost over $130 billion. . . . The Union of Concerned Scientists estimates that just launching enough Spaced-Based Interceptors (SBI) to ensure full global coverage could cost $40 billion to $60 billion. All of this expenditure might be justified if the ballistic missile defense system could be shown to work and if the ballistic missile threat were the most urgent danger facing the United States. But neither of these propositions is true."

15. Having contributed slightly more than $4 million total to just thirty members of Congress in 2001–2006, mostly to members of the Armed Services Committee or the Defense Appropriations Subcommittee, the missile defense industry ensured strong advocacy. Campaign-finance reformers see such an arrangement as strongly favorable

to the industry: $4 million spent in campaign funding, $50 billion received in missile defense program acquisition costs, yielding a 12,500 percent return on investment. The top two recipients of missile defense–related contributions in the Senate 2001–2006 were Alabama Republicans Richard Shelby and Jeff Sessions. Many members of the House outdid members of the Senate, especially Pennsylvania Democrat Jack Murtha, ranking Democrat on the House Defense Appropriations Subcommittee during the post-Clinton Republican administrations. The six most generous donations to House members ranged from $73,000 down to $41,000, three to Democrats and three to Republicans: (1) Lockheed Martin to Jim Saxton (R–NJ), chair, House Armed Services Subcommittee on Terrorism, Unconventional Threats and Capabilities, and member, House Armed Services Subcommittee on Projection Forces; (2) BAE Systems to Jack Murtha (D–PA), ranking Democrat, House Appropriations Subcommittee on Defense; (3) Northrop Grumman to Jane Harman (D–CA), member, House Permanent Select Committee on Intelligence, and member, House Committee on Homeland Security; (4) Boeing to James Moran (D–VA), member, House Appropriations Subcommittee on Defense; (5) L-3 to Jerry Lewis (R–CA), chair, House Appropriations Committee, and former chair, House Appropriations Subcommittee on Defense; and (6) Titan to Duncan Hunter (R–CA), chair, House Armed Services Committee. Hartung et al., *Tangled Web;* Brandon Michael Carius, "Procuring Influence: An Analysis of the Political Dynamics of District Revenue from Defense Contracting" (MPP thesis, Georgetown University, 2009), 3–6.

16. Budget totals change retroactively as emergency spending gets added later and funds get reappropriated. Here, "2001" and "2004" refer to fiscal years. For the figures for FY2001 and FY2004, see Table 1-1, "National Defense Budget Summary," in Office of the Under Secretary of Defense (Comptroller), National Defense Budget Estimates for FY2003, March 2002, and FY2007, March 2006, 4. For the explanation of budget authority and outlays, see "Overview" in, e.g., National Defense Budget Estimates for FY2009, 1. Beginning in FY2005, the total of budget authority plus outlays has exceeded $1 trillion. Unclassified defense budget information is summarized in the yearly "Green Books," available at Under Secretary of Defense (Comptroller), "DoD Budget Request," comptroller.defense.gov/budgetmaterials .aspx (accessed Apr. 3, 2016). For Iraq spending, see Donald L. Barlett and

James B. Steele, "Billions Over Baghdad," *Vanity Fair,* Oct. 2007; Matt Kelley, "Rebuilding Iraq: Slow but Steady Progress," *USA Today,* Mar. 22, 2010. Through 2004, the US had spent $6.8 billion on reconstruction in Iraq; through 2009, it had spent $44.6 billion.

17. American Security Project, "About: Vision–Strategy–Dialogue," www.american securityproject.org/about/ (accessed July 1, 2014; Apr. 10, 2017).

18. When viewed in July 2014, the ACLU National Security Project's self-description had a soaring tone: "Our way forward lies in decisively turning our backs on the policies and practices that violate our greatest strength: our Constitution and the commitment it embodies to the rule of law. Liberty and security do not compete in a zero-sum game; our freedoms are the very foundation of our strength and security." When viewed in Apr. 2017, the banner on the redesigned National Security page had become quite matter-of-fact: "The ACLU's National Security Project is dedicated to ensuring that U.S. national security policies and practices are consistent with the Constitution, civil liberties, and human rights." ACLU National Security Project, "National Security: What's at Stake," www.aclu.org/national-security.

19. National Security Agency/Central Secu-

rity Service, www.nsa.gov; "Mission and Strategy," www.nsa.gov/about/mission -strategy (accessed Apr. 10, 2017). In April 2016, the mission was expressed as "Global Cryptologic Dominance Through Responsive Presence and Network Advantage."

20. Re Snowden, see, e.g., *Citizenfour,* documentary film (dir. Laura Poitras, 2014); video interviews by the *New Yorker*'s Jane Mayer (2014), www.youtube.com/watch ?v=fidq3jow8bc; and Harvard Law School professor Lawrence Lessig (2014), www .youtube.com/watch?v=o_Sr96TFQQE (accessed Apr. 10, 2017). Following seven months of revelations and accusations by and against Snowden, a major US newspaper contended: "The shrill brigade of his critics say Mr. Snowden has done profound damage to intelligence operations of the United States, but none has presented the slightest proof that his disclosures really hurt the nation's security." Editorial Board, "Edward Snowden, Whistle-Blower," op-ed, *New York Times,* Jan. 1, 2014. In an Oct. 2014 interview published in *The Nation,* Snowden said that certain phrases "parroted" by the media are intended "to provoke a certain emotional response — for example, 'national security.' . . . But it is not national security that they're concerned with; it is state security. And that's a key distinction. We don't like to use the phrase

'state security' in the United States because it reminds us of all the bad regimes. But it's a key concept, because when these officials are out on TV, they're not talking about what's good for you. They're not talking about what's good for business. They're not talking about what's good for society. They're talking about the protection and perpetuation of a national state system." Katrina vanden Heuvel and Stephen F. Cohen, "Edward Snowden: A 'Nation' Interview," *The Nation,* Nov. 17, 2014.

21. National Priorities Project, "Cost of National Security," www.nationalpriorities .org/cost-of (accessed Apr. 3, 2016).
22. Re antibiotic resistance, see, e.g., Sabrina Tavernise, "U.S. Aims to Curb Peril of Antibiotic Resistance," *New York Times,* Sept. 18, 2014; Gardiner Harris, " 'Superbugs' Kill India's Babies and Pose an Overseas Threat," *New York Times,* Dec. 3, 2014. Re DoD and climate change, see Department of Defense, *2014 Climate Change Adaptation Roadmap,* www.acq.osd .mil/ie/download/CCARprint.pdf (accessed Dec. 4, 2014). The first sentence on page 1 reads: "Climate change will affect the Department of Defense's ability to defend the Nation and poses immediate risks to U.S. national security." Secretary of Defense James Mattis reiterated this position in early 2017: "Climate change is impact-

ing stability in areas of the world where our troops are operating today. . . . It is appropriate for the Combatant Commands to incorporate drivers of instability that impact the security environment in their areas into their planning." Andrew Revkin (ProPublica), "Trump's Defense Chief Cites Climate Change as National Security Challenge," *Science,* Mar. 14, 2017, DOI: 10 .1126/science.aal0911 (accessed Apr. 10, 2017).

23. European Commission, "Horizon 2020 Programme: Security," ec.europa.eu/ programmes/horizon2020/en/area/security (accessed July 8, 2017).

24. Canadian astronaut Chris Hadfield observed while gazing down on war-torn Syria from inside the International Space Station, "We're all in this together. . . . And so when we do look down on a place that is currently in great turmoil or strife, it's hard to reconcile the inherent patience and beauty of the world with the terrible things that we can do to each other as people and can do to the Earth itself." "Canadian Astronaut Appeals for Peace from Space," Phys.org, Jan. 10, 2013, phys.org/news/ 2013-01-canadian-astronaut-appeals-peace -space.html. Another of the many examples is the Indian-American astronaut Sunita Williams, who said in Jan. 2007, live by satellite from the ISS, "It is hard to imagine

anybody arguing down there." "Peace Is the Message of Sunita Williams," *OneIndia,* Jan. 11, 2007, www.oneindia.com/2007/01/11/peace-is-the-message-of-sunita-williams-1168510495.html (accessed Apr. 10, 2017).

25. *The Space Report 2006,* 1.

26. From 1984 through 2013 the event was called the National Space Symposium. It was renamed in 2014 "to reflect the event's truly global profile." Space Foundation, "About the Space Symposium: History," www.spacesymposium.org/about/space-symposium. Between 2003 and 2009 the foundation also put on a separate, specifically military Strategic Space Symposium, co-sponsored by the DoD's US Strategic Command and Space News. More recently, the foundation has begun staging "a boutique investment conference" called the Space Technology and Investment Forum, www.spacetechforum.com/ (accessed Apr. 10, 2017).

27. CNN.com/WORLD, "War in Iraq: U.S. Launches Cruise Missiles at Saddam," Mar. 20, 2003, www.cnn.com/2003/WORLD/meast/03/19/sprj.irq.main/ (accessed Apr. 10, 2017).

28. According to the Space Foundation, the 19th National Space Symposium "grew 20 percent as compared to the 2002 event . . . more than 5,200 total participants . . . more

than 1,400 symposium registrants were joined by over 1,000 students and teachers and an estimated 2,800 exhibitors, volunteers, customer support representatives, media, guests and others. . . . More than 120 companies, agencies and organizations participated in the exhibition center — also a new record." "Space Foundation Reports National Space Symposium Growth," news release, Apr. 29, 2003, available at Space Ref, www.spaceref.com/news/viewpr.html ?pid=11401 (accessed Apr. 3, 2016).

29. Leonard David, "Military Space Operations in Transformation," Space.com, Apr. 8, 2003, www.space.com/news/nss _warfighter_030408.html (link disabled).

30. For the timetable, see "War in Iraq: War Tracker / Archive," CNN.com, www.cnn .com/SPECIALS/2003/iraq/war.tracker/ index.html; "Struggle for Iraq — War in Iraq: Day by Day Guide," BBC News, news .bbc.co.uk/2/hi/in_depth/middle_east/2002/ conflict_with_iraq/day_by_day_coverage/ default.stm (accessed Apr. 3, 2016).

31. Commission to Assess US National Security Space Management, *Report,* xviii.

32. National Science Board, *S & E Indicators 2016* (Arlington, VA: National Science Foundation, 2016), O–4, O–5, 3–6, 3–7, 3–18, 3–19, Fig. 3–33, 3–77, 3–103, 4–55, fig. 6–3, 6–20, www.nsf.gov/statistics/2016/ nsb20161/uploads/1/nsb20161.pdf; Na-

tional Science Board, *S & E Indicators 2014* (Arlington, VA: National Science Foundation, 2014), Appendix table 2–33, 2–34, www.nsf.gov/statistics/seind14/content/etc/nsb1401.pdf; Space Foundation, *The Space Report 2016*, 16, 24–25, 64–68; *Space Report 2017*, 47–48. See also Neil deGrasse Tyson, "Science in America," Facebook, Apr. 21, 2017, www.facebook.com/notes/neil-degrasse-tyson/science-in-america/10155202535296613/ (accessed July 8, 2017).

33. Northrop Grumman, "2016 Annual Report," www.northropgrumman.com/AboutUs/AnnualReports/Documents/pdfs/2016_noc_ar.pdf, 21–22, 1, 45; "Starshade," Northrop Grumman, www.northropgrumman.com/Capabilities/Starshade/Pages/default.aspx; "Capabilities," Northrop Grumman, www.northropgrumman.com/Capabilities/Pages/default.aspx (accessed Apr. 11, 2017).

34. Eric Schmitt with Bernard Weinraub, "A Nation at War: Military; Pentagon Asserts the Main Fighting Is Finished in Iraq," *New York Times*, Apr. 15, 2003; CNN.com, "Inside Politics: Commander in Chief Lands on USS Lincoln," May 2, 2003, www.cnn.com/2003/ALLPOLITICS/05/01/bush.carrier.landing/; Jarrett Murphy, AP, "Text of Bush Speech," CBS News, May 1,

2003, www.cbsnews.com/news/text-of-bush -speech-01-05-2003/ (accessed Apr. 4, 2016).

35. According to a 2006 Zogby International poll of troops serving in Iraq, "Three quarters of the troops had served multiple tours and had a longer exposure to the conflict: 26% were on their first tour of duty, 45% were on their second tour, and 29% were in Iraq for a third time or more." www.zogby.com/NEWS/ReadNews.dbm ?ID=1075 (link disabled).

36. "By the end of Rumsfeld's tenure in late 2006, there were an estimated 100,000 private contractors on the ground in Iraq — an almost one-to-one ratio with active-duty American soldiers." Jeremy Scahill, "Bush's Shadow Army," *The Nation,* Apr. 2, 2007. According to a Congressional Research Service report on a slightly later time frame, "the number of troops in Iraq dropped from a high of 169,000 in September 2007 to a low of 95,900 in March 2010, a decrease of 43%. The total number of contractors dropped from a high of 163,000 in September 2008 to 95,461 in March 2010, a decrease of 42%. The number of PSCs [private security contractors] peaked at 13,232 in June 2009." Moshe Schwartz, *The Department of Defense's Use of Private Security Contractors in Iraq and Afghanistan: Background, Analysis, and Options for Con-*

gress, report, Congressional Research Service, June 22, 2010, 7, fpc.state.gov/documents/organization/145576.pdf (accessed Apr. 4, 2016).

37. On April 23, 2006, the Iraqi death count was 34,511 minimum / 38,660 maximum, according to Iraq Body Count (characterized by the BBC News in October 2004 as "a respected database run by a group of academics and peace activists"), www.iraq bodycount.net (accessed Apr. 23, 2006). A much higher number for "excess Iraqi deaths as a consequence of the war" — between 392,979 and 942,636 — is found in a noted study by Gilbert Burnham et al., "Mortality After the 2003 Invasion of Iraq: A Cross-sectional Cluster Sample Survey," *The Lancet* 368:9545 (Oct. 21, 2006), 1421–28, www.ncbi.nlm.nih.gov/pubmed/17055943 (accessed Apr. 4, 2016). As of the first week of Apr. 2006, 17,469 US troops had been wounded, according to "U.S. Wounded by Month," at Iraq Coalition Casualty Count, icasualties.org/oif/woundedchart.aspx (accessed Apr. 23, 2006).

38. Estimates of the true costs by two eminent economists, Nobel laureate Joseph E. Stiglitz and Linda Bilmes, increased steadily from $1–2 trillion in early 2006, to a minimum of $2.267 trillion in late 2006 — factoring in long-term and macroeconomic

costs such as medical care and disability for veterans; capital expenditures to replace or restore military equipment destroyed or depleted during the war; the true costs of recruitment, activation, lost earnings, disability, and death of the troops; and rising oil prices. Interest payments on the money borrowed to fight the war would add another $264–308 billion. There would be intangible additional costs in, e.g., reduced US capability to respond to national security threats in other regions, increasing anti-American sentiment abroad, decline in US influence on issues ranging from trade negotiations to criminal justice. In early 2008 the estimate reached $3 trillion; in late 2015 it rose to $5–7 trillion (Stiglitz); in late 2016, nearly $5 trillion (Bilmes). Bilmes and Stiglitz, "A Careless War of Excessive Cost — Human and Economic," *San Francisco Chronicle,* Jan. 22, 2006, www.sfgate.com/opinion/article/A-careless-war -of-excessive-cost-human-and-2542816 .php; Bilmes and Stiglitz, "Encore: Iraq Hemorrhage," *Milken Institute Review* (4th Q, 2006), 76–83, www8.gsb.columbia.edu/ faculty/jstiglitz/sites/jstiglitz/files/2006_Iraq _War_Milken_Review.pdf war; Stiglitz and Bilmes, *The Three Trillion Dollar War: The True Cost of the Iraq Conflict* (New York: W. W. Norton, 2008); see also the continually updated website Three Trillion Dollar

War: The True Cost of the Iraq and Afghanistan Conflicts, threetrilliondollarwar.org. By early 2010, the monthly costs of the war in Afganistan were exceeding those of the war in Iraq; see Richard Wolf, "Afghan War Costs Now Outpace Iraq's," *USA Today,* May 13, 2010, usatoday30.usatoday.com/news/military/ 2010-05-12-afghan_N.htm (accessed Apr. 4, 2016).

39. Space Foundation, " 'One Industry — Go for Launch!' at the 22nd National Space Symposium," press release, Apr. 3, 2006, www.nss.org/pipermail/isdc2006/2006-April/000239.html: "More than 135 companies and organizations will showcase exhibits in the Lockheed Martin Exhibit Center, representing an increase in the square footage of exhibits from last year by 40 percent." See also Space Foundation, "Space Foundation Declares 22nd National Space Symposium a Huge Success," press release, Apr. 8, 2006, www.spacefoundation.org/media/press-releases/space-foundation-declares-22nd-national-space-symposium-huge-success (accessed Apr. 4, 2016).

40. American Institute of Physics, "House Appropriators Want More Money for NASA," *FYI: The American Institute of Physics Bulletin of Science Policy News* 47, Apr. 13, 2006.

41. According to a nationwide 2005 Harris Poll of 1,833 US adults, "In 10 years, seven

in 10 (70%) U.S. adults think China will be a superpower. Forty-one percent think Japan will be as well, followed by the European Union (31%), United Kingdom (25%), India (20%) and Russia (15%)." See PRNewswire, "U.S. Public Less Concerned about China's Potential to Grow Economically than Militarily in the Next Ten Years," Nov. 15, 2005, www.prnewswire .com/news-releases/us-public-less-con cerned-about-chinas-potential-to-grow -economically-than-militarily-in-the-next -ten-years-55627132.html (accessed Apr. 4, 2016).

42. NASA, "NASA Names Worden New Ames Center Director," press release 06-193, Apr. 21, 2006, www.nasa.gov/home/ hqnews/2006/apr/HQ_06193_Worden _named_director.html (accessed Apr. 4, 2016).

43. University Communications, "Scientists Polled on Solar System Exploration Program Priorities," news release, *UA News,* University of Arizona, Apr. 24, 2006, uanews.arizona.edu/story/scientists-polled -on-solar-system-exploration-program -priorities (accessed Apr. 4, 2016).

44. See Space Foundation, *Space Report 2012,* 109; *Space Report 2014,* 104; *Space Report 2017,* 43, "Exhibit 3b: Space Work-force Trends in the United States, Europe,

Japan and India" and "Exhibit 3c: U.S. Space Industry Core Employment, 2005–2016." See also Bureau of Labor Statistics, "Databases, Tables & Calculators by Subject: Employment, Hours, and Earnings from the Current Employment Statistics Survey (National) — All employees, thousands, total nonfarm, seasonally adjusted, 2007–2017," US Department of Labor, data.bls.gov/timeseries/CES0000000001 (accessed Oct. 2, 2017). The period covered by Exhibit 3b in the *Space Report 2017* is 2005 through 2015; by Exhibit 3c, 2005 through the second quarter of 2016; and by the BLS total nonfarm chart and table, Jan. 2007 through June 2017.

45. Mike Wall, "NASA to Pay $70 Million a Seat to Fly Astronauts on Russian Spacecraft," Space.com, Apr. 30, 2013, www.space.com/20897-nasa-russia-astronaut-launches-2017.html; "NASA: Seats on Russian Rockets Will Cost U.S. $490 Million," CBS/AP, Aug. 6, 2015, www.cbsnews.com/news/nasa-seats-on-russian-rockets-will-cost-u-s-490-million/ (accessed Apr. 3, 2016).

46. See, e.g., William J. Broad, "Physicists Compete for the Biggest Project of All," *New York Times,* Sept. 20, 1983; Associated Press, "Legislation Introduced to Spur Super Collider," *New York Times,* Aug. 10, 1987; Ben A. Franklin, "Texas Is Awarded

Giant U.S. Project on Smashing Atom," *New York Times,* Nov. 11, 1988; David Appell, "The Super Collider That Never Was," *Scientific American,* Oct. 15, 2013; Trevor Quirk, "How Texas Lost the World's Largest Super Collider," *Texas Monthly,* Oct. 21, 2013, www.texasmonthly.com/articles/how-texas-lost-the-worlds-largest-super-collider/ (accessed Jan. 10, 2018).

47. US General Accounting Office, "Federal Research — Super Collider Is Over Budget and Behind Schedule," Report to Congressional Requesters, GAO/RCED-93-87, Feb. 1993, babel.hathitrust.org/cgi/pt?id=uiug.30112033998011;view=1up;seq=1 (accessed Jan. 12, 2018).

48. Hearing Before the Subcommittee on Oversight and Investigations of the Committee on Energy and Commerce, House of Representatives, 103rd Congress, "Mismanagement of DOE's Super Collider," Serial 103-76, June 30, 1993 (Washington, DC: US Government Printing Office, 1994), babel.hathitrust.org/cgi/pt?id=ucl.31210013511959 (accessed Jan. 12, 2018). The committee chair, John Dingell (D–MI), stated at the outset of the hearing, "While the science of this project is fascinating, that is not the focus of today's hearing. . . . The Subcommittee on Oversight and Investigations has examined dozens of defense acquisitions in-depth. Many of

them were seriously mismanaged. But the SSC ranks among the worst projects that we have seen in terms of contract mismanagement and failed government oversight" (1).

49. President Bill Clinton, personal communication to Tyson; Michael Wines, "House Kills the Supercollider, And Now It Might Stay Dead," *New York Times,* Oct. 19, 1993.

50. "Contributions to Growth of Worldwide R & D Expenditures, by Selected Region, Country, or Economy: 2000–15," pie chart, in National Science Board, *Science & Engineering Indicators 2018 Digest,* NSB-2018-2, Jan. 2018, 5, fig. D, www.nsf.gov/statistics/2018/nsb20181/assets/1407/digest.pdf (accessed Jan. 23, 2018). More generally, in 2000 the United States accounted for 31 percent of the world economy, and China 4 percent. By 2015 that figure dropped to slightly more than 24 percent for the United States and almost 15 percent for China. See Robbie Gramer, "Infographic: Here's How the Global GDP Is Divvied Up," *Foreign Policy,* Feb. 24, 2017, foreignpolicy.com/2017/02/24/infographic-heres-how-the-global-gdp-is-divvied-up/ (accessed Jan. 23, 2018); Evan Osnos, "Making China Great Again," *New Yorker,* Jan. 8, 2018, 38.

51. Scott Simon, "Razor Technology, On the Cutting Edge," *Weekend Edition Saturday,* July 17, 2010, www.npr.org/templates/transcript/transcript.php?storyId=12858 3887 (accessed Apr. 11, 2017).

52. Re desirability of American preeminence, see, e.g., a report by the conservative Washington, DC, think tank Project for the New American Century: "The United States is the world's only superpower, combining preeminent military power, global technological leadership, and the world's largest economy. Moreover, America stands at the head of a system of alliances which includes the world's other leading democratic powers. At present the United States faces no global rival. America's grand strategy should aim to preserve and extend this advantageous position as far into the future as possible" (*Rebuilding America's Defenses: Strategy, Forces and Resources For a New Century,* Sept. 2000, i, www.informationclearinghouse.info/pdf/RebuildingAmericasDefenses.pdf, accessed Apr. 4, 2016).

53. United States Commission on National Security/21st Century, *Road Map for National Security: Imperative for Change* — Phase III Report, Feb. 15, 2001, 30, govinfo.library.unt.edu/nssg/PhaseIIIFR.pdf (accessed Apr. 4, 2016).

54. Council on Competitiveness, *Competitive Index: Where America Stands,* 2007, 15, 67, www.compete.org/storage/images/uploads/File/PDF%20Files/Competitiveness_Index_WhereAmerica_Stands_March_2007.pdf (accessed Apr. 4, 2016).

55. Joan Johnson-Freese, *Heavenly Ambitions: America's Quest to Dominate Space* (Philadelphia: University of Pennsylvania Press, 2009), ix.

56. Marc Kaufman and Dafna Linzer, "China Criticized for Anti-Satellite Missile Test," *Washington Post,* Jan. 19, 2007; Johnson-Freese, *Heavenly Ambitions,* 9–10, 15.

57. Translation displayed at the British Museum, room 7. Another translation of the Standard Inscription, adapted from that by Samuel M. Paley, reads, in part: "the divine weapon of the Great Gods, the potent king, the king of the world, the king of Assyria; . . . the powerful warrior who always lived by [his] trust in Assur, his lord; who has no rival among the princes of the four quarters of the earth; [who is] the shepherd of his people, fearless in battle, the overpowering tidewater who has no opponent; [who is] the king, subjugator of the unsubmissive, who rules the total sum of all humanity; [who is] the potent warrior, who tramples his enemies, who crushes all the adversaries; [who is] the disperser of the

host of the haughty; [who is] the king who always lives by [his] trust in the Great Gods, his lords; and captured all the lands himself, ruled all their mountainous districts, [and] received their tribute; who takes hostages, who establishes victory over all their lands." Vaughn E. Crawford, Prudence O. Harper, and Holly Pittman, *Assyrian Reliefs and Ivories in the Metropolitan Museum of Art: Palace Reliefs of Assurnasirpal II and Ivory Carvings from Nimrud* (New York: Metropolitan Museum of Art, 1980), full text at archive.org/stream/Assyrian ReliefsandIvoriesinTheMetropolitanMu seumofArtPalaceReliefsofAssurnasirpalIIan/ AssyrianReliefsandIvoriesinTheMetropoli tanMuseumofArtPalaceReliefsofAssurnasir palIIan_djvu.txt (accessed Apr. 5, 2016).

58. J. H. Parry, *Trade and Dominion: The European Overseas Empires in the Eighteenth Century* (New York: Praeger, 1971), 3, 5–6.

59. J. M. Coetzee, *Waiting for the Barbarians* (New York: Penguin, 1982), 133.

60. Ron Suskind, "Without a Doubt" [print title] or "Faith, Certainty and the Presidency of George W. Bush" [online title], *New York Times Magazine,* Oct. 17, 2004.

61. Maureen Dowd, "Are We Rome? Tu Betchus!" op-ed, *New York Times,* Oct. 11, 2008.

62. Mick Weinstein, "Ben's Bid to Boost Buck,"Yahoo Finance, June 6, 2008, finance .yahoo.com/expert/article/stockblogs/86614 (link disabled).

2. Star Power

1. See discussion of biological timekeeping in organisms from blue-green algae to humans in Roger G. Newton, *Galileo's Pendulum* (Cambridge, MA: Harvard University Press, 2004), 4–23.
2. For instance, the ca. 35,000 BC Lebombo bone from the mountains bordering South Africa and Swaziland; the ca. 20,000 BC Ishango bone from the border between Zaire and Uganda; the ca. 18,000 BC cave drawings at Lascaux, France. It is speculated that many of the early notched "calendar bones" were kept by women in order to monitor menstrual cycles.
3. Ronald A. Wells, "Astronomy in Egypt," in *Astronomy Before the Telescope,* ed. Christopher B. F. Walker (London: British Museum Press, 1996), 33–34; James Henry Breasted, "The Beginnings of Time-Measurement and the Origins of Our Calendar," *Scientific Monthly* 41:4 (Oct. 1935), 294.
4. In *Astronomy Before the Telescope,* ed. Walker, see David Pingree, "Astronomy in India," 129, for a discussion of the *kalpa*

and associated units, and Anthony F. Aveni, "Astronomy in the Americas," 272–73, for the Maya.

5. Nicholas Goodrick-Clarke, *The Occult Roots of Nazism: Secret Aryan Cults and Their Influence on Nazi Ideology: The Ariosophists of Austria and Germany, 1890–1935* (New York: New York University Press, 1992), 104, 192–97. Lanz von Liebenfels, publisher of *Ostara,* was preoccupied with all things occult, strange, spiritual, and Aryan. The subtitle of *Ostara* translates as something like "the blond ones and the rights of man." Hitler discovered this publication as a young man and went to the *Ostara* office to buy back issues, but he was so obviously poor that Liebenfels simply gave him the copies.

6. Lillian Lan-ying Tseng, *Picturing Heaven in Early China* (Cambridge, MA: Harvard University Asia Center, 2011), 45–47, 238, 316–19, 335–36. The Han view of Heaven conflated the sky and the supreme deity; a tomb being a microcosm, its ceiling represents the celestial realm. Furthermore, squareness refers to Earth, roundness to Heaven. Regarding the lunar lodges, writes Tseng, the ancient Chinese divided the sky into twenty-eight segments. Within each segment, the brighter stars were seen as a lodge — a place to rest or reside — because

the Moon moved cyclically from one segment to the next. Jessica Rawson, in "The Eternal Palaces of the Western Han: A New View of the Universe," *Artibus Asiae* 59:1/2 (1999), notes that the more elaborate of the Han rock-cut tombs "were complete settings for the afterlife. Each tomb was an entire universe centered on its occupant" (13). Our thanks to Jessica Rawson of Oxford University and John P. C. Moffett, librarian of the Needham Research Institute, Cambridge University, for their assistance.

7. Clive Ruggles, "Archaeoastronomy in Europe," in *Astronomy Before the Telescope,* ed. Walker, 21–23.

8. Ron Cowen, "Peru's Sunny View," *Science News* 171:18 (May 5, 2007), 280–81; J. McK Malville et al., "Astronomy of Nabta Playa," *African Skies/Cieux Africains* 11 (July 2007), 2–7.

9. As an example of detailed Chinese celestial chronicles, Gang Deng, *Chinese Maritime Activities and Socioeconomic Development, c. 2100 BC–1900 AD: Contributions in Economics and Economic History* 188 (Westport, CT: Greenwood Press, 1997), 36, excerpts a Zhou Dynasty (ca. 1046–256 BC) book on the positions of Taurus at night: "It shows up from the east at sea level in the Sixth Month, / . . . Reaches the zenith

in the Eighth Month, / . . . Drops to sea level in the Tenth Month." Colin Ronan, "Astronomy in China, Korea and Japan," in *Astronomy Before the Telescope,* ed. Walker, 247, stresses the exclusivity of astronomical, as well as meteorological and astrological, activity and the potential disruptiveness of the findings. Everyone who was not part of the emperor's staff of expert astronomers was discouraged from engaging in astronomy-related activities. Astronomical records were closely guarded — in effect, classified documents. Also see F. Richard Stephenson, "Modern Uses of Ancient Astronomy," in *Astronomy Before the Telescope,* ed. Walker, 331–32.

10. Noel Barnard, "Astronomical Data from Ancient Chinese Records: The Requirements of Historical Research Methodology," *East Asian History* 6 (Dec. 1993), 47–74; David S. Nivison, Kevin Pang, et al., "Astronomical Evidence for the *Bamboo Annals'* Chronicle of Early Xia," *Early China* 15 (1990), 87–95, 97–196; Salvo De Meis and Jean Meeus, "Quintuple Planetary Groupings — Rarity, Historical Events and Popular Beliefs," *J. Brit. Astronomical Assoc.* 104:6 (1994), 293–97.

11. Alexander Jones, "The Antikythera Mechanism and the Public Face of Greek Science," *Proceedings of Science,* PoS(An-

tikythera & SKA)038, 2012, pos.sissa.it/cgi
-bin/reader/conf.cgi?confid=170; Tony
Freeth and Alexander Jones, "The Cosmos
in the Antikythera Mechanism," *ISAW Papers* 4, Feb. 2012, dlib.nyu.edu/awdl/isaw/
isaw-papers/4/ (accessed Apr. 7, 2017);
Tony Freeth et al., "Decoding the Ancient
Greek Astronomical Calculator Known as
the Antikythera Mechanism," *Nature* 444
(Nov. 30, 2006), 587–91. Was the Mechanism unique? One notable, though nonsurviving, related object is the *sphaera* of Posidonius, described in a firsthand account by
Cicero, "in which single turnings have the
same effect for the Sun and Moon and the
five planets as occurs in single days and
nights." In addition, historians of technology contend that nobody in the ancient
world would have cast a complex device
full-scale in "expensive and intractable"
bronze unless it was already clear from
previous wooden models that it would
work, and wooden versions and components do not survive; the Mechanism would
have been "part of a long period of technical evolution largely hidden from us."
Stephanie Dalley and John Peter Oleson,
"Sennacherib, Archimedes, and the Water
Screw: The Context of Invention in the
Ancient World," *Technology and Culture*
44:1 (Jan. 2003), 16. Re ruling out the possibility of a later date for the Mechanism,

Jones writes, in part, "Also the Egyptian calendar scale on the front was designed to be removable so that the beginning of the Egyptian year could be lined up with any position of the Sun in the zodiac. That was necessary because the Egyptian calendar year was always 365 days long, with no leap years, so that the calendar year gradually shifted backwards relative to the natural seasons and the Sun's apparent motion through the zodiac. But after Egypt came under Roman rule in 30 BC, leap years were instituted every 4 years, so after that reform there would have been no need to make the scale adjustable in this way. More generally, the state of knowledge of astronomy built into the Mechanism makes good sense for around the 2nd and 1st centuries BC, whereas by say Ptolemy's time (2nd century AD) it would have seemed quite crude and archaic" (email to Avis Lang, Apr. 7, 2017). For context, see "Time and Cosmos in Greco-Roman Antiquity," exhibition curated by Alexander R. Jones, Oct. 2016–Apr. 2017, Institute for the Study of the Ancient World, New York University, isaw.nyu.edu/exhibitions/ time-cosmos. The advanced technologies used in recent work on the object were provided by X-Tek Systems and Hewlett-Packard.

12. See, e.g., D. L. Simms, "Archimedes and the Burning Mirrors of Syracuse," *Technol-*

ogy and Culture 18:1 (Jan. 1977), 1–24; Wilbur Knorr, "The Geometry of Burning-Mirrors in Antiquity," *Isis* 74:1 (Mar. 1983), 53–73. In the fall of 2005 David Wallace, a mechanical engineer at MIT, and his students staged two simulations of this incident; see detailed accounts at "Archimedes Death Ray: Idea Feasibility Testing," web.mit.edu/2.009/www/experiments/deathray/10_ArchimedesResult.html, and "2.009 Archimedes Death Ray: Testing with MythBusters," web.mit.edu/2.009/www/experiments/deathray/10_Mythbusters.html (accessed Dec. 17, 2006).

13. John Noble Wilford, "Homecoming of Odysseus May Have Been in Eclipse," *New York Times,* June 24, 2008.

14. In Herodotus, *Histories,* 440 BC, trans. George Rawlinson, available as *The History of Herodotus,* Internet Classics Archive, classics.mit.edu/Herodotus/history.html (accessed Apr. 4, 2017), we read: "As, however, the balance had not inclined in favour of either nation, another combat took place in the sixth year, in the course of which, just as the battle was growing warm, day was on a sudden changed into night. This event had been foretold by Thales, the Milesian, who forewarned the Ionians of it, fixing for it the very year in which it actually took place. The Medes and Lydians,

when they observed the change, ceased fighting, and were alike anxious to have terms of peace agreed on" (1.74).

15. Book 9.12–21, *The Histories of Polybius,* Loeb Classical Library Edition, vol. 4, 1922–27, text in the public domain, penelope.uchicago.edu/Thayer/E/Roman/Texts/Polybius/9*.html (accessed Apr. 4, 2017).

16. *Histories of Polybius,* 9.15.1–5.

17. *Histories of Polybius,* 9.19.1–3. The "men acquainted with astronomy" would have included, most notably, Anaxagoras, who may have seen the eclipses of both 463 BC and 478 BC and who as a young man had hypothesized that the Moon is opaque and thus capable of casting a shadow on Earth; see Dana Mackenzie, "Don't Blame It on the Gods," *New Scientist,* June 14, 2008, 50–51.

18. Alan C. Bowen, "The Art of the Commander and the Emergence of Predictive Astronomy," in *Science and Mathematics in Ancient Greek Culture,* ed. C. J. Tuplin and T. E. Rihll (Oxford: Oxford University Press, 2002), 76, 87–89.

19. Edward Cavendish Drake, *A New Universal Collection of Authentic and Entertaining Voyages and Travels* (London: J. Cooke, 1768), 32.

20. For overviews of astrology in history, see,

e.g., S. J. Tester, *A History of Western Astrology* (Woodbridge, UK: Boydell Press, 1987); Anthony Grafton, "Girolamo Cardano and the Tradition of Classical Astrology: The Rothschild Lecture, 1995," *Proc. Amer. Philosophical Society* 142:3 (Sept. 1998), 323–33; Ellic Howe, *Astrology and Psychological Warfare During World War II* (London: Rider, 1972). Referring to Rudolf II, Kepler wrote in a letter on Easter 1611, "Astrology does the emperor endless harm when a shrewd astrologer wishes to play lightly with the credulity of people. I must see to it that it does not happen to our emperor. . . . The ordinary astrology is a mass of garbage and can easily be twisted and its messages recited out of both sides of the mouth." Mark Graubard, "Astrology's Demise and Its Bearing on the Decline and Death of Beliefs," *Osiris* 13 (1958), 239. See also Richard Kremer's review of Tester's *History of Western Astrology* in *Speculum* 65 (1990), 209; Sheila J. Rabin, "Kepler's Attitude Toward Pico and the Anti-Astrology Polemic," *Renaissance Quarterly* 50:3 (Autumn 1997), 759, 764.

21. Ptolemy, *Tetrabiblos* I.1, ed. and trans. F. E. Robbins (Cambridge: Harvard University Press, 1940), 3–4.

22. Grafton, "Girolamo Cardano," 326.

23. Ptolemy, *Tetrabiblos* I.4: "Of the Power of

the Planets"; I.5: "Of Beneficent and Maleficent Planets"; I.6: "Of Masculine and Feminine Planets"; II.3: "Of the Familiarities Between Countries and the Triplicities and Stars."

24. This is echoed in Goodrick-Clarke, *Occult Roots of Nazism,* 103: "[Otto] Pöllner's first work, *Mundan-Astrologie* [Mundane Astrology] (1914), laid the basis of political astrology by casting the horoscopes of state, people and cities, in order to determine their future destiny, while his second work, *Schicksal und Sterne* [Destiny and the Stars] (1914), traced the careers of European royalty according to the dictates of their natal horoscopes. [Ernst] Tiede gave an analysis of all belligerent state-leaders' horoscopes, before declaring that there was a two to one chance of victory for the Central Powers."

25. Often, of course, a celestial configuration does have a clear connection to an earthly event. Major earthquakes are more likely to take place at new and full Moons, when tides are at their highest and lowest and especially huge loads of water are straining plate boundaries. But such phenomena are a matter of ordinary physics and seismology, not astrology. The opposing view is stated by a well-known Indian astrologer, founder of the *Astrological Magazine,* who contended in his book *Astrology in Forecast-*

ing Weather and Earthquakes, "There is a need on the part of the seismologists and meteorologists to shed their prejudices and embark on the investigation of the ingenious methods used for thousands of years with such success." Michael T. Kaufman, "Bangalore Venkata Raman, Indian Astrologer, Dies at 86," *New York Times,* Dec. 23, 1998.

26. Grafton, "Girolamo Cardano," 326.

27. "[T]he mistakes of those who are not accurately instructed in its practice, and they are many, as one would expect in an important and many-sided art, have brought about the belief that even its true predictions depend upon chance. . . . Secondly, most, for the sake of gain, claim credence for another art in the name of this, and deceive the vulgar, because they are reputed to foretell many things, even those that cannot naturally be known beforehand. . . . Nor is this deservedly done; it is the same with philosophy — we need not abolish it because there are evident rascals among those that pretend to it." Ptolemy, *Tetrabiblos* I:2: "That Knowledge by Astronomical Means is Attainable, and How Far."

28. George Sarton, "Astrology in Roman Law and Politics," *Speculum* 31:1 (Jan. 1956), 160; Tester, *History of Western Astrology,* 110.

29. During the Thirty Years' War, at the Spanish court, diplomat Diego de Saavedra Fa-

jardo argued against the recommendation of Philip IV's Council of State that the king should stop paying attention to astrological predictions. For Saavedra, both astrology and history provided knowledge and relevant models of action; as one historian writes, "[D]ivine and natural laws could not be known by humanity without the use of somewhat trustworthy learned disciplines. . . . Without these means, the entire paradigm of order and organic linkage between macrocosm and human microcosm collapsed." Abel A. Alves, "Complicated Cosmos: Astrology and Anti-Machiavellianism in Saavedra's *Empresas Políticas,*" *Sixteenth Century J.* 25:1 (Spring 1994), 67–68.

30. Tester, *History of Western Astrology,* 220.
31. William D. Stahlman, "Astrology in Colonial America: An Extended Query," *William and Mary Quarterly,* 3rd ser., 13:4 (Oct. 1956), 557.
32. See also, e.g., N. M. Swerdlow, "Galileo's Horoscopes," *J. History of Astronomy* 35 (pt. 2):119 (2004), 135–41; Mario Biagioli, "Galileo the Emblem Maker," *Isis* 81:2 (June 1990), 232–36; Richard S. Westfall, "Science and Patronage: Galileo and the Telescope," *Isis* 76:1 (Mar. 1985), 11–30; Nick Kollerstrom, "Galileo's Astrology," in *Largo Campo di Filosofare, Eurosymposium Galileo 2001,* ed. J. Montesinos and C. Solís

(Puerto de la Cruz, 2001), 421–31, also at www.skyscript.co.uk/galast.html; *Galileo's Astrology,* ed. Nicholas Campion and Nick Kollerstrom, special issue of *Culture and Cosmos* 7:1 (Spring/Summer 2003).

33. See, e.g., Stahlman, "Astrology in Colonial America," 561; Ellic Howe, *Urania's Children: The Strange World of the Astrologers* (London: William Kimber, 1967), 21–67; Howe, *Astrology and Psychological Warfare,* 14–17. In the City of London the Stationers' Company, a craft guild, had been granted the monopoly on almanac publishing as early as 1603, when it began to publish an annual almanac. The *Vox Stellarum,* written by Francis Moore for many years (even well after his death), had a print order of 393,750 copies in 1803. The first weekly astrological periodical in any language, published in London every Saturday beginning in 1824, was called *The Straggling Astrologer of the Nineteenth Century; Or, Magazine of Celestial Intelligences;* the first almanac to have daily predictions was *The Prophetic Messenger for 1827, An Original, Entertaining, and Interesting Melange* — also published in London (Howe, *Urania's Children,* plates 1 and 2, following p. 36). Stahlman writes that in the American colonies between 1639 and 1799, more than a thousand almanacs were published,

a few of them with considerable input from the colonies' earliest astronomers (561). Writing in 1931, Carl Jung criticized the common notion that "astrology had been disposed of long since and was something that could safely be laughed at. But today, rising out of the social deeps, it knocks at the doors of the universities from which it was banished some three hundred years ago." C. G. Jung, "The Spiritual Problem of Modern Man," in *Civilisation in Transition*, quoted in Howe, *Astrology and Psychological Warfare*, 12–13.

34. Polls taken in 1978, 1985, 1990, 2005, and 2012. Shoshana Feher, "Who Looks to the Stars? Astrology and Its Constituency," *J. Scientific Study of Religion* 31:1 (Mar. 1992), 88; Stephanie Rosenbloom, "Today's Horoscope: Now Unsure," *New York Times*, Aug. 28, 2005; Pew Research Center, "Many Americans Mix Multiple Faiths," Dec. 9, 2009, www.pewforum.org/2009/12/09/ many-americans-mix-multiple-faiths/ (accessed Aug. 3, 2017); National Science Foundation, "Science and Technology: Public Attitudes and Understanding," *Science and Engineering Indicators 2014*, www.nsf.gov/statistics/seind14/index.cfm/ chapter-7/c7h.htm (accessed Apr. 5, 2017); Joan Quigley, *"What Does Joan Say?" My Seven Years as White House Astrologer to*

Nancy and Ronald Reagan (New York: Pinnacle, 1991), 9–14, 19; Snopes.com, "Urban Legends Reference Pages: Rumors of War (False Prophecy)," www.snopes.com/rumors/predict.htm (accessed Dec. 3, 2006); Brooks Hays, "Majority of Young Adults Think Astrology Is a Science," UPI, Feb. 12, 2014, www.upi.com/Science _News/2014/02/11/Majority-of-young -adults-think-astrology-is-a-science/5201 392135954/ (accessed Apr. 5, 2017).

35. Khushwant Singh, *The Collected Novels: Train to Pakistan* (New Delhi: Penguin, 1999), 64; Maseeh Rahman, "Wedding Frenzy Hits India as Every Sphere of Life Comes under Influence of Planets," *Guardian,* Nov. 29, 2003; Agence France Presse–English, "Indian Couples Rush to Marry on Luckiest Day of Wedding Season," Nov. 26, 2005; Press Trust of India, "30,000 Couples Tie Knot in Delhi," Nov. 27, 2005; Indo-Asian News Service, "Flower Business Soars with Delhi's Marriage Season," Dec. 2, 2005; Amrit Dhillon, "Down the Aisle," *South China Morning Post,* Nov. 7, 2006. Quotation re satellites from "India's Space Science," *Statesman (India),* Dec. 31, 2005. See also "A Havan Kund in the Laboratory?" *The Hindu,* May 22, 2001; "Master of Business Astrology," *Economist,* May 1, 2004; "India's Supreme Court Ap-

proves University Instruction in Astrology,"
Agence France Presse–English, May 5,
2004.

36. Vikram Chandra, *Sacred Games* (New
York: HarperCollins, 2007), 547.

37. Brian Diemert, "The Trials of Astrology
in T. S. Eliot's *The Waste Land:* A Gloss on
Lines 57–59," *J. Modern Literature* 22:1
(Fall 1998), 178.

38. Joanne Kaufman, "Profiting from the
Positions of Planets," *New York Times,* Nov.
3, 1985; N. R. Kleinfield, "Seeing Dollar
Signs in Searching the Stars," *New York
Times,* May 15, 1988; Gary Weiss, "When
Scorpio Rises, Stocks Will Fall," *Business
Week,* June 14, 1993, 106; Anne Matthews,
"Markets Rise and Fall, but He's Always
Looking Up," *New York Times,* Mar. 12,
1995; Reid Kanaley, "Astrological Web Sites
Predict Market Movements," *Philadelphia
Inquirer,* Oct. 15, 1999; "Investrend Co-
Sponsors Astrologers Fund Triple Gold
Investment Conference February 1," *Finan-
cial Times Information,* Jan. 30, 2006; David
Roeder, "Some Large-cap Deals Hide in
Plain Sight," *Chicago Sun Times,* Apr. 30,
2006.

39. Ilia D. Dichev and Troy D. Janes, "Lunar
Cycle Effects in Stock Returns," Social Science Research Network, Aug. 2001, 3–4,
papers.ssrn.com/sol3/papers.cfm?abstract

_id=281665 (accessed Apr. 5, 2017).

40. Theodore White, "The Challenging Transits of Autumn 2007: How to Survive & Prosper," posted Aug. 2007, www.internationalastrologers.com/astro_meteorologist.htm; "Transits and the Economy," Sept. 20, 2007, www.internationalastrologers.com/transits_and_the_economy.htm. Also see "Theo's 2009–2010 World Economic Astrological Report: How to Survive the 2010s in a New Global Structure," Nov. 20, 2008, skyscript.co.uk/forums/viewtopic.php?t=3962&sid=d4b80774bb6e39ca2d48a16faa7d7aa8 (accessed Feb. 7, 2009).

41. For Howe's wartime activities as a forger, see Herbert A. Friedman, "Conversations with a Master Forger," *Scott's Monthly Stamp Journal,* Jan. 1980, www.psywar.org/forger.php (accessed Apr. 5, 2017).

42. Howe, *Astrology and Psychological Warfare,* 27–28.

43. Howe, *Astrology and Psychological Warfare,* 29–30, 36, 66, 197. The astronomer was H. H. Kritzinger, who wrote a book titled *Mysteries of the Sun and Soul,* in which he interpreted a quatrain of Nostradamus's as foretelling a crisis in 1939 between Britain and Poland; see Geoffrey Ashe, *Encyclopedia of Prophecy* (Santa Barbara: ABC-CLIO, 2001), 126. After the war, books with names such as *The Stars of*

War and Peace, by Louis de Wohl (1952), and *Zodiac and Swastika,* by Wilhelm Wulff (1968), appeared. See also Goodrick-Clarke, *Occult Roots of Nazism.*

44. Howe, *Astrology and Psychological Warfare,* chap. 3. Nothing in which people are interested can be completely silenced. On p. 49, n. 1, Howe quotes a passage from the autobiography of Hans Blüher, a founder of the Wandervögel youth movement, who in 1934 asked an astrologer friend to tell him about Hitler's horoscope: "My friend . . . leaned towards me and whispered into my ear through a cupped hand: 'He's a homicidal maniac!' "

45. Howe, *Astrology and Psychological Warfare,* 145–46; Jo Fox, "Propaganda and the Flight of Rudolf Hess, 1941–45," *J. Modern History* 83:1 (Mar. 2011), 78–110; James Edgar, "Rudolf Hess Plane Wreckage Hidden by Scottish Farmers, Letter Reveals," *Telegraph,* May 30, 2014. Numerous British political, military, and medical persons interrogated Hess over the course of days and weeks, but "[t]he interrogations revealed that Hess was so infused with the spirit of working toward the Führer that he had undertaken his mission in order to please him, not to undermine him. During his interviews, he insisted that Britain was a spent force, that capitulation was inevitable,

and that a negotiated peace was the only way to save British cities from the havoc wreaked on Rotterdam or Warsaw" (Fox, "Propaganda and the Flight," 87).

46. The Propaganda Ministry found two adjacent quatrains in Nostradamus especially suitable for the French psy-ops campaign: "Brabant, Flanders, Ghent, Bruges and Boulogne / are temporarily united with the great Germany. / But when the passage of arms is finished / the great Prince of Armenia will declare war. / Now begins an era of humanity of divine origin, / the age of peace is founded by unity, / war, now captive, sits on half the world, / and peace will be preserved for a long time." As one attendee reported, Goebbels discussed the matter as follows: "This is a thing we can exploit for a long time. [T]hese forecasts by Monsieur Nostradamus . . . must be disseminated only by handbills, handwritten, or at most typed, secretly. . . . The thing must have an air of being forbidden. . . . Interpretation: Introduction of the new order in Europe by Greater Germany, occupation of France only temporary, Greater Germany ushers in the thousand years' Reich and a thousand years' peace. Naturally, all this silly rubbish must also go out to France over the [secret] transmitters." Quoted in *The Secret Conferences of Dr. Goebbels: The Nazi Propaganda War*

1939–43, ed. Willi A. Boelcke (New York: E. P. Dutton, 1970), 6. The topic of how best to spread the news according to Nostradamus recurred throughout 1940. Also see Howe, *Astrology and Psychological Warfare,* chap. 10, "Nostradamus and Psychological Warfare," 133–44.

47. Besides a selection of relevant passages, this commission yielded 299 copies of a facsimile of the posthumous 1568 compilation *Les Prophéties de M. Michel Nostradamus,* with a thirty-two-page commentary by Krafft tucked inside its front cover. As Howe writes, "There was never any question of the book being freely available in the bookshops. A widespread public interest in Nostradamus and his prophecies was the last thing that the authorities desired, for they realised that no author was a better source for potential rumours, and if there were to be any rumours at all, then they would prefer to invent their own." Krafft died in 1945 while being moved from the Oranienburg concentration camp to Buchenwald. Howe, *Astrology and Psychological Warfare,* 190–91; Ashe, *Encyclopedia of Prophecy,* 125–27.

48. Howe, *Astrology and Psychological Warfare,* 197–99, 177–91.

49. According to Howe, *Astrology and Psychological Warfare,* 158, a report that circulated

among Nazi Party officials in the summer of 1944 referred to security operatives having seen a "considerable increase in all possible forms of prophecy about the future course of the war." Another slant on this appears in Anthony Heilbut, *Exiled in Paradise: German Refugee Artists and Intellectuals in America from the 1930s to the Present* (New York: Viking, 1983), 131: "As the war draws to a close, [l]ight entertainment is introduced, in Goebbels's words, 'to unobtrusively take the mind off the working day.' The schmaltzy music, the chats with crystal-gazers, astrologers, and palmists, are all signs of despair." A general corroboration appears in Ernst Kris and Hans Speier, *German Radio Propaganda: Report on Home Broadcasts During the War* (London: Oxford University Press, 1944), 103, n. 1: "Astrologers are more likely to be consulted during war than in peacetime. In the London press, astrological advertisements have increased since the beginning of the war, despite the fact that the size of newspapers has generally diminished."

50. Kris and Speier, *German Radio Propaganda,* 107 (fig. III), 109.

51. Kris and Speier, *German Radio Propaganda,* 103–10.

52. Howe, *Astrology and Psychological Warfare,* 191–96.

53. Perhaps the most important exception to this was Heinrich Himmler, rather than Rudolf Hess. See, e.g., Goodrick-Clarke, *Occult Roots of Nazism,* 5–6, 192, and Hugh Trevor-Roper, *The Last Days of Hitler,* 6th ed. (Chicago: University of Chicago Press, 1992), 71–74, 127–31. Trevor-Roper's statement "It is true that his personal court contained some strange figures; . . . that (like Hitler and Wallenstein) he was unduly influenced by his astrologer, Wulf" is, however, implictly negated by Howe.

54. Goodrick-Clarke, *Occult Roots of Nazism;* Howe, *Astrology and Psychological Warfare;* Trevor-Roper, *Last Days of Hitler.*

55. Trevor-Roper, *Last Days of Hitler,* 138–44; Howe, *Astrology and Psychological Warfare,* 200–204.

56. From the account of a secretary who worked in the same room as Goebbels's secretary, quoted in Trevor-Roper, *Last Days of Hitler,* 142–43.

57. Fritz Brunhübner, *Pluto,* trans. Julie Baum (Washington, DC: American Federation of Astrologers, n.d. [preface dated December 1934]). We thank Louise S. Sherby and the Archives & Special Collections of the Hunter College Libraries for making available Hunter's copy of the manuscript. See pp. 16, 67, 81, and passim for numerous additional astrological evoca-

tions of Pluto's characteristics.

58. Brunhübner, *Pluto,* 75.

3. Sea Power

1. Modern human remains 40,000 years old have been found in Tianyuan Cave near Beijing; remains from Fa Hien and Batadomba Lena caves in Sri Lanka are 35,000 years old. Some archaeologists contend that modern humans, equipped with new technology, began to emigrate from Africa along coastlines and reached southern Asia some 65,000 years ago. DNA evidence, for example the prevalence of high proportions of unusual mutations among New Guineans, Australians, and Andaman Islanders, supports this contention. See Dan Jones, "Going Global," *New Scientist,* Oct. 27, 2007, 36–41. See also Heather Pringle, "Follow That Kelp," *New Scientist* Aug. 11, 2007, 41; J. F. O'Connell and J. Allen, "Dating the Colonization of Sahul (Pleistocene Australia–New Guinea): A Review of Recent Research," *J. Archaeological Science* 31:6 (June 2004), abstract at doi:10.1016/j .jas.2003.11.005. Citing the work of Jon Erlandson, Pringle says that modern humans had arrived in the Willandra Lakes region "some 50,000 years ago." O'Connell and Allen "conclude that while the continent was probably occupied by 42–45,000

BP, earlier arrival dates are not well-supported." Archaic humans trekked far from Africa as well: almost 40,000-year-old bones of archaic hominids (possibly Neanderthals) have been found as far east as southern Siberia. See, e.g., Roxanne Khamsi, "Neanderthals Roamed as Far as Siberia," *New Scientist,* Sept. 30, 2007.

2. David Lewis, *We, the Navigators: The Ancient Art of Landfinding in the Pacific,* ed. Derek Oulton, 2nd ed. (Honolulu: University of Hawaii Press, 1994), 205ff., 21; E. G. R. Taylor, *The Haven-Finding Art: A History of Navigation from Odysseus to Captain Cook* (London: Hollis & Carter, 1956), 72–78; Barry Cunliffe, *The Extraordinary Voyage of Pytheas the Greek* (New York: Walker, 2002), 120–21. Lewis points out that boobies and frigate birds avoid alighting on the sea because their feathers become water-logged. Cunliffe relates another tale from the Icelandic saga of Flóki, who released three ravens in the course of his westward journey from Norway. The first flew east, back to land; the second circled the ship; the third headed straight westward, to Iceland. Cunliffe also mentions the starring role of land birds in the Portuguese navigator Pedro Alvares Cabral's encounter with the unknown land of Brazil in 1500: while sailing very wide of

the West African coast on a journey intended to reach India, Cabral saw the birds, followed them, and ended up in the harbor now known as Porto Seguro. Pelican sightings on Sept. 19 and 20, 1492: "Saw a pelican coming from W. N. W. and flying to the S. W; an evidence of land to the westward, as these birds sleep on shore, and go to sea in the morning in search of food, never proceeding twenty leagues from the land." Christopher Columbus, *Personal Narrative of the First Voyage of Columbus to America: From a Manuscript Recently Discovered in Spain,* trans. Samuel Kettell (Boston: T. B. Wait, 1827), archive.org/details/personalnarrativ00colu (accessed Apr. 6, 2017).

3. This phrase dates back to Elizabethan times; see Taylor, *Haven-Finding Art,* xii.

4. Excerpt from Michiel Coignet, *Instruction nouvelle des poincts plus excellents & necessaires, touchant l'art de naviguer* (1581), quoted in J. H. Parry, *The Age of Reconnaissance* (London: Phoenix Press, 1963), 83.

5. Charles H. Cotter, *A History of Nautical Astronomy* (New York: American Elsevier, 1968), 1. This hydrographer gave lessons in "gunnery," navigation, and several branches of mathematics and ran a map shop in London. See, e.g., "John Seller [ca. 1630–

1697], New York Public Library, www.nypl.org/research/chss/epo/mapexhib/seller.html (accessed Apr. 6, 2017).

6. The text of *Shangshu,* an anonymous Zhou Dynasty work, is as follows: "It shows up from the east at sea level in the Sixth Month, / Arises up half way from the zenith in the Seventh Month, / Reaches the zenith in the Eighth Month, / Declines to the west half way to the sea level in the Ninth Month, / Drops to sea level in the Tenth Month." Gang Deng, *Chinese Maritime Activities and Socioeconomic Development, c. 2100 B.C.–1900 A.D.: Contributions in Economics and Economic History 188* (Westport, CT: Greenwood Press, 1997), 36. For *Kitab al-Fawa'id,* see G. R. Tibbetts, *Arab Navigation in the Indian Ocean Before the Coming of the Portuguese* (London: The Royal Asiatic Society of Great Britain and Ireland, 1971), 130–31. This work includes both a translation of *Kitab al-Fawa'id* and a detailed discussion of Arab navigation.

7. For Homer and 1955, see Taylor, *Haven-Finding Art,* 9–13; for 1492, see J. E. D. Williams, *From Sails to Satellites: The Origin and Development of Navigational Science* (Oxford: Oxford University Press, 1992), 32. Taylor says "more than 12 degrees away from the pole" for Homer's day but includes a chart comparing Polaris at 1000

BC (showing it at 72/73 degrees N) and at 1955 (showing it at 90 degrees N). Williams is referring to the "true celestial pole." For the position of Polaris in AD 15,000 (declination 44° 27'), see Starry Night Pro, Simulation Curriculum Corp., v. 6.4.3.

8. Ursa Major is also called the Plough. In *Haven-Finding Art,* Taylor points out that the Greek name for the constellation, Arctos, "Bear," was also the word for "north" and that the Latin name, Septentrio, "north," derives from *septem triones,* seven plough-oxen (9). William B. Gibbon, in "Asiatic Parallels in North American Star Lore: Ursa Major," *J. Amer. Folklore* 77:305 (July–Sept. 1964), 236, points out that many pre-Columbian peoples of the United States and Canada also called Ursa Major the Bear, while some called it the Seven Brothers. The dipper imagery was also widespread: slaves fleeing the American South were told to "follow the drinking gourd" to make their way north. In the Arab world, the stars in the handle of the Plough were seen as having the form of a boat, and ibn Mājid proposes that Noah was inspired by the form of the Bear/Plough to build a ship and become the first navigator (Tibbetts, *Arab Navigation,* 69).

9. Homer, *Odyssey,* V.278–80, trans. A. T. Murray, at www.theoi.com/Text/HomerOdyssey5.html (accessed Apr. 6,

2017); Taylor, *Haven-Finding Art,* 9, 40, 43. Taylor uses the translation quoted here, while the Murray translation reads: "the Bear, which men also call the Wain, which ever circles where it is and watches Orion, and alone has no part in the baths of Ocean."

10. The Vikings' difficulty resulted from the obliquity of their course in relation to the horizon, as well as the drastic seasonal changes in the duration of day and night. See Taylor, *Haven-Finding Art,* chap. 4, "The Irish and the Norsemen," 65–85. For the Pacific Islanders, see Lewis, *We, the Navigators.*

11. Images dating to ca. 3100 BC of boats with sails have been found in Egypt, also a model dated to ca. 3400 BC in Mesopotamia. Lionel Casson, *The Ancient Mariners: Seafarers and Sea Fighters of the Mediterranean in Ancient Times* (Princeton: Princeton University Press, 1991), 4.

12. Casson, *Ancient Mariners,* 30–32.

13. Casson, *Ancient Mariners,* 6–21, 170–73; Deng, *Chinese Maritime Activities,* 113; Andrew Lawler, "Indus Script: Write or Wrong," *Science* 306:5704 (Dec. 17, 2004), 2027; *The Indian Ocean: Explorations in History, Commerce, and Politics,* ed. Satish Chandra (New Delhi: Sage, 1987), 30–31, 153–57; Lionel Casson, *Travel in the Ancient*

World (Baltimore: Johns Hopkins University Press, 1994), 369. Grain, olive oil, and wine are the top three items found in amphoras from shipwrecks in the Mediterranean.

14. Inscriptions relating to the casting of Athena Promachos, a towering bronze statue that once stood on the Acropolis, indicate that in the mid-fifth century BC one talent of tin sold for 233 drachmas and one talent of copper sold for 35 drachmas. James D. Muhly, "Sources of Tin and the Beginnings of Bronze Metallurgy," *Amer. J. Archaeology* 89:2 (Apr. 1985), 276–77.

15. Muhly writes, "Contact between the Aegean (and lands to the east) and Iberia goes back no earlier than the ninth century BC and the onset of Phoenician expansion/colonization of the western Mediterranean" ("Sources of Tin," 286). Javier G. Chamorro seems to disagree: "The archaeological and metallurgical evidence points to Iberian exploitation of the [silver] mines [of Tartessos] prior to the arrival of the Phoenicians and Greeks in the eighth to sixth centuries B.C.E. The mines remained in Tartessian hands; Phoenicians and Greeks simply provided new markets." Chamorro, "Survey of Archaeological Research on Tartessos," *Amer. J. Archaeology* 91:2 (Apr. 1987), 200.

16. A review of S. Bianchetti's *Pitea di Mas-*

salia: L'Oceano by K. Zimmerman mentions earlier journeys to northwestern Europe by Colaeus (blown off course), Midacritus, and Himilco (*Classical Review,* new ser. 50:1 [2000], 29). But some scholars apparently think there are even fewer bits of unarguable evidence concerning these than concerning Pytheas's. One account that ascribes greater credibility to Himilco's voyage than to Pytheas's is Luis A. García Moreno, "Atlantic Seafaring and the Iberian Peninsula in Antiquity," *Mediterranean Studies* 8 (1999), 1–13.

17. Casson, *Ancient Mariners,* 75.
18. Christina Horst Roseman, *Pytheas of Massalia: On the Ocean — Text, Translation and Commentary* (Chicago: Ares, 1994). C. F. C. Hawkes, in *The Eighth J. L. Myres Memorial Lecture — Pytheas: Europe and the Greek Explorers* (Oxford: Blackwell, 1977) mentions that Virgil's phrase was *"ultima Thule";* he contends that "Thule as Iceland seems to me clear" (34–35). Another scholar on Thule: "[T] here are three schools of thought: that Thule is indeed Iceland, that it was Norway, and that it was Shetland. It will be clear from the way I have presented the evidence that I belong to the Iceland school. To me the evidence seems unassailable" (Cunliffe, *Extraordinary Voyage,* 131–32).

19. Cunliffe, *Extraordinary Voyage,* 95–97, 102–103, 128–31; Hawkes, *Eighth Myres Lecture,* 37; Roseman, *Pytheas,* 121. Quoted descriptions are taken from Geminus, *Introduction to Celestial Phenomena* (first century AD) and Polybius. Cunliffe points out that Pytheas would have been measuring in the unit known as the stade, equivalent to 125 paces and amounting to 8.0 or 8.3 stades to the Roman mile, depending on who was doing the measuring. Ignoring fractals, the long and the short of it is that Britain's coastline, as given in *Encyclopaedia Britannica* and cited by Cunliffe, measures 4,548 miles — and Pytheas's approximation amounted to something in the neighborhood of 4,400 miles. Today (presumably taking into account only some of the ins and outs), according to the Ordnance Survey, Britain's national mapping agency, "The coastline length around mainland Great Britain is 11,072.76 miles," www.ordnancesurvey.co.uk/osweb site/freefun/didyouknow/ (accessed May 17, 2010). But as Benoit Mandelbrot famously proposed in "How Long Is the Coastline of Britain?" there is no end of different numbers. Nevertheless, Pytheas was certainly on the right track compared with contemporaries of his, such as the disbelieving Strabo.

20. Roseman, *Pytheas,* 7–20, writes that

eighteen known ancient writers referred to Pytheas by name between 300 BC and AD 550, notably Eratosthenes, Hipparkhos, Polybius, Strabo, and Pliny the Elder. Two more — Poseidonios and Diodoros — likely used his information but did not name him in their extant works. For a discussion of reasons not to credit Pytheas with this voyage, see Moreno, "Atlantic Seafaring and the Iberian Peninsula."

21. Taylor, *Haven-Finding Art,* 44; Casson, *Ancient Mariners,* 124; Cunliffe, *Extraordinary Voyage,* 99–100; Hawkes, *Eighth Myres Lecture,* 27–28, 30, 35–37.

22. Roseman, *Pytheas,* 117ff.

23. There is much doubt as to whether the Necho expedition was completed, though no doubt that it was undertaken. Herodotus, in *Histories* 4.42, reported that "[o]n their return, they declared — I for my part do not believe them, but perhaps others may — that in sailing round [Africa] they had the sun upon their right hand." In fact, only by going south of the equator can one see the Sun in such a position, so the very assertion Herodotus rejects is the one that best argues for the journey's actually having taken place. See discussion of Necho's seventh-century BC and Carthaginian king Hanno's fifth-century BC voyages in Casson, *Ancient Mariners,* 116–24.

24. Dava Sobel, *Longitude: The True Story of a Lone Genius Who Solved the Greatest Scientific Problem of His Time* (New York: Walker, 2005), 4.

25. Taylor, *Haven-Finding Art,* 12–13; Williams, *Sails to Satellites,* 8–9; Tibbetts, *Arab Navigation,* 129–32, 314; B. Arunachalam, "Traditional Sea and Sky Wisdom of Indian Seamen and Their Practical Applications," in *Tradition and Archaeology: Early Maritime Contacts in the Indian Ocean,* ed. Himanshu Prabha Ray and Jean-François Salles (New Delhi: Manohar, 1996), 264 and nn. 6–8. Tibbetts includes ibn Maājid quoting the Syrian rationalist philosopher and poet al-Ma'arri (129):

Suhail is the cheek of the beloved in colour
As the heart of the lover with its throbbing
Standing alone like the leading horseman
Clearly visible before the cavalry's ranks.

26. Tibbetts, *Arab Navigation,* 125; Alfred Clark, "Medieval Arab Navigation on the Indian Ocean: Latitude Determinations," *J. Amer. Oriental Society* 113:3 (July–Sept. 1993), 360, 363.

27. Taylor, *Haven-Finding Art,* 129, 161, x.

28. Deng, *Chinese Maritime Activities,* 37; Abdul Sheriff, "Navigational Methods in the Indian Ocean," in *Ships and the Development of Maritime Technology on the Indian Ocean,* ed. Ruth Barnes and David Parkin

(New York: Routledge Curzon, 2002), 216–18.

29. Deng, *Chinese Maritime Activities,* 39; Taylor, *Haven-Finding Art,* 92, 96, and generally 89–97. See also Barbara M. Kreutz, "Mediterranean Contributions to the Medieval Mariner's Compass," *Technology and Culture* 14:3 (July 1973), 367–83.

30. Guyot of Provins, quoted in Taylor, *Haven-Finding Art,* 95–96.

31. Taylor, *Haven-Finding Art,* 111–16, 140; Parry, *Age of Reconnaissance,* 1–16, 38–40, 77, 88–89.

32. Gail Vines, "The Other Side of Ohthere," *New Scientist,* June 28, 2008, 52–53; Taylor, *Haven-Finding Art,* 97, 155–56. As Williams puts it, "Plutarch, writing in the first century AD, had a clearer idea of African geography than any West European had in, say, AD 1400" (*Sails to Satellites,* 6; see also 13). Re Zheng He, Deng notes, "Zheng He's round-Asia marathon was by no means unprecedented in Chinese diplomacy: twelve centuries earlier, during the Three Kingdoms Period, Sun Quan, king of Wu, sent Zhu Ying and Kang Tai overseas for a twenty-year-long diplomatic mission, visiting southeast Asia, the Asian subcontinent, the Arabian sea region, and even the eastern Roman Empire" (*Chinese Maritime Activities,* 12). See also "Tuan Ch'eng-shih:

Chinese Knowledge in the Ninth Century," in G. S. P. Freeman-Grenville, *The East African Coast: Select Documents from the First to the Earlier Nineteenth Century* (Oxford: Clarendon Press, 1962), 10. Deng also contends that "Zheng He's maritime activities were military, or at least semimilitary," pointing out that "the great majority of the passengers were soldiers to 'show off China's wealth and strength to overseas by presenting arms' " (10). Mark Denny, by contrast, presents Zheng He's voyages as a version of the peaceable potlatch in *How the Ocean Works: An Introduction to Oceanography* (Princeton: Princeton University Press, 2008): "Unlike the Portuguese, who arrived in India with thoughts of spices and the slave trade, the Chinese desired to demonstrate the superiority of China by handing out gifts. Thus, the Treasure Ship carried treasure from China to the rest of the world to dazzle natives with the majesty of the Middle Kingdom." Chapter 1, "Discovering the Oceans," available at press .princeton.edu/chapters/s8693.html (accessed Apr. 7, 2017).

33. Translation of Azurara's chronicle in Emilia Viotti da Costa, "The Portuguese–African Slave Trade: A Lesson in Colonialism," *Latin American Perspectives* 12:1 (Winter 1985), 44; also see Parry, *Age of*

Reconnaissance, 35–36.

34. William E. Burrows, *This New Ocean: The Story of the First Space Age* (New York: Random House, 1998), 435.

35. Jorge Cañizares-Esguerra, *Nature, Empire, and Nation: Explorations in the History of Science in the Iberian World* (Stanford: Stanford University Press, 2006), 10–11, 20–21. The author cites a number of images that reinforce this interpretation: Captain John Smith, colonizer of Virginia, appears in his *Generall Historie of Virginia* (1624) "as a fully armored knight standing right next to a globe." *America Retectio* (ca. 1589), a series of engravings by a Flemish artist, shows Amerigo Vespucci using a quadrant for astronomical observation; beside him is "a banner bearing the cross, a reminder that Vespucci first described the constellation of the Southern Cross. A broken mast reminds the viewer that the knight-cosmographer has survived a tempest." In another engraving, Ferdinand Magellan is "depicted as a knight clad in full armor who charts the heavens by means of an armillary sphere, a lodestone, and a compass."

36. Arthur Davies, "Prince Henry the Navigator," *Transac. and Papers (Institute of Brit. Geographers)* 35 (Dec. 1964), 119–27; Taylor, *Haven-Finding Art,* 159; Viotti da Costa, "Portuguese–African Slave Trade," 45–46;

Spanish conquistador Bernal Díaz del Castillo on the reason to undertake voyages of conquest, quoted in Parry, *Age of Reconnaissance,* 19.

37. Kenneth Pomeranz and Steven Topik, *The World That Trade Created: Society, Culture, and the World Economy, 1400 to the Present,* 2nd ed. (Armonk, NY: M. E. Sharpe, 2006), 3–40. See also "al-Idrisi: The First Western Notice of East Africa," in Freeman-Grenville, *East African Coast,* 19–20: "The Zanj of the East African coast have no ships to voyage in, but use vessels from Oman and other countries which sail to the islands of Zanj which depend on the Indies . . . They own and exploit iron mines; for them iron is an article of trade and the source of their largest profits."

38. Parry, *Age of Reconnaissance,* 15; Pomeranz and Topik, *World That Trade Created,* 142–43.

39. Parry, *Age of Reconnaissance,* 80–81, 83–99; Williams, *Sails to Satellites,* 27; Taylor, *Haven-Finding Art,* 160–61.

40. Viotti da Costa, "Portuguese–African Slave Trade," 44–45, 47. See also Garrett Mattingly, "No Peace beyond What Line?" *Transac. Royal Historical Society,* 5th ser., 13 (1963), 147; Parry, *Age of Reconnaissance,* 32.

41. John Law, "On the Methods of Long

Distance Control: Vessels, Navigation, and the Portuguese Route to India," *Sociological Rev.* 32:S1 (May 1984), 234–63; Parry, *Age of Reconnaissance,* 94–96; Taylor, *Haven-Finding Art,* 162–66. The first extant terrestrial globe (1492), the "Erdapfel," made by Martin Behaim, was executed on parchment and then stretched over a globe.

42. Parry, *Age of Reconnaissance,* 11–15.

43. The author of *Breve compendio de la sfera y de la arte de navegar* (1551) noted, "If two points on the equator are 60 leagues apart, two points on the same meridians at latitude 60° are only 30 leagues apart, but the chart, *being plane,* shows them still to be 60 leagues apart." The FRS was John Wood, who attempted the Northeast Passage in 1676. See Williams, *Sails to Satellites,* 42, 45.

44. Parry, *Age of Reconnaissance,* 72–73.

45. David B. Quinn, "Columbus and the North: England, Iceland, and Ireland," *William and Mary Quarterly,* 3rd ser., 49:2 (Apr. 1992), 278–97; P. E. H. Hair, "Columbus from Guinea to America," *History in Africa* 17 (1990), 115. There is some debate about the Iceland trip. The trip to what is now Ghana was to a recently established Portuguese fort on the Gold Coast, São Jorge da Mina de Ouro.

46. Columbus's incorrect calculations are

discussed in detail in, e.g., W. G. L. Randles, "The Evaluation of Columbus' 'India' Project by Portuguese and Spanish Cosmographers in the Light of the Geographical Science of the Period," *Imago Mundi* 42 (1990), 50–64; Williams, *Sails to Satellites,* 15–16. See also "Privileges and Prerogatives Granted by Their Catholic Majesties to Christopher Columbus: 1492," Avalon Project, Yale Law School, avalon.law.yale.edu/15th_century/colum.asp (accessed Apr. 8, 2017).

47. Parry, *Age of Reconnaissance,* 69–70; Williams, *Sails to Satellites,* 9, 16, 18; Randles, "Evaluation of Columbus' 'India' Project," 54–55. Seventeen centuries earlier, Eratosthenes had raised the idea of heading west from Lisbon to reach China. For Mandeville, see C. W. R. D. Moseley, "Behaim's Globe and 'Mandeville's Travels,'" *Imago Mundi* 33 (1981), 89–91. The e-book of *The Travels of Sir John Mandeville* is available at www.gutenberg.org/ebooks/782 (accessed Apr. 8, 2017). Two contradictory passages give the flavor: "And therefore in the Septentrion, that is very north, is the land so cold, that no man may dwell there. And, in the contrary, toward the south it is so hot, that no man ne may dwell there, because that the sun, when he is upon the south, casteth his beams all straight upon that

part" (chap. 14) vs. "In Ethiopia be many diverse folk; and Ethiope is clept Cusis. In that country be folk that have but one foot, and they go so blyve that it is marvel. And the foot is so large, that it shadoweth all the body against the sun, when they will lie and rest them. In Ethiopia, when the children be young and little, they be all yellow; and, when that they wax of age, that yellowness turneth to be all black" (chap. 17). For details of the Columbus brothers' intrigue, see Arthur Davies, "Behaim, Martellus and Columbus," *Geographical J.* 143:3 (Nov. 1977), 451–59.

48. Parry, *Age of Reconnaissance,* 70, 83–84, 90–96.

49. Williams, *Sails to Satellites,* 26–27.

50. Re Hipparchus, see David Royster, "Mathematics and Maps," Carolinas Mathematics Conference, Oct. 17, 2002, 2–3, www.ms.uky.edu/~droyster/talks/NCCTM _2002/Mapping.pdf (accessed Apr. 8, 2017). For an extensive analysis of the map as state secret, see J. B. Harley, "Silences and Secrecy: The Hidden Agenda of Cartography in Early Modern Europe," *Imago Mundi* 40 (1988), 57–76. Re grids, see Parry, *Age of Reconnaissance,* 101.

51. See Gomes Eannes de Azurara (Portugal's royal chronicler), *The Chronicle of the Discovery and Conquest of Guinea,* trans.

C. R. Beazley (London: Hakluyt Society, 1899), 84–85, e-text at archive.org/details/chroniclediscov00presgoog (accessed Apr. 8, 2017).

52. Antonio Pigafetta, *Magellan's Voyage: A Narrative Account of the First Circumnavigation* (1534), trans. and ed. R. A. Skelton (New York: Dover, 1969), 1, 5–8, 148.

53. J. B. Harley, "Rereading the Maps of the Columbian Encounter," *Annals of the Association of American Geographers* 82:3 (Sept. 1992), 529–30.

54. Denis Cosgrove, "Globalism and Tolerance in Early Modern Geography," *Annals of the Assoc. of Amer. Geographers* 93:4 (Dec. 2003), 854; 852–70.

55. E. G. R. Taylor, "Gerard Mercator: A.D. 1512–1594," *Geographical J.* 128:2 (June 1962), 202; Mark Monmonier, *Rhumb Lines and Map Wars: A Social History of the Mercator Projection* (Chicago: University of Chicago Press, 2004), chap. 3, "Mercator's Résumé," www.press.uchicago.edu/Misc/Chicago/534316.html (accessed Apr. 8, 2017); David Turnbull, "Cartography and Science in Early Modern Europe: Mapping the Construction of Knowledge Spaces," *Imago Mundi* 48 (1996), 14, 23 nn. 46, 48. Mercator was not only a mapmaker but also an instrument maker, engraver, and author of a comprehensive project, published in

three phases, called *Atlas sive Cosmographi-cae Meditationes de Fabrica Mundi et Fabri-cati Figura* (Atlas, or Cosmographic Medita-tions on the Fabric of the World and the Figure of the Fabrick'd). One of his specu-lations was that magnetic variation in compasses had a terrestrial cause. His work did not, however, result in the comprehen-sive scientization of cartography, whether oceanic or terrestrial: as late as the 1750s, on a major map of Germany that included two hundred locations, a mere thirty-three were fixed by astronomical determinations of latitude, and none well fixed by longi-tude. Mercator, by the way, was imprisoned in 1544 for four months on grounds of heresy. Re the questionnaires, Turnbull writes that the Consejo Real y Supremo de las Indias issued them between 1569 and 1577: "On the whole this attempt to as-semble the empire failed because of the lack of trained and disciplined personnel. Many failed to answer the questionnaire; those who did often misunderstood the questions or the instructions on making observations or gave inaccurate responses."

56. Wright published his mathematical prin-ciples and his world map in 1599, in *Cer-taine Errors in Navigation,* and is sometimes credited with creating the first world map based on the Mercator projection after

Mercator's own of 1569. But Wright had given a Dutch acquaintance, Jocodus Hondius, an early version of the manuscript of *Certaine Errors* sometime before 1593, and Hondius, although having promised not to publish any of the information, seems to have preempted Wright by creating a world map of his own in 1598, as well as maps of individual regions. As the Dutch were ramping up their maritime exploration in the 1590s, presumably Hondius had much to gain from stealing Wright's ideas. See Brian Hooker, "New Light on Jodocus Hondius' Great World Mercator Map of 1598," *Geographical J.* 159:1 (Mar. 1993), 45–46. Even more interesting is the expedition that ultimately resulted in *Certaine Errors:* having become a noted mathematician and cosmographer, Wright was in 1589 " 'called forth to the public business of the nation, by the Queen' " and asked to take part in an "expedition" — a thieving party, a pirates' voyage — to the Azores, led by the Earl of Cumberland. See E. J. S. Parsons and W. F. Morris, "Edward Wright and His Work," *Imago Mundi* 3:1 (1939), 61. "The object of Cumberland's expedition," write Parsons and Morris, "was to prey upon Spanish commerce." Cumberland and his men took several ships: one full of spices, three full of sugar, and a fifth, the most valuable, full of hides, silver, and cochineal,

which was wrecked off Cornwall.
57. See, e.g., Parry, *Age of Reconnaissance,*
100–127. In chap. 2 Parry contends, "It was
not the sixteenth century, but the seventeenth, which saw the eclipse of the Mediterranean. The heirs of Italian mercantile
predominance were not the Portuguese, but
the English and the Dutch. . . . The Span-
iards and the Portuguese, however, lacked
the capital . . . and the financial organization to exploit commercially the discoveries
which they made" (48).
58. Joyce E. Chaplin, "The Curious Case of
Science and Empire," *Rev. in Amer. History*
34:4 (Dec. 2006), 436–37. She frames the
general issue thus: "Those busy Europeans
of the early modern era. Among the many
things they managed to do in the centuries
that stretched from 1500 to 1800, they
defined modern science and they created
modern empires. How did they do it? Did
the work depend on an efficient synergy,
the two projects somehow supporting each
other, science as handmaid to empire? Or
did Europeans' accomplishment depend on
a division of labor, in which different people
did different things in parallel, nevertheless
contributing to a larger program of European definition and control of the globe?
Or were the revolutions in science and in
global domination merely coincidental, hav-

ing nothing, really, to do with each other?" (434).

59. State Library of New South Wales, "The Crew on the *Endeavour,*" Papers of Sir Joseph Banks, www2.sl.nsw.gov.au/banks/series_03/crew_01.cfm (accessed Sept. 26, 2017).
60. See generally Donald H. Menzel, "Venus Past, and the Distance of the Sun," *Proc. Amer. Philosophical Society* 113:3 (June 16, 1969), 197–202; Donald A. Teets, "Transits of Venus and the Astronomical Unit," *Mathematics Magazine* 76:5 (Dec. 2003), 335–48; "James Cook and the Transit of Venus," NASA Science, May 27, 2004, science.nasa.gov/science-news/science-at-nasa/2004/28may_cook (accessed Apr. 8, 2017); Neil deGrasse Tyson, "The Long and the Short of It," *Natural History* 114:3 (Apr. 2005), 26–27. Early in the third century b c Aristarchus estimated that the Sun was at least nineteen times farther from Earth than was the Moon. In the following century, Hipparchus agreed. So did Ptolemy. That estimate is about twenty times too small. Copernicus didn't change that estimate, but Kepler did, proposing a distance of 3469 Earth radii, which is about seven times too small. In 1671–73, three French astronomers, based on their observations of Mars from Paris and from French Guiana, calcu-

lated the Earth–Sun distance as 87,000,000 miles, which is only about 7 percent too small. In 1771, basing his calculation on the findings of the 1769 Venus transit, a British astronomer named Thomas Hornsby came up with 93,726,900 English miles. Today the average Earth–Sun distance, which constitutes a standard unit called the AU, or astronomical unit — our very own yardstick — is 92,955,807 miles (exactly 149,597,870,700 meters). Reverend Hornsby was off by only eight-tenths of one percent. But because the Sun daily loses mass (which is carried away in the solar wind), the value formally assigned to one AU is, in fact, a quantity that slowly changes with time.

61. E. G. R. Taylor, "Position Fixing in Relation to Early Maps and Charts," *Bull. Brit. Society for the History of Science* 1:2 (Aug. 1949), 27; Turnbull, "Cartography and Science," 6–7, 21 nn. 19, 20; Vikram Chandra, *Sacred Games* (New York: HarperCollins, 2007), 293.

62. Taylor, *Haven-Finding Art,* 51–52, 140. For inland waters, Herodotus distinguished a day's sail from a day's voyage in an oared boat.

63. Tyson, "Long and the Short," 24–26; Taylor, *Haven-Finding Art,* 49.

64. Museum labels, Royal Observatory Greenwich; Sobel, *Longitude,* 56; Robert

Howard, "Psychiatry in Picture. Psychiatry (2002), A10. The final degradation and downfall in H(1730s series *Rake's Progress,* "The 1 Bedlam," includes a "longitude lun telescope in hand, pursuing to the poi insanity his personal solution to "the lu tive puzzle of the age" (Howard).

65. Parry, *Age of Reconnaissance,* 118–2 quote from Williams, *Sails to Satellites,* 80

66. For an overview, see, e.g., Cotter, *Nauti cal Astronomy,* 180–267 (marred by a few errors in names and dates). For the odder options, see Sobel, *Longitude,* 41–49.

67. Quoted in Cotter, *Nautical Astronomy,* 188.

68. Sobel, *Longitude,* 35. Re Gemma, two dates — 1522 and 1530 — crop up in different authors. The 1530 reference is specified as "De usu globi" in D. J. Struik, "Mathematics in the Netherlands During the First Half of the XVIth Century," *Isis* 25:1 (May 1936), 47, in which Gemma "showed how to determine geographical longitudes with the aid of a watch."

69. Sobel, *Longitude,* 7, 58–59; Williams, *Sails to Satellites,* 80.

70. Quoted in Sobel, *Longitude,* 106.

71. Sobel, *Longitude,* 128–45, 149.

72. Although Maskelyne and his Board would not permit the real thing to accompany

Cook, they hired a clockmaker admired even by Harrison himself to copy H-4 for £500. See Sobel, *Longitude,* 138–45, 152–53.

73. Sobel, *Longitude,* 152–64.

74. Williams, *Sails to Satellites,* 79.

75. Text of full conference proceedings and related documents at "1884 International Meridian Conference," www.ucolick.org/~sla/leapsecs/scans-meridian.html (accessed Apr. 9, 2017). Re inviting astronomers to speak, see Session 2, Oct. 2, 1884, 15–21.

76. "The Meridian Conference," *Science* 4:89 (Oct. 17, 1884), 376–77.

77. "The Meridian Conference," *Science* 4:91 (Oct. 31, 1884), 421.

78. Stephen Malys, John H. Seago, Nikolaos K. Pavlis, P. Kenneth Seidelmann, and George H. Kaplan, "Prime Meridian on the Move: Pre-GPS Techniques Actually Responsible for the Greenwich Shift," GPS World, Jan. 13, 2016, gpsworld.com/prime-meridian-on-the-move/ (accessed Sept. 29, 2017); for an in-depth discussion, see Stephen Malys, John H. Seago, Nikolaos K. Pavlis, P. Kenneth Seidelmann, and George H. Kaplan, "Why the Greenwich Meridian Moved," *J. Geodesy* 89:12 (Dec. 2015), 1263–72. Universal Time, or UT1, was formerly known as Greenwich Mean Time.

4. Arming the Eye

1. Fred Watson, *Stargazer: The Life and Times of the Telescope* (Cambridge, MA: Da Capo Press, 2004), 49–50, 296–97; Albert Van Helden, "The Invention of the Telescope," *Transac. Amer. Philosophical Society* 67:4 (June 1977), 9, n. 4. The pope referred to was Gerbert d'Aurillac, who reigned as Pope Sylvester II from 999 to 1003.
2. From Galileo's *Sidereus Nuncius:* "Afterward I made another more perfect one for myself that showed objects more than sixty times larger. Finally, sparing no labor or expense, I progressed so far that I constructed for myself an instrument so excellent that things seen through it appear about a thousand times larger and more than thirty times closer than when observed with the natural faculty only." In *Archives of the Universe: A Treasury of Astronomy's Historic Works of Discovery,* ed. Marcia Bartusiak (New York: Pantheon Books, 2004), 81. As to the availability of ready-made lenses, Van Helden marshals extensive evidence that by the middle of the sixteenth century, the shops of spectacle sellers across Europe commonly offered a selection of both concave and convex lenses of varying strengths. The explosion of book publication in mid-fifteenth-century Europe, spurred by Johannes Gutenberg's invention

of the printing press with movable metal type, had led to rapid increases in myopia. The solution — concave spectacle lenses — was for sale in Florence by 1451 (Van Helden, "Invention of the Telescope," 10–11).

3. See Watson, *Stargazer,* 71–73, re observations prior to Galileo's. As with the rest of his scientific discoveries, Harriot did not publish his results, note J. J. O'Connor and E. F. Robertson in "Thomas Harriot," Mac-Tutor History of Mathematics Archive, University of St. Andrews, Scotland, www-gap.dcs.st-and.ac.uk/history/Biographies/Harriot.html (accessed Apr. 13, 2017).

4. Watson, *Stargazer,* 55–62; Van Helden, "Invention of the Telescope," 25–26, 36–42.

5. Engel Sluiter, "The Telescope Before Galileo," *J. History of Astronomy* 28:92 (Aug. 1997), 225–26.

6. The self-characterization of Galileo is from his book *Sidereus Nuncius, or The Sidereal Messenger* (1610), trans. Albert Van Helden (Chicago: University of Chicago Press, 1989), 1. The Venetian connection was the Republic's chief theologian, Fra Paolo Sarpi, who had been tasked with inspecting and testing the earlier petitioner's telescope and who might well have provided Galileo with extremely detailed information about it. See Mario Biagioli, "Did Galileo Copy

the Telescope? A 'New' Letter by Paolo Sarpi," in *The Origins of the Telescope,* ed. Albert Van Helden, Sven Dupré, Rob van Gent, and Huib Zuidervaart (Amsterdam: KNAW Press/Royal Netherlands Academy of Arts and Sciences, 2010), 203–30, innovation.ucdavis.edu/people/publications/biagioli-did-galileo-copy-the-telescope (accessed July 26, 2017).

7. Letter to Leonardo Donato, Doge of Venice, Aug. 24, 1609, in Galileo, *Sidereus Nuncius,* 7–8. Like most creative individuals in need of money, Galileo was thoroughly familiar with having to plead for support. In his discussion of a December 1605 letter from Galileo to Cosimo II, the Medici prince who would soon become Grand Duke of Tuscany, the historian of science Richard S. Westfall writes, "He [Galileo] had prepared his instructions for the use of the geometric and military compass for publication as a pamphlet, and in the spring of 1605 he formally sought permission to dedicate it to the crown prince, Cosimo. Galileo parlayed acceptance of the dedication into an invitation to instruct the prince in mathematics during his summer vacation and did not thereafter relent in his quest. He wrote the prince the flattering letters that an absolute ruler expected of a client, declaring himself 'one of his most faithful and devoted servants,'

and proclaiming his desire to demonstrate 'by how much I prefer his yoke to that of any other Master, since it seems to me that the suavity of his manner and the humanity of his nature are able to make anyone desire to be his slave.' The terms of Galileo's address . . . would not have seemed sycophantic to Galileo's contemporaries. Almost no one challenged the legitimacy of a hierarchically ordered society, the precondition of the patronage that supported Galileo in an economically unproductive occupation." Richard S. Westfall, "Science and Patronage: Galileo and the Telescope," *Isis* 76:1 (Mar. 1985), 14.

8. All quotes from Van Helden, "Invention of the Telescope," 15, 28–30.

9. See, e.g., Van Helden, "Invention of the Telescope," 11, 26; Engel Sluiter, "The First Known Telescopes Carried to America, Asia and the Arctic, 1614–39," *J. History of Astronomy* 28:91 (May 1997), 141; Engel Sluiter, "The Telescope Before Galileo," *J. History of Astronomy* 28:92 (Aug. 1997), 224–29.

10. *Las Lanzas,* also known as *Surrender at Breda* (1634–35), 307 × 367 cm., Museo Nacional del Prado, Madrid.

11. Robert K. Merton writes that there were more wars in seventeenth-century Europe than in any preceding century or any succeeding century until the twentieth ("Sci-

ence, Technology and Society in Seventeenth Century England," *Osiris* 4 [1938], 564). Geoffrey Parker later writes, "Hardly a decade [in European history] can be found before 1815 in which at least one battle did not take place. . . . In the sixteenth century there were less than ten years of complete peace; in the seventeenth there were only four" (*The Military Revolution: Military Innovation and the Rise of the West, 1500–1800,* 2nd ed. [Cambridge: Cambridge University Press, 1996], 1). On commercialization, see William H. McNeill, *The Pursuit of Power: Technology, Armed Force, and Society since A.D. 1000* (Chicago: University of Chicago Press, 1982), chap. 4. On military technology, see Merton, "Science, Technology," 543–57.

12. William Molyneux, *Dioptrica Nova: A treatise of dioptricks in two parts, wherein the various effects and appearances of spherick glasses, both convex and concave, single and combined, in telescopes and microscopes, together with their usefulness in many concerns of humane life, are explained* (London: Benj. Tooke, 1692), 243. Quoted in Peter Abrahams, "When an Eye Is Armed with a Telescope: The Dioptrics of William and Samuel Molyneux," paper, Antique Telescope Society, Sept. 2002, home.europa .com/~telscope/ molyneux.txt (accessed

Apr. 15, 2017).

13. Merton, "Science, Technology," 372 n. 8, 373 n. 9, 543–44; Parker, *Military Revolution,* 177 n. 2, quoting J. R. Hale, *War and Society in Renaissance Europe 1450–1620* (1985). Merton quotes John W. Fortescue, *A History of the British Army* (1899): "It is hardly too much to say that for, at any rate, the four years from 1642 to 1646 the English went mad about military matters. Military figures and metaphors abounded in the language and literature of the day."

14. Samuel Butler, "The Elephant in the Moon" (1676, published posthumously).

15. Re lenses vs. mirrors: Early investigators tried out a wide variety of lens shapes and combinations of lenses. René Descartes, the seventeenth-century French mathematician and philosopher, proposed an especially complicated lens whose shape was a combination of an elliptical solid and a hyperbolic solid, set at right angles to each other — perhaps a fine idea, but the technology to make such a lens didn't exist in his day. Other investigators tried lengthening the tube that held the lenses; one tube, the creation of a brewer, Johannes Hevelius, was so long that it had to be propped up with ropes and pulleys, and would drift off target with the slightest puff of wind.

For those who chose to work with mirrors rather than lenses, there were other problems: How do you arrange the mirrors so that your head doesn't get in the way? How do you balance the virtues and drawbacks of reflective metal versus transparent glass? What do you use to polish the mirrors? Can you combine mirrors and lenses in the same telescope? Isaac Newton's solution was to use a concave primary mirror to collect and reflect the light onto a flat, angled secondary mirror that redirected the now-converging beam of light out the side of the tube, where the viewer looked at the image through an eyepiece. William Herschel, fresh from his discovery of Uranus in 1781, built himself a forty-foot-long telescope — the world's largest at the time — fitted with a four-foot-wide mirror made mostly of polished copper. The mirror was so large that the viewer, standing on an angled platform above it, would block only a small fraction of the mirror's total collecting area. From that vista you would observe the direct reflection of light through an eyepiece rather than reflected from a secondary mirror.

16. Albert Van Helden, "The Telescope in the Seventeenth Century," *Isis* 65:1 (Mar. 1974), 42; Robert Hooke, *Micrographia* (1665), preface, quoted in Van Helden, "Invention of the Telescope," 27–28 n. 23; letter from Galileo to Giuliano de Medici,

Nov. 13, 1610, quoted in Westfall, "Science and Patronage," 23.

17. One such was a gentleman acting as Portugal's official observer during a sea battle between the French and the Portuguese off the coast of Brazil in late 1614. Noticing that the Brazilian-born Creole commander of the Portuguese forces took advantage of a pause in the fighting to pick up his spyglass, the observer told him he was wasting everybody's time — that looking through a telescope "will neither lessen our task nor make our enemies fewer." Sluiter, "First Known Telescopes," 141–45.

18. Sluiter, "First Known Telescopes," 141–45; Yasuaki Iba, "Fragmentary Notes on Astronomy in Japan (Part III)," *Popular Astronomy* 46 (1938), 94.

19. Martin van Creveld, *Command in War* (Cambridge, MA: Harvard University Press, 1985), 10–11, 115; Frederick the Great, "The King of Prussia's Military Instructions to His Generals," Articles V, VIII, www.au.af.mil/au/awc/awcgate/readings/fred_instructions.htm (accessed Apr. 15, 2017).

20. Silvio A. Bedini, "Of 'Science and Liberty': The Scientific Instruments of King's College and Eighteenth Century Columbia College in New York," *Annals of Science* 50:3 (May 1993), 214; Edward

Redmond, "George Washington: Surveyor and Mapmaker — Washington as Land Speculator," Library of Congress, www.loc.gov/collections/george-washington-papers/articles-and-essays/george-washington-survey-and-mapmaker/washington-as-land-speculator/ (accessed Apr. 15, 2017).

21. Benjamin Franklin, *Proposals Relating to the Education of Youth in Pensilvania*, 1749, 30, facsimile at sceti.library.upenn.edu/pages/index.cfm?so_id=7430&pageposition=30&level=2 (accessed Apr. 15, 2017).

22. Meanwhile, the Royal Society of London for the Improvement of Natural Knowledge — founded in 1660, a quarter century after the founding of Harvard College — continued to support and recognize the work of colonial scientists. Frederick E. Brasch, "John Winthrop (1714–1799), America's First Astronomer, and the Science of His Period," *Publications of the Astronomical Society of the Pacific* 28:165 (Aug.–Oct. 1916), 156.

23. See Bedini, " 'Science and Liberty,' " 214–15.

24. Letter from George Washington to William Heath, Sept. 5, 1776, in Henry P. Johnston, *The Campaign of 1776 Around New York and Brooklyn . . . Containing Maps, Portraits, and Original Documents* (Brooklyn: Long Island Historical Society, 1878),

www.gutenberg.org/files/21990/21990.txt (accessed Apr. 16, 2017).

25. The original painting was done in 1850 but was damaged by fire; a full-size copy painted by the artist in 1851 now hangs at the Metropolitan Museum of Art in New York City.

26. Deborah Jean Warner, *Alvan Clark & Sons: Artists in Optics* (Washington, DC: Smithsonian Institution Press, 1968), 33.

27. Letter from George Washington to Brigadier General Anthony Wayne, July 10, 1779, National Digital Library Program, Library of Congress, cdn.loc.gov/service/mss/mgw/mgw3b/009/009.xml (accessed Apr. 16, 2017).

28. Van Creveld, *Command in War,* 12; in his slightly later work *Technology and War, From 2000 B.C. to the Present,* rev. and exp. (New York: The Free Press, 1991), van Creveld writes that in the sphere of military intelligence during the period 1500–1830, "technological developments were minimal" (120). Two classics written at about the same time that simply omit mention of the telescope are McNeill, *Pursuit of Power,* and Parker, *Military Revolution.*

29. Van Creveld, *Command in War,* 281 n. 23; van Creveld, *Technology and War,* 117–20.

30. Frederick the Great, "Military Instructions," Article I; van Creveld, *Technology in*

War, 107, 123; McNeill, *Pursuit of Power,* 126–29. Frederick the Great writes of "taking care that the troops be furnished with bread, flesh, beer, brandy, &c." Van Creveld estimates the daily food requirement for an army besieging a fortress, say 50,000 troops and 33,000 horses, at 1.5 kg per day for each man and 15 kg per day for each horse, or a total of 475 tons of food each day.

31. See, e.g., van Creveld, *Technology in War,* 86, 96, 106, and generally "The Age of Machines, 1500–1830," 81–149.
32. For extensive discussion and documentation of the earliest means of signaling, see Gerard J. Holzmann and Björn Pehrson, *The Early History of Data Networks* (Los Alamitos, CA: IEEE Computer Society Press, 1995), 1–29, 43–44; condensed pdf available at people.seas.harvard.edu/~jones/cscie129/papers/Early_History_of_Data_Networks/ The_Early_History_of_Data_Networks.html (accessed Apr. 16, 2017); Alexander J. Field, "French Optical Telegraphy, 1793–1855: Hardware, Software, Administration," *Technology and Culture* 35:2 (Apr. 1994), passim; George B. Dyson, *Darwin Among the Machines: The Evolution of Global Intelligence* (Reading, MA: Addison-Wesley Longman, 1997), 131–39. Jamie Morton, *The Role of the Physical Environment in Ancient Greek Sea-*

faring (Leiden: Brill, 2001), discusses beacons and fires, including the legend, repeated by Euripides, that Nauplius, king of Euboea, intentionally set misleading beacon fires on a dangerous, rocky promontory so as to cause the wrecking of the Greek fleet returning from Troy, because the fires would be interpreted by the Greeks as the lights of a safe harbor (210–12). Polybius, *Histories,* 10.45.5, 10.43.2.

33. Holzmann and Pehrson, *Early Data Networks,* 35–38; Dyson, *Darwin Among the Machines,* chap. 8, "On Distributed Communications," 133–34, 137–38; Field, "French Optical Telegraphy," 332.

34. The root of "tachygraph" (French *tachygraphe*) is the Greek *tachys,* meaning "swift" — also found in "tachometer" (an instrument for measuring the velocity of a machine) as well as "tachyon" (a hypothetical faster-than-light particle).

35. Quoted in Holzmann and Pehrson, *Early Data Networks,* 56–57.

36. See detailed technical descriptions in, e.g., Field, "French Optical Telegraphy," 320–22, 331–38; figs. 1–2, pp. 334–35. Field likens the Chappes' system to American Sign Language — "Sign language is, in a sense, optical telegraphy over short distances, with hand, arm, and finger signals the analogue to the positions of the Chappe

apparatus. Both systems use a large and complex transmission set, since both can rely on the acuity of visual recognition; both are constrained by the time required to compose individual signals" — and points out that ASL "is in fact an offshoot of, and linguistically similar to, a code originally developed in France in the 18th century" (329).

37. Andy Martin, "Mentioned in Dispatches: Napoleon, Chappe and Chateaubriand," *Modern & Contemporary France* 8:4 (2000), 446–47; van Creveld, *Command in War,* 60.

38. In French, the complete transcription reads:

> Le Directeur de la Correspondance Télégraphique de Strasbourg au Citoyen Commissaire du pouvoir exécutif près l'administration municipale de Strasbourg.
>
> Transmission télégraphique de Paris à Strasbourg le 21 Brumaire.
>
> Le corps législatif est transporté à St. Cloud. Bonaparte est nommé Commandant de Paris. Tout est tranquille et content.
>
> Le Directoire a donné sa démission. Moreau, général, commande au palais du Directoire.
>
> Pour copie, Durant

We are grateful to Maryline Simler of the Société d'Histoire de la Poste et de France Telecom en Alsace for providing us with the text of the original communication. For a photo of the transcribed telegram, written on the custom letterhead of the time, see musee.ptt.alsace.pagesperso-orange.fr/page%20tour.htm (accessed Oct. 7, 2017).

39. *Aeneid,* 1.278–79.

40. US military historians take the position that the United States was the first country to have a signal corps, though perhaps the dedicated operators of the Chappe telegraph should collectively be considered a close forerunner. Not until Feb. 1863 did Congress vote to establish the Union signalers as a separate corps. But the training of the men who became the US Signal Corps had begun in June 1861, and later that year Congress authorized $21,000 for their activities. The Confederate Congress voted to authorize such a corps in Apr. 1862. Rebecca Robbins Raines, *Getting the Message Through: A Branch History of the U.S. Army Signal Corps* (Washington, DC: Center of Military History, US Army, 1996), 3, 8–12, 29; Paul J. Scheips, "Union Signal Communications: Innovation and Conflict," *Civil War History* 9:4 (Dec. 1963), 402–403.

41. Raines, *Getting the Message Through,* 8, 23–24, 29; Scheips, "Union Signal Communications," 401–402; George Raynor Thompson, "Civil War Signals," *Military Affairs* 18:4 (Winter 1954), 189–90; Edwin C.

Fishel, *The Secret War for the Union: The Untold Story of Military Intelligence in the Civil War* (Boston: Houghton Mifflin, 1996), 38ff. Fishel quotes Alexander's account many years later: "I was watching the flag of our station at Stone Bridge [where the Warrenton Turnpike crosses over Bull Run] when in the distant edge of the field of view of my glass, a gleam caught my eye. It was the reflection of the sun (which was low in the east behind me), from a polished brass field piece." He immediately signaled to the nearby commanders, and Fishel contends that "the two commanders' action in response to the enemy advance was more timely than it could have been in the absence of the signal service. It would be too much to say that Alexander's intelligence won the Bull Run battle, but it certainly helped the Confederates from losing it. For that they had to thank an inventive young Yankee physician" (39–40).

42. In *Secret War,* Fishel argues that the immobility of the European "networks of semaphore towers . . . made them virtually useless by a marching and fighting army" (37–38).

43. Although Myer had (temporarily, as it turned out) lost his post in November 1863, he remained deeply preoccupied with the well-being and development of the Signal Corps, which was, after all, his

creation. He had begun *A Manual of Signals for the Use of Signal Officers in the Field* prior to his dismissal, and a sympathetic clerk in the Washington headquarters of the Signal Corps arranged for its printing. The title page does not mention Myer; instead it says, "Published by Order of the War Department / Washington: Government Printing Office." The next, enlarged edition, published in 1868 by D. Van Nostrand, has a lengthier title and lists Bvt. Brig. Genl. Albert J. Myer as author. Scheips, "Union Signal Communications," 413–14.

44. Initially, leftward meant "1" and rightward meant "2"; later the reverse became standard. In some accounts, therefore, "A" is listed as "2-2" and "B" as "2-1-1-2." See Major General A. W. Greely, "The Signal Corps," in *Photographic History of The Civil War in Ten Volumes,* vol. 8, ed. Francis Trevelyan Miller and Robert Sampson Lanier (New York: Review of Reviews, 1912), 312–40. Alphabet, numerals, and code signals listed on pp. 314 and 316.

45. Fishel, *Secret War,* 4; Albert J. Myer, *A Manual of Signals: For the Use of Signal Officers in the Field, and for Military and Naval Students, Military Schools, etc.* (New York: D. Van Nostrand, 1868), 231.

46. Myer, *Manual of Signals,* 232.

47. In the fall of 1863, Myer's "cipher disk" — a device for selecting preset variations on the regular code — came into general use; there is some evidence that the Confederate side was henceforth unable to read Union signals but that the Union side continued to be able to read the Confederates'. Scheips, "Union Signal Communications," 407 n. 32. Re casualties, the Union corps had a ratio of killed to wounded of 150 percent, and one Medal of Honor. Major General Greely, in "The Signal Corps," writes, "Did a non-combatant corps ever before suffer such disproportionate casualties — killed, wounded, and captured? Sense of duty, necessity of exposure to fire, and importance of mission were conditions incompatible with personal safety — and the Signal Corps paid the price. While many found their fate in Confederate prisons, the extreme danger of signal work, when conjoined with stubborn adherence to outposts of duty, is forcefully evidenced by the fact that the killed of the Signal Corps were one hundred and fifty per cent. of the wounded, as against the usual ratio of twenty per cent" (318). See also Raines, *Getting the Message Through,* 29.

48. For discussions of some of the battles as experienced by the Signal Corps, see, e.g., Greely, "The Signal Corps"; Thompson,

"Civil War Signals"; Raines, *Getting the Message Through,* 23–28. Battles whose outcomes are considered to have been shaped in part by signalers include Bull Run, Antietam, Chancellorsville, and Allatoona, as well as Gettysburg.

49. For accounts of signaling at Gettysburg, see, e.g., J. Willard Brown, *The Signal Corps, U.S.A. in the War of the Rebellion* (Boston: US Veteran Signal Corps Association, 1896), 359–72; Alexander W. Cameron, "The Signal Corps at Gettysburg," *Gettysburg* 3 (July 1990), 9–15; Raines, *Getting the Message Through,* 25–27; Thompson, "Civil War Signals," 197–98. General Lee's post-encounter report included the statement "The advance of the enemy to [Gettysburg] was unknown" (Fishel, *Secret War,* 522).

50. Official Records XXVII, Part III, 488, cited in, e.g., Raines, *Getting the Message Through,* 25.

51. This was one of several messages sent in quick succession, quoted in Brown, *Signal Corps, U.S.A.,* 360–61.

52. Quoted in Brown, *Signal Corps, U.S.A.,* 367–68.

53. Report by Capt. E. C. Pierce, quoted in Brown, *Signal Corps, U.S.A.,* 361–62. The July 3 entry in the diary of one of Pierce's flagmen, Sergeant Luther C. Furst, concurs

in vivid detail: "Were up before daylight. Began to signal in direction of Gettysburg at daybreak. Held our station all day, but were much annoyed by the enemy's sharpshooters in and near the Devil's Den. Have to keep under cover to protect ourselves. The large rocks piled up all around us serve as good protection. Today there have been seven men killed and wounded near our station by the enemy's sharpshooters; hundreds on all sides of us by the enemy's severe cannonading. Up to near noon there has been considerable skirmishing along line. A little later the whole of the artillery on both sides opened up and shell flew fast and thick. A good many have been struck near our station, but we are able to keep up communication. The fight upon the right is said to have been very severe, but our troops have held their positions and repulsed the enemy at every point." Quoted in Brown, *Signal Corps, U.S.A.*, 362–64.

54. Report by L. B. Norton, chief signal officer, Army of the Potomac, quoted in Brown, *Signal Corps, U.S.A.*, 372.

55. Raines, *Getting the Message Through*, 45–47, 53–54.

56. Raines, *Getting the Message Through*, 131, 145, chap. 5 passim. In the Signal Corps's 1914 annual report, the chief signal officer, while expressing uncertainty and fear about the dropping of bombs by air-

craft, predicted that "if the future shows that attack from the sky is effective and terrible, as may prove to be the case, it is evident that, like the rain, it must fall upon the just and upon the unjust, and it may be supposed will therefore become taboo to all civilized people; and forbidden at least by paper agreements."

57. Joseph W. Slade, "Review: *Getting the Message Through: A Branch History of the U.S. Army Signal Corps,*" *Technology and Culture* 39:3 (July 1998), 592.

58. Raines, *Getting the Message Through,* 169–70, 190, chap. 5.

59. For the beginnings, see, e.g., Hugh Barty-King, *Eyes Right: The Story of Dollond & Aitchison Opticians, 1750–1985* (London: Quiller Press, 1986), 15–53.

60. Barty-King, *Eyes Right,* 34.

61. Barty-King, *Eyes Right,* 53.

62. Warner, *Alvan Clark & Sons,* 99: "[Between 1863 and 1865] the Clarks sold the Navy at least 165 spyglasses at prices ranging from \$25.75 to \$35.00 apiece." Using 1863 dollars (the Civil War resulted in a significant lowering of prices by 1865), that would be about \$500 to \$700 in 2016 dollars, based on purchasing power as tracked by the Consumer Price Index. Calculator at Measuring Worth, www.measuringworth.com (accessed Jan. 16, 2018).

63. An acquaintance who had seen Clark in 1885 mentioned that the optician's thumbs had burst open from his having used them as polishers. Warner, *Alvan Clark & Sons,* 27.

64. Heber D. Curtis, "Optical Glass," *Publications of the Astronomical Society of the Pacific* (Apr. 1919), 77, archive.org/stream/publica tionsast30pacigoog/publicationsast30paci goog_djvu.txt (accessed Apr. 17, 2017).

65. Fuel usage alone presents a daunting challenge. In the US during World War I, for instance, "the question of fuel and gas for the glass-melting furnaces and for other operations became serious during the coal shortage of the winter 1917–18. When it is realized that the glass plant at the Bausch & Lomb factory alone consumed 33,000,000 cubic feet of illuminating gas per month, a quantity sufficient to meet the needs of a city of 80,000 inhabitants, the scale of its fuel consumption and of the difficulty of meeting the situation adequately is apparent." US Army Ordnance Department/Lt. Col. F. E. Wright, *The Manufacture of Optical Glass and of Optical Systems: A Wartime Problem* (Washington, DC: Government Printing Office, 1921), 288, archive.org/details/manufactureof opt00unitrich (accessed Apr. 17, 2017). For a clear explanation of the process, see Cur-

tis, "Optical Glass," 77–85.

66. In about 1820, for instance, a London instrument maker paid eight guineas for "a rude piece of flint glass about five inches in diameter" (Fred Watson, *Stargazer: The Life and Times of the Telescope* [Cambridge, MA: Da Capo, 2005], 183–85). A guinea being one pound and one shilling, eight guineas would have a 2016 purchasing power of more than £600, based on the retail price index (www.measuringworth .com) — about US$1,000. But window glass, too, was an expensive item, and Britain imposed a "deeply unpopular" window tax from 1696 through 1851; see "About Parliament: Living Heritage: Window Tax," www.parliament.uk/about/living -heritage/transformingsociety/towncountry/ towns/tyne-and-wear-case-study/about-the -group/housing/window-tax/ (accessed Apr. 17, 2017). In the US soon after the Civil War, coal fields along with natural gas were developed, and glass manufacture took place in exactly those areas: Pennsylvania, Ohio, West Virginia.

67. Zeiss unveiled the first optical planetarium projector in 1923; New York's Hayden Planetarium has had a Zeiss "projection planetarium" (Zeiss's own term) ever since its founding in 1935.

68. Antje Hagen, "Export versus Direct Investment in the German Optical Indus-

try: Carl Zeiss, Jena and Glaswerk Schott & Gen. in the UK, from Their Beginnings to 1933," *Business History* 38:4 (1996), 4, 17 n. 23. During the Boer War, Zeiss supplied the British army with binoculars; during the Russo-Japanese War, it supplied both sides with them.

69. Zeiss, "The Carl Zeiss Foundation in Jena, 1885–1945: Expansion of the Product Portfolio," www.zeiss.com/corporate/int/history/company-history/at-a-glance.html#inpagetabs-1 (accessed Apr. 18, 2017).

70. Zeiss, "History of Zeiss in Oberkochen," www.zeiss.com/ corporate/int/history/locations/oberkochen.html; Zeiss, "Background Story: The Development of Carl Zeiss Between 1945 and 1989," www.zeiss.com/corporate/int/history/company-history/20-years-of-reunification/background-story.html; Zeiss, "Lens in a Square — The Zeiss Logo," www.zeiss.com/corporate/int/history/company-history/the-zeiss-logo.html (accessed Nov. 15, 2017).

71. William Tobin, "Evolution of the Foucault–Secretan Reflecting Telescope," *J. Astronomical History and Heritage* 19:2 (2016), 106–84, available at SAO/NASA ADS Astronomy Abstract Service, adsabs.harvard.edu/abs/2016JAHH. . .19..106T (accessed Oct. 19, 2017).

72. Stephen C. Sambrook, "The British

Armed Forces and Their Acquisition of Optical Technology: Commitment and Reluctance, 1888–1914," in *Year Book of European Administrative History* 20 (Baden-Baden: Nomos Verlagsgesellschaft, 2008), 2.

73. US Army Ordnance Department, *Manufacture of Optical Glass,* chaps. 1, 7.

74. Curtis, "Optical Glass," 77; Stephen C. Sambrook, "No Gunnery Without Glass: Optical Glass Supply and Production Problems in Britain and the USA, 1914–1918" (working paper, Sept. 2000), n.p., home.europa.com/~telscope/glass-ss.txt; Stewart Wills, "How the Great War Changed the Optics Industry," *Optics & Photonics News* 27 (Jan. 2016), www.osa-opn.org/home/articles/volume_27/january_2016/features/how_the_great_war_changed_the_optics_industry/ (accessed Apr. 18, 2017).

75. As described by Sambrook, although Britain and the US had a heavy reliance on Schott, the exact nature of that reliance, both prior to and during World War I, is complex: "Sometimes, perhaps many times, it was as much the result of Schott's 'publicity' and growing reputation as the 'ne plus ultra' in glass manufacture, as a real need to use Schott glass. Many of the 'new Jena glasses' introduced in the 1890s were duplicated by Parra Mantois and to a

lesser degree by Chance prior to 1914. The problems of dependency which arose in Britain after 1914 were usually caused by a maker having previously designed an optical system which incorporated (say) one single element using a Schott glass which wasn't replicated by another maker. Solving that problem meant either copying the Schott glass or re-designing the rest of the system around whatever glass WAS available" (Sambrook, email to Avis Lang, Dec. 6, 2009).

76. Hagen, "Export versus Direct Investment," 11.

77. One such warning is found in NAK ADM 116/3458, Aug. 27, 1915, Admiralty (ADM) Report on the state of glass supplies, which describes a meeting on July 13, 1912, between Richard Glazebrook, director of the National Physical Laboratory, the Third Sea Lord, and the Director of Naval Ordnance. A "leading optician" had written to Glazebrook in 1911, saying that Schott glasses were widely used in optical instruments supplied by British makers to the Admiralty and that "[i]n the event of a war with Germany . . . a stoppage of optical glass would simply paralyse the optical trade." A year later, Glazebrook consulted "seven or eight leading opticians" who considered that it was "essential" to use German glasses in "most" of the Admiral-

ty's instruments and that both British and French glass was "unreliable" in transparency and homogeneity. It is possible that the opticians were overstating the case and angling for greater government support, but the warning was taken seriously enough to result in the formation of a committee to articulate domestic research requirements (Sambrook, email to Avis Lang, Dec. 7, 2009).

78. Economic historians differentiate "defense share" (military expenditures expressed as a percentage of total expenditures by a central government) from "military burden" (military expenditures expressed as a percentage of GDP, a far broader category that includes all goods and services in the nation-state as a whole). Economic historian Jari Eloranta presents a number of examples of military spending in terms of defense share: In England from 1535 to 1547, for instance, the defense share averaged 29 percent; from 1685 to 1813, the English average was 75 percent and in any given year did not drop below 55 percent. During the early nineteenth century the English average was about 39 percent and from 1870 to 1913 about 37 percent. During World War I the average annual defense share was massive: England 1914–18, 49 percent; France 1914–18, 74 percent; Germany 1914–18, 91 percent; US

1917–18, 47 percent. Jari Eloranta, "Military Spending Patterns in History," EH.Net Encyclopedia, ed. Robert Whaples (Sept. 16, 2005), eh.net/encyclopedia/military -spending-patterns-in-history (accessed Apr. 18, 2017).

79. Raines, *Getting the Message Through,* 172, 191; Curtis, "Optical Glass," 81. Curtis emphasizes the role of the federal government's Bureau of Standards in this effort: "In times of peace a busy place engaged in scientific and industrial researches nearly as numerous as those of the combined physical and chemical laboratories of the universities of the country; under the stress of war, the Bureau perforce expanded into a personnel of nearly fifteen hundred, gathered from all over the United States, working and experimenting in every phase of the war's scientific needs."

80. Stephen C. Sambrook, "The British Optical Munitions Industry Before the Great War," in *Proceedings,* Economic History Society Annual Conference, Royal Holloway College, University of London, Apr. 2004, 52, www.ehs.org.uk/events/ehs -annual-conference-archive.html (accessed Apr. 18, 2017).

81. Stephen C. Sambrook, The Optical Munitions Industry in Great Britain, 1888–1923, PhD diss., University of Glasgow, 2005, 156.

82. Sambrook, "No Gunnery Without Glass"; Sambrook, "British Armed Forces and Acquisition."
83. In 1913 Britain's exports totaled $3.1 billion, Germany's $2.4 billion. All figures were converted to USD by the authors, using gold standard par value. Hugh Neuburger and Houston H. Stokes, "The Anglo-German Trade Rivalry, 1887–1913: A Counterfactual Outcome and Its Implications," *Social Science History* 3:2 (Winter 1979), 187–88, 191–92.
84. Treaty of Versailles, Articles 168–70, 202.
85. The final list of verboten items to be surrendered was presented in the Blue Book's thirty-three chapters. Many were dual use, and soon Germany challenged the breadth of the list, "arguing that the inclusion of items such as cooking utensils, and more importantly transportation, would not only hurt the German economy but would hinder reparations deliveries to the Allies and make German political conditions conducive to Bolshevism." Richard J. Schuster, *German Disarmament After World War I: The Diplomacy of International Arms Inspection* (London: Routledge, 2006), 41.
86. J. H. Morgan, *Assize of Arms: The Disarmament of Germany and Her Rearmament (1919–1939)* (New York: Oxford University Press, 1946), 37–38.

87. Schuster, *German Disarmament,* 63–64; Morgan, *Assize of Arms,* 35, 40. Germany asked that some eighty factories be permitted to engage in production of war material; the IAMCC eventually allowed fourteen smallish firms to each produce a specific type of weapon. See Schuster, *German Disarmament,* 42–45, for details on the destruction of arms. While a German writer, Hans Seeger, emphasizes the widespread destruction rather than the confiscation of precision optics, some historians, such as Michael Buckland, contend that such *Feinmechanik* would have been attractive booty for the Allied conquerors (email to Avis Lang, Dec. 2009), and Morgan points out that within a few months after the German government declared it had surrendered all required arms, "hundreds of newly manufactured howitzers [were found] walled up in a single factory," and a "vast 'gun park' of heavy artillery" was found to have been "secreted in the forts of Königsberg." Seeger, *Militärische Ferngläser und Fernrohre in Heer, Luftwaffe und Marine* (1996), 32, English translation at www.europa.com/∼telscope/trsg2.txt (accessed Apr. 18, 2017); Morgan, *Assize of Arms,* 35.

88. Hagen, "Export versus Direct Investment," 1 and n. 6, 4–7, 11–12, 17–18 n. 25; US Army Ordnance Department, *Manu-*

facture of Optical Glass, chap. 1; Sambrook, "British Optical Munitions Industry," 54. According to the Ordnance Department, "Educational and research institutions obtained a large part of their equipment from Germany and offered no special inducement for American manufacturers to provide such apparatus. Duty-free importation favored and encouraged this dependence on Germany for scientific apparatus."

89. The Zeiss website states: "Although the production of instruments for civilian use had been dominant in the 1920s and early 1930s, Jena never lost sight of the development of military instruments, as the advances being achieved at that time in the field of precision engineering and optics were equally suitable for civilian and military purposes." "The Carl Zeiss Foundation in Jena," www.zeiss.com/corporate/int/history/company-history/at-a-glance.html#inpagetabs-1 (accessed Apr. 18, 2017).

90. P. G. Nutting, "The Manufacture of Optical Glass in America," letter, Science 46:1196 (Nov. 30, 1917), 539. In 2016 purchasing power, half a million 1913 dollars would be more than $12 million, according to Measuring Worth, www.measuringworth.com/ppowerus (accessed July 26, 2017).

91. Hagen, "Export versus Direct Invest-

ment," 17 n. 25; M. Herbert Eisenhart and Everett W. Melson, "Development and Manufacture of Optical Glass in America," *Scientific Monthly* 50:4 (Apr. 1940), 323; Raines, *Getting the Message Through,* 174.

92. The War Industries Board mobilized, co-ordinated, and regulated production; headed by Bernard Baruch for the final eight months of the war, it converted about a quarter of US industrial production to military purposes. The National Bureau of Standards developed a new crucible material that could survive the assault of barium crown glass melts; it also assisted with the testing of glass and of finished optical instruments. The US Geological Survey located new sources of raw materials, such as sufficiently pure silica sand. See US Army Ordnance Deparartment, *Manufacture of Optical Glass,* introduction and table 1.

93. The frontispiece of *Sidereus Nuncius* reads: "MAGNA, LONGEQVE ADMIRA-BILIA / Spéctacula pandens, sus-piciendáque proponens vniquique" (trans. Albert Van Helden). For Galileo's drawings, see "Sidereus Nuncius, Galileo Galilei (Facsimile)," Museo Galileo VirtualMuseum, catalogue.museogalileo.it/object/GalileoGalileiSidereusNunciusFacsimile.html (accessed Apr. 13, 2017). Until Galileo went blind in the late 1630s, his

eyesight as well as his instruments were taken as authoritative. However, the seventeenth-century astronomer and telescope maker Johannes Hevelius was not cowed by Galileo's reputation. In *Selenographia,* Hevelius's book on the Moon, he critiques Galileo's depiction of the Moon in *Sidereus Nuncius:* "Galileo did not have a sufficiently good telescope or could not devote sufficient care to those observations of his or, what is most likely, was ignorant of the art of picturing and drawing that serves this work very well, and no less than acute vision, patience, and toil." Albert Van Helden, "Telescopes and Authority from Galileo to Cassini," *Osiris,* 2nd ser. (1994), 15–18.

94. The writer also says, "Painters need not despair; their labours will be as much in request as ever, but in a higher field: the finer qualities of taste and invention will be called into action more powerfully; and the mechanical process will be only abridged and rendered more perfect. What chemistry is to manufactures and the useful arts, this discovery will be to the fine art; improving and facilitating the production, and lessening the labour of the producer; not superseding his skill, but assisting and stimulating it." Spectator, "Self-Operating Processes of Fine Art. The Daguerotype," *The Mu-*

seum of Foreign Literature, Science and Art 35 (Mar. 1839), 341–43, formerly available at "Daguerreian Texts: The First Two Years (1839–1840)," Daguerrian Society, www .daguerre.org/resource/texts/self_op.html (link disabled).

95. Fox Talbot was also the author of the first book to be illustrated with photographs, The Pencil of Nature (1844–46).

96. François Arago, "Fixation des images qui se forment au foyer d'une chambre obscure" (1839), in Oeuvres complètes de François Arago, vol. 7, ed. Jean Augustin Barral (Paris: Gide et J. Baudry, 1854–62), 4–5, trans. Stéphan Reebs and Avis Lang.

97. Arago, "Fixation des images," 6; Gérard de Vaucouleurs, Astronomical Photography: From the Daguerreotype to the Electron Camera, trans. R. Wright (New York: Macmillan, 1961), 13–16. Arago's two collaborators in the Moon-imaging experiment were Pierre-Simon Laplace and Étienne-Louis Malus (who was part of Napoleon's invasion of Egypt). The urging of Daguerre was done by a triad — Arago, Jean-Baptiste Biot, and Alexander von Humboldt — whom de Vaucouleurs describes as "his three renowned physicist-astronomer confidants in the Academy."

98. François Arago, "Report" (1839), in Classic Essays in Photography, ed. Alan Trach-

tenberg (New Haven: Leete's Island Books, 1981), 21–22. Similar statements are made a month later in François Arago, "Le daguerréotype: Rapport fait à l'Académie des Sciences de Paris le 19 août 1839" (Caen: L'Échoppe, reprint 1987), 18–22.

99. For the British side of this story, see R. Derek Wood, "The Daguerreotype Patent, the British Government, and the Royal Society," *History of Photography* 4:1 (Jan. 1980), 53–59.

100. "Despite these early successes, most professional astronomers shunned the photographic process. Photography, as it was practiced then, was noxious, imprecise, and inefficient." Alan W. Hirshfeld, "Picturing the Heavens: The Rise of Celestial Photography in the 19th Century," *Sky & Telescope* (Apr. 2004), 38. Two of the many excellent overviews of early astronomical photography are de Vaucouleurs, *Astronomical Photography,* and John Lankford, "The Impact of Photography on Astronomy," in *Astrophysics and Twentieth-Century Astronomy to 1950, Part A — The General History of Astronomy,* vol. 4, ed. Owen Gingerich (Cambridge, UK: Cambridge University Press, 1984), 16–39. On reproducibility, see Walter Benjamin's widely reprinted 1937 essay "The Work of Art in the Age of Mechanical Reproduction."

101. Thomas Melvill, "Observations on Light and Colours (1752)," reprinted in *J. Royal Astronomical Society of Canada* 8 (Aug. 1914), 242–43.

102. See Ian Howard-Duff, "Joseph Fraunhofer (1787–1826)," *J. Brit. Astronomical Assoc.* 97:6 (1987), 339–47.

103. Letter from Bunsen to Sir Henry Roscoe, Nov. 15, 1859, quoted in Mary E. Weeks and Henry M. Leicester, *Discovery of the Elements* (Easton, PA: J. Chemical Education, 1968), 598.

104. As translated in John Hearnshaw, "Auguste Comte's Blunder: An Account of the First Century of Stellar Spectroscopy and How It Took One Hundred Years to Prove That Comte Was Wrong," *J. Astronomical History and Heritage* 13:2 (2010), 90.

105. De Vaucouleurs, *Astronomical Photography,* 35, 49. The Henrys' instrument had a thirteen-inch aperture. Tenth-magnitude stars took them twenty seconds; sixteenth-magnitude took eighty minutes. In 1885, by using long exposures, the Henrys discovered a hitherto unobserved nebula surrounding the Pleiades, even though that region of the sky had been scrutinized by other astronomers for decades. See also "Obituary Notices: Associate: Prosper, Henry," *Monthly Notices of the Royal Astronomical Society* 64 (Feb. 1904), 296–98.

106. Lankford, "Impact of Photography," 29.

107. Samuel P. Langley was the first winner of the Henry Draper Medal and the founder of the Smithsonian Astrophysical Laboratory. Re newness and the quotation (from Sir William Huggins), see A. J. Meadows, "The New Astronomy," in *Astrophysics and Twentieth-Century Astronomy to 1950,* ed. Gingerich, 59, 70.

108. Harwit resigned under political pressure in May 1995 because of objections to the museum's planned exhibition marking the fiftieth anniversary of the atomic bomb dropped on Hiroshima by the American B-29 bomber *Enola Gay.* The controversial exhibition, meant to go beyond a commemoration of the end of World War II, was to include material on the consequences of the bombing. Because of strong advance criticism from groups including the American Legion and the Air Force Association, it was canceled. See Edward J. Gallagher, "History on Trial: The Enola Gay Controversy," Lehigh University, www.lehigh.edu/%7Eineng/enola (accessed Apr. 12, 2017).

109. Martin Harwit, *Cosmic Discovery: The Search, Scope, and Heritage of Astronomy,* 1st ed. (New York: Basic Books, 1981), 13–17, 20; Michael J. Sheehan, *The International*

Politics of Space — Space Power and Politics series (London/New York: Routledge, 2007), 2.

110. Isaac Newton, *Opticks: or, a Treatise of the Reflections, Refractions, Inflections, and Colours of Light,* 4th ed. (London, 1730), bk. 1, pt. 1, prop. viii, prob. 2; see p. 110 of the Project Gutenberg ebook at sirisaacnewton.info/writings/opticks-by-sir -isaac-newton (accessed Jan. 13, 2018).

111. Robert W. Duffner, *The Adaptive Optics Revolution: A History* (Albuquerque: University of New Mexico Press, 2009), ix.

112. For more detail, see, e.g., Neil deGrasse Tyson, "Star Magic," *Natural History* 104:9 (Sept. 1995), 18–20, digitallibrary.amnh .org/handle/2246/6501 (accessed Jan. 14, 2018). Adaptive optics systems are cheaper and simpler for infrared than for visible light, because differences in temperature and density among atmospheric patches are less destructive to infrared wavelengths. As a result, the effective size of an atmospheric patch is correspondingly larger, the mirror need not be as heavily segmented, and finding a nearby guide star is more likely. In addition, the moment-to-moment changes in atmospheric conditions are less severe and occur at a slower rate, so the guide star need not be monitored as rapidly and need not be as bright.

113. See Ann Finkbeiner, *The Jasons: The Secret History of Science's Postwar Elite* (New York: Viking, 2006).

114. Duffner, *Adaptive Optics Revolution,* 14–15; Robert Q. Fugate, quoted in Robert W. Duffner, "Revolutionary Imaging: Air Force Contributions to Laser Guide Star Adaptive Optics," Historical Perspectives — *ITEA Journal* 29:4 (Dec. 2008), 341.

115. For more on the military contributions, see, e.g., Duffner, *Adaptive Optics Revolution,* passim; John W. Hardy, *Adaptive Optics for Astronomical Telescopes* (New York: Oxford University Press, 1998), 16–25, 217–21, 378–79; Robert W. Smith, review of Duffner, *Adaptive Optics Revolution, Isis* 101:3 (2010), 673–74; N. Hubin and L. Noethe, "Active Optics, Adaptive Optics, and Laser Guide Stars," *Science* 262:5138 (Nov. 26, 1993), 1390–94; Ann Finkbeiner, "Astronomy: Laser Focus," *Nature* 517:7535 (Jan. 27, 2015), www.nature.com/news/astronomy-laser-focus-1.16741; GlobalSecurity.org, "Airborne Laser Laboratory," www.globalsecurity.org/space/systems/all.htm# (accessed Jan. 14, 2018). Also see Hardy, *Adaptive Optics,* 11–16, for early efforts at compensating for atmospheric turbulence.

116. Hardy, *Adaptive Optics,* 378–79. For more on Hardy and his work, see Duffner,

Adaptive Optics Revolution, 31ff.

. Johnson quoted in US Air Force, *Space Operations: Air Force Doctrine Document 2-2,* Nov. 27, 2006, 1, fas.org/irp/doddir/usaf/afdd2_2.pdf (accessed Apr. 12, 2017).

. John F. Kennedy, "President Kennedy's Special Message to the Congress on Urgent National Needs, May 25, 1961," John F. Kennedy Presidential Library and Museum, www.jfklibrary.org/Research/Research-Aids/JFK-Speeches/United-States-Congress-Special-Message_19610525.aspx (accessed Apr. 12, 2017). This is the same speech in which Kennedy declared, "I believe that this nation should commit itself to achieving the goal, before this decade is out, of landing a man on the moon and returning him safely to the earth. No single space project in this period will be more impressive to mankind, or more important for the long-range exploration of space; and none will be so difficult or expensive to accomplish."

. The Union of Concerned Scientists maintains a database of all orbiting satellites at www.ucsusa.org/nuclear-weapons/space-weapons/satellite-database#.WPELGqK1tnJ, updated "roughly quarterly." As of Dec. 31, 2016, there were 1,459. As of Aug. 31, 2017, there were 1,738.

. Corona was called Discoverer; Zenit was called Kosmos. For a comprehensive case study of the Defense Support Program satellites that also addresses the overall political dynamics and implications of military programs, see Jeffrey T. Richelson, *America's Space Sentinels: The History of the DSP and SBIRS Satellite Systems,* 2nd ed. (Lawrence: University Press of Kansas, 2012).

121. Joan Johnson-Freese, *Heavenly Ambitions: America's Quest to Dominate Space* (Philadelphia: University of Pennsylvania Press, 2009), 81.

122. T. S. Subramanian, "An ISRO Landmark," *Frontline* 18:23, Nov. 10–23, 2001, www.frontline.in/static/html/fl1823/18230780.htm; Habib Beary, "India's Spy Satellite Boost," BBC News, Nov. 27, 2001, news.bbc.co.uk/2/hi/south_asia/1679321.stm; PTI, "India to Launch Spy Satellite on April 20," *Times of India,* Apr. 8, 2009, timesofindia.indiatimes.com/india/India-to-launch-spy-satellite-on-April-20/articleshow/4374544.cms (accessed Apr. 12, 2017).

123. European Global Navigation Satellite Systems Agency, "Galileo Is the European Global Satellite-based Navigation System," www.gsa.europa.eu/european-gnss/galileo/galileo-european-global-satellite-based-navigation-system; "European Parliament Resolution of 10 July 2008 on Space and Security (2008/2030/INI)," www.europarl

.europa.eu/sides/getDoc.do?pubRef=-//EP//
TEXT+TA+P6-TA-2008-0365+0+DOC
+XML+V0//EN; Galileo GNSS, "European Satellite Systems in Service of European Security," June 14, 2016, galileognss
.eu/european-satellite-systems-in-service-of
-european-security; see also Vincent Reillon
and Patryk Pawlak, "EU Space Policy:
Industry, Security and Defence," Galileo
GNSS, June 13, 2016, galileognss.eu/
eu-space-policy-industry-security-and
-defence. (All accessed Apr. 14, 2017.)

124. Trudy E. Bell and Tony Phillips, "A
Super Solar Flare," NASA Science, May 6,
2008, science.nasa.gov/science-news/
science-at-nasa/2008/06may_carrington
flare (accessed Sept. 9, 2017).

125. John Dos Passos, "The House of Morgan," *Nineteen Nineteen,* book 2 of *U.S.A.*
(Boston: Houghton Mifflin, 1932/1960),
293–94.

126. Dwight D. Eisenhower, "Farewell Address: Transcript," Jan. 17, 1961, sec. IV,
University of Virginia Miller Center,
millercenter.org/the-presidency/presidential
-speeches/january-17-1961-farewell-address
(accessed Apr. 18, 2017).

127. Matthew Weiner, creator, *Mad Men,*
AMC, season 2, episode 10, "The Inheritance."

5. Unseen, Undetected, Unspoken

1. Much more can be said about invisibility. Writers of fairy tales, devisers of magical potions, devotees of various faiths, practitioners of disembodied communication, fearful children, novelists, string theorists, mathematicians, musicians, the elderly, the homeless, the poor, those on the receiving end of prejudice and discrimination — all sorts of people have had other, multiple experiences of invisibility. For a far-ranging approach to the topic, see Philip Ball, *Invisible: The Dangerous Allure of the Unseen* (Chicago: University of Chicago Press, 2015). The opening chapters examine concurrent developments in mysticism, magic, modern technology, and modern science; in a later chapter, "The People Who Can't Be Seen," Ball describes the invisibility of people who are deemed marginal as "a condition imposed by the selective vision of others, a manufactured blind spot painted over by the mind's eye" (191). See also the discussion of multifarious invisibilities in the review of Ball's book by Kathryn Schulz, "Sight Unseen," *New Yorker,* Apr. 13, 2015, 75–79. Schulz makes the point that "[a]lmost everything around us is imperceptible, almost all the rest is maddeningly difficult to perceive, and what

remains scarcely amounts to anything. . . . As for the part that exists in or near our own planet, the stuff that is visible to us in any literal sense: that is a decimal attenuating out almost to nothing, a speck of dust in the cosmic hinterlands" (78).

2. Antony van Leeuwenhoeck, "Observations communicated to the Publisher by Mr. Antony van Leeuwenhoeck, in a Dutch Letter of the 9th of Octob. 1676. here English'd: Concerning little Animals by him observed in Rain-Well-Sea- and Snow water as also in water wherein Pepper had lain infused," *Philosophical Transac. Royal Society* 1677:12 (Mar. 25, 1677), 828–29, digitized pages at rstl.royalsocietypublishing .org/content/12/133/ 821.full.pdf+html (accessed Jan. 17, 2018).

3. The number 7 had carried rich associations since well before Newton's day, including the seven notes in the heptatonic scale, the seven "classical planets," and the seven days of the week. See, e.g., Robert Finlay, "Weaving the Rainbow: Visions of Color in World History," *J. World History* 18:4 (2007), 387; June W. Allison, "Cosmos and Number in Aeschylus' Septem," *Hermes* 137:2 (2009), 130.

4. "Refrangibility" simply means refraction. A light ray that is refracted is being deflected, or bent, from the straight path it

would otherwise travel on. What causes the bending is some kind of surface that the light ray encounters, or some change in the medium through which it has been passing. Here is Newton on the necessity of experiments: "You know, the proper Method for inquiring after the properties of things is, to deduce them from Experiments. And I told you, that the Theory, which I propounded, was evinced to me, not by inferring 'tis thus because not otherwise, that is, not by deducing it only from a confutation of contrary suppositions, but by deriving it from Experiments concluding positively and directly. The way therefore to examin it is, by considering, whether the Experiments which I propound do prove those parts of the Theory, to which they are applyed; or by prosecuting other Experiments which the Theory may suggest for its examination." Isaac Newton, "A Serie's of Quere's propounded by Mr. Isaac Newton, to be determin'd by Experiments, positively and directly concluding his new Theory of Light and Colours; and here recommended to the Industry of the Lovers of Experimental Philosophy, as they were generously imparted to the Publisher in a Letter of the said Mr. Newtons of July 8. 1672," *Philosophical Transac. Royal Society* 85 (July 15, 1672), 5004.

5. ". . . for about a 1/4 or 1/3 of an Inch at

either end of the Spectrum the Light of the Clouds seemed to be a little tinged with red and violet, but so very faintly, that I suspected that Tincture might either wholly, or in great Measure arise from some Rays of the Spectrum scattered irregularly by some Inequalities in the Substance and Polish of the Glass . . ." "Exper. 3" in Sir Isaac Newton, Knt., *Opticks: or, A Treatise of the Reflexions, Refractions, Inflexions and Colours of Light,* 4th ed. corr. (London: William Innys, 1730), Project Gutenberg Ebook 33504 (2010), 30, www.gutenberg.org/files/33504/33504-h/33504-h.htm (accessed Apr. 19, 2017).

6. Newton, *Opticks,* Qu. 25, first sentence.

7. William Herschel, "Investigation of the Powers of the prismatic Colours, to heat and illuminate Objects; with Remarks, that prove the different Refrangibility of radiant Heat . . . ," *Philosophical Transac. Royal Society* 90 (1800), 272. Writing several decades later, another Englishman described Herschel's finding in Victorian language: "The experiment proved that, besides its luminous rays, the sun emitted others of low refrangibility, which possessed great calorific power, but were incompetent to excite vision." J. Tyndall, "On Calorescence," *Philosophical Transac. Royal Society* 156 (1866), 1. There is also evidence that

seventeenth-century French and Italian researchers had begun, in a much less organized way, to investigate the invisible rays that produced heat; see James Lequeux, "Early Infrared Astronomy," *J. Astronomical History and Heritage* 12:2 (2009), 125–26. The term "infrared" did not come into use until about 1880; see S. D. Price, *History of Space-based Infrared Astronomy and the Air Force Infrared Celestial Backgrounds Program,* AFRL-RV-HA-TR-2008-1039 (Hanscom AFB, MA: Air Force Research Laboratory, 2008), 36.

8. In early 2016 a detector called Advanced LIGO first measured a kindred phenomenon: gravitational waves, made up not of photons but of gravitons, with wavelengths about the size of the system that generated them — up to a thousand kilometers. LIGO was built to detect the workings of gravity, not light, on a cosmic scale, and it marks a totally new era in astrophysical detection.

9. As the first invisible form of electromagnetic radiation to be discovered, radio waves — enabling "communications [to be] borne on the pervasive wireless ether" — became, like the newly discovered invisible magic of electricity, proof and beneficiary of centuries of spiritualist leanings and occult assumptions. Ball, *Invisible,* 101 and chap. 4, "Rays That Bridge Worlds," 90–134.

10. C. J. Seymour Baker, "Correspondence: Camouflage," *J. Royal Society of Arts* (Mar. 19, 1920), 298; Michael Taussig, "Zoology, Magic, and Surrealism in the War on Terror," *Critical Inquiry* 34:S2 (Winter 2008), S98–S116.

11. Sun Tzu, *The Art of War,* trans. Lionel Giles, chap. 1, "Laying Plans," sec. 18–19, in *The Strategy Collection: The Art of War, On War, The Prince* (Waxkeep Publishing, 2013), loc. 11794.

12. Paul Daniel Emanuele, "Vegetius and the Roman Navy: Translation and Commentary, Book Four," 31–46, "Part II: Translation, XXXVII," 28 (MA thesis, Department of Classics, University of British Columbia, 1974).

13. Ball, *Invisible,* 241.

14. Claudia T. Covert, "Art at War: Dazzle Camouflage," *Art Documentation: J. Art Libraries Society of North America* 26:2 (Fall 2007), 50–51. The French created their camouflage service in 1915, the British in 1916, and the Americans in 1917. Charlie Chaplin adopted the tree disguise for the 1918 movie *Shoulder Arms,* in which "he ran around in a tree costume knocking down German soldiers at the front" (Ball, *Invisible,* 242).

15. Ball, *Invisible,* 244–50.

16. The first three examples are cited by Ball

in *Invisible:* A British stage magician who worked with the British Army during World War I, in an attempt to hide aircraft from searchlights, painted the planes with varnish that he then covered over with black felt powder before it had dried (249); a Japanese engineer, Susumu Tachi, has created a material called "retro-reflectum," made up of small light-reflecting beads, which transmits to the front of an object the exact view from its back (229–30); a skyscraper being designed for South Korea is to be surrounded by outward-facing cameras and coated with LEDs that project improved versions of what the cameras record (231–32). The multiple-lens approach to disappearing was developed at the University of Rochester; it relies on four standard lenses of different focal lengths arranged in a line and separated by carefully calculated distances, www.rochester.edu/newscenter/ watch-rochester-cloak-uses-ordinary-lenses -to-hide-objects-across-continuous-range-of -angles-70592/ (accessed July 19, 2015).

17. Chen-Pang Yeang, "The Study of Long-Distance Radio-Wave Propagation, 1900–1919," *Historical Studies in the Physical and Biological Sciences* 33:2 (2003), 369–403.

18. The initial cost of a call to London was $75 for the first three minutes; after seven more years of intensive R & D, the initial cost of a call to Tokyo was $39 for three

minutes. See AT&T, "The History of AT&T," www.corp.att.com/history (accessed Apr. 19, 2017).

19. Karl G. Jansky, "Directional Studies of Atmospherics at High Frequencies," *Proc. Institute of Radio Engineers* 20 (1932), 1920; Karl G. Jansky, "Electrical Disturbances Apparently of Extraterrestrial Origin," *Proc. IRE* 21:10 (Oct. 1933), 1387–98.

20. Addressing the 94th meeting of the American Astronomical Society on March 23, 1956, Cyril M. Jansky Jr., also a radio engineer, called his brother Karl's work on noise "in effect a wedding ceremony" between pure science and applied science, pointing out that he himself (and by implication, most people working in S & T) "used to define a pure scientist as one who if he saw a practical application of what he was doing somehow felt contaminated by commercialism and an applied scientist as one who if he could not see a practical application of his work would lose interest." C. M. Jansky Jr., "My Brother Karl Jansky and His Discovery of Radio Waves from Beyond the Earth," *Cosmic Search* 1:4, www.bigear.org/vol1no4/jansky.htm (accessed Nov. 3, 2015).

21. Grote Reber, "A Play Entitled the Beginning of Radio Astronomy," *J. Royal Astro-*

nomical Society of Canada 82:3 (June 1988), 94, adsabs.harvard.edu/full/1988JRASC..82 . . .93R (accessed Apr. 19, 2017).

22. For Jansky's own detailed description of this apparatus, see Jansky, "Directional Studies," 4–7.

23. Lisa Grossman, "New Questions about Arecibo's Future Swirl in the Wake of Hurricane Maria," *ScienceNews,* Sept. 29, 2017, www.sciencenews.org/blog/science -public/new-questions-about-arecibos -future-swirl-wake-hurricane-maria (accessed Oct. 28, 2017).

24. Cheng Yingqi and Yang Jun, "Massive Telescope's 30-ton 'Retina' Undergoes Final Test," *China Daily,* Nov. 23, 2015, www.chinadaily.com.cn/china/2015-11/23/ content_22509826.htm (accessed Apr. 19, 2017).

25. Initially the British called their version RDF, or radio direction finding. See the brief, eloquent explanation of radar by the UK's WWII radar maven Robert Watson-Watt, in J. T. Randall, "Radar and the Magnetron," *J. Royal Society of Arts* 94:4715 (Apr. 12, 1946), 304.

26. "[B]y the use of [a] generator of stationary waves and receiving apparatus properly placed . . . , it is practicable to transmit intelligible signals or to control or actuate at will any one or all of such apparatus for many other important and valuable pur-

poses, as for . . . ascertaining the relative position of a body or distance of the same with reference to a given point or for determining the course of a moving object, such as a vessel at sea, the distance traversed by the same or its speed, or for producing many other useful effects at a distance dependent on the intensity, wave length, direction or velocity of movement." Nikola Tesla, "Art of Transmitting Electrical Energy Through the Natural Medium," US Patent 787,412; application filed May 16, 1900; renewed June 17, 1902; specification dated Apr. 18, 1905, www.teslauniverse .com/nikola-tesla/patents/us-patent-787412 -art-transmitting-electrical-energy-through -natural-mediums (accessed Apr. 19, 2017). In 1917 Tesla proposed that a submarine could be detected by the same wireless invention of his that had already been used to detect underground ore — a statement similar to Marconi's in 1922. "Nikola Tesla Tells of Country's War Problems," *New York Herald,* Apr. 15, 1917, www.teslauniverse .com/nikola-tesla/articles/nikola-tesla-tells -countrys-war-problems (accessed Apr. 19, 2017). Re early efforts in multiple countries, see Louis Brown, *A Radar History of World War II: Technical and Military Imperatives* (Bristol, UK, and Philadelphia, PA:

Institute of Physics Publishing, 1999), 40–49.

27. Quoted in Andrew J. Butrica, *To See the Unseen: A History of Planetary Radar Astronomy,* NASA History Series: NASA SP-4218 (Washington, DC: NASA, 1996), 1, ntrs.nasa.gov/archive/nasa/casi.ntrs.nasa.gov/19960045321.pdf (accessed Apr. 19, 2017).

28. Butrica, *See the Unseen,* 1–2.

29. Brown, *Radar History of WWII,* ix.

30. Brown, *Radar History of WWII,* xi.

31. For a detailed look at the many roadblocks in the Soviet Union as well as the competing technologies, see John Erickson, "Radio-location and the Air Defence Problem: The Design and Development of Soviet Radar 1934–40," *Science Studies* 2:3 (July 1972), 241–63. Re sound detection in early-warning systems as early as 1917, including "acoustical mirrors" post–World War I, see David Zimmerman, *Britain's Shield: Radar and the Defeat of the Luftwaffe* (Stroud, UK: Amberly, 2013), 23–50; the book also extensively addresses the political and scientific background to radar development in Britain. Roadblocks to the development of shortwave radar in Britain as well as Germany are discussed in Bernard Lovell, "The Cavity Magnetron in World War II: Was the Secrecy Justified?" *Notes*

and Records of the Royal Society of London 58:3 (Sept. 2004), 286–91. Also see Brown, *Radar History of World War II,* 40–91, including this grim mid-1930s example from Germany: The Kriegsmarine ordered engineers working on an early radar system to abandon the cathode-ray tube because the tube was too delicate for shipboard use. Not long afterward, a ship carrying a prototype radar set with such a tube went down; the entire crew died, but the cathode-ray tube continued to function (75). Re infrared, Brown points out that although infrared detection became widespread after the war, its wartime applications were severely limited by the fact that good semiconductors did not yet exist and the photoelectric effect had not yet been fully mastered (41).

32. Brown, *Radar History of World War II,* 33–49. See also Zimmerman, *Britain's Shield,* 53–55.

33. Zimmerman, *Britain's Shield,* 65–70, presents detailed information re defensive measures.

34. A German historian of technology writes that "there was no intensive interaction between the scientists and the services, the level of integration of components into a system was low, and operational efficiency was weak"; Walter Kaiser, "A Case Study in the Relationship of History of Technology and of General History: British Radar

Technology and Neville Chamberlain's Appeasement Policy," *Icon* 2 (1996), 38. Re the extreme secrecy, Brown notes, in his caption to a 1938 photograph originally included in the 1939 edition of a German compilation of the world's navy ships, that the ship conspicuously displayed a Seetakt antenna (a large, pale, rather flat box) in front of the foremast, but that "the photograph was passed for publication by naval authorities, all kept in the dark about the new technique and, of course, unable to recognize the apparent mattress as the mark of a secret weapon" (*Radar History of World War II,* 32).

35. Brown, *Radar History of World War II,* 40–96, 280–81.

36. Zimmerman, *Britain's Shield,* 184, 186–88; Kaiser, "Case Study: British Radar," 34–35, 37; Brown, *Radar History of World War II,* 64, 82–83. Kaiser writes, "The reasons for the extraordinary achievements of British radar technology lie preeminently in an alert military policy and a far-sighted strategy." Britain's approach was to form "organized cadres of scientists to assist the services." In 1937–38 the government compiled a list of skilled workers suitable for employment in war production; in addition, with input from the Royal Society, universities, and technical institutions, it

developed a registry of highly qualified volunteers for war service. "The creation and successful use of institutional structures to direct the difficult process of transforming science into technology was an essential," Kaiser argues.

37. Robert Watson-Watt was the Radio Research Board's most visible actor in the effort to develop Britain's radar effort. Much of the work of his Radio Research Station at Slough had to do with the ionosphere. On Feb. 12, 1935, just two weeks after being contacted by the Air Ministry's director of scientific research, he sent a secret memorandum to the ministry titled "Detection of Aircraft by Radio Methods" and noted in his cover letter, "It turns out so favourably that I am still nervous as to whether we have not got a power of ten wrong, but even that would not be fatal." The final draft was titled "Detection and Location of Aircraft by Radio Methods." One of Watson-Watt's biographers called the memorandum "the political birth of radar"; Watson-Watt himself went so far as to claim it as "marking the birth of radar." Butrica, *See the Unseen,* 3 n.9. For the cover letter, see "Radar Personalities: Sir Robert Watson-Watt," www.radarpages.co .uk/people/images/wwfig3.jpg (accessed Apr. 19, 2017). After the war, Watson-Watt himself recast the memo into plain English

at the beginning of a very readable article, "Radar Defense Today — and Tomorrow," *Foreign Affairs* 32:2 (Jan. 1954), 230–43, esp. 231–34. For a later technical, but also readable, analysis of the memorandum, see B. A. Austin, "Precursors to Radar: The Watson-Watt Memorandum and the Daventry Experiment," *Int. J. Electrical Engineering Education* 36 (1999), 364–72.

38. Zimmerman, *Britain's Shield,* 208–35, 263–79.

39. Brown, *Radar History of World War II,* 49, 56, 287.

40. See, e.g., Lovell, "Cavity Magnetron in World War II," 283–94; J. T. Randall, "Radar and the Magnetron," *J. Royal Society of Arts* 94:4715 (Apr. 12, 1946), 313; Butrica, *See the Unseen,* 3–6. Raytheon made 80 percent of them, according to "Raytheon Company History," www.raytheon.com/ourcompany/history/ (accessed Jan. 17, 2016).

41. The failure of intelligence in connection with the attack of Pearl Harbor, as well as questions of the role played by insufficient communication between branches of the armed services, by President Roosevelt, and by "technical surprise," are fraught topics. Specifically regarding radar, however, Butrica states in an extensively footnoted paragraph: "A mobile SCR-270, placed on Oahu as part of the Army's Aircraft Warn-

ing System, spotted incoming Japanese airplanes nearly 50 minutes before they bombed United States installations. . . . The warning was ignored, because an officer mistook the radar echoes for an expected flight of B-17s" (Butrica, *See the Unseen*, 5). Elsewhere, citing other sources, historian Alvin Coox states: "Two Army enlisted men tinkering with a new radar set detected the first sizable air squadrons, but that crucial intelligence was dismissed because unarmed B-17 Flying Fortresses were expected from California that morning" ("The Pearl Harbor Raid Revisited," *J. Amer.–East Asian Relations* 3:3 — *Special Issue: December 7, 1941: The Pearl Harbor Attack* [Fall 1994], 220).

More recently, a staff historian for the US Army's Communications–Electronics Command states in the *CERDEC Monthly View* of July 2009, published by the Army's Communications–Electronics Research, Development and Engineering Center, in an article honoring the "flawless" performance of the radar itself:

On Dec. 7, 1941, three SCR-270 radar sets in operation on the northern shore of Oahu recorded impulses between 4 a.m. and 7 a.m., indicating the approach of what would turn out to be two Japanese reconnaissance planes. . . .

One of the radar stations reported the

findings to a Navy lieutenant on duty at the Information Center at Fort Shafter, Hawaii. The lieutenant reported it to another Navy lieutenant, who determined that the Navy "had a reconnaissance flight out, and that's what this flight was."

At 7:02 a.m., the radar detected an aircraft approaching Oahu at a distance of about 130 miles. The Signal Corps radar operators telephoned the Information Center at Fort Shafter and reported a "large number of planes coming in from the north, three points east." The operator at Fort Shafter informed his superior that the radar operator said he had never seen anything like it, and it was "an awful big flight." (Floyd Hertweck, " 'It was the largest blip I'd ever seen': Fort Monmouth Radar System Warned of Pearl Harbor Attack," cecom.army.mil/historian/pubAr tifacts/Articles/2010-01-01_0900-FILE -CERDEC%20Monthly%20View%20July %202009%20-%20SCR%20 270.pdf, accessed Dec. 11, 2015; link disabled)

42. Brown, *Radar History of World War II,* x, 5–6.
43. Brown, *Radar History of World War II,* 279–80. Martin Harwit, former director of the National Air and Space Museum, also views technology as strongly determinative: "The most important observational discoveries result from substantial technological

innovation in observational astronomy[, and a] novel instrument soon exhausts its capacity for discovery." Harwit, *Cosmic Discovery,* 18–19.

44. T. R. Kennedy Jr., "Theory of Radar: More Information on Radio Detection Device Is Made Public," *New York Times,* Apr. 29, 1945; William S. White, "Secrets of Radar Given to World: Its Role in War and Uses for Peacetime Revealed in Washington and London," *New York Times,* Aug. 15, 1945.

45. Randall, "Radar and the Magnetron," 314.

46. Quoted in Kaiser, "Case Study: British Radar," 38. Women were crucial to that "operational efficiency," because it was they who took on the task of closely monitoring nuances in the incoming signals — the "small wiggles in oscilloscope traces" — at the Chain Home early-warning stations. Radar, declared Watson-Watt, was "the secret that was kept by a thousand women" (Watson-Watt, "Radar Defense Today," 230). Brown quotes an Australian explanation as to why women proved so valuable: "Women did make the best radar operators, because they watched the screen" (Brown, *Radar History of World War II,* 2, 64). Kaiser, too, acknowledges the role played by women: "Due to a sort of unconscious pattern recognition, operators, par-

ticularly the Women's Auxiliary Air Force (the WAAFs), acquired the skill to detect signals even below noise level" (38).

47. Brown, *Radar History of World War II*, x, 6.

48. In 1946 Watson-Watt painted a sunny portrait of UK wartime cooperation across sectors in the effort to develop and improve radar: "a co-operation which, I believe, was unsurpassed and unequalled in any part of the war effort. It was a co-operation in which the natural philosopher and the engineer in the university worked with the physicists and the mathematicians and the workers of all kinds in the industry, with the men in the Government establishments and with the uniformed forces, from top to bottom rank, making an extraordinarily reassuring and happy story of full interplay between all the contributory factors necessary to the winning of the war." Randall, "Radar and the Magnetron," 314.

49. For details on early research in planetary radar astronomy, see Butrica, *See the Unseen*, 7–27.

50. Lovell belatedly discovered this fact in 1977, while visiting the radio telescope at Effelsberg, near Bonn, and discussing collaboration between Jodrell Bank and the German facility, whose director was Otto Hachenberg. At dinner, Hachenberg raised the topic of doing science during the war

years and said to his counterpart Lovell, "I am well aware of your wartime occupation because as a young man then working in Telefunken I was sent to investigate the equipment in a bomber that crashed near Rotterdam in 1943." Lovell, "Cavity Magnetron in WWII," 288.

51. Butrica, *See the Unseen,* 21–26.

52. William E. Burrows, *This New Ocean: The Story of the First Space Age* (New York: Random House, 1998), 67–68.

53. See, e.g., Burrows, *This New Ocean,* 94–123; David H. DeVorkin, *Science with a Vengeance: How the Military Created the US Space Sciences after World War II* (New York: Springer-Verlag, 1992), 34–57. Hitler, in fact, had ordered the destruction of Germany's research facilities and research records, and Wernher von Braun and his colleagues had been ordered off the main V-2 facility at Peenemünde. As Burrows writes, "Where Peenemünde's rocketeers were concerned, however, getting rid of the only card they had with which to barter their futures would have been unthinkably stupid." Von Braun understood that those records and those rocketeers "were a treasure trove of data on the world's operational ballistic missile technology and the starter set for going to space." So his assistant and a group of unhappy soldiers crated fourteen

805

tons of irreplaceable, indescribably precious documents, carried the crates into a vaulted room inside an abandoned mine shaft, and dynamited the entrance, sealing the room. Meanwhile, the rocketeers maneuvered to turn themselves over to the Americans, in an operation that came to be known as Paperclip (Burrows, *This New Ocean,* 108–16).

54. Remark to a colleague, as quoted in Jonathan Allday, *Apollo in Perspective: Spaceflight Then and Now* (Bristol and Philadelphia: Institute of Physics Publishing, 2000), 85 n.1.

55. The first round of one hundred US-produced V-2s were to be built from 360-plus metric tons of V-2 parts rushed out of Germany's underground Mittelwerk factory in late summer 1945 and shipped to the United States by the US Army's Special Mission V-2 prior to the Soviet army's takeover of the area. However, by January 1946 it had become clear that many components were either damaged or simply missing. It appeared that only twenty-five V-2s could be assembled with the available parts and that they would have to be assembled quickly, because some components were deteriorating as they sat in the desert Southwest. DeVorkin, *Science With a Vengeance,* 48, 61–62.

56. DeVorkin, *Science with a Vengeance,* 154, 67. DeVorkin contends that war rather than science held the reins, writing, "Military goals had indeed become scientific goals in the warhead of a V-2 missile."

57. Watson-Watt, "Radar Defense Today," 240.

58. Letter from Secretary of the Navy James Forrestal to Merwyn Bly, Senior Engineer in the Bureau of Ships, Dec. 4, 1945, on the occasion of Bly's receipt of the Distinguished Civilian Service Award for his role in the development of chaff, wikipedia .org/wiki/Chaff_%28countermeasure% 29#/media/File:Letter_from_Secretary_of _the_Navy,_James_Forrestal,_to_Merwyn _Bly.jpg (accessed Apr. 20, 2017).

59. Brown, *Radar History of World War II,* 295–97. Contemporary chaff is often made of thin, aluminum-coated wire or glass fiber.

60. "Counter Radar Devices," *Science News Letter for December 8, 1945,* 355; Col. Arthur P. Weyermuller, USAF, "Stealth Employment in the Tactical Air Force (TAF) — A Primer on Its Doctrine and Operational Use" (Carlisle, PA: US Army War College, 1992), 2, nsarchive.gwu.edu/ NSAEBB/NSAEBB443/docs/area51_18 .PDF (accessed Apr. 20, 2017).

61. Brown, *Radar History of World War II,* 288–98.

62. USAF, "Air Force Stealth Technology

Review," June 10–14, 1991, "Tab A: Value of Stealth," nsarchive.gwu.edu/NSAEBB/ NSAEBB443/docs/ area51_14.PDF (accessed Apr. 20, 2017).

63. P. Ya. Ufimtsev, *Method of Edge Waves in the Physical Theory of Diffraction* (Izd-Vo Sovetskoye Radio, 1962), trans. Foreign Technology Division, Air Force Systems Command (Dayton, OH: Wright-Patterson Air Force Base, 1971), viii, v. Re Skunk Works, currently about 90 percent of their projects are classified, and most are "so secret that employees can't tell one another what they're working on." W. J. Hennigan, " 'Chief Skunk' at a Hush-Hush Weapons Complex," *Los Angeles Times,* May 13, 2012. But see "Skunk Works Critique of Secrecy and Security Policies," Federation of American Scientists: Project on Government Secrecy, fas.org/sgp/othergov/ skunkworks.html (accessed Apr. 20, 2017).

64. "The Area 51 File: Secret Aircraft and Soviet MiGs — Declassified Documents Describe Stealth Facility in Nevada: National Security Archive Electronic Briefing Book No. 443," ed. Jeffrey T. Richelson, National Security Archive, George Washington University, Oct. 29, 2013, nsarchive .gwu.edu/NSAEBB/NSAEBB443/ (accessed Apr. 20, 2017). Another, recent approach to stealth aircraft revisits the possibilities of cloaking. Engineers at Iowa

State University have developed a flexible radar-trapping "meta-skin": small split rings filled with a liquid metal alloy and embedded in multiple layers of silicon that can be stretched/tuned to capture different wavelengths. When an object — such as, hypothetically, a successor to the B-2 bomber — is wrapped in the meta-skin, it suppresses radar from all directions and angles. Siming Yang, Peng Liu, Mingda Yang, Qiugu Wang, Jiming Song, and Liang Dong, "From Flexible and Stretchable Meta-Atom to Metamaterial: A Wearable Microwave Meta-Skin with Tunable Frequency Selective and Cloaking Effects," *Scientific Reports* 6 (2016), 21921, doi: 10 .1038/srep21921; "Iowa State engineers develop flexible skin that traps radar waves, cloaks objects," Iowa State University, Mar. 4, 2016, news release, www.news.iastate .edu/news/2016/03/04/meta-skin (accessed Apr. 20, 2017). Re the design differences between the F-117A and the B-2, Moore's law over ten years gives 6.67 doubling cycles, which equals a factor of 100 increase in computing power.

65. At a meeting held by SETI (the Search for Extraterrestrial Intelligence) in 1976, the vice president for R & D at Hewlett Packard, Bernard Oliver, quoted from a 1971 report by NASA's Project Cyclops (in which he participated) that introduced

the term "water hole": "Nature has provided us with a rather narrow band in this
best part of the spectrum that seems especially marked for interstellar contact. It lies
between the spectral lines of hydrogen
(1420 MHz) and the hydroxyl radical
(1662 MHz). Standing like the Om and the
Um on either side of a gate, these two emissions of the disassociation products of water
beckon all water-based life to search for its
kind at the age-old meeting place of all species: the water hole." Oliver then remarks,
"It is easy to dismiss this as romantic,
chauvinistic nonsense, but is it? We suggest
that it is chauvinistic and romantic but that
it may not be nonsense." See Bernard M.
Oliver, "Colloquy 4 — The Rationale for a
Preferred Frequency Band: The Water
Hole," SP-419 SETI: The Search for Extra-
terrestrial Intelligence, history.nasa.gov/SP
-419/s2.4.htm (accessed Apr. 20, 2017).

66. For a historical overview of what are
generally termed nonlethal weapons, see
Ando Arike, "The Soft-Kill Solution: New
Frontiers in Pain Compliance," *Harper's*
(Mar. 2010), 38–47. Re the USAF's ADS
system, Arike writes, "Active denial works
like a giant, open-air microwave oven, using
a beam of electromagnetic radiation to heat
the skin of its targets to 130 degrees and
force anyone in its path to flee in pain —
but without injury, officials insist, making it

one of the few weapons in military history to be promoted as harmless to its targets" (38). The extent to which the US military wishes to stress the nonlethality and limited impact is evident in the DoD's Non-Lethal Weapons Program's webpage titled "Active Denial System FAQs," on which the DoD states that fifteen years of research and more than thirteen thousand volunteer exposures have demonstrated that the weapon (ADS) "is safe." In Q9, it addresses the question of whether the system works like a microwave oven (answer: no) and emphasizes the difference in impact of millimeter waves vs. microwaves: "The ADS, a non-lethal directed-energy weapon, projects a very short duration (on the order of a few seconds) focused beam of millimeter waves at a frequency of 95 gigahertz (GHz). A microwave oven operates at 2.45 GHz. At the much higher frequency of 95 GHz, the associated directed energy wavelength is very short and only physically capable of reaching a skin depth of about 1/64 of an inch. A microwave oven operating at 2.45 GHz has a much longer associated wave length, on the order of several inches, which allows for greater penetration of material and efficiency in heating food. The ADS provides a quick and reversible skin surface heating sensation that does not penetrate into the target." jnlwp.defense.gov/About/

FrequentlyAskedQuestions/ActiveDenial SystemFAQs.aspx (accessed Apr. 20, 2016).

67. Like its predecessors, the KH-11 was a secret affair. Space technology sleuth Craig Covault, veteran of nearly four decades of journalism at *Aviation Week,* recently told the exciting tale of how he, his magazine, and the chairman of the Joint Chiefs of Staff arranged to preserve that secrecy — an arrangement that held until the late summer of 1978, when the arrest of a CIA employee who had sold the KH-11 manual to the Soviets for a laughable $3,000 opened the door for Covault to write about something that was already partly public. So as not to blow the cover off the program completely, he says, he agreed to "dribble in the details across many issues of the magazine, not trumpet the whole program at once." Craig Covault, "Anatomy of a Scoop," *Aviation Week & Space Technology,* May 9, 2016, 32–33.

68. One early USAF experiment in spy satellites, SAMOS, which got under way shortly after Sputnik, was not of the standard film-return type. It took pictures on film, developed and scanned the film in orbit, and relayed the data via a radio link. But only a few dozen images a day could be transmitted, owing to the slowness of the system. Because this yield was regarded as insufficient to be of value, SAMOS was canceled

in the early 1960s.

69. "Lockheed Martin Honors Pioneers of Recently Declassified National Reconnaissance Satellites," press release, Jan. 25, 2012, Lockheed Martin, www.lockheed martin.com/us/news/press-releases/2012/ january/0125_ss_satellite.html (accessed Apr. 21, 2017).

70. Figures for resolution and other features vary from source to source. Among the sources consulted were the fact sheets, reports on declassification, and other materials available at Center for the Study of National Reconnaissance, "The Gambit and Hexagon Programs," www.nro.gov/ history/csnr/gambhex/index.html, including "Hexagon: America's Eyes in Space," Sept. 2011, www.nro.gov/history/csnr/gambhex/ Docs/Hex_fact_sheet.pdf. Other sources were T.-W. Lee, *Military Technologies of the World,* vol. 1 (Westport, CT: Greenwood/ Praeger Security International, 2009), 142–49; "U.S. Satellite Imagery 1960–1999: National Security Archive Electronic Briefing Book No. 13," ed. Jeffrey T. Richelson, National Security Archive, George Washington University, Apr. 1999, nsarchive.gwu .edu/NSAEBB/NSAEBB13/#26; Dwayne Day, "Reconnaissance and Signals Intelligence Satellites," US Centennial of Flight Commission, 2003, www.centennialofflight .net/essay/SPACEFLIGHT/recon/SP38

.htm; Craig Covault, "Titan, Adieu," *Aviation Week & Space Technology* 163:16 (Oct. 24, 2005), 28–29; John Pike, "Eyes in the Sky: Satellite Reconnaissance," *Harvard Int. Rev.* 10:6 (Aug./Sept. 1988), 21–23, 26; Jeffrey Richelson, "Monitoring the Soviet Military," *Arms Control Today* 16:7 (Oct. 1986), 14–15; Jeffrey T. Richelson, "The NRO Declassified: National Security Archive Electronic Briefing Book No. 33," National Security Archive, George Washington University, Sept. 2000, nsarchive.gwu.edu/NSAEBB/NSAEBB35/index.html; "Military Surveillance Sat," Encyclopedia Astronautica, www.astronautix.com/fam/milcesat.htm#chrono; National Reconnaissance Office, "Released Records," www.nro.gov/foia/declass/collections.html. (All online sources accessed Mar. 25–26, 2016.)

71. William E. Burrows, *The Survival Imperative: Using Space to Protect Earth* (New York: Forge/Tom Doherty Associates, 2006), 141ff.

72. "Mission to Comet Tempel 1: Deep Impact: About the Mission," Jet Propulsion Laboratory, NASA, www.jpl.nasa.gov/missions/deep-impact/ (accessed Apr. 21, 2017).

73. NASA, "The Deep Impact Spacecraft: Overview" (with links to "Flight System," "Impactor," and "Instruments"), May 11,

2005, www.nasa.gov/mission_pages/
deepimpact/spacecraft/index.html#;
NASA, "Deep Impact Kicks Off Fourth of
July with Deep Space Fireworks," July 4,
2005, www.nasa.gov/mission_pages/deep
impact/media/deepimpact-070405-1.html;
Shyam Bhaskaran, "Autonomous Naviga-
tion for Deep Space Missions," American
Institute of Aeronautics and Astronautics
SpaceOps 2012 Conference, Stockholm,
www.spaceops2012.org/proceedings/
documents/id1267135-Paper-001.pdf (ac-
cessed Apr. 21, 2017); P. Thomas et al.,
"The Nucleus of Comet 9P/Tempel 1:
Shape and Geology from Two Flybys,"
Icarus 222 (2013), 458.

6. Detection Stories

1. Bernard Lovell, *The Story of Jodrell Bank*
   (New York: Harper & Row, 1968), xii, 170,
   29.
2. Lovell, *Jodrell Bank,* 196.
3. Lovell, *Jodrell Bank,* 197–208, 217–29. The
   suddenly formidable military value of the
   Mark I was that it "totally unexpectedly appeared as the only instrument that could be
   used as a long distance radar capable of
   giving warning of the launch of an ICBM
   in the USSR"; it was instantly clear "that
   Britain had built a unique instrument which
   was in great demand for commanding and

receiving the telemetry from US and USSR satellites, particularly those venturing far out into the Solar System." Francis Graham-Smith and Bernard Lovell, "Diversions of a Radio Telescope," *Notes & Records of the Royal Society* 62 (2008), 197; Jodrell Bank Centre for Astrophysics, "The 250ft Mk I Radio Telescope," www.jb.man.ac.uk/history/mk1.html (accessed Apr. 20, 2017). See also Tim O'Brien, "When Was the Lovell Telecope at Jodrell Bank First Switched On?" Jodrell Bank Discovery Centre, University of Manchester, Oct. 29, 2014, www.jodrellbank.net/lovell-telescope-jodrell-bank-first-switched/ (accessed Apr. 20, 2017).

4. Lovell, *Jodrell Bank,* 200.
5. Lovell, *Jodrell Bank,* 230–34, 235 n.1, 239. The author writes that Vice President Richard Nixon said, "None of us know that it is really on the moon," and that former President Harry Truman "said the Russian feat was 'a wonderful thing — if they did it.' " He writes further that a US radio telescope recorded the final minute of Luna 2's signals, but that the "scientists worked in an establishment with military associations and were not allowed to announce their success. I was informed that any such release would have been incompatible with the official American reserve on the success of Lunik II" (235 n.1).

6. Lovell, *Jodrell Bank,* 209–16. For more discussion of Jodrell Bank's key role during the early years of the space race, see Graham-Smith and Lovell, "Diversions of a Radio Telescope," 197–204; Lovell, *Jodrell Bank,* 230–244, 250–52. Also see, by a Swedish Space Corporation rocket engineer, Sven Grahn, "Jodrell Bank's Role in Early Space Tracking Activities," Jodrell Bank Centre for Astrophysics, www.jb.man .ac.uk/history/tracking/ (accessed Apr. 20, 2017).

7. Lovell, *Jodrell Bank,* 240–42.

8. Strangely, not all bursts of gamma rays are equally lethal, nor are they all of cosmic origin. A terrestrial team of investigators found that at least fifty of them pop off daily near the tops of thunderclouds, a split second before ordinary lightning bolts. How did they figure this out? With ground-based detectors tuned to the lowest wavelengths of the radio band. Rebecca E. Kessler, "Flash of Insight," *Natural History* 114:7 (Sept. 2005), 16. See also Neil deGrasse Tyson, "Knock 'Em Dead," in Tyson, *Death by Black Hole and Other Cosmic Quandaries* (New York: W. W. Norton, 2007), 278–81.

9. Two multifaceted discussions of gamma-ray bursts are Jonathan I. Katz, *The Biggest Bangs: The Mystery of Gamma-Ray Bursts, the Most Violent Explosions in the Universe*

(New York: Oxford University Press, 2002), and Govert Schilling, *Flash! The Hunt for the Biggest Explosions in the Universe,* trans. Naomi Greenberg-Slovin (Cambridge: Cambridge University Press, 2002). Atomic bombs of the fission variety, like those dropped on Hiroshima and Nagasaki at the end of World War II, did produce gamma radiation as a by-product of uranium atoms breaking apart, but the much more powerful thermonuclear fusion bomb, developed postwar as a "deterrent" and fortunately not dropped anywhere yet, would produce a much more intense explosion of gamma rays.

10. Quoted in *The New Quotable Einstein,* ed. Alice Calaprice (Princeton: Princeton University Press, 2005), 173.

11. Karen C. Fox, "NASA's Van Allen Probes Spot an Impenetrable Barrier in Space," NASA, Nov. 26, 2014, www.nasa.gov/content/goddard/van-allen-probes-spot-impenetrable-barrier-in-space (accessed Apr. 20, 2017).

12. For an in-agency account of Vela, see Sidney G. Reed, Richard H. Van Atta, and Seymour J. Deitchman, *DARPA Technical Accomplishments: An Historical Review of Selected DARPA Projects,* vol. 1, IDA paper P-2192, Institute for Defense Analyses, Nov. 1990, 11-1–11-10, www.dod.mil/pubs/

foi/Reading_Room/DARPA/301.pdf (accessed Apr. 20, 2017). Besides the Vela Hotel program, which focused on satellite detection of atmospheric and space-based nuclear explosions, there was the Vela Sierra program to develop ground-based methods for detecting those same explosions as well as the Vela Uniform program for detecting underground nuclear explosions (11-1).

13. For the graph, see Schilling, *Flash!*, 12.

14. Ray W. Klebesadel, Ian B. Strong, and Roy A. Olson, "Observations of Gamma-Ray Bursts of Cosmic Origin," *Astrophysical Journal* 182 (June 1, 1973), L86.

15. See Trevor Weekes, "Very High Energy Gamma Ray Astronomy 101," Harvard–Smithsonian Center for Astrophysics, June 2012, fermi.gsfc.nasa.gov/science/mtgs/summerschool/2012/week2/ACT_Weekes.pdf, for a PowerPoint tutorial/history by one of the field's pioneers (accessed Apr. 20, 2017).

16. Keith A. Shrock, "Space-Based Infrared Technology Center of Excellence," fact sheet, AFRL Space Vehicles Directorate, Space Technology Division, Infrared Technologies Center of Excellence Branch, Kirtland AFB and Hanscom AFB, Apr. 3, 2007, www.kirtland.af.mil/About-Us/Fact-Sheets/Display/Article/826053/space-based-infrared-technology-center-of-excellence/ (accessed Apr. 20, 2017).

17. E. A. Davis, ed., *Science in the Making: Scientific Development as Chronicled by Historic Papers in the* Philosophical Magazine — *With Commentaries and Illustrations,* vol. 1: 1798–1850 (London: Taylor & Francis, 1995), 165.

18. "Edison and the Unseen Universe," *Scientific Amer.* 39:8, suppl. 138 (Aug. 24, 1878), 112. For a discussion of the tasimeter, see Thomas A. Edison, "On the Use of the Tasimeter for Measuring the Heat of the Stars and of the Sun's Corona," *Amer. J. Science* 17:97 (Jan. 1879), 52–54. Our thanks to librarian Mai Reitmeyer of the American Museum of Natural History for locating these sources.

19. G. Neugebauer and R. B. Leighton, *Two-Micron Sky Survey: A Preliminary Catalog,* NASA SP-3047 (Washington, DC: NASA, 1969), ntrs.nasa.gov/archive/nasa/casi.ntrs .nasa.gov/19690028611.pdf (accessed Apr. 20, 2017).

20. Russell G. Walker and Stephan D. Price, *AFCRL Infrared Sky Survey, Vol. 1: Catalog of Observations at 4, 11, and 20 Microns,* ADA 016397 (Hanscom AFB, MA: Optical Physics Laboratory, Air Force Cambridge Research Laboratories, July 1975), www .dtic.mil/dtic/tr/fulltext/u2/a016397.pdf (accessed Apr. 20, 2017). A revised version, by Price and Murdock, with the addition of

observations at 27 microns, was published in 1983.

21. 2MASS: 2 Micron All Sky Survey, "Introduction: 1. 2MASS Overview," Dec. 20, 2006, www.ipac.caltech.edu/2mass/releases/allsky/doc/sec1_1.new.html (accessed Apr. 20, 2017).

22. S. D. Price, *History of Space-Based Infrared Astronomy and the Air Force Infrared Celestial Backgrounds Program,* AFRL-RV-HA-TR-1008-1039 (Hanscom AFB, MA: Air Force Research Laboratory — Space Vehicles Directorate, Apr. 2008), xi, 11ff.

23. "Mansfield Amendment: Research Restriction Diluted," *Science News* 98:17 (Oct. 24, 1970), 332; Philip M. Boffey, "Mansfield Amendment Not Yet Dead," *Science* 170:3958 (Nov. 6, 1970), 613; "Mansfield Amendment: Defense Research Curbs Eased," *Science News* 99:3 (Jan. 16, 1971), 50.

24. Martin Harwit, "The Early Days of Infrared Space Astronomy," in *The Century of Space Science,* ed. J. A. Bleeker, J. Geiss, and M. Huber (Dordrecht: Kluwer, 2002), 304. Harwit goes on to say that, based on the pattern of expenditures on infrared work during the later 1980s and early 1990s, "infrared astronomers may expect to inherit even more powerful techniques than those currently available provided they

patiently stand by the closed door that normally separates military from academic infrared space astronomy. Occasionally that door opens a crack, and an arm hands out some highly desirable piece of technology. [But w]here the military has had no apparent interests . . . , astronomers have had to develop instrumentation on their own and progress has been far slower" (327). Antoni Rogalski, a Polish infrared astronomer, concurs: "After World War II, infrared detector technology development was and continues to be primarily driven by military applications." A. Rogalski, "History of Infrared Detectors," *Opto-Electronics Review* 20:3 (2012), 279.

A brief look at infrared detector technology: The 1930s were an active period for militarily useful advances in infrared detection, including portable detectors. In the United States, for instance, RCA came up with an IR tube that, with the advent of war, turned into the RCA 1P25 image converter, which was used for America's Sniperscopes and Snooperscopes beginning in 1942. Meanwhile, in Germany in the early 1930s, Edgar Kutzscher, a professor of physics at the University of Berlin, discovered that lead sulphide had excellent photoconductive properties; based on that discovery, the German army undertook a secret wartime program to manufacture infrared detectors beginning in 1943. In January 1945 that factory was captured by the Soviet Union. After the end of hostilities, Kutzscher was sent to Britain and, like so many other valued German scientists, eventually ended up in the United States — in southern California, working for Lockheed Aircraft Corporation. Rogalski, "History," 283–84. See also D. J. Lovell, "The Thirty-third Anecdote — Wartime Incentive: Robert Joseph Cashman," *Optical Anecdotes* (Bellingham: SPIE — The International Society for Optical Engineering, 1981), 115–18.

25. Ronald E. Doel, *Solar System Astronomy in America: Communities, Patronage, and Interdisciplinary Science, 1920–1960* (New York: Cambridge University Press, 1996), 77.

26. Alexander Szalay, personal communications with Neil deGrasse Tyson, Jan. 31, 2018.

27. National Science and Technology Medals Foundation, "2003 National Medal of Science Laureate Biopic: Riccardo Giacconi," 2003, www.nationalmedals.org/laureates/riccardo-giacconi# (accessed Oct. 30, 2017).

28. Riccardo Giacconi, "The Dawn of X-ray Astronomy," Nobel lecture, Dec. 8, 2002, 112–14, www.nobelprize.org/nobel_prizes/

physics/laureates/2002/giacconi-lecture.pdf (accessed Oct. 31, 2017).

29. Der-Ann Hsu and Richard E. Quandt, "Statistical Analyses of Aircraft Hijackings and Political Assassinations," research memo, Econometric Research Program, Princeton University, Feb. 1976, 1–3, 9, 12, 14–16, www.princeton.edu/~erp/ERParchives/archivepdfs/M194.pdf; Aviation Safety Network, "Airliner Hijackings: 1942– ," aviation-safety.net/statistics/period/stats.php?cat=H2; US Department of Transportation, "Aircraft Hijackings and Other Criminal Acts Against Civil Aviation Statistical and Narrative Reports," May 1983, www.ncjrs.gov/pdffiles1/ Digitization/91941NCJRS.pdf (accessed Oct. 31, 2017).

30. AS&E, "Company: History," www.as-e .com/company/history/# (accessed Oct. 31, 2017).

31. 93rd Congress (1973–1974), S.39 — An Act to amend the Federal Aviation Act of 1958 to implement the Convention for the Suppression of Unlawful Seizure of Aircraft; to provide a more effective program to prevent aircraft piracy; and for other purposes, Title II: Air Transportation Security Act, www.congress.gov/bill/93rd-congress/senate-bill/39 (accessed Oct. 31, 2017).

32. Eric J. Chaisson, *The Hubble Wars: Astrophysics Meets Astropolitics in the Two-Billion-Dollar Struggle over the Hubble Space Tele-*

scope (New York: HarperCollins, 1994), xi.

33. "Classification of TALENT and KEY-HOLE Information," Special Center Notice: Security, No. 6-64, Jan. 16, 1964, nsarchive.gwu.edu/NSAEBB/NSAEBB225/doc23.pdf (accessed Apr. 20, 2017). See also Burrows, *This New Ocean,* 241–42.

34. See, e.g., Dwayne A. Day, "The Flight of the Big Bird (parts 1–4)," *Space Review,* Jan. 17–Mar. 28, 2011, www.thespacereview.com/article/1761/1–www.thespacereview.com/article/1809/2; Roger Guillemette, "Declassified US Spy Satellites Reveal Rare Look at Cold War Space Program," Space.com, Sept. 18, 2011, www.space.com/12996-secret-spy-satellites-declassified-nro.html (accessed Apr. 21, 2017); Chaisson, *Hubble Wars,* 208; Philip Chien, "High Spies," *Popular Mechanics* 173:2 (Feb. 1996), n.p.; John M. Doyle, "Big Bird, Uncaged," *Air & Space Smithsonian,* Dec. 2011/Jan. 2012, 10.

35. Chaisson, *Hubble Wars,* 88–93.

36. Chaisson, *Hubble Wars,* 96.

37. National Research Council — Committee for a Decadal Survey of Astronomy and Astrophysics, *New Worlds, New Horizons in Astronomy and Astrophysics* (Washington, DC: National Academies Press, 2010). For an accessible summary of the committee's issues and priorities, see "2020 Vision: An

Overview of *New Worlds, New Horizons in Astronomy and Astrophysics,*" www.nap.edu/resource/12951/bpa_064932.pdf (accessed Feb. 13, 2018).

38. Dennis Overbye, "Ex-Spy Telescope May Get New Identity as a Space Investigator," *New York Times,* June 4, 2012.

39. Office of Management and Budget, *An American Budget: Major Savings and Reforms — Fiscal Year 2019* (Washington, DC: US Government Publishing Office, 2018), 92, www.whitehouse.gov/wp-content/uploads/2018/02/msar-fy2019.pdf (accessed Feb. 14, 2018). See also, e.g., Amina Khan, "Trump's Budget Would Kill NASA's WFIRST Telescope. Astronomers Say That Would Be a Mistake," *Los Angeles Times,* Feb. 12, 2018; Dennis Overbye, "Astronomers' Dark Energy Hopes Fade to Gray," *New York Times,* Feb. 19, 2018.

7. Making War, Seeking Peace

1. See the UCS Satellite Database, compiled and updated approximately four times per year by the Union of Concerned Scientists, www.ucsusa.org/nuclear-weapons/space-weapons/satellite-database# (accessed Mar. 15, 2018).

2. On May 19, 1998, a satellite named Galaxy IV, operated by PanAmSat, suddenly failed, cutting off tens of millions — the vast

majority — of the pagers in the United States, as well as local affiliates of National Public Radio, certain kinds of credit-card processing, and other forms of communication. A couple of weeks of coronal mass ejections and solar flares preceded the event; some scientists attribute the failure to the effects of the solar storms, others to peculiarities of the tin solder used in key components. "PanAmSat Satellite Outage Interrupts Pager, Television Service in the U.S.," *Wall Street Journal,* May 20, 1998; Lawrence Zuckerman, "Satellite Failure Is Rare, and Therefore Unsettling," *New York Times,* May 21, 1998; "A Week of Solar Blasts: The Space Weather Event of May 1998," pwg.gsfc.nasa.gov/istp/outreach/events/98/ (accessed Apr. 22, 2017). As Zuckerman wrote, "Workers around the country who had come to depend on their beepers for everything from emergency calls to the price of soybeans were suddenly in the dark. As in a major electricity blackout or the disruption of telephone service, users suddenly realized how much they had taken technology for granted."

3. Joint Chiefs of Staff, *Space Operations: Joint Publication* 3–14, May 29, 2013, www.dtic.mil/doctrine/new_pubs/jp3_14.pdf (accessed June 19, 2016). This unclassified document "provides military guidance for

use by the Armed Forces in preparing their appropriate plans" but leaves open the possibility for contrary action "when, in the judgment of the commander, exceptional circumstances dictate otherwise" (i).

4. Air Force Space Command, "Commander's Strategic Intent," May 6, 2016, 5, www.afspc.af.mil/Portals/3/Commander%20Documents/AFSPC%20Commander%E2%80%99s%202016%20Strategic%20Intent.pdf?ver=2016-05-09-094135-810 (accessed Apr. 22, 2017).

5. Office of the Secretary of Defense, *Annual Report to Congress: Military and Security Developments Involving the People's Republic of China 2016,* 36, www.defense.gov/Portals/1/Documents/pubs/2016%20China%20Military%20Power%20Report.pdf (accessed June 26, 2016). Per a 2010 amendment to the National Defense Authorization Act for Fiscal Year 2000, these reports are to be issued annually through 2030.

6. "III. Strategic Guideline of Active Defense" and "Force Development in Critical Security Domains," in "Full Text: China's Military Strategy," *China Daily,* May 26, 2015, www.chinadaily.com.cn/china/ 2015-05/26/content_20820628.htm (accessed June 26, 2016). A Chinese defense "white paper" has been issued every two years since 1998

(Secretary of Defense, *Military and Security Developments,* 3).

7. A 2015 study found that in the context of any confrontation that might arise in or near the South China Sea, US and Chinese counterspace capabilities are now on a par. Eric Heginbotham et al., *The U.S.–China Military Scorecard: Forces, Geography, and the Evolving Balance of Power 1996–2017* (Santa Monica, CA: RAND, 2015), 257–58.

8. The foreword to *Counterspace Operations: Air Force Doctrine Document 2–2.1* (Aug. 2, 2004), the first US Air Force position paper wholly devoted to counterspace, states: "Counterspace operations have defensive and offensive elements, both of which depend on robust space situation awareness. These operations may be utilized through the spectrum of conflict and may achieve a variety of effects from temporary denial to complete destruction of the adversary's space capability" (i). Among the document's statements of foundational doctrine are "defensive counterspace (DCS) operations preserve US/friendly ability to exploit space to its advantage via active and passive actions to protect friendly space-related capabilities from enemy attack or interference" and "offensive counterspace (OCS) preclude an adversary from

exploiting space to their advantage" (vii). "Space superiority" is defined in *AFDD 2–2.1* as "the degree of control necessary to employ, maneuver, and engage space forces while denying the same capability to an adversary" (55). For "agility" and "resilience capacity," see Air Force Space Command, "Commander's Strategic Intent." For current US and Chinese counterspace measures, see Heginbotham et al., *U.S.–China Military Scorecard,* 227–58.

9. David Axe, "The Great Debate: When It Comes to War in Space, U.S. Has the Edge," Reuters, Aug. 10, 2015, blogs.reuters.com/great-debate/2015/08/09/the-u-s-military-is-preparing-for-the-real-star-wars; Lee Billings, "War in Space May Be Closer Than Ever," *Scientific American,* Aug. 10, 2015, www.scientificamerican.com/article/war-in-space-may-be-closer-than-ever (accessed Apr. 22, 2017).

10. See NASA Astromaterials Research and Exploration Science: Orbital Debris Program Office, "Orbital Debris Graphics," www.orbitaldebris.jsc.nasa.gov/photo-gallery.html (accessed Apr. 22, 2017).

11. Sun Tzu, "The Attack by Fire," from *The Art of War,* trans. Lionel Giles, 1910, chap. XII, secs. 1–13, in *The Strategy Collection: The Art of War, On War, The Prince* (Waxkeep Publishing, 2013), loc. 12219–31.

12. Homer, *The Iliad,* trans. Caroline Alexander (New York and London: HarperCollins, 2015), introduction, 4.460–62, 5.66–68, 5.301–308.

13. Leonardo da Vinci, "Letter to Il Moro" (1493), in *A Documentary History of Art,* vol. 1, ed. Elizabeth G. Holt (Garden City, NY: Doubleday Anchor, 1957), 273–75.

14. Carl von Clausewitz, "What Is War?" from *On War,* trans. Col. James J. Graham, 1873, bk. I, chap. 1, sec. 24, in *Strategy Collection,* loc. 2501. The longer formulation here is "a real political instrument, a continuation of political commerce, a carrying out of the same by other means."

15. Despite its similar-sounding name, the taser — developed in the 1970s by a nuclear physicist who spent part of his work life at NASA — is not based on directed energy; it is a gun that fires not bullets but electrodes, which then conduct electric current through the victim's muscles. The word "taser" is an acronym for "Thomas A. Swift's Electric Rifle."

16. H. G. Wells, *The War of the Worlds* (1898; Amazon Digital Services, Public Domain Book, 2012), 20, 25, 52, 73, 104.

17. William J. Fanning Jr., "The Historical Death Ray and Science Fiction in the 1920s and 1930s," *Science Fiction Studies* 37:2 (July 2010), 253–74; David Zimmerman, *Britain's Shield: Radar and the Defeat of the*

Luftwaffe (Stroud, Gloucestershire: Amberly, 2013), 72–75. Winston Churchill, "Shall We All Commit Suicide?" *Nash's Pall Mall Magazine,* Sept. 24, 1924, quoted (though dated as 1921) in Zimmerman, *Britain's Shield,* 61.

18. A. P. (Albert Percival) Rowe, *One Story of Radar* (Cambridge: Cambridge University Press, 1948/2015), 6.
19. Zimmerman, *Britain's Shield,* 76; Rowe, *One Story of Radar,* 6–7; B. A. Austin, "Precursors to Radar: The Watson-Watt Memorandum and the Daventry Experiment," *Int. J. Electrical Engineering Education* 36 (1999), 366–67, www.bawdseyradar .org.uk/wp-content/uploads/2012/12/Wilkins-Calculations.pdf (accessed Apr. 22, 2017); David E. Fisher, *A Summer Bright and Terrible: Winston Churchill, Lord Dowding, Radar, and the Impossible Triumph of the Battle of Britain* (Berkeley, CA: Shoemaker & Hoard, 2005), 66–68.
20. Rowe, *One Story of Radar,* 6.
21. Giovanni de Briganti, "2015 Ushers In the Era of Laser Weapons," Defense-Aerospace.com, Jan. 5, 2014, www.defense -aerospace.com/articles-view/feature/5/1599 75/2015-ushers-in-era-of-laser-weapons .html; Aaron Mehta, "Laser Weapons Ready for Use Today, Lockheed Executives Say," *Defense News,* Mar. 16, 2016, www

.defensenews.com/story/defense/innovation/ 2016/03/15/laser-weapons-directed-energy -lockheed-pewpew/81826876/ (accessed Apr. 22, 2017).

22. Bob Preston, Dana J. Johnson, Sean J. A. Edwards, Michael Miller, and Calvin Shipbaugh, *Space Weapons Earth Wars* (Santa Monica, CA: RAND, 2002), 25, 30.

23. David Wright, Laura Grego, and Lisbeth Gronlund, *The Physics of Space Security: A Reference Manual* (Cambridge, MA: American Academy of Arts and Sciences, 2005), 2, 5.

24. See, e.g., Yasmin Tadjdeh, "Directed Energy Weapons Gaining Acceptance Across U.S. Military," *National Defense,* Aug. 2016, 38–39, digital.nationaldefense magazine.org/i/708228-aug-2016/39; Sydney J. Freedberg Jr., "Lasers Vs. Drones: Directed Energy Summit Emphasizes the Achievable," *Breaking Defense,* June 23, 2016, breakingdefense.com/2016/06/lasers -vs-drones-directed-energy-summit-empha sizes-the-achievable; Sydney J. Freedberg Jr., "The Laser Revolution: This Time It May Be Real," *Breaking Defense,* July 28, 2015, breakingdefense.com/2015/07/the -laser-revolution-this-time-it-may-be-real. Freedberg's 2015 piece quotes Frank Kendall, undersecretary of defense for acquisition, technology, and logistics, who "lived through Reagan's Strategic Defense Initia-

tive and the Airborne Laser — but this time, he thinks, lasers really are becoming reality." In the following paragraph, however, the author says that's "not an easy case to make" and quotes a former head of the Missile Defense Agency, retired Lieutenant General Trey Obering, saying, "DE [directed energy] was right around the corner in 1976. It was right around the corner in 1986. It was right around the corner in 1996." A clearinghouse for responsible information on directed-energy technologies for mostly military but also civilian use is *Wave Front: The Directed Energy Newsletter,* published by DEPS, the Directed Energy Professional Society, www.deps.org/DEPSpages/newsletter.html. (All accessed Apr. 22, 2017.)

25. Preston et al., *Space Weapons Earth Wars,* 128.

26. The futurist Herman Kahn (Dr. Strangelove) used this phrase as a book title in 1962 and again in 1985.

27. In 1946 the Air Force was still the Army Air Force, and LeMay was its R & D director, but the National Security Act of 1947 created the Air Force as an independent branch. In 1948 LeMay became the first commander of Strategic Air Command. Many Americans who lived through the era of the Vietnam War will remember General LeMay for his advice about North Vietnam:

"My solution to the problem would be to tell them frankly that they've got to draw in their horns and stop their aggression, or we're going to bomb them back into the Stone Age." From LeMay, *Mission with LeMay: My Story* (New York: Doubleday, 1965), quoted in, e.g., Alfonso Narvaez, "Gen. Curtis LeMay, an Architect of Strategic Air Power, Dies at 83," *New York Times,* Oct. 2, 1990. Less widely known is his statement about war in general: "But all war is immoral and if you let that bother you, you're not a good soldier." "Race for the Superbomb: People & Events: General Curtis E. LeMay (1906–1990)," *American Experience,* www.pbs.org/wgbh//amex/bomb/peopleevents/pandeAMEX61.html (accessed Apr. 22, 2017).

28. Buchheim articulates the Eisenhower administration's dualistic view of space efforts: "The statesman, endeavoring to promote world peace, can see both a hope and a threat in astronautics. International cooperation in space enterprises could help to promote trust and understanding. Astronautics can provide physical means to aid international inspection, and thereby, can help in the process toward disarmament and the prevention of surprise attack. Astronautics can also lead to military systems which, once developed and de-

ployed, may make hopes of disarmament, arms control, or inspection more difficult to fulfill." Quoted in Sean N. Kalic, *US Presidents and the Militarization of Space 1946–1967* (College Station: Texas A&M University Press, 2012), 44.

30. Ronald Reagan, "Address to the Nation on Defense and National Security," Mar. 23, 1983, transcript, Ronald Reagan Presidential Library and Museum, www.reagan library.archives.gov/archives/speeches/1983/ 32383d.htm (accessed Apr. 22, 2017).

30. Steven R. Weisman, "Reagan Proposes U.S. Seek New Way to Block Missiles," *New York Times,* Mar. 24, 1983.

31. "Boost-phase intercept has the big advantage, especially for small states, that is for North Korea, that you can get close. You can intercept before the missile has stopped burning — maybe 250 seconds, four minutes after it ignites — and you can do that really quite comfortably with the warning that we've had since 1970 from our defense support program satellites. You could launch an interceptor very comfortably 100 seconds after the ICBM starts burning. The collision, [or] the intercept, would be made comfortably before the missile got up to full speed, so it wouldn't fall anywhere near the United States. If you intercepted 10 seconds before the end of powered flight, it would fall 5,000 kilometers short." Sherry

Jones, "Missile Wars — Interview: Richard Garwin," *Frontline,* PBS, 2002, www.pbs.org/wgbh/pages/frontline/shows/missile/interviews/ garwin.html (accessed Apr. 22, 2017).

32. Burton Richter, "It Doesn't Take Rocket Science," *Washington Post,* July 23, 2000, quoted in Mary H. Cooper, "Missile Defense: Should the U.S. Build a Missile Defense System," *CQ Researcher,* Sept. 8, 2000 (accessed Apr. 22, 2017).

33. John M. Broder, " 'Brilliant Pebbles' a Last Hope?: 'Star Wars' Stakes Future on Mini-Missile Concept," *Los Angeles Times,* Apr. 29, 1989. See also William J. Broad, "What's Next for 'Star Wars'? 'Brilliant Pebbles,' " *New York Times,* Apr. 25, 1989.

34. James A. Abrahamson and Henry F. Cooper, "What Did We Get for Our $30-Billion Investment in SDI/BMD?" National Institute for Public Policy, Sept. 1993, 8, 2, 5, www.nipp.org/wp-content/uploads/2014/11/What-for-30B_.pdf (accessed Apr. 22, 2017).

35. "Strategic Defense Initiative (SDI) Budget Slashed, Funds Earmarked," *CQ Almanac 1990,* 46th ed. (Washington, DC: Congressional Quarterly, 1991), 619–93, library.cqpress.com/cqalmanac/cqal90-1111525 (accessed July 22, 2016); Michael R. Gordon, "Pentagon Curbing Public Data on 'Star Wars,' " *New York Times,*

Jan. 26, 1987.

36. Barry Grass, "CISER Survey: Top Scientists Oppose SDI 8–1," *Cornell Chronicle* 18:11 (Nov. 6, 1986); Steven Soter, "SDI Survey," *Science* 235:4791 (Feb. 20, 1987), 831; Philip W. Anderson et al., "Open Letter to Congress," Mar. 12, 2986, in Marshall W. Nirenberg Papers, Profiles in Science, National Library of Medicine, profiles .nlm.nih.gov/ps/access/JJBBSJ.pdf (accessed Apr. 22, 2017); John Kogut, "Say No to a 'Dumb, Dangerous' Program," *The Scientist* 1:7 (Feb. 23, 1987); William Sweet, "Science Wars over Star Wars," *Editorial Research Reports 1986,* vol. 2 (Washington, DC: CQ Press, 1986). For a nontechnical narrative about the people and stories behind Star Wars, see William J. Broad, *Star Warriors: A Penetrating Look into the Lives of the Young Scientists Behind Our Space Age Weaponry* (New York: Simon and Schuster, 1985).

37. "SDI Debate: Is the Strategic Defense Initiative in the National Interest," Nov. 18, 1987, C-SPAN, Program 532-1, www.c -span.org/video/?532-1/sdi-debate (accessed Apr. 22, 2017).

38. Dimitri K. Simes, *After the Collapse: Russia Seeks Its Place as a Great Power* (New York: Simon and Schuster, 1999), chap. 1 at www.nytimes.com/books/first/s/simes

-collapse.html (accessed Oct. 23, 2017). This is Simes's translation/paraphrase of a statement made in private conversation by Marshal Sergei Akhromeyev, a former chief of the General Staff as well as Gorbachev's military advisor.

39. Bradley Graham, "Rumsfeld Pares Oversight of Missile Defense Agency," *Washington Post,* Feb. 16, 2002; Missile Defense Agency, "Airborne Laser Test Bed Successful in Lethal Intercept Experiment," news release, Feb. 11, 2010, www.mda.mil/news/10news0002.html; Jim Wolf and David Alexander, "U.S. Successfully Tests Airborne Laser on Missile," Reuters, Feb. 12, 2010, www.reuters.com/article/usa-arms-laser-idUSN111660620100212?type=marketsNews (accessed Apr. 22, 2017). Also see generally Laura Grego, George N. Lewis, and David Wright, *Shielded from Oversight: The Disastrous US Approach to Strategic Missile Defense,* Union of Concerned Scientists, July 2016.

40. There is much variation in terminology used for orbiting space rocks, based on size, location, and material constituents. For an overall look at asteroids, comets, and impactors, see Neil deGrasse Tyson, "Killer Asteroids," in Tyson, *Space Chronicles: Facing the Ultimate Frontier,* ed. Avis Lang (New York: W. W. Norton, 2012), 45–54. Many

asteroids explode in the upper atmosphere. For instance, the Nuclear Test Ban Treaty Organization, which relies on a network of sensors to pick up the "infrasound signature" of nuclear detonations, detected twenty-six explosions between 2000 and 2013 that were caused by asteroids rather than nuclear detonations. Most exploded high above Earth, but several did hit. There was a six-hundred-kiloton impact in Chelyabinsk, Russia, in 2013, and asteroid impacts greater than twenty kilotons in Indonesia in 2009, in the Southern Ocean in 2004, and in the Mediterranean Sea in 2002. None of these was detected in advance. B612 Foundation, "B612 Foundation Releases Video at Museum of Flight Earth Day Event Showing Evidence of 26 Atomic Bomb Scale Asteroid Impacts Since 2000," news release, Apr. 22, 2014, b612foundation.org/wp-content/uploads/2016/06/B612_PR_042214.pdf (accessed Apr. 10, 2018).

41. Center for Near Earth Object Studies, "CNEOS Is NASA's Center for Computing Asteroid and Comet Orbits and Their Odds of Earth Impact," Jet Propulsion Laboratory, cneos.jpl.nasa.gov; NASA, "Planetary Defense Frequently Asked Questions," www.nasa.gov/planetary defense/faq (accessed Oct. 23, 2017).

42. Carl Sagan and Steven J. Ostro, "Dangers

of Asteroid Deflection," *Nature* 368 (Apr. 7, 1994), 501.

43. Preston et al., *Space Weapons Earth Wars,* 41–42, 173–83.

44. See Mika McKinnon and Mia Risra, "A Scientist Responds . . . to *Deep Impact,"* io9, June 10, 2015, io9.gizmodo.com/ a-scientist-responds-to-deep-impact -1709206458 (accessed Oct. 24, 2017).

45. At the Asteroid Deflection Research Center (ADRC) at Iowa State University, under the aegis of NASA, work has been proceeding on the Hypervelocity Asteroid Intercept Vehicle (HAIV), a two-part craft that would be launched by rocket, like any other missile. Upon nearing the target, it would rely on cameras and sensors for a reliable aim. First, the "leader spacecraft" would hit the surface of the object, creating a shallow crater and destroying itself in the process; then the "follower spacecraft" would make its way into the crater and detonate its nuclear explosive device deep within, where it would have the maximum effect. The yields under consideration range from 300 kilotons to 2 megatons. See Bong Wie, "Optimal Fragmentation and Dispersion of Hazardous Near-Earth Objects: NIAC Phase I Final Report," Sept. 25, 2012, 14, www.nasa.gov/pdf/718394main _Wie_2011_PhI_NEO_Mitigation.pdf (accessed Apr. 22, 2017).

46. Strana.Ru, "Russia K-19 Nuclear Submarine Saved the Globe from Third World War," trans. Leila Wilmers, Pravda.Ru, July 6, 2006, www.pravdareport.com/history/06 -07-2006/83000-submarine-0/ (accessed Apr. 22, 2017).

47. *Command and Control,* dir. Robert Kenner, American Experience Films, PBS, 2016, www.commandandcontrolfilm.com, based on Eric Schlosser, *Command and Control: Nuclear Weapons, the Damascus Accident, and the Illusion of Safety* (New York: Penguin, 2013), see esp. 225–26, 325–34, 425–27. See also, e.g., Neil Denny, "Interview: Eric Schlosser's Command and Control," *Little Atoms* 1, Jan. 17, 2016, littleatoms.com/ interview-eric-schlossers -command-and-control; Scott D. Sagan, "On the Brink? How Safe Are Our Nukes?" *American Scholar* (Autumn 2013), theamericanscholar.org/on-the-brink/#.V -gW5CRoBdk; Eric Schlosser interviewed by Amy Goodman and Nermeen Shaikh, "How the U.S. Narrowly Avoided a Nuclear Holocaust 33 Years Ago, and Still Risks Catastrophe Today," *Democracy Now!,* Sept. 18, 2013, transcript at www.democracynow .org/2013/9/18/how_the_us_narrowly _avoided_a (all accessed Apr. 22, 2017). Between 1950 and March 1968 alone, there were at least twelve hundred "significant" US nuclear incidents, according to a study

by Sandia Labs (Schlosser, *Command and Control,* 327).

48. NASA, "New Desktop Application Has Potential to Increase Asteroid Detection, Now Available to Public," Asteroid Redirect Mission, release 15-041, Mar. 15, 2015, www.nasa.gov/press/2015/march/new -desktop-application-has-potential-to -increase-asteroid-detection-now-available; B612, "Our Mission: Dedicated to the Discovery and Deflection of Asteroids," b612foundation.org/our-mission/#sentinel -mission (accessed Apr. 22, 2017); Edward T. Lu and Stanley G. Love, "Gravitational Tractor for Towing Asteroids," *Nature* 438 (Nov. 10, 2005), 177–78.

49. Michael Krepon with Michael Katz-Hyman, "Space Weapons and Proliferation," *Nonproliferation Review* 12:2 (July 2005), 325. Krepon stresses the need to distinguish between "residual" or "latent" capabilities and developed, "dedicated" space warfare capabilities, hence his reference to "specifically designed and flight-tested."

50. Alexei Arbatov, "Preventing an Arms Race in Space," in *Outer Space: Weapons, Diplomacy, and Security,* ed. Alexei Arbatov and Vladimir Dvorkin (Washington, DC: Carnegie Endowment for International Peace, 2010), 87.

51. Andrew Cockburn, "The New Red Scare: Reviving the Art of Threat Inflation," *Harper's* 333:1999 (Dec. 2016), 25. Cockburn cites a quote from Ivan Selin, director of the Strategic Forces Division in the Pentagon's Office of Systems Analysis, who was fond of saying, "Welcome to the world of strategic analysis, where we program weapons that don't work to meet threats that don't exist."

52. Vladimir Dvorkin, "Space Weapons Programs," in *Outer Space,* ed. Arbatov and Dvorkin, 31–45; Viktor Mizin, "Non-Weaponization of Outer Space: Lessons from Negotiations," in *Outer Space,* ed. Arbatov and Dvorkin, 52–53; Matthew Evangelista, "The Paradox of State Strength: Transnational Relations, Domestic Structures, and Security Policy in Russia and the Soviet Union," *Int. Organization* 49:1 (Winter 1995), 14–17; Joan Johnson-Freese, *Heavenly Ambitions: America's Quest to Dominate Space* (Philadelphia: University of Pennsylvania Press, 2009), 39; Matthew Mowthorpe, *The Militarization and Weaponization of Space* (Lanham, MD: Lexington Books/Rowman & Littlefield, 2004), 70. Of interest in the present context is Evangelista's discussion of physicist Richard Garwin's role in the drafting of treaty language re the limitation of ASAT weap-

ons, following Garwin's having met Soviet physicist Evgenii Velikhov at an early 1983 meeting between a Soviet delegation and the US National Academy of Sciences' Committee on International Security and Arms Control (14–16).

53. Dvorkin, "Space Weapons Programs," in *Outer Space,* ed. Arbatov and Dvorkin, 35.
54. For more on dazzling, blinding, jamming, and spoofing, see Wright et al., *Physics of Space Security,* 117–30.
55. Samuel R. Delaney, *Babel-17* (1966; repr. Open Road Media, 2014), loc. 1256–57.
56. The complete text of all UN treaties, principles, and resolutions regarding space law is available through the UN Office for Outer Space Affairs, www.unoosa.org/oosa/en/ourwork/spacelaw/index.html. The foundational Outer Space Treaty came into force in 1967; see www.unoosa.org/oosa/en/ourwork/spacelaw/treaties/introouterspacetreaty.html. Efforts against weaponization came later; see, e.g., UN General Assembly, "Resolution 62/20: Prevention of an Arms Race in Outer Space," Dec. 5, 2007, www.oosa.unvienna.org/pdf/gares/ARES_62_020E.pdf. (All accessed Apr. 23, 2017.)
57. The UN Office for Outer Space Affairs began to address concerns about space debris officially in 1994 and eventually produced a set of recommendations titled "Space Debris Mitigation Guidelines,"

which were endorsed by the General Assembly in 2007 but remain voluntary. For UNOOSA and the UN's approach to space debris, see www.unoosa.org/oosa/en/ourwork/topics/space-debris.html; www.un.org/en/events/tenstories/08/spacedebris.shtml. For "Space Debris Mitigation Guidelines," see www.unoosa.org/pdf/publications/st_space_49E.pdf. See also NASA's publication *Orbital Debris Quarterly News,* orbitaldebris.jsc.nasa.gov/quarterly-news/newsletter.html/. See UN Office at Geneva, Conference on Disarmament: Introduction to the Conference, www.unog.ch/80256EE600585943/(httpPages)/BF18ABFEFE5D344DC1256F3100311CE9?OpenDocument; CD Documents Related to Prevention of an Arms Race in Outer Space, www.unog.ch/80256EE600585943/(httpPages)/D4C4FE00A7302FB2C12575E4002DED85?OpenDocument. (All accessed Apr. 23, 2017.)

58. James Clay Moltz, *Crowded Orbits: Conflict and Cooperation in Space* (New York: Columbia University Press, 2014), 148, 151.

59. David S. F. Portree, "NASA's Origins and the Dawn of the Space Age," Monograph 10, NASA History Division, 2005, www.hq.nasa.gov/office/pao/History/40thann/nasaorigins.htm. For the Soviet approach to IGY and the Rome meeting,

see Asif A. Siddiqi, "Korolev, Sputnik, and the International Geophysical Year," in *Reconsidering Sputnik: Forty Years Since the Soviet Satellite,* ed. Roger D. Launius, John M. Logsdon, and Robert W. Smith (Australia: Harwood Academic Publishers, 2000), 47, history.nasa.gov/sputnik/siddiqi .html (accessed Apr. 23, 2017). See generally Walter A. McDougall, . . . *the Heavens and the Earth: A Political History of the Space Age* (Baltimore: Johns Hopkins University Press, 1997), 118–24, 134.

60. In the 1920s and 1930s, Russian rocket pioneer Konstantin Tsiolkovsky had considered two versions of the multistage rocket. One was sequential — a "rocket train," with the successive stages one behind the other. The other was an array — a "rocket squadron." Soviet space exploration advocate Mikhail Tikhonravov, who edited Tsiolkovsky's writings in the late 1920s, believed that the multistage rocket had numerous advantages over a single giant rocket and strongly supported the "rocket package" approach: a cluster of identical, not-giant rockets that could easily be mass-produced and, depending on how many were in the cluster, could provide various levels of thrust. See Asif A. Siddiqi, *The Red Rockets' Glare: Spaceflight and the Soviet Imagination, 1857–1957* (New York: Cambridge

University Press, 2010), 252–54.

61. In 2003, NASA's Space Infrared Space Telescope Facility, tuned specifically to the infrared part of the spectrum, would be renamed the Spitzer Space Telescope in his honor.

62. As the very first Director of Defense Research and Engineering, Herbert York, wrote in *Making Weapons, Talking Peace:* "Unnecessary duplication was rife, and vicious interservice struggles over rules and missions were creating confusion." Quoted in James Clay Moltz, *The Politics of Space Security: Strategic Restraint and the Pursuit of National Interests* (Stanford, CA: Stanford University Press, 2008), 95. The Air Force also began to attract lavish funding: in 1948, shortly after the fall of Czechoslovakia, "a supposedly penny-proud Congress" voted for 25 percent more funding for the Air Force than had been requested by the secretary of defense. This vote took place during the "war scare" of 1948 and the resultant call for a military buildup — a buildup that was partly a response to active lobbying by US aircraft manufacturers. Walter LaFeber, *America, Russia, and the Cold War, 1945–2006,* 10th ed. (New York: McGraw Hill, 2006), 79–80.

63. Kalic, *Presidents and Militarization of Space,* 19–25.

64. Siddiqi, *Red Rockets' Glare,* 244–46; Stalin quote originally from David Holloway, *Stalin and the Bomb: The Soviet Union and Atomic Energy, 1939–1956.*

65. Siddiqi, *Red Rockets' Glare,* 201–206, 241–89. For Korolev's modus operandi within the constraints of Soviet bureaucracy under Stalin and Khrushchev, see Slava Gerovitch, "Stalin's Rocket Designers' Leap into Space: The Technical Intelligentsia Faces the Thaw," *Osiris* 23:1 (2008), 189–209.

66. Siddiqi, *Red Rockets' Glare,* 290–331.

67. James Clay Moltz, *Asia's Space Race: National Motivations, Regional Rivalries, and International Risks* (New York: Columbia University Press, 2012), 46–48, 73–75; Evan Osnos, "The Two Lives of Qian Xuesen," *New Yorker,* Nov. 3, 2009, www.newyorker.com/news/evan-osnos/ the-two-lives-of-qian-xuesen; Michael Wines, "Qian Xuesen, Father of China's Space Program, Dies at 98," *New York Times,* Nov. 3, 2009 (accessed July 5, 2016). See also Iris Chang's biography, *Thread of the Silkworm* (New York: BasicBooks, 1995), in which the scientist's name is spelled Tsien Hsueshen. During Qian's Immigration and Naturalization Service hearings in 1950–51, when interrogated about his loyalty in the event of a potential US–China conflict,

Qian said, "My essential allegiance is to the people of China. If a war were to start between the United States and China, and if the United States war aim was for the good of the Chinese people, and I think it will be, then, of course, I will fight on the side of the United States" (Chang, 170). Also, as Chang writes, "INS officers failed to see the irony of deporting a scientist accused of Communist leanings to a Communist country — especially when this scientist was a world-renowned expert in ballistic missile design" (193).

68. LaFeber, *America, Russia,* 62, 73ff. Quote re American consensus is from hearings of the Committee on Foreign Relations (92).

69. LaFeber, *America, Russia,* 45, 62; Vojtech Mastny, *The Cold War and Soviet Insecurity: The Stalin Years* (New York: Oxford University Press, 1996), 27–28, 41, 110–33.

70. National Security Council — Executive Secretary, "Report to the National Security Council on United States Objectives and Programs for National Security," NSC 68, Apr. 14, 1950, Harry S. Truman Library and Museum, www.trumanlibrary.org/whistlestop/study_collections/coldwar/documents/pdf/10-1.pdf (accessed Apr. 23, 2017).

71. NSC, "Report on US Objectives," NSC 68, 5–6, 35, 11, 39.

72. NSC, "Report on US Objectives," NSC

68, 54ff. For an analysis of NSC 68 in historical context, see LaFeber, *America, Russia,* 103–105 and generally chap. 4, "The 'Different World' of NSC-68 (1948–1950)," 83–105.

73. Office of the Historian, "Milestones: 1945–1952 — NSC 68, 1950," US Department of State, history.state.gov/milestones/1945-1952/NSC68 (accessed Apr. 23, 2017); LaFeber, *America, Russia,* 147.

74. National Science Foundation and National Academy of Sciences, "Plans for Construction of Earth Satellite Vehicle Announced," press release, July 29, 1955, www.eisenhower.archives.gov/research/online_documents/ igy/1955_7_29_NSF_Release.pdf; "The White House, statement by James C. Hagerty," press release, July 29, 1955, www.eisenhower.archives.gov/research/online_documents/igy/1955_7_29_Press_Release.pdf (accessed Apr. 23, 2017). However, as Kalic points out in *Presidents and the Militarization of Space,* while Eisenhower touted the scientific and civilian character of the Vanguard satellite program in public, the Department of Defense carried out most of the work (33).

75. Kalic, *Presidents and Militarization of Space,* 31–34. Wernher von Braun was at the Army's Redstone Arsenal in Alabama and continued to push for support. In 1954

he requested $100,000 to launch a satellite using existing technology — "a tiny price to pay given that 'a man-made satellite, no matter how humble (five pounds), would be a scientific achievement of tremendous impact'" (McDougall, *Heavens and Earth*, 118–19). Everett Dolman, in *Astropolitik: Classical Geopolitics in the Space Age* (London and Portland, OR: Frank Cass, 2002), cites a claim from the respected *Spaceflight Directory* that "von Braun was fully prepared to launch a satellite into orbit on a Redstone rocket in September 1957" (109). William E. Burrows, in *This New Ocean: The Story of the First Space Age* (New York: Random House, 1998), writes that von Braun "was not engaging in idle promises" — that in September 1956 his team at the Redstone Arsenal had test-launched an eighty-four-pound payload more than three thousand miles over the Atlantic; had the trajectory been upward rather than downrange, von Braun maintained, the Redstone rocket could have launched a satellite into orbit (188).

76. Bernard Lovell, *The Story of Jodrell Bank* (New York: Harper & Row, 1968), 187. Lovell goes on to say that John Hagen, director of Project Vanguard, told him in August 1957 that no Vanguard launch could possibly be attempted until several

months in the future, whereupon Lovell quotes himself saying, "Then you will certainly be beaten by the Russians." Hagen replied that "he did not believe there was the slightest chance of this: the Russians were known to be encountering severe difficulties and were attending a conference in the U.S. in early October to discuss them." That same week, news of a successful Soviet test of an ICBM appeared. Lovell then assumed that Hagen's statement amounted to a form of secrecy, and that "Vanguard was much more nearly ready than Hagen had publicly indicated. Unfortunately, as was soon to be revealed, Vanguard was not only late but nearly a total failure. The U.S. had failed to give the project the priority and support which was necessary" (190–91).

77. Dolman, *Astropolitik,* 106; Siddiqi, *Red Rockets' Glare,* 290. Re the unprecedented nature of Korolev's presentation, Siddiqi writes, "With few exceptions, in the 1940s and 1950s, almost no one from the defense industry — from the lowest mechanic to the highest chief designer — was allowed to write publicly under his or her own name or to reveal the place of his or her employment. Weapons research institutes or design bureaus were openly identified only with a post office box number" (293). Nor was Korolev's work ascribed to him by name;

instead he was generally referred to as the Chief Designer; as one journalist put it, "he was never named in state communiqués because of official disapproval of 'the cult of personalities.' " Robin McKie, "Sergei Korolev: The Rocket Genius Behind Yuri Gagarin," *Guardian,* Mar. 12, 2011.

78. Siddiqi, "Korolev, Sputnik," 51; Kalic, *Presidents and Militarization of Space,* 33, 92.

79. Dolman, *Astropolitik,* 107–109; Burrows, *This New Ocean,* 187; Curtis Peebles, *High Frontier: The U.S. Air Force and the Military Space Program* (Washington, DC: Air Force History and Museums Program, 1997), 10; McDougall, *Heavens and Earth,* 123–24, 134.

80. For details on the R-7, see Siddiqi, "Korolev, Sputnik," 45–56.

81. "A few days ago a super-long-range, multistage intercontinental ballistic missile was launched. The tests of the missile were successful; they fully confirmed the correctness of the calculations and the selected design. The flight of the missile took place at a very great, hitherto unattained, altitude. Covering an enormous distance in a short time, the missile hit the assigned region. The results obtained show that there is the possibility of launching missiles into any region of the terrestrial globe." Quoted in Siddiqi, "Korolev, Sputnik," 58. Siddiqi

notes that it was highly unusual for the Soviet Union to publicize military successes.

82. Siddiqi, *Red Rockets' Glare*, 2.

83. An earlier Vanguard launch attempt, on Dec. 2, 1958, took place during post-Sputnik hearings by Lyndon Johnson's Senate Armed Services subcommittee on preparedness. Journalists variously named the failure "Kaputnik," "Stayputnik," and "Flopnik." See Thomas M. Gaskin, "Senator Lyndon B. Johnson, the Eisenhower Administration and U.S. Foreign Policy, 1957–60," *Presidential Studies Quarterly* 24:2 (Spring 1994), 348. Also of note: five days after the launch of Sputnik 1, top officials at the Department of Defense who were involved in Vanguard rather boldly misrepresented the situation in a briefing for Senator Johnson's subcommittee, saying that Vanguard and the US satellite program had very little to do with the US missile program and that they themselves could not offer an assessment of Sputnik's "military significance." Kalic, *Presidents and Militarization of Space,* 92–93.

84. Burrows, *This New Ocean,* x, 201. One of the innumerable despairing contemporary responses, this one from an aide to Senator Lyndon Johnson, inadvertently reveals the absence of science education via its use of "floating around" to mean "orbiting" and

its reference to "universe": "It is unpleasant to feel that there is something floating around in the air which the Russians can put up and we can't. . . . It really doesn't matter whether the satellite has any military value. The important thing is that the Russians have left the earth and the race for the control of the universe has started" (quoted in Peebles, *High Frontier,* 9).

85. Deborah D. Stine, "U.S. Civilian Space Policy Priorities: Reflections 50 Years After Sputnik," Congressional Research Service, Feb. 2, 2009, 2–5, fas.org/sgp/crs/space/RL34263.pdf (accessed Apr. 23, 2017).

86. ARPA's four space programs provide a good look at national priorities in the spring of 1958, specifically an emphasis on military goals and a de-emphasis of scientific goals: Missile Defense Against ICBMs, Military Reconnaissance Satellites, Developments for Application to Space Technology, and Advanced Research for Scientific Purposes. See Bruno W. Augenstein, "Evolution of the U.S. Military Space Program, 1945–1960: Some Key Events in Study, Planning, and Program Development," Paper P-6814, RAND, Sept. 1982, 13, www.rand.org/content/dam/rand/pubs/papers/2008/P6814.pdf (accessed Apr. 23, 2017).

87. "National Security Council Report: Statement of Preliminary U.S. Policy on Outer Space," NSC 5814/1, Aug. 18, 1958, doc.

442, para. 26, Office of the Historian, US Department of State, history.state.gov/ historicaldocuments/frus1958-60v02/d442 (accessed Apr. 23, 2017). See discussion at McDougall, *Heavens and Earth,* 180–83.

88. Vannevar Bush, *Science, The Endless Frontier: A Report to the President,* July 1945, www.nsf.gov/about/history/nsf50/ vbush1945_content.jsp#sect6_6 (accessed Oct. 26, 2017).

89. Dwight D. Eisenhower, *Waging Peace: The White House Years 1956–1961* (Garden City, NY: Doubleday, 1965), 257, quoted in Preston et al., *Space Weapons Earth Wars,* 9. By the early 1950s, American policymakers had come to regard both satellite reconnaissance and ballistic-missile defense systems as essential tools. Nevertheless, even though the United States initiated covert development programs for a number of military space projects, both nondestructive and weaponized, hot on the heels on Sputnik, Eisenhower's science advisors "judged space to be an unsuitable arena for weapons, labeling space weapons 'clumsy and ineffective ways of doing a job' " (10). Re Nixon, email from James Clay Moltz, Apr. 27, 2018.

90. William J. Broad, " 'Star Wars' Traced to Eisenhower Era," *New York Times,* Oct. 28,

1986; Johnson-Freese, *Heavenly Ambitions,* 4.

91. "Appendix II: Current Attitudes and Activities Regarding Biological Contamination of Extraterrestrial Bodies," in Leonard Reiffel, *A Study of Lunar Research Flights,* vol. 1, AD 425380/AFSWC TR-59-39 (Kirtland AFB, NM: Air Force Special Weapons Center, June 19, 1959), 292, oai .dtic.mil/oai/oai?verb=getRecord&meta dataPrefix=html& identifier=AD0425380 (accessed Apr. 23, 2017). Vol. 1 is now unclassified; vol. 2 remains unavailable and may have been destroyed. This study was one of a series of projects collectively labeled A119 that Reiffel directed between 1949 and 1962.

92. Antony Barnett, "US Planned One Big Nuclear Blast for Mankind," *Guardian,* May 13, 2000. Compare the more neutral language in the 1959 report's introduction:

Rapidly accelerating progress in space technology clearly requires evaluation of the scientific experiments or other human activities which might be carried out in the vicinity of the earth's natural satellite. Among various possibilities, the detonation of a nuclear weapon on or near the moon's surface has often been suggested. The motivation for such a detonation is clearly threefold: scientific, military and political.

The scientific information which might be obtained from such detonations is one of the major subjects of inquiry of the present work. On the other hand, it is quite clear that certain military objectives would be served since information would be supplied concerning the environment of space, concerning detection of nuclear device testing in space and concerning the capability of nuclear weapons for space warfare. . . . Obviously[,] specific positive effects would accrue to the nation first performing such a feat as a demonstration of advanced technological capability. (Reiffel, *Lunar Research Flights,* 2)

93. For text of disarmament resolutions 1148 and 1149, see UN General Assembly, Resolutions Adopted by the General Assembly During Its Twelfth Session, www.un.org/documents/ga/res/12/ares12.htm (accessed Apr. 23, 2017).

94. Raymond L. Garthoff, "Banning the Bomb in Outer Space," *International Security* 5:3 (Winter 1980–81), 25–40.

95. In mid-October 1962, President Kennedy was given photographic proof, collected by an American U-2 spy plane, that — contrary to US official assumptions — the Soviet Union was constructing launch sites in Cuba for missiles with ranges of both one thousand and more than two thousand

miles. Kennedy publicly announced this news on Oct. 22 as "provid[ing] a nuclear strike capability against the Western Hemisphere," and the United States instituted a maritime blockade of "all offensive military equipment." The actual launch of a nuclear missile from Cuba would be regarded, declared the president, "as an attack by the Soviet Union on the United States, requiring a full retaliatory response upon the Soviet Union." He demanded the removal of the weapons, the two sides went on full nuclear alert, letters were exchanged and ignored and examined, bargains involving US missiles in Turkey were proposed, America organized around an air strike to take place on Oct. 30, and everybody remained terrified until Oct. 28, when Khrushchev agreed to the removal of Soviet missiles from Cuba along with American missiles from Turkey. In 1992, Russia disclosed previously secret information showing that, contrary to the 1962 US assumption that nuclear warheads would not yet have been installed on the Cuban missiles, they in fact were: forty-two intermediate-range and nine short-range missiles, all with their warheads in place, guarded by forty thousand troops and ready for launch. It would have been possible to obliterate any site or city in America except those in Washington state. As Secretary of

Defense Robert McNamara said in 1992 upon learning this information, "It meant that had a U.S. invasion been carried out . . . there was a 99 percent probability that nuclear war would have been initiated." See LaFeber, *America, Russia,* 221, 231–37. See also the detailed account by the Central Intelligence Agency, "Cuban Missile Crisis, 1962, Value of Photo Intelligence," May 8, 2007, www.cia.gov/library/center-for-the-study-of-intelligence/kent-csi/docs/v44i4a09p_0002.htm to –0015.htm (accessed Sept. 13, 2016; by Apr. 23, 2017, photos had been removed).

96. See, e.g., Edward R. Finch Jr., "Outer Space for 'Peaceful Purposes,' " *American Bar Association Journal* 54:4 (Apr. 1968), 365–67. Finch argues that "peaceful" signifies "nonaggressive" rather than "nonmilitary," and states: "In Russian the word for 'military' essentially means warlike rather than pertaining to the armed services of a country, while in English 'peaceful' is not regarded as the opposite of 'military.' " The report of the Rumsfeld Space Commission concurs, saying that most nations agree that "peaceful" means "nonaggressive," but goes much further: "There is no blanket prohibition in international law on placing or using weapons in space, applying force from space to earth or conducting military operations in and through space. . . . The U.S.

must be cautious of agreements . . . that, when added to a larger web of treaties or regulations, may have the unintended consequences of restricting future activities in space," which is of course precisely what the Outer Space Treaty is intended to do. See *Report of the Commission to Assess United States National Security Space Management and Organization,* Jan. 11, 2001, xviii. See also Andrew D. Burton, "Daggers in the Air: Anti-satellite Weapons and International Law," *Fletcher Forum of World Affairs* (Winter 1988), 151–53; Kalic, *Presidents and Militarization of Space,* 81–82. Kalic quotes a Kennedy administration undersecretary of state who alerted the president in 1962 to the " 'widespread confusion over the distinction between peaceful and aggressive, and military and civilian' " uses of space, and advised him to make clear that "peaceful" was not synonymous with "civilian" nor "aggressive" synonymous with "military."

97. National Security Council — Executive Secretary, "Report to the National Security Council on Basic National Security Policy," NSC 162/2, Oct. 30, 1953; paras. 2, 13c, fas.org/irp/offdocs/nsc-hst/nsc-162-2.pdf (accessed Apr. 24, 2017).

98. Kalic, *Presidents and Militarization of*

Space, 3–6; McDougall, *Heavens and Earth,* 335ff.

99. Kalic, *Presidents and Militarization of Space,* 88; Sinclair Lewis, *It Can't Happen Here* (New York: Doubleday, 1935; Signet Classics, 1970, 2014), 138.

100. Quoted in LaFeber, *America, Russia,* 204; McDougall, *Heavens and Earth,* 138.

101. Quoted in McDougall, *Heavens and Earth,* 114.

102. Dwight D. Eisenhower, "Annual Message to the Congress on the State of the Union," Jan. 10, 1957, at Gerhard Peters and John T. Woolley, The American Presidency Project, www.presidency.ucsb.edu/ws/?pid=11029 (accessed Apr. 24, 2017). McDougall contends that Eisenhower profoundly hoped to "open up" the Soviet Union, and that "[i]f it could be done voluntarily in the context of arms control, Eisenhower was even willing to forego a purely national space program" (*Heavens and Earth,* 127–28).

103. "Report by the Technological Capabilities Panel of the Science Advisory Committee," Feb. 14, 1955, S/S–RD Files: Lot 71 D 171; Top Secret; Restricted Data; available with omissions; Office of the Historian, US Department of State, history.state.gov/historicaldocuments/frus1955-57v19/d9 (accessed Apr. 24, 2017); McDougall, *Heav-*

ens and Earth, 115–18.

104. Dwight D. Eisenhower, "Radio and Television Address to the American People on Science in National Security," Nov. 7, 1957, at Gerhard Peters and John T. Woolley, The American Presidency Project, University of California, Santa Barbara, www.presidency.ucsb.edu/ws/?pid=10946 (accessed Apr. 24, 2017); National Security Council, "Preliminary Policy on Outer Space," NSC 5814/1, para. 30.

105. See full text of "National Aeronautics and Space Act of 1958 (Unamended)," NASA, history.nasa.gov/spaceact.html.

106. For detailed discussion of "spies in the skies" and "the reconnaissance war," see Burrows, *This New Ocean,* 225–58; Jeffrey T. Richelson, *America's Space Sentinels: The History of the DSP and SBIRS Satellite Systems,* 2nd ed. (Lawrence: University Press of Kansas, 2012).

107. Earth takes twenty-three hours and fifty-six minutes to complete one rotation relative to the stars. This is the precise orbital period of a geostationary (also called geosynchronous) satellite. The higher the orbit, the longer it takes; satellites that live in low Earth orbit (LEO), such as the Hubble Telescope and the International Space Station, take ninety minutes to complete one orbit. Arthur C. Clarke first made a brief, offhand mention of the possibility of geo-

synchronous communications satellites in a letter to the editor, "V2 for Ionosphere Research," *Wireless World,* Feb. 1945, 58. Less than a year later came his fully formulated proposal: "Extra-Terrestrial Relays: Can Rocket Stations Give World-wide Radio Coverage?," *Wireless World,* Oct. 1945, 305–308. Facsimiles of the letter and the article are at lakdiva.org/clarke/1945ww/ (accessed Nov. 7, 2017). Clarke would later collaborate with director Stanley Kubrick on the 1968 science fiction film classic *2001: A Space Odyssey,* released the year before the first Moon landing. "Live via satellite" TV broadcasts also began in 1968, the onscreen note to viewers a proud indicator that space was enabling the future. By the late 1970s the note had disappeared: live satellite broadcasts had become the new normal. Robert Yowell, "Splashdown, Live Via Satellite," AirSpaceMag.com, Apr. 13, 2016, www.airspacemag.com/daily-planet/splashdown-live-satellite-180958760/ (accessed Nov. 7, 2017).

108. See Kalic, *Presidents and Militarization of Space,* chap. 2 passim, 47–57, 71–73; Preston et al., *Space Weapons Earth Wars,* 9–12; Burrows, *This New Ocean,* 226–68. Also relevant is Augenstein, "Evolution of the U.S. Military Space Program," an early overview of the multiplicity of postwar military space projects, by a key figure in

the RAND and Lockheed Corporations as well as the DoD. For Dyna-Soar, see Mc-Dougall, *Heavens and Earth,* 339–41, and Chris Bergin, "The Story of the Dyna-Soar," NASASpaceflight.com, Jan. 7, 2006, www.nasaspaceflight.com/2006/01/the-story-of-the-dyna-soar/ (accessed Nov. 7, 2017); for the X-15 and Dyna-Soar, see Burrows, *This New Ocean,* 249–55. A photo of the X-15 is at Burrows, fourth page of illustrations following p. 206; the accompanying caption states that "USAF and NASA markings reflect the symbiotic relationship between the military and civilian space programs."

109. One historian contends, "All actions of Congress with regard to space between 1957 and 1961 can be attributed to Johnson, who worked hard to demonstrate that the Democrats in general, and he in particular, were the leading forces in Congress for stepped-up efforts in space exploration." He also characterizes Johnson as "probably the first who understood that space was the ideal 'battleground' of the cold war, that by competing with the Soviet Union for technical leadership and peaceful dominance in space, the United States could show that it was the superior nation." Andreas Reichstein, "Space — the Last Cold War Frontier?" *Amerikastudien/American Studies* 44:1 (1999), 115–16. A more nuanced, and

lively, exploration of this moment in US history can be found in Gaskin, "Senator Lyndon B. Johnson," 341–61, esp. 347ff. One political insider relayed to Johnson advice from another political insider that the Sputnik hearings "if properly handled, would blast the Republicans out of the water, unify the Democratic Party and elect you President." The strategic recommendation was that Johnson "should plan to plunge heavily into this one" (348).

110. Gaskin, "Senator Lyndon B. Johnson," 348.
111. For the maneuvers that created Johnson's opportunity to speak before the United Nations, see Gaskin, "Senator Lyndon B. Johnson," 349–51.
112. This quotation is a composite of extracts from Eilene Galloway, "Organizing the United States Government for Outer Space: 1957–1958," conference paper presented at "Reconsidering Sputnik: Forty Years Since the Soviet Satellite," Washington, DC, Sept. 30–Oct. 1, 1997, gos.sbc.edu/g/galloway2.html (accessed Apr. 24, 2017); "In Essentials, Unity," op-ed, *New York Times,* Nov, 18, 1958, 36; and Thomas J. Hamilton, "Johnson Tells the U.N. Nation Is United on Space," *New York Times,* Nov. 18, 1958, 1, 10.
113. Moltz, *Politics of Space Security,* 71.
114. Dolman, *Astropolitik,* 87.

115. Moltz, *Politics of Space Security,* 121, 90.

116. Committee on Aeronautical and Space Sciences, US Senate, *Staff Report: Documents on International Aspects of the Exploration and Use of Outer Space, 1954–1962,* May 9, 1963, 182, www.spacelaw.olemiss .edu/library/space/US/Legislative/Congress/ 88/Senate/reports/docno18.pdf (accessed Apr. 24, 2017).

117. During the election, Kennedy attacked the Eisenhower administration for allowing a missile gap to develop, even though US intelligence sources had already revealed the nonexistence of such a gap. "In fact . . . overflights, still highly classified at the time, showed that the United States had a commanding lead in deployed nuclear-tipped missiles." See Moltz, *Politics of Space Security,* 105–106, 106 n. 157; Peebles, *High Frontier,* 4, 9–10. LaFeber writes that Khrushchev called the ICBM the "ultimate weapon" and exploited the imaginary Soviet lead by quoting "the West's own greatly exaggerated views of Soviet missile capacity, thereby reinforcing the exaggerations." He also notes that the US nuclear arsenal tripled, from six thousand to eighteen thousand weapons, between 1958 and 1960 — including fourteen Polaris nuclear submarines that each carried sixteen missiles

(*America, Russia,* 202–205). See also detailed discussion in McDougall, *Heavens and Earth,* 226–31. Khrushchev quote from *Pravda,* Jan. 28, 1959, quoted in McDougall, *Heavens and Earth,* 240.

118. McDougall, *Heavens and Earth,* 346–48, 335; Burrows, *This New Ocean,* 241.

119. Moltz, *Politics of Space Security,* 111–12, 107; Kalic, *Presidents and Militarization of Space,* 69–71, 76–79; Office of Management and Budget, Historical Tables, Table 1.1 ("Summary of Receipts, Outlays, and Surpluses or Deficits (–): 1789–2021") and Table 4.1 ("Outlays by Agency: 1962–2021"), www.whitehouse.gov/omb/budget/Historicals (accessed Jan. 2, 2017; by Apr. 24, 2017, link was disabled; by Apr. 10, 2018, link was restored and updated).

120. John F. Kennedy, "Special Message to Congress on Urgent National Needs," May 25, 1961, John F. Kennedy Presidential Library and Museum, www.jfklibrary.org/Research/Research-Aids/JFK-Speeches/United-States-Congress-Special-Message_19610525.aspx (accessed Apr. 24, 2017).

121. Kalic, *Presidents and Militarization of Space,* 97–100.

122. John F. Kennedy, "Address at Rice University on the Nation's Space Effort," Sept. 12, 1962, transcript, www.jfklibrary.org/Asset-Viewer/MkATdOcdU06X5

uNHbmqm1Q.aspx (accessed Apr. 29, 2017).

123. Office of the Historian, US Department of State, "Draft Proposals for US–USSR Cooperation," Apr. 13, 1961, *Foreign Relations of the United States, 1961–1963*, vol. 25, doc. 387, history.state.gov/historicaldoc uments/frus1961-63v25/d387 (accessed Apr. 24, 2017). Quote from the general in McDougall, *Heavens and Earth*, 342.

124. The two large 1962 US nuclear tests were Sedan (ground-based) and Starfish Prime (high-altitude). For a discussion of the AEC's project in Alaska, see Douglas L. Vandegraft, "Project Chariot: Nuclear Legacy of Cape Thompson," *Proceedings of the US Interagency Arctic Research Policy Committee Workshop on Arctic Contamination,* Session A: Native People's Concerns about Arctic Contamination II: Ecological Impacts, May 6, 1993, Anchorage, Alaska, arcticcircle.uconn.edu/VirtualClassroom/ Chariot/ vandegraft.html (accessed Apr. 24, 2017). See also Ronald E. Doel and Kristine C. Harper, "Prometheus Unleashed: Science as a Diplomatic Weapon in the Lyndon B. Johnson Administration," in "Global Power Knowledge: Science and Technology in International Affairs," *Osiris* 21:1 (2006), 70 n. 22. It was already clear that extensive, durable harm could result

from a range of advanced technologies, and efforts at prevention were under way. In April 1963, for instance, the president issued National Security Action Memorandum 235/1, requiring agencies such as the CIA and the State Department to do an advance review of any large-scale, potentially controversial experiments that could conceivably have adverse effects on the environment.

125. One commonly acknowledged positive component of America's conduct during the missile crisis was the Kennedys' decision to respond favorably to a first letter from Khrushchev and to ignore a second, problematic one that arrived shortly afterward. One relatively unheralded positive component on the Soviet side was an intervention on Oct. 27 by a Soviet officer on submarine B-59, who urged his fellow officers not to execute an order to fire their ten-kiloton nuclear-tipped torpedo at a US warship that was bombarding them with depth charges. The firing required approval from three officers; Vasili Arkhipov refused to be the third. This piece of the story became generally known only forty years afterward, at a 2002 conference at Brown University on the crisis; an oft-repeated quote from that conference comes from Thomas Blanton, director of the National Security Archive: "a guy called Vasili Arkhi-

pov saved the world." Marion Lloyd, "Soviets Close to Using A-bomb in 1962 Crisis, Forum Is Told," *Boston Globe,* Oct. 13, 2002; Edward Wilson, "Thank You Vasili Arkhipov, the Man Who Stopped Nuclear War," *Guardian,* Oct. 27, 2012; Neil Genzlinger, "Same Cuba Crisis, Different Angles: 50 Years Later — Cuban Missile Crisis Revisited on PBS," *New York Times,* Oct. 22, 2012.

126. Quoted in McDougall, *Heavens and Earth,* 331. Canadian Nobel laureate and prime minister Lester Pearson may have originated the phrase "balance of terror" in 1955. President Kennedy used it in his inaugural address in 1961:

> Finally, to those nations who would make themselves our adversary, we offer not a pledge but a request: that both sides begin anew the quest for peace, before the dark powers of destruction unleashed by science engulf all humanity in planned or accidental self-destruction. . . . [O]nly when our arms are sufficient beyond doubt can we be certain beyond doubt that they will never be employed. But neither can two great and powerful groups of nations take comfort from our present course — both sides overburdened by the cost of modern weapons, both rightly alarmed by the steady spread of the deadly atom, yet both racing to alter that

Militarization of Space, 95.

137. See, e.g., E.L., "Barry Goldwater on Space: GOP Candidate Wants Military, Not Civilians, to Run Space Program," *Science* 145 (July 31, 1964), 470–71; George B. Kistiakowsky, "Johnson or Goldwater — Two Scientists Explain Their Choice: The Case for Johnson," *Science* 146 (Oct. 16, 1964), 380–82; NASA Historical Staff, Office of Policy Planning, *Astronautics and Aeronautics, 1964: Chronology on Science, Technology and Policy* (Washington, DC: NASA, 1964), 232–33, history.nasa.gov/AAchronologies/1964.pdf (accessed Apr. 25, 2017); Richard Dean Burns and Joseph M. Siracusa, *A Global History of the Nuclear Arms Race: Weapons, Strategy, and Politics,* 2 vols. (Santa Barbara, CA: Praeger/ABC-CLIO, 2013), 379; Kalic, *Presidents and Militarization of Space,* 102–105.

138. The Gulf of Tonkin Resolution — the free hand in Southeast Asia — passed the House 416–0 and the Senate 88–2; see discussion in LaFeber, *America, Russia,* 251–52. See also Burns and Siracusa, *Global History,* 379–80: "[Johnson] poured considerable effort into reducing tensions between Washington and Moscow by pursuing measures that lessened the prospect of nuclear conflict. . . . Bureaucrats from the Joint Chiefs of Staff to the State Depart-

ment often fought Johnson's arms control proposals; moreover, his policies often ran counter to the views of many Americans who endorsed more strident policies[, such as] the American Security Council." For an overview of Johnson-era legislation, see Joseph A. Califano Jr., "Seeing Is Believing — The Enduring Legacy of Lyndon Johnson," keynote address, centennial celebration, May 19, 2008, www.lbjlibrary.org/lyndon-baines-johnson/perspectives-and-essays/seeing-is-believing-the-enduring-legacy-of-lyndon-johnson (accessed Nov. 9, 2017).

139. As the very first Director of Defense Research and Engineering, Herbert York, wrote in *Making Weapons, Talking Peace:* "Unnecessary duplication was rife, and vicious interservice struggles over rules and missions were creating confusion." Quoted in Moltz, *Politics of Space Security,* 95.

140. Burrows, *This New Ocean,* 238–41, 246. In addition, beginning in 1961, the Air Force found itself competing for primacy in space reconnaissance with the new National Reconnaissance Office and the Central Intelligence Agency.

141. Kalic, *Presidents and Militarization of Space,* 105–107. One source of opposition was the Critical Issues Council of the Republican Citizens Committee of the United States, which saw "no reason to

believe that putting a man on the moon could contribute to our military strength" and urged a multinational rather than solely US lunar-landing program so as to share the prodigious costs.

142. The remaining one-quarter of federal space funding received by the military during the Johnson years went into programs both successful and aborted, including KEYHOLE and CORONA photoreconnaissance satellites; the Manned Orbiting Laboratory project, a nonwarfighting reconnaissance space station for the Air Force that was defunded well before achieving orbit; and the Defense Meteorological Satellite Program, set up by the National Reconnaissance Office. Satellites and heavy-lift booster rockets took up most of the money, while a much smaller amount went toward antisatellite weapons and ballistic missile defense, in tandem with the Soviets' continuing work on bombardment satellites. Communications satellites were rapidly altering the way information could be transmitted and exchanged, and in 1968 a constellation of twenty-six US satellites called the Defense Satellite Communication System began operations — a mere half-decade after nineteen nations, including the United States, had formed the civilian International Telecommunications Satellite Consortium (Intelsat) and Bell

Labs/AT&T had generated their first live television transmission via commercial satellite. See, e.g., Burrows, *This New Ocean,* 241–71; Kalic, *Presidents and Militarization of Space,* 101–102, 107, 110–11.

143. Moltz, *Politics of Space Security,* 143, 152–54.

144. Quoted in Kalic, *Presidents and Militarization of Space,* 117.

145. See, e.g., letter from Robert C. Seamons Jr., NASA Deputy Administrator, to James E. Webb, NASA Administrator, Feb. 25, 1967, and Statement by James E. Webb, Feb. 25, 1967, in NASA, "Report of Apollo 204 Review Board," NASA-TM-84-105, Apr. 5, 1967, 3-57, 3-58, 3-61, history.nasa.gov/Apollo204/summary.pdf (accessed Nov. 11, 2017).

146. McDougall, *Heavens and Earth,* 344–45.

147. Kalic, *Presidents and Militarization of Space,* 109–114; McDougall, *Heavens and Earth,* 344–45.

148. Moltz, *Politics of Space Security,* 125–26. Moltz contends that the movement toward space agreements was bilateral, not unilateral. He describes the gradual movement toward US–Soviet space cooperation and arms deceleration that followed the signing of the Limited Test Ban Treaty as the emergence of "cooperative restraint."

149. Lyndon B. Johnson, "Remarks at the Signing of the Treaty on Outer Space," Jan. 27, 1967, Peters and Woolley, *American Presidency Project,* www.presidency.ucsb.edu/ws/?pid=28205 (accessed Apr. 25, 2017). On February 7, 1967, the President transmitted the treaty to the Senate, which ratified it in late April. Urging ratification, Johnson addressed the senators: "Today, outer space is free. It is unscarred by conflict. No nation holds a concession there. It must remain this way. We of the United States do not acknowledge that there are landlords of outer space who can presume to bargain with the nations of the Earth on the price of access to this domain. We must not — and we need not — corrupt this great opportunity by bringing to it the very antagonism which we may, by courage, overcome and leave behind forever if we proceed with this joint adventure into this new realm." Quoted in Kalic, *Presidents and Militarization of Space,* 115.

150. Mizin, "Non-Weaponization of Outer Space," in *Outer Space,* ed. Arbatov and Dvorkin, 50–51. In 1985 the United States presented a "broad interpretation" of the treaty, asserting that it did not "prohibit the testing of BMD [ballistic missile defense] components in space" (57).

151. On a related 1994 UN resolution,

Enlargement of the Committee on the Peaceful Uses of Outer Space, the United States was the sole vote in opposition; there were no abstentions. See United Nations Office for Outer Space Affairs, "Documents and Resolutions Database," www.unoosa.org/oosa/documents-and-resolutions/search.jspx?&view=resolutions (accessed Apr. 25, 2017).

152. Mizin, "Non-Weaponization of Outer Space," 54–56; Tim Weiner, "Lies and Rigged 'Star Wars' Test Fooled the Kremlin, and Congress," *New York Times,* Aug. 18, 1993; Sergei Oznobishchev, "Codes of Conduct for Outer Space," in *Outer Space,* ed. Arbatov and Dvorkin, 69–77. Mizin's assessment, shared by many, is that SDI "was really not only a grandiose new technological project to revamp the U.S. armed forces, but also a kind of active measure designed to lure the USSR into an exhausting competition that it was destined to lose" (56). For the Soviet submissions to the General Assembly, see documents A/36/192 (Aug. 20, 1981), A/38/194 (Aug. 23, 1983), and A/39/243 (Sept. 27, 1984) at "Documents by Symbol," General Assembly of the United Nations, www.un.org/en/ga/documents/symbol.shtml (accessed Apr. 25, 2017).

153. Letter dated February 12, 2008, from the Permanent Representative of the Rus-

sian Federation and the Permanent Representative of China to the Conference on Disarmament, CD/1839 (incorporating Draft: Treaty on Prevention of the Placement of Weapons in Outer Space and of the Threat or Use of Force Against Outer Space Objects), UN Conference on Disarmament, Feb. 29, 2008, documents-dds-ny.un.org/doc/UNDOC/GEN/G08/604/02/PDF/G0860402.pdf. Letter dated Sept. 11, 2015, from the Permanent Representative of China to the Conference on Disarmament and the Chargé d'affaires a.i. of the Russian Federation addressed to the Secretary-General of the Conference, CD/2042 (incorporating Follow-up comments by the Russian Federation and China on the analysis submitted by the United States of America of the updated Russian-Chinese draft PPWT), Conference on Disarmament, Sept. 14, 2015, documents-dds-ny.un.org/doc/ UNDOC/GEN/G15/208/38/PDF/G1520838.pdf (accessed July 31, 2016; by Apr. 25, 2017, links were disabled).

154. Detlev Wolter, *Common Security in Outer Space and International Law* (Geneva: United Nations Institute for Disarmament Research, 2006), 157–58, www.files.ethz.ch/isn/122089/2006_CommonSecuritySpace_en.pdf; condensed version at Wolter, "Legal Foundations and Essential Treaty Elements for a System of Common Security

883

in Outer Space," Global Security Institute, 2007, www.worldacademy.org/files/System _of_Common_Security_in_Outer_Space .pdf (accessed Apr. 25, 2017).

155. Burton, "Daggers in the Air," 147. Elaborating why the OST is the "high-water mark," the author states that "the marked decline in the regard to the United Nations paid by the United States in recent years makes it unlikely that a significant repeat or extension of the Outer Space Treaty will be negotiated under the aegis of that body."

156. Ratification by 105 member states as of January 1, 2017, www.unoosa.org/docu ments/pdf/spacelaw/treatystatus/AC105_C2 _2017_CRP07E.pdf (accessed Nov. 13, 2017). Here the term "ratification" includes not only actual ratification but also what the UN calls acceptance, approval, accession, or succession. The status of international agreements relating to outer space is updated annually and is available through the United Nations Office for Outer Space Affairs at www.unoosa.org/oosa/en/ SpaceLaw/treatystatus/index. html.

157. Some statements of space doctrine do not mention the Outer Space Treaty or the existence of an international legal framework. For example, *Space Operations: Air Force Doctrine Document 2-2,* Aug. 23, 1998, lists in its "Suggested Readings" a UN website on the status of space treaties

but does not otherwise discuss space law, www.globalsecurity.org/jhtml/jframe .html#http://www.globalsecurity.org/ military/library/policy/usaf/afdd/2-2/afdd2-2 .pdf|||AFDD%202-2:%20Space%20 Operations. Eight years later, the updated doctrine document includes multiple references to treaty provisions and space law (*Space Operations: Air Force Doctrine Document 3-14,* Nov. 27, 2006/July 28, 2011, www.globalsecurity.org/jhtml/jframe .html#http://www.globalsecurity.org/ military/library/policy/usaf/afdd/3-14/afdd3 -14_2011.pdf|||AFDD%203-14:%20 Space%20Operations). Other documents may briefly refer to legal compliance in extremely general terms — for example, the national space policy of the Barack Obama administration, dated June 28, 2010, at www.au.af.mil/au/awc/awcgate/whitehouse/ national_space_policy_28june2010.pdf, which states, "All nations have the right to explore and use space for peaceful purposes, and for the benefit of all humanity, in accordance with international law," but then hastens to note, in the following sentence, "Consistent with this principle, 'peaceful purposes' allows for space to be used for national and homeland security activities."

On the other hand, the national space policy (unclassified) of the George W. Bush administration, dated Aug. 31, 2006, at history.nasa.gov/ostp_space_policy06.pdf, refers only to "treaty compliance" as it relates to the obligations of the Director of National Intelligence, and preemptively rules out any additional legal instruments: "The United States will oppose the development of new legal regimes or other restrictions that seek to prohibit or limit U.S. access to or use of space. Proposed arms control agreements or restrictions must not impair the rights of the United States to conduct research, development, testing, and operations or other activities in space for U.S. national interests." The DoD's Obama-era *Quadrennial Defense Review 2014,* archive.defense.gov/ pubs/ 2014_Quadrennia_Defense_Review.pdf, by contrast, is more receptive to the development of additional agreements: "All of the Department's [DoD's] initiatives in space will continue to be underpinned by U.S. Government efforts to work with industry, allies, and other international partners to shape rules of the road in this domain," although here, too, the very next sentence reasserts the centrality of conflict: "We will retain and strengthen our power projection capabilities so that we can deter conflict, and if deterrence fails, win decisively against aggressors" (20). (All accessed Apr. 26, 2017).

158. US Air Force, *Space Operations: Air Force Doctrine Document 2.2,* Nov. 27, 2006, 26–27, www.globalsecurity.org/jhtml/ jframe.html#http://www.globalsecurity.org/ military/library/policy/usaf/afdd/2-2/afdd2-2 -2006.pdf|||AFDD%202-2:%20Space %20Operations (accessed Apr. 26, 2017).

159. Preston et al., *Space Weapons Earth Wars,* excludes from consideration a number of categories of weapons: those that pass through space without achieving orbit, those in space that simply improve the efficacy of ground-based weapons, cyberweapons dependent on space-based communications systems, and space-based weapons of use only against space-based targets. It also excludes classified weapons.

160. Preston et al., *Space Weapons Earth Wars,* 1.

161. I. F. Stone, "First Call for a Test Ban" (Nov. 1, 1954), in I. F. Stone, *The Best of I. F. Stone* (New York: PublicAffairs, 2006), 117.

162. Preston et al., *Space Weapons Earth Wars,* 2.

163. Preston et al., *Space Weapons Earth Wars,* 3, 94.

164. Preston et al., *Space Weapons Earth Wars,* 17–18, 37.

165. Johnson-Freese, *Heavenly Ambitions,* 6. The "highly intrusive forms" quote is from

Nancy Gallagher and John D. Steinbruner, *Reconsidering the Rules for Space Security* (Cambridge, MA: American Academy of Arts and Sciences, 2008), 2. Another military scholar, John J. Klein, in *Space Warfare: Strategy, Principles and Policy* (London and New York: Routledge, 2006) — part of Routledge's *Space Power and Politics* series — acknowledges that while an aggressive US policy of space weaponization, "whether for offensive or defensive purposes," would be useful in protecting national security, it would also paint the United States "as becoming more powerful too quickly" and could well lead to "collective attempt[s] to contest this space hegemony through diplomatic, economic, information, and perhaps even military endeavors." Ultimately, "after the completion of any expensive multi-year weapons program there is no guarantee that national security and the ability to command space will be improved in the end" (145–46).

166. The USA supplied 33 percent of all arms transfers 2012–16; Russia was second, at 23 percent. Aude Fleurant, Pieter D. Wezeman, Siemon T. Wezeman, and Nan Tian, "Trends in International Arms Transfers, 2016," fact sheet, Stockholm International Peace Research Institute, 2, Feb. 2017, www.sipri.org/sites/default/files/Trends-in-international-arms-transfers

-2016.pdf (accessed Apr. 27, 2017).

167. Preston et al., *Space Weapons Earth Wars*, 74–75.

168. There is an extensive literature on the shift in international law away from the venerable principle of *cuius est solum eius est usque ad coelum et ad sidera,* variously translated but, in essence, meaning "he who owns the land owns it up to the heavens and down to hell." With the ascent of Sputnik, the issue of sovereignty up to the heavens shifted to the practical issue of where to establish the boundary between earth, sky, and outer space. See, for instance, Burton, "Daggers in the Air," 143, 149–50, 153; Dolman, *Astropolitik,* 115–20; Philip W. Quigg, "Open Skies and Open Space," *Foreign Affairs* 37:1 (Oct. 1958).

169. Preston et al., *Space Weapons Earth Wars*, 101–106.

170. Joint Chiefs of Staff, *Joint Vision 2020: America's Military — Preparing for Tomorrow,* Summer 2000. A headline elucidates the Joint Chiefs' vision: "Dedicated individuals and innovative organizations transforming the joint force for the 21st century to achieve full spectrum dominance: Persuasive in peace; Decisive in war; Preeminent in any form of conflict" (58).

171. Scott A. Weston, "Examining Space Warfare: Scenarios, Risks, and US Policy

Implications," *Air & Space Power Journal* 23:1 (Spring 2009), 74, www.au.af.mil/au/afri/aspj/airchronicles/apj/apj09/spr09/weston.html (accessed May 3, 2017).

172. Moltz, *Politics of Space Security,* 119–21, 130–32. Starfish Prime was ninety times more powerful than "Little Boy" (Hiroshima) and seventy times more powerful than "Fat Man" (Nagasaki).

173. "Interview: Walter LaFeber, Historian," *American Experience,* PBS, www.pbs.org/wgbh/americanexperience/features/interview/truman-lafeber/ (accessed Aug. 28, 2016).

174. James Forrestal, *The Forrestal Diaries,* quoted in LaFeber, *America, Russia,* 86; Mastny, *Cold War,* 49, 101, 123, 127.

175. LaFeber, *America, Russia,* 27–28.

176. Reports of yield vary. Figures given here are from the Preparatory Commission for the Comprehensive Test Ban Treaty Organization, www.ctbto.org/nuclear-testing/ (accessed Apr. 27, 2017). The Soviet device detonated in October 1961 is called Tsar Bomba and was by far the largest-ever nuclear bomb, its power equivalent to about 3,800 Hiroshima-type bombs detonated simultaneously.

177. American Security Council, "Top-Level Civilian Committee Urges President Kennedy to Stop Geneva Test-Ban Negotiations, Resume Atomic Underground Tests,"

press release, May 16, 1961, archive.org/ stream/AmericanSecurityCouncil/American %20Security%20Council-2#page/n19/ mode/2up (accessed Apr. 27, 2017).

178. Richard H. Kohn and Joseph P. Harahan, "U.S. Strategic Air Power, 1948–1962: Excerpts from an Interview with Generals Curtis E. LeMay, Leon W. Johnson, David A. Burchinal, and Jack J. Catton," *Int. Security* 12:4 (Spring 1988), 85–86. Re "merciless," William E. Burrows characterizes LeMay as a man "who never lost sight of the fact that his service's core mission was destroying the enemy" (*This New Ocean,* 237).

179. National Security Council — Executive Secretary, "National Security Policy," NSC 162/2; 2, 13, 19, 22. During the mid-1950s, key segments of both the Eisenhower and Khrushchev administrations contended that simultaneously increasing expenditures on nuclear weapons and cutting conventional forces was a sensible strategy, though key segments of the military disagreed. See Matthew Evangelista, "Cooperation Theory and Disarmament Negotiations in the 1950s," *World Politics* 42: 4 (July 1990), 510–12.

180. Robert S. McNamara, "The Military Role of Nuclear Weapons: Perceptions and Misperceptions," *Foreign Affairs* 62:1 (Fall

1983), 63. Montgomery's statement comes from a speech to the Royal United Services Institute, London.

181. Scott Shane, "1950s U.S. Nuclear Target List Offers Chilling Insight," *New York Times,* Dec. 22, 2015.

182. North Atlantic Military Committee, "Final Decision on MC 14/2 (Revised): A Report by the Military Committee to the North Atlantic Council on Overall Strategic Concept for the Defense of the North Atlantic Treaty Organization Area," declassified, May 23, 1957, 9[289], 13[293], in Gregory W. Pedlow, ed., "NATO Strategy Documents 1949–1969," NATO International Staff Central Archives, n.d., www.bits.de//NRANEU/nato-strategy/ MC14-2.pdf (accessed Apr. 27, 2017).

183. Burns and Siracusa, *Global History,* 377.

184. This assessment comes from Thomas Graham Jr., an attorney and disarmament diplomat who served with the US Arms Control and Disarmament Agency from 1970 to 1997. Burns and Siracusa, *Global History,* 426, 431.

185. LaFeber, *America, Russia,* 208–12, 223–24; Burns and Siracusa, *Global History,* chap. 14, "Reagan, Gorbachev, and Nuclear Arms: Ending the Cold War," 413–45.

186. Michael S. Gerson, "No First Use: The Next Step for U.S. Nuclear Policy," *Int.*

Security 35:2 (Fall 2010), 7; US Air Force, *Nuclear Operations: Air Force Doctrine Document 3-72*, May 7, 2009 (incorporating Change 2, Dec. 14, 2011), 17–18, www .globalsecurity.org/military/library/policy/ usaf/afdd/3-72/afdd3-72_2011.pdf; US Department of Defense, *Nuclear Posture Review Report*, Apr. 2010, v, www.defense .gov/Portals/1/features/defenseReviews/ NPR/2010_Nuclear_Posture_Review _Report.pdf (accessed Apr. 27, 2017).

187. "Trump Repeats Call for US Nuclear Supremacy," BBC News, Feb. 24, 2017, www.bbc.com/news/world-us-canada -39073303 (accessed Apr. 27, 2017).

188. James E. Cartwright and Bruce G. Blair, "End the First-Use Policy for Nuclear Weapons," op-ed, *New York Times*, Aug. 14, 2016.

189. Michael Krepon, "Not Just Yet for No First Use," blog post, Arms Control Wonk: Leading Voices on Arms Control, Disarmament and Non-Proliferation, July 31, 2016, www.armscontrolwonk.com/archive/ 1201722/not-just-yet-for-no-first-use/ (accessed Apr. 27, 2017). Krepon argued here that although there were almost no good arguments against US adoption of a no-first-use policy, one argument — timing — had some merit. He wrote that the actions of Soviet leader Vladimir Putin and the statements of candidate Donald Trump as

of mid-2016 made it a bad time to pressure President Obama to declare, before the end of his term in office, an American commitment to no first use.

190. NATO, "Defence and Deterrence: Clause 17," *Active Engagement, Modern Defence: Strategic Concept for the Defence and Security of the Members of the North Atlantic Treaty Organization,* Nov. 19–20, 2010, 14, www.nato.int/nato_static_fl2014/assets/pdf/pdf_publications/20120214_strategic-concept-2010-eng.pdf. Re global developments, see, e.g., *Perspectives on the Evolving Nuclear Order,* ed. Toby Dalton, Togzhan Kassenova, and Lauryn Williams (Washington, DC: Carnegie Endowment for International Peace, 2016), carnegie endowment.org/files/NuclearPerspectives_final.pdf; "Pakistan: Nuclear," Nuclear Threat Initiative, Apr. 2016, www.nti.org/learn/countries/pakistan/nuclear/ (accessed Aug. 22, 2016); Rick Gladstone, "A Treaty Is Reached to Ban Nuclear Arms. Now Comes the Hard Part," *New York Times,* July 7, 2017; UN General Assembly, "Draft Treaty on the Prohibition of Nuclear Weapons," A/CONF.229/2017/L.3/Rev.1, limited distribution, July 6, 2017, www.undocs.org/en/a/conf.229/2017/L.3/Rev.1 (accessed Aug. 7, 2017).

191. Ramesh Thakur, "Why Obama Should

Declare a No-First-Use Policy for Nuclear Weapons," *Bulletin of the Atomic Scientists,* Aug. 19, 2016, thebulletin.org/why-obama -should-declare-no-first-use-policy-nuclear -weapons9789 (accessed Aug. 21, 2016). Thakur is a co-convener of the Asia Pacific Leadership Network for Nuclear Non-Proliferation and Disarmament.

192. "Timeline," *Bull. Atomic Scientists,* thebulletin.org/timeline; Science and Security Board, "It is two and a half minutes to midnight: 2017 Doomsday Clock Statement," *Bull. Atomic Scientists,* Jan. 26, 2017, thebulletin.org/sites/default/files/Final% 202017%20Clock%20Statement.pdf (accessed Apr. 27, 2017); Science and Security Board, "Statement from the President and CEO: It Is Now Two Minutes to Midnight," *Bull. Atomic Scientists,* Jan. 25, 2018, thebulletin.org/2018-doomsday-clock -statement (accessed Jan. 25, 2018).

193. Bruce M. DeBlois, "The Advent of Space Weapons," *Astropolitics* 1:1 (Spring 2003), 36.

194. Mizin, "Non-Weaponization of Outer Space," 58.

195. Johnson-Freese, *Heavenly Ambitions,* 35; see also 119–32.

196. See, e.g., *Outer Space,* ed. Arbatov and Dvorkin, 72–110: Oznobishchev, "Codes of Conduct," 72–73; Alexei Arbatov, "Preventing an Arms Race in Space," 79–102; Alexei

Arbatov and Vladimir Dvorkin, "Conclusion," 103–10.

197. See the presentation of the treaty, preceded by a lengthy discussion of issues from the US point of view, at www.congress.gov/105/cdoc/tdoc28/CDOC-105tdoc28.pdf (accessed Sept. 29, 2016).

198. See Mary Beth D. Nikitin, "Comprehensive Nuclear-Test-Ban Treaty: Background and Current Developments," Congressional Research Service RL-33458, Sept. 1, 2016, www.fas.org/sgp/crs/nuke/RL33548.pdf (accessed Sept. 29, 2016); "Senate Holds First Hearing on Comprehensive Nuclear-Test-Ban Treaty Since 1999," *FYI: The AIP Bulletin of Science Policy News* 106, Sept. 8, 2016; "UN Resolution on Nuclear-Test-Ban Treaty Spurs Debate on Treaty's Merits," *FYI: The AIP Bulletin of Science Policy News* 120, Sept. 27, 2016. As of early 2018, India, North Korea, and Pakistan have neither signed nor ratified the CTBT; China, Egypt, Iran, Israel, and the United States have signed but not ratified.

199. Bruce Cumings interviewed by Amy Goodman, "On Asia Trip, Trump Met by Protests Calling on U.S. to Open Diplomatic Relations with North Korea," *Democracy Now!,* Nov. 10, 2017, transcript at www.democracynow.org/2017/11/10/on_asia

_trip_trump_met_by (accessed Mar. 6, 2018).

200. Lewis, *It Can't Happen Here,* 7, 9.

8. Space Power

1. George Orwell, *Nineteen Eighty-Four* (New York: Houghton Mifflin Harcourt, 1949), Kindle loc. 3747–52; Kimiko de Freytas-Tamura, "George Orwell's '1984' Is Suddenly a Best-Seller," *New York Times,* Jan. 25, 2017. Three weeks later, the book was still number one on Amazon's lists of both classic and contemporary literature and fiction, political fiction, satire, and dystopian science fiction. Former president George W. Bush weighed in on the nature of power in February 2017 on national TV, one month into the Trump presidency, as the new administration began to ignore long-standing practices of press access and freedom and to repeatedly vilify journalism as "fake news" and an "enemy of the people": "We need the media to hold people like me to account. I mean, power can be very addictive and it can be corrosive and it's important for the media to call to account people who abuse their power, whether it be here or elsewhere." Peter Baker, "Former President George W. Bush Levels Tacit Criticism at Trump," *New York Times,* Feb. 27, 2017.

2. Commission on the Future of the United States Aerospace Industry, *Anyone, Anything, Anywhere, Anytime: Final Report,* Dec. 2002, 3–1, history.nasa.gov/AeroCommis sionFinalReport.pdf (accessed Dec. 17, 2016).

3. Kevin Pollpeter, Eric Anderson, Jordan Wilson, and Fan Yang, *China Dream, Space Dream: China's Progress in Space Technologies and Implications for the United States* (Washington, DC: IGCC/US–China Economic and Security Review Commission, 2015), 5, 7, and generally 1–7 for China's approach to power, www.uscc.gov/sites/ default/files/Research/China%20Dream% 20Space%20Dream_Report.pdf (accessed Nov. 18, 2016). Also see James Clay Moltz, "China's Space Technology: International Dynamics and Implications for the United States," testimony at US–China Economic and Security Review Commission hearing, May 11, 2011, www.uscc.gov/sites/default/ files/5.11.11Moltz.pdf (accessed Nov. 21, 2016). Re white papers, see Information Office of the State Council, People's Republic of China, "China's Space Activities in 2006 — Preface," Oct. 2006, www.china .org.cn/english/features/book/183672.htm; Information Office of the State Council, People's Republic of China, "China's Space Activities in 2011 — Preface," Dec. 2011,

news.xinhuanet.com/english/china/2011 -12/29/c_131333479.htm; Information Office of the State Council, People's Republic of China, "China's Military Strategy — I. National Security Situation," May 2015, *China Daily,* www.chinadaily.com.cn/china/ 2015-05/26/content_20820628.htm (accessed Dec. 16, 2016); State Council Information Office of the People's Republic of China, "China's Space Activities in 2016 — Preamble," *Global Times*/Xinhua, Dec. 27, 2016, www.globaltimes.cn/content/ 1025893.shtml (accessed Jan. 8, 2017).

4. John F. Kennedy, "Address at Rice University on the Nation's Space Effort," transcript, Sept. 12, 1962, John F. Kennedy Presidential Library and Museum, www .jfklibrary.org/Asset-Viewer/MkATdOcd U06X5uNHbmqm1Q.aspx (accessed Apr. 29, 2017).

5. Robert C. Seamans Jr., *Project Apollo: The Tough Decisions,* Monographs in Aerospace History 37, SP-2005-4537 (Washington, DC: NASA History Division, 2007), 45, history.nasa.gov/monograph37.pdf (accessed Apr. 29, 2017).

6. Richard W. Orloff, *Apollo by the Numbers: A Statistical Reference* (Washington, DC: NASA History Division, 2005), history.nasa .gov/SP-4029/Apollo_18-16_Apollo _Program _Budget_Appropriations.htm.

The 1961 budget for the Apollo Moon landing program was $1 million; the next year it jumped to $160 million and then quadrupled in both of the next two years (accessed Apr. 29, 2017). In constant 2010 dollars NASA funding has reached $20 billion three times since the 1970s: 1991–93. See "Appendix C: A Half Century of NASA Spending 1959–2010: NASA Outlays in Relation to Total U.S. Federal Government Outlays and to GDP," in Neil deGrasse Tyson, *Space Chronicles: Facing the Ultimate Frontier,* ed. Avis Lang (New York: W. W. Norton, 2012), 331–32. During the George W. Bush and Barack Obama presidencies (2001–16), NASA's funding amounted to about a third less than the allocation to the Department of Energy and about twice that to the Environmental Protection Agency. For details on funding by agency, see Office of Management and Budget, "Table 4.1 — Outlays by Agency: 1962–2022" and "Table 4.2 — Percentage Distribution of Outlays by Agency: 1962–2022," n.p., in ' "Introduction to the Historical Tables: Structure, Coverage, and Concepts," www.whitehouse.gov/sites/whitehouse.gov/files/omb/budget/fy2018/hist.pdf (accessed Aug. 13, 2017).

7. Space Foundation, *The Space Report 2013: The Authoritative Guide to Global Space Ac-*

tivity (Colorado Springs: Space Foundation, 2013), 1; *Space Report 2016,* 37; *Space Report 2017,* 16. See Exhibit "Global Space Activity Revenues and Budgets" in *Space Reports 2013* through *2016* and "A Snapshot: The Global Space Economy in 2016 —Total $329.306 B," in *Space Report 2017,* 16. See also *Space Report 2010,* 50; *Space Report 2011,* 55. For years prior to 2010 the estimates for non-US military space spending are based on fewer sources and do not include China; for more recent years, China is included, but the dollar amounts are estimates, as China does not make exact figures public. Nevertheless, it is worth noting the Space Foundation's statement that in 2008 an estimated "95% of the worldwide government spending on defense-related space programs occurred in the United States" (*Space Report 2010,* 50).

8. Formerly it was common for documents and institutions to pair space with air, suggesting a continuum, as in *Air and Space Power in the New Millennium,* ed. Daniel Gouré and Christopher M. Szara (Washington, DC: Center for Strategic & International Studies, 1997), or the Smithsonian National Air and Space Museum.

9. US Air Force, *Space Operations: Air Force Doctrine Document 2-2,* Nov. 27, 2006, 1, 6, 35, www.globalsecurity.org/jhtml/jframe

.html#http://www.globalsecurity.org/
military/library/policy/usaf/afdd/2-2/afdd2-2
-2006pdf|||AFDD%202-2:%20Space%
20Operations (accessed Apr. 29, 2017).

10. US Air Force, *Counterspace Operations:
Air Force Doctrine Document 2-2.1,* Aug. 2,
2004, vii, 27, 33–34, 40, www.globalsecurity
.org/jhtml/jframe.html#http://www.global
security.org/military/library/policy/usaf/
afdd/2-2-1/afdd2-2-1.pdf|||AFDD%202-2
.1:%20Counterspace%20Operations (accessed Apr. 29, 2017).

11. US Air Force, *Space Operations,* 7; Joint
Chiefs of Staff, "The National Military
Strategy of the United States of America
2015," June 2015, 3, www.jcs.mil/Portals/
36/Documents/Publications/2015_National
_Military_Strategy.pdf; Tyrone C. Marshall
Jr., "Officials Update Congress on Military
Space Policy, Challenges," American Forces
Press Service, DoD News, Mar. 12, 2014,
archive.defense.gov/news/newsarticle.aspx
?id=121826; Jim Garamone, "Stratcom
Chief: U.S. Must Maintain Space Dominance," DoD News, Feb. 6, 2015, archive
.defense.gov/news/newsarticle.aspx?id
=128130 (accessed Nov. 13, 2016).

12. General John E. Hyten, "Space Mission
Force: Developing Space Warfighters for
Tomorrow," white paper, US Air Force
Space Command, June 29, 2016, 2–3, 5,
www.afspc.af.mil/Portals/3/documents/

White%20Paper%20-%20Space%20
Mission%20Force/AFSPC%20SMF%20
White%20Paper%20-%20FINAL%20-%
20AFSPC%20CC%20Approved%20on%
20June%2029.pdf?ver=2016-07-19-09
5254-887 (accessed Apr. 29, 2017).

13. Marcia S. Smith, "Top Air Force Officials: Space Now Is a Warfighting Domain," SpacePolicyOnline.com, May 17, 2017, www.spacepolicyonline.com/news/top-air-force-officials-space-now-is-a-warfighting-domain. See also US Government Accountability Office, "DOD Space Acquisition Management and Oversight: Information Presented to Congressional Committees," GAO-16-592R, July 27, 2016, www.gao.gov/assets/680/678697.pdf (accessed Nov. 21, 2017).

14. Council of the European Union, "Implementation Plan on Security and Defence," Nov. 14, 2016, 14, 30; "Shared Vision, Common Action: A Stronger Europe — A Global Strategy for the European Union's Foreign and Security Policy," June 2016, 4, 44; both at www.consilium.europa.eu/en/press/press-releases/2016/11/14-conclusions-eu-global-strategy-security-defence/ (accessed Apr. 29, 2017).

15. European Commission, "Space Strategy for Europe," COM(2016) 705, Oct. 26, 2016, 5, 11, ec.europa.eu/docsroom/documents/19442 (accessed Apr. 29, 2017).

16. Michael Sheehan, *International Politics of Space* (London: Routledge, 2007), 72–90; Joan Johnson-Freese, *Space as a Strategic Asset* (New York: Columbia University Press, 2007), 169–96.

17. European Defence Agency, "Latest News: EU and US Government Defense Spending," news release, Jan. 25, 2012, www.eda.europa.eu/info-hub/press-centre/latest-news/12-0125/EU_and_US_government_Defence_spending; Zoe Stanley-Lockman and Katharina Wolf, "European Defence Spending 2015: The Force Awakens," European Union Institute for Security Studies — *Brief Issue* 10 (Mar. 2016), 1–2, www.iss.europa.eu/uploads/media/Brief_10_Defence_spending. pdf (accessed Apr. 29, 2017). In 2015, when the EU countries' military spending totaled just over €203 billion, Asia spent €277 billion.

18. See, e.g., Glenn Kessler, "Fact Checker: Trump's Claim That the U.S. Pays the 'Lion's Share' for NATO," *Washington Post,* Mar. 30, 2016; Michael R. Gordon and Niraj Chokshi, "Trump Criticizes NATO and Hopes for 'Good Deals' with Russia," *New York Times,* Jan. 15, 2017.

19. European Commission, "Space Strategy for Europe."

20. European Commission, "New Commission Space Policy Puts Focus on Improving

People's Daily Lives and Boosting Europe's Competitiveness," fact sheet/press release, Oct. 26, 2016, europa.eu/rapid/press-release _MEMO-16-3531_en.htm (accessed Apr. 30, 2017).

21. See, e.g., European Space Agency (ESA), "Ministerial Council 2016: What Is Space 4.0?" www.esa.int/About_Us/Ministerial _Council_2016/ What_is_space_4.0; ESA, "Media Backgrounder: ESA's Ministerial 2016 in Lucerne," press release, Nov. 14, 2016, www.esa.int/For_Media/Press _Releases/Media_backgrounder_ESA_s _Ministerial_2016_in_Lucerne; Jan Wörner, " 'Space 4.0' Can Help EU Overcome Its Challenges," *Parliament Magazine,* Mar. 4, 2016, www.theparliamentmagazine .eu/articles/opinion/ space-40-can-help-eu -overcome-its-challenges; ESA, "Council Meeting Held at Ministerial Level on 1 and 2 December 2016: Resolutions and Main Decisions," Dec. 2, 2016, esamultimedia .esa.int/docs/corporate/ For_Public_Release _CM-16_Resolutions_and_Decisions.pdf (accessed Apr. 30, 2017).

22. US Central Command, "Operation Desert Shield/Desert Storm: Executive Summary," July 11, 1991, 1–2, nsarchive.gwu .edu/NSAEBB/NSAEBB39/document6 .pdf. During the first night of Desert Storm, a total of 668 aircraft attacked Iraq, 530 of which were from the US Air Force

and another 90 from US Navy carriers and the US Marine Corps. Britain contributed two dozen aircraft to the attack, France and Saudi Arabia one dozen each. See Airpower Research Institute: College of Aerospace Doctrine, Research and Education, "Airpower in the Gulf War," *Essays on Air and Space Power,* vol. II (Maxwell AFB, AL: Air University Press, 1997), 69, 72. See also Everett C. Dolman, *Astropolitik: Classical Geopolitics in the Space Age* (London: Frank Cass, 2002), 152; Steven J. Bruger, "Not Ready for the 'First Space War,' What About the Second?" Operations Department, Naval War College, May 17, 1993, 1, ii, www.dtic.mil/dtic/tr/fulltext/u2/a266557 .pdf; Department of Defense, *Conduct of the Persian Gulf War: Final Report to Congress,* Apr. 1992, 18, 227–28, 642–61 [reprint pagination], www.ssi.army.mil/ !Library/Desert%20Shield-Desert% 20Storm%20Battle%20Analysis/Conduct% 20of%20the%20Persian%20Gulf%20 War%20-%20Final%20Rpt%20to%20 Congress.pdf (accessed Apr. 30, 2017).

23. US Space Command, "Operations Desert Shield and Desert Storm: Assessment," Jan. 1992, 2, nsarchive.gwu.edu/NSAEBB/ NSAEBB39/document10.pdf (accessed Apr. 30, 2017).

24. Sir Peter Anson and Dennis Cummings,

"The First Space War: The Contribution of Satellites to the Gulf War," *RUSI Journal* 136:4 (Winter 1991), 45; US Department of Defense, *Conduct of the Persian Gulf War,* 26.

25. Prior to Iraq's 1990 invasion and temporary annexation of Kuwait, which triggered the 1991 attack by the US-led Coalition, Iraq had attempted "aggressive, provocative campaigns to attach Kuwait to Iraq" in 1938–41 and 1961–63; see Robert G. Landen, "Review: *Kuwait and Iraq: Historical Claims and Territorial Disputes,* by Richard Schofield," *Middle East Studies Association Bulletin* 26:2 (Dec. 1992), 221–22. For a fuller historical account, see Peter Sluglett, "The Resilience of a Frontier: Ottoman and Iraqi Claims to Kuwait, 1871–1990," *Int. History Rev.* 24:4 (Dec. 2002), 783–816.

26. See GPS.gov, "GPS Accuracy," www.gps .gov/systems/ gps/performance/accuracy/; "Augmentation Systems," www.gps.gov/ systems/augmentations/. In late April 2016, Aerospace Corp. recorded the most accurate positioning to date: thirty-eight centimeters, or fifteen inches. Julius Delos Reyes, "GPS Registers Most Accurate Signal Yet," US Air Force News, www.af .mil/News/ArticleDisplay/tabid/223/Article/ 757533/gps-registers-most-accurate-signal -yet.aspx (accessed Oct. 29, 2016).

27. Space and Missile Systems Center and SMC History Office, "Evolution of GPS: From Desert Storm to Today's Users," US Air Force News, Mar. 24, 2016, www.af.mil/ News/ArticleDisplay/tabid/223/Article/ 703894/evolution-of-gps-from-desert-storm -to-todays-users.aspx; "Desert Storm: The First SpaceWar," Gray Space and the War fighter, Project 1997-0563, www.au.af.mil/ au/awc/awcgate/grayspc/dstorm/dstorm .htm; James Drew, "Boeing B-52 Evolves Again with Guided Weapons Launcher," FlightGlobal.com, Jan. 15, 2016, www .flightglobal.com/news/articles/boeing-b-52 -evolves-again-with-guided-weapons-launch -420874/; "AGM-86C/D Conventional Air Launched Cruise Missile," Federation of American Scientists, fas.org/nuke/guide/usa/ bomber/calcm.htm; Benjamin Raughton, "Desert Storm: 2nd Bomb Wing Leads the Air War," Barksdale AFB News, Jan. 14, 2016, www.barksdale.af.mil/News/ ArticleDisplay/tabid/2668/Article/641881/ desert-storm-2nd-bomb-wing-leads-the-air -war.aspx; Kris Osborn, "Stealth, GPS, 'Smart Bomb' and More: How Desert Storm Changed Warfare Forever," *National Interest,* Nov. 21, 2016, nationalinterest.org/ blog/the-buzz/stealth-gps-smart-bombs -more-how-desert-storm-changed-war -18477. (All accessed May 1, 2017.)

Regarding the hit rate, the historian of

the 37th Tactical Fighter Wing writes, "Statistically the [37th TFW] compiled a record that is unparalleled in the chronicals [*sic*] of air warfare: the 'Nighthawks' achieved a 75 percent hit rate on pinpoint targets (1669 direct hits and 418 misses) while crippling nearly 40 percent of enemy strategic targets." See Harold P. Myers, "Nighthawks over Iraq: A Chronology of the F-117A Stealth Fighter in Operations Desert Shield and Desert Storm — Special Study 37FW/HO-91-1," Office of History, Headquarters 37th Fighter Wing, Twelfth Air Force, Tactical Air Command, Jan. 9, 1992, 3–4, nsarchive.gwu.edu/NSAEBB/NSAEBB39/document9.pdf (accessed Jan. 17, 2017). The Department of Defense, in its title V report to Congress, put the hit rate at 80 percent. In 1993 the contractor, Lockheed Martin, went so far as to say "one bomb = one kill." As a follow-up to such claims, in mid-1997 the National Security and Internal Affairs Division of the Government Accountability Office produced an unclassified version of its report to Congress in which it summarizes its overall finding regarding the F-117A as follows: "[T]he F-117 bomb hit rate ranged between 41 and 60 percent — which is considered to be highly effective, but is still less than the 80-percent hit rate reported after the war by DOD, the Air Force, and the primary contractor." The report subsequently presents an extremely detailed breakdown of claims, corroborations, and conclusions. The 41–60 percent figure was derived as follows: "[T]he expected probability of a target's being damaged to the desired level would be based on the number of bombs tasked, reduced by the proven probability of bomb release (75 percent), and reduced further by the demonstrated hit rate (between 55 and 80 percent). Therefore, in Desert Storm, the probability of a target's receiving damage from a scheduled F-117 strike (that is, the probability of bomb release times the demonstrated hit rate) was between 41 and 60 percent." See Government Accountability Office, *Operation Desert Storm: Evaluation of the Air Campaign,* GAO/NSAID-97-134, June 1997, 1, 110, 125–38, 225–26, and passim, www.gao.gov/archive/1997/ns97134 .pdf (accessed Jan. 17, 2017).

For a recent, on-the-ground investigation of other issues associated with hit rates — such as civilian deaths and the identification of targets — see Azmat Khan and Anand Gopal, "The Uncounted," *New York Times Magazine,* Nov. 16, 2017, which examines Coalition air strikes intended to push ISIS out of Iraq and Syria during 2014–17. While a spokesman for Central

Command characterized the campaign as "one of the most precise air campaigns in military history," Khan and Gopal found that 20 percent of the 103 air strikes they exhaustively investigated resulted in civilian deaths, a rate more than thirty times that acknowledged by the Coalition. They suggest two primary sources of civilian deaths: proximity to an actual ISIS target and questionable intelligence. "[I]n about half of the strikes that killed civilians," they write, "we could find no discernible ISIS target nearby. Many of these strikes appear to have been based on poor or outdated intelligence."

28. Jamming adds noise to the already weak signals. Iraqi forces installed jammers atop palaces and other landmarks. Larry Greenemeier, "GPS and the World's First 'Space War,' " *Scientific American,* Feb. 8, 2016, www.scientificamerican.com/article/gps-and -the-world-s-first-space-war/ (accessed Apr. 30, 2017).

29. Figures for deployed GPS receivers are from US Department of Defense, *Conduct of the Persian Gulf War,* 678. Other figures are from US Central Command, "Operation Desert Shield/Desert Storm: Executive Summary," 1, and US General Accounting Office, "Desert Storm: Air Campaign," 14. Source of quotation: Space and Missile Systems Center, "Evolution of GPS."

30. Greenemeier, "GPS and First 'Space War' "; Sam Jones, "Satellite Wars," *Financial Times,* Nov. 20, 2015, www.ft.com/cms/ s/2/637bf054-8e34-11e5-8be4-3506bf 20cc2b.html#ixzz3tDtUkpkq; Marcia S. Smith, "Military and Civilian Satellites in Support of Allied Forces in the Persian Gulf War," Congressional Research Service, Feb. 27, 1991, www.hsdl.org/?view&did= 712697 (accessed Apr. 30, 2017); Bruger, "Not Ready for 'First Space War,' " 13; Andrew Pollack, "War Spurs Navigation by Satellite," *New York Times,* Feb. 6, 1991.

31. Malcolm W. Browne, "New Space Beacons Replace the Compass," *New York Times,* Nov. 8, 1988.

32. Extended quotation from Colin S. Gray, "The Influence of Space Power upon History," *Comparative Strategy* 15:4 (1996), 303. Several years later, Gray published a response, as it were, to our comment about weakening: "Clausewitz Rules, OK? The Future Is the Past: With GPS," *Rev. Int. Studies* 25 (Dec. 1999), 161–82. Here he asserted the primacy of strategy over technology and thus the minor contribution of all technological innovation: "[T]he game of polities (or security communities) does not change from age to age, let alone from decade to decade. . . . Whether humans navigate by the stars or via the satellites of

the US Global Positioning System (GPS), and whether they communicate by smoke signals or via space vehicles, matters not at all for the permanent nature of strategy" (163, 182).

33. "[George W.] Bush's Speech on the Start of War," transcript, *New York Times,* Mar. 20, 2003. Notwithstanding Bush's statement of purpose, no evidence for so-called WMDs had been detected, even after extensive satellite reconnaissance and extensive searches by UN weapons inspectors. "Nuclear Inspection Chief Reports Finding No New Weapons," transcript, *New York Times,* Jan. 28, 2003.

34. GPS.gov, "Selective Availability," www.gps.gov/systems/ gps/modernization/sa/; "Data from the First Week Without Selective Availability: GPS Fluctuations Over Time on May 2, 2000," www.gps.gov/systems/gps/modernization/sa/data/. Reported maximum accuracy varies from 2.66 meters (just under nine feet) to 2.2 meters (just over seven feet); Bob Brewin, "Pentagon Tweaked GPS Accuracy to Within Three Meters During Iraq War," *Computerworld,* June 24, 2003, www.computerworld.com/article/2569842/mobile-wireless/pentagon-tweaked-gps-accuracy-to-within-three-meters-during-iraq-war.html (accessed Apr. 30, 2017); William B. Scott

and Craig Covault, "High Ground over Iraq," *Aviation Week & Space Technology* 158:23 (June 9, 2003), 44–48.

35. Phillip Swarts, "SpaceX's Low Cost Won GPS 3 Launch, Air Force Says," *Space-News,* Mar. 15, 2017, spacenews.com/spacexs-low-cost-won-gps-3-launch-air-force-says; GPS.gov, "Space Segment," www.gps.gov/systems/gps/space; GPS.gov, "Program Funding," www.gps.gov/policy/funding (accessed Apr. 30, 2017). The Consolidated Appropriations Act of 2016 allocated $937 million to the DoD (for procurement and development) and some $130 million to the DoT (for support of civilian R&D programs such as the Wide Area Augmentation System and Alternative Positioning, Navigation, and Timing).

36. The angle of GLONASS is 64.8 degrees; that of GPS is 55 degrees. Up-to-date GLONASS status is at Information and Analysis Center for Position, Navigation and Timing, "GLONASS Constellation Status," www.glonass-iac.ru/en/GLONASS/ (accessed Apr. 30, 2017).

37. RIA Novosti, "GLONASS: Dispelling the Myths Around Russia's GPS," interview, Jan. 10, 2014, Russia Behind the Headlines, rbth.com/science_and_tech/2014/01/10/glonass_dispelling_the_myths_around_russias_gps_33183.html. The US position on denial is publicly stated as follows: "It is

U.S. policy to prevent hostile use of GPS through localized denial (i.e., military jamming) that does not unduly disrupt civil and commercial GPS access outside the battlefield." GPS.gov, "United States Policy," www.gps.gov/policy. See also Beebom, "What Is GLONASS and How It Is Different from GPS," Aug. 25, 2016, beebom.com/what-is-glonass-and-how-it-is-different-from-gps (accessed Nov. 7, 2016).

38. Defense Advanced Research Projects Agency, "About DARPA," www.darpa.mil/about-us/about-darpa; Robert Lutwak, "Atomic Clock with Enhanced Stability (ACES)," DARPA, www.darpa.mil/program/atomic-clock-with-enhanced-stability (accessed Apr. 30, 2017).

39. European Space Agency, "Atomic Clock Ensemble in Space (ACES)," flyer, Sept. 2011, wsn.spaceflight.esa.int/docs/others/aces_flyer.pdf; European Space Agency, "Atomic Clock Ensemble in Space (ACES)," fact sheet, wsn.spaceflight.esa.int/docs/Factsheets/20%20ACES%20LR.pdf (accessed Apr. 30, 2017); Greenemeier, "GPS and First 'Space War.'"

40. Anson and Cummings, "The First Space War," 45.

41. Bruger, "Not Ready for 'First Space War,'" 7; Anson and Cummings, "The First Space War," 45–48. See also, e.g.,

"Desert Storm: The First Space War — Gray Space"; US Space Command, "Desert Shield and Desert Storm," 47–54; Smith, "Military and Civilian Satellites," CRS-1–3; US Department of Defense, *Conduct of the Persian Gulf War,* 873–75; James A. Walker, Lewis Bernstein, and Sharon Lang, Historical Office, US Army Space and Missile Defense Command, *Seize the High Ground: The Army in Space and Missile Defense* (Washington, DC: US Government Printing Office, 2003), 156–57; Stephen Cass, "Legendary U.S. Satellite Put Out to Pasture," *MIT Technology Review,* Oct. 14, 2009, www.technologyreview.com/s/415716/legendary-us-satellite-put-out-to-pasture/ (accessed Feb. 7, 2017). Cass makes the point that NASA was not in fact the Tracking and Data Relay Satellite system's priority user: "Although it was never advertised, the biggest users of the TDRS constellation weren't NASA astronauts and scientists, but the military and the National Reconnaisance Office, who had priority use of the system for keeping in touch with their spy satellites. This occasionally caused frustration for scientific users of the system, especially during tense geopolitical moments [such as] the first Gulf War."

42. Smith, "Military and Civilian Satellites,"

CRS-10.

43. US Department of Defense, *Conduct of the Persian Gulf War,* 219–20, 871–73; US Space Command, "Desert Shield and Desert Storm," 33–38; Smith, "Military and Civilian Satellites," CRS-6; Anson and Cummings, "The First Space War," 51–52; Walker et al., *Seize the High Ground,* 153.

44. US Space Command, "Desert Shield and Desert Storm," 40.

45. US Space Command, "Desert Shield and Desert Storm," 39–46; Craig Covault, "Recon Satellites Lead Allied Intelligence Effort," *Aviation Week & Space Technology,* Feb. 4, 1991, 25–26; Anson and Cummings, "The First Space War," 50–53; Smith, "Military and Civilian Satellites," CRS-7–10; US Department of Defense, *Conduct of the Persian Gulf War,* 877–78, 652–53; "Desert Storm: The First Space War — Gray Space"; Walker et al., *Seize the High Ground,* 154; Alan Riding, "After the War; France Concedes Its Faults in War," *New York Times,* May 8, 1991.

46. Craig Covault, "Desert Storm Reinforces Military Space Directions," *Aviation Week & Space Technology,* Apr. 8, 1991, 42–47; Vice President's Space Policy Advisory Board, "The Future of the U.S. Space Industrial Base: A Task Group Report," Nov. 1992, vi, history.nasa.gov/33081.pt1

.pdf (accessed May 1, 2017).

47. Anthony H. Cordesman, *The Iraq War: Strategy, Tactics, and Military Lessons* (Westport, CT: Praeger/Center for Strategic and International Studies, 2003), 8, 184, 199–200; William B. Scott and Craig Covault, "High Ground over Iraq," *Aviation Week & Space Technology* 158:23 (June 9, 2003), 44; Paul Wolfowitz, "Testimony on U.S. Military Presence in Iraq: Implications for Global Defense Posture," as prepared for delivery to the House Armed Services Committee, Washington, DC, June 18, 2003, available at GlobalSecurity.org, www.globalsecurity.org/wmd/library/news/iraq/2003/06/iraq-030618-dod03.htm (accessed Feb. 18, 2017).

48. Cordesman, *Iraq War,* 199, 195–96.

49. Fred Kaplan, "The End of the Age of Petraeus: The Rise and Fall of Counterinsurgency," *Foreign Affairs* 92:1 (Jan.–Feb. 2013), 85, 88; Phil Klay, "Money as a Weapons System," in *Redeployment* (New York: Penguin, 2014; winner of the 2014 National Book Award for Fiction), 78; Cordesman, *Iraq War,* 235, 217. Re "victory," see, e.g., Max Boot, "The New American Way of War," *Foreign Affairs* 82:4 (July–Aug. 2003), 41–58. For Boot as for Rumsfeld and so many others at the time, the entire war was over and done with. The

Western allies "won so quickly"; it was "one of the signal achievements in military history . . . a spectacular success" (44). His article ends on a celebratory note: "the victory in Iraq shows that the military is making impressive progress toward making the American way of war more effective and more humane" (58). For an ongoing account of documented violent incidents, see Iraq Body Count, www.iraqbodycount.org/. The issues re Iraq's schools are exemplified by the situation in the city of Mosul: a decade and a half after the supposed victory, National Public Radio and UNICEF had grounds to celebrate the reopening of seventy — less than one-fifth — of Mosul's four hundred public schools, all of which had been closed for two years; report by Alice Fordham, *All Things Considered,* NPR, Feb. 16, 2017. For archaeological uses of satellite data on Iraq, see, e.g., Kristin Romey, "Iconic Ancient Sites Ravaged in ISIS's Last Stand in Iraq," *National Geographic,* Nov. 10, 2016, news.national geographic.com/2016/11/iraq-mosul-isis -nimrud-khorsabad-archaeology/ (accessed Feb. 16, 2017).

50. Craig Covault, "Fade to Black," *Aviation Week & Space Technology* 164:20, May 15, 2006, 24–26; David Talbot, "How Technology Failed in Iraq," *MIT Technology Review,*

Nov. 2004, www.technologyreview.com/s/403319/how-technology-failed-in-iraq/ (accessed Feb. 19, 2017).

51. *Space Report 2017*, 1–2, 9, 14; *Space Report 2016*, 1–24; World Bank, "World Development Indicators Database: Gross Domestic Product 2015," Oct. 11, 2016, 1, 4, databank.worldbank.org/data/download/GDP.pdf (accessed Nov. 20, 2016). Examples of 2016 civil space spending by the space agencies of countries with struggling economies: Argentina ($125 million), Brazil ($50 million), Bolivia ($31 million), Nigeria ($31 million).

52. Johnson-Freese, *Space as a Strategic Asset*, 232.

53. Pollpeter et al., *China Dream*, 4, 8; Sheehan, *International Politics of Space*, 142, 147–52, 161; Wu Ji et al., "Prospect for Chinese Space Science in 2016–2030," *Bull. of Chinese Academy of Sciences* 30:6 (2015), English abstract at www.bulletin.cas.cn/publish_article/2015/6/20150601.htm. The abstract elaborates on the agenda thus: "a series of scientific satellite programs and missions in frontier scientific fields, such as the formation and evolution of the universe, the exploration of exoplanets and extraterrestrial life, the formation and evolution of the solar system, solar activities and their impact on the earth's space environment,

the development and evolution of the earth's system, new physics beyond the current basic physics theories, the law of matter motion and the law of life activity in space environment, etc; and to drive the great-leap-forward of aerospace and related high technologies." See also, e.g., Edward Wong, "China Launches Quantum Satellite in Bid to Pioneer Secure Communications," *New York Times,* Aug. 16, 2016; Mike Wall, "China Launches Pioneering 'Hack-Proof' Quantum-Communications Satellite," Space.com, Aug. 16, 2016, www .space.com/33760-china-launches-quantum -communications-satellite.html (accessed May 1, 2017).

54. "Dr. Vikram Ambalal Sarabhai," Indian Space Research Organisation, www.isro.gov .in/about-isro/dr-vikram-ambalal-sarabhai; T. S. Subramanian, "An ISRO Landmark," *Frontline* 18:23 (Nov. 10–23, 2001), www .frontline.in/navigation/?type=static&page =flonnet&rdurl=fl1823/18230780.htm (accessed Nov. 23, 2016); Ellen Barry, "India Launches 104 Satellites from a Single Rocket, Ramping Up a Space Race," *New York Times,* Feb. 15, 2017; Kai Schultz and Hari Kumar, "India Tests Ballistic Missile, Posing New Threat to China," *New York Times,* Jan. 18, 2018.

55. United Nations Development Programme, "Table 1: Human Development

Index and Its Components," *Human Development Report 2015: Work for Human Development,* 2015, 208–211, hdr.undp.org/sites/default/files/2015_human_development_report.pdf. Life expectancy at birth: Japan 83.5, Canada 82.0, China 75.8, India 68.0. Average years of schooling: Canada 13.0, Japan 11.5, China 7.5, India 5.4. For military spending, see SIPRI Military Expenditure Databases, "Military expenditure by country as percentage of government spending, 1988–2016" and "Military expenditure by country as a share of GDP, 2003–2016," Stockholm International Peace Research Institute, 2017, www.sipri.org/databases/milex (accessed Nov. 26, 2017). SIPRI's exact figures for 2016 military spending as a percentage of government spending were Canada 2.4%, Japan 2.6%, China 6.2%, India 8.9%, USA 9.3%, and Russia 15.5%. Re US 36% share of global military spending, see Niall McCarthy, "The Top 15 Countries for Military Expenditure in 2016 [Infographic]," *Forbes,* Apr. 24, 2017, www.forbes.com/sites/niallmccarthy/2017/04/24/the-top-15-countries-for-military-expenditure-in-2016-infographic/#2036c07843f3 (accessed Nov. 27, 2017).

56. Canadian Space Agency, "Canadian Space Milestones," www.asc-csa.gc.ca/eng/about/milestones.asp, "Canadarm and

Canadarm2 — Comparative Table," www
.asc-csa.gc.ca/eng/iss/canadarm2/c1-c2.asp,
"History of the Canadian Astronaut
Corps," www.asc-csa.gc.ca/eng/astronauts/
canadian/ history-of-the-canadian-astronaut
-corps.asp, "Canadian Science on the
International Space Station," www.asc-csa
.gc.ca/eng/iss/science/default.asp (accessed
Dec. 7, 2016). The Canadian Space Agency
falls within the portfolio of the Minister of
Industry. More physicians than military
personnel have served as Canada's astronauts. Among Canadarm's notable achievements were retrieving the Hubble Space
Telescope for five servicing missions between 1993 and 2009 and the connection
of the first two modules of the International
Space Station in 1998. In 2008, Cana-
darm2 transferred one module of the Japanese ISS space lab Kibo from the space
shuttle to the space station, a maneuver
celebrated in a popular poster.

57. "NORAD History," North American
Aerospace Defense Command, www.norad
.mil/About-NORAD/NORAD-History/;
Colonel T. J. Grant, "Space Policy," Canadian Forces College, Nov. 26, 1998, 3, 19,
21, www.cfc.forces.gc.ca/259/260/261/
grant2.pdf; Max Paris, "Canadian Forces
Put Their 1st Satellite in Orbit," CBC
News, Feb. 25, 2013, www.cbc.ca/news/
politics/canadian-forces-put-their-1st

-satellite-in-orbit-1.1338715; Andre Dupuis, "An Overview of Canadian Military Space in 2014," pt. 1 Feb. 9, pt. 2 Feb. 17, 2015, SpaceRef Canada, spaceref.ca/military-space/an-overview-of-canadian-military-space-in-2014—part-2.html (accessed May 1, 2017).

58. Space Foundation, *Space Report 2017,* 10, 15. The top five space spenders in dollar amounts in 2016 were the United States, the European Space Agency, China, Japan, and Russia. Percentage of GDP spent on space in 2016: Japan .062%, Canada .021% (US .239%; Russia .122%; China .039% avg.).

59. The cabinet-level Office of National Space Policy was created in July 2012. Before then, JAXA was supervised by the Minister of Education, Culture, Sports, Science, and Technology and the Minister of Public Management, Home Affairs, Posts, and Telecommunications. Office of National Space Policy, "Planning Policy of Development and Utilization of Space and the Headquarters for Japanese Space Policy," www.cao.go.jp/en/pmf/pmf_20.pdf; Japan Aerospace Exploration Agency, "JAXA History," global.jaxa.jp/about/history/index.html; "ISAS History," global.jaxa.jp/about/history/isas/index_e.html; "SS-520 Sounding Rockets," ISAS, www.isas.jaxa.jp/e/enterp/rockets/sounding/ss520.shtml;

"Catalogue of ISAS Missions," ISAS, www .isas.jaxa.jp/e/enterp/missions/catalogue .shtml; "Missions: About Our Projects," global.jaxa.jp/projects/; "Japanese Experimental Module (KIBO)," iss.jaxa.jp/en/ kiboexp/ef/ (accessed Dec. 8, 2016).

60. James Clay Moltz, *Asia's Space Race: National Motivations, Regional Rivalries, and International Risks* (New York: Columbia University Press, 2012), 43–69; Paul Kallender, "Japan's New Dual-Use Space Policy: The Long Road to the 21st Century," *Notes de l'Ifri: Asie.Visions* 88 (Nov. 2016), www.ifri.org/sites/default/files/atoms/ files/ japan_space_policy_kallender.pdf; Maeda Sawako, "Transformation of Japanese Space Policy: From the 'Peaceful Use of Space' to 'the Basic Law on Space,' " *Asia-Pacific Journal: Japan Focus* 7:44:1 (Nov. 2009), 1–7, apjjf.org/-Maeda-Sawako/ 3243/ article.pdf; Steven Berner, "Japan's Space Program: A Fork in the Road?" RAND, 2005, www.rand.org/content/dam/ rand/ pubs/technical_reports/2005/RAND _TR184.pdf (accessed May 1, 2017).

61. For Cold War 2.0, see Evan Osnos, David Remnick, and Joshua Yaffa, "Active Measures," *New Yorker,* Mar. 6, 2017, 40–55. "For nearly two decades, U.S.–Russian relations have ranged between strained and miserable," write the authors. "Many Russian and American policy experts no longer

hesitate to use phrases like 'the second Cold War' " (44). For a best-selling in-depth investigation, see Michael Isikoff and David Corn, *Russian Roulette: The Inside Story of Putin's War on America and the Election of Donald Trump* (New York: Twelve/Hachette, 2018).

62. For the saga of Apollo–Soyuz and the decades leading up to it, see Edward Clinton Ezell and Linda Neuman Ezell, *The Partnership: A History of the Apollo–Soyuz Test Project* (Washington, DC: NASA, 1978), history.nasa.gov/SP-4209.pdf. See also the Nixon–Kosygin "Cooperation in Space: Agreement Between the United States of America and the Union of Soviet Socialist Republics Concerning Co-operation in the Exploration of the Use of Outer Space for the Peaceful Purposes, May 24, 1972," www.archives.gov/files/presidential-libraries/events/centennials/nixon/images/exhibit/agreement-of-cooperation.pdf (accessed May 1, 2017). Article I of the agreement commits the two parties to "develop cooperation in the fields of space meteorology; study of the natural environment; exploration of near earth space, the moon and the planets; and space biology and medicine." Article 3 sets the stage for Apollo–Soyuz, specifically the development of "compatible rendezvous

and docking systems of United States and Soviet manned spacecraft and stations in order to enhance the safety of manned flights in space and to provide the opportunity for conducting joint scientific experiments in the future." The agreement was renewable at five-year intervals; President Carter renewed it in 1977, but in 1982 President Reagan let it lapse.

63. William E. Burrows, *This New Ocean: The Story of the First Space Age* (New York: Random House, 1998), 585. For his portrayal of the vicissitudes of the Gorbachev and Yeltsin space programs, see chap. 15, "Downsizing Infinity," 551–90.

64. For a minute-by-minute account of Nov. 9, 1989, see, e.g., Laurence Dodds, "Berlin Wall: How the Wall Came Down, As It Happened 25 Years Ago," *Telegraph,* Nov. 9, 2014, www.telegraph.co.uk/history/11219434/Berlin-Wall-How-the-Wall-came-down-as-it-happened-25-years-ago-live.html (accessed Feb. 27, 2017).

65. Igor Filatochev and Roy Bradshaw, "The Soviet Hyperinflation: Its Origins and Impact Throughout the Former Republics," *Soviet Studies* 44:5 (1992), 739–59; Walter LaFeber, *America, Russia, and the Cold War, 1945–2006,* 10th ed. (Boston: McGraw-Hill, 2008), 366–67, 391–93. GNP fell 17 percent in 1991, according to Filatochev

and Bradshaw (742).

66. James Clay Moltz, *The Politics of Space Security: Strategic Restraint and the Pursuit of National Interests* (Stanford: Stanford University Press, 2008), 205, 208, 212.

67. Moltz, *Politics of Space Security,* 204–18. The original source of the locomotive quotation is a 1987 article, "Space Exploration and New Thinking," in *International Affairs* (Moscow). For the tale of Skif, see Dwayne A. Day and Robert G. Kennedy III, "Soviet Star Wars," *Air & Space Smithsonian,* Jan. 2010, www.airspacemag.com/space/soviet-star-wars-8758185/?all (accessed May 1, 2017). Also see Sheehan, *International Politics of Space,* 55–66, for an overview of Soviet space efforts, with an emphasis on space as a domain for diplomacy and cooperation, especially with the Communist bloc and nonaligned nations, during the 1970s and 1980s. Of the name Mir, Sheehan writes that it was intended to create a contrast to "the American effort to militarise and 'weaponise' space through the Strategic Defense Initiative." He further argues that "SDI needed to be challenged symbolically in this way because Gorbachev was aware that a Soviet effort simply to match the American programme would not only be strategically destabilising, but was likely to expose the economic weaknesses

and technological limitations of the USSR" (66).

68. "The Gorbachev Visit; Excerpts from Speech to U.N. on Major Soviet Military Cuts," trans. Soviet Mission, *New York Times*, Dec. 8, 1988. For the disasters of Soyuz TM-5 and Phobos 1, see Burrows, *This New Ocean*, 573–75.

69. Roald Sagdeev, *The Making of a Soviet Scientist: My Adventures in Nuclear Fusion and Space from Stalin to Star Wars* (New York: John Wiley & Sons, 1994), ix, 186–91, 321–24.

70. Figures cited at p. 74 in Kathleen J. Hancock, "Russia: Great Power Image Versus Economic Reality," *Asian Perspective* 31:4 (2007), 71–98. See also LaFeber, *America, Russia*, 388–95; Burrows, *This New Ocean*, 585.

71. Burrows, *This New Ocean*, 586–88. See also Francis X. Clines, "Going-Out-of-Business Sale for Soviets' Space Program," *New York Times*, Aug. 8, 1993.

72. Burrows, *This New Ocean*, 601–609; Moltz, *Politics of Space Security*, 230–33, 240–45, 250–52; Richard Stone, "A Renaissance for Russian Space Science," *Science*, Apr. 7, 2016, www.sciencemag.org/news/2016/04/ renaissance-russian-space-science (accessed May 1, 2017). Burrows includes an especially grim detail re a $300-million,

twenty-nation Mars-bound spacecraft, launched by Russia, that fell into the Pacific because of the failure of the Proton rocket's fourth stage. He writes that "parts of Mars 96 had been integrated at Tyuratam in the glow of kerosene lamps because the Kazakhs had cut off the electricity in exasperation over a pile of unpaid bills" (601).

73. "GLONASS Constellation Status"; Jason Davis, "What's the Matter with Russia's Rockets?" blog, Planetary Society, Dec. 2, 2016, www.planetary.org/blogs/jason-davis/2016/20161201-whats-the-matter-russias-rockets.html; Emma Grey Ellis, "Russia's Space Program Is Blowing Up. So Are Its Rockets," *Wired,* Dec. 7, 2016, www.wired.com/2016/12/russias-space-program-blowing-rockets; Michael Weiss and Pierre Vaux, "How a U.S.-Russian Space Rocket Deal Funds Putin's Cronies," *Daily Beast,* May 31, 2016, www.thedailybeast.com/articles/2016/05/31/the-u-s-violates-its-own-sanctions-to-buy-russian-space-rockets.html; Anatoly Zak, "A Rare Look at the Russian Side of the Space Station," *Air & Space Smithsonian,* Sept. 2015, www.airspacemag.com/space/ rare-look-russian-side-space-station-180956244; Stone, "Renaissance for Russian Space Science"; Space Research Institute of the Russian Academy of Sciences (IKI), www.iki.rssi.ru/eng; Anatoly Zak, "Spektr-RG to Expand

Horizons of X-ray Astronomy," Russian Space Web, Jan. 2017, www.russianspace web.com/spektr_rg.html (accessed Mar. 12, 2017).

74. *Space Report 2016,* 37, 48; *Space Report 2017,* 8; Anatoly Zak, "Russia Approves Its 10-Year Space Strategy," blog post, Planetary Society, Mar. 23, 2016, www.planetary .org/blogs/guest-blogs/2016/0323-russia -space-budget.html; Davis, "What's the Matter with Russia's Rockets?"

75. Vladimir Putin, "Russian President Vladimir Putin State of the Nation Address," C-SPAN, Mar. 1, 2018, simultaneous translation, 1:22: 04–1:38:18, www.c-span.org/ video/?441907-1/russian-president-vladimir -putin-state-nation-address (accessed Mar. 8, 2017). See also, e.g., Andrew Roth, "Putin Threatens US Arms Race with New Missiles Declaration," *Guardian,* Mar. 1, 2018; Neil MacFarquhar and David E. Sanger, "Putin's 'Invincible' Missile Is Aimed at U.S. Vulnerabilities," *New York Times,* Mar. 1, 2018; Anton Troianovski, "Putin Claims Russia Is Developing Nuclear Arms Capable of Avoiding Missile Defenses," *Washington Post,* Mar. 1, 2018; Vladimir Isachenkov, AP, "Putin Shows New Russian Nuclear Weapons: 'It Isn't a Bluff,' " *Washington Post,* Mar. 1, 2018.

76. NASA Advisory Council, *Task Force on*

International Relations in Space, International Space Policy for the 1990s and Beyond (1987), quoted in Johnson-Freese, *Space as a Strategic Asset,* 180.

77. Johnson-Freese, *Space as a Strategic Asset,* 179–82; Zak, "Rare Look at Russian Side."

78. Burrows, *This New Ocean,* 139–46 (quotation at 143), 508. The *Collier's* series (Mar. 1952–Apr. 1954) was collectively titled "Man Will Conquer Space *Soon.*" Burrows writes, "With [Wernher] von Braun as its architect and credible specialists filling in details based on real science and engineering rather than fanciful speculation, the articles constituted a blueprint for the U.S. space program" (144).

79. Ronald Reagan, "Address Before a Joint Session of the Congress on the State of the Union," Jan. 25, 1984, at Gerhard Peters and John T. Woolley, American Presidency Project, www.presidency.ucsb.edu/ws/?pid=40205 (accessed May 2, 2017).

80. See, e.g., Philip M. Boffey, "Higher Cost Predicted for Space Station," *New York Times,* July 7, 1987; William J. Broad, "How the $8 Billion Space Station Became a $120 Billion Showpiece," *New York Times,* June 10, 1990; US General Accounting Office, "Space Station: NASA's Search for Design, Cost, and Schedule Stability Continues,"

GAO/NSAID-91-125, Mar. 1991, www.gao
.gov/assets/160/150248.pdf; J. R. Minkel,
"Is the International Space Station Worth
$100 Billion?" Space.com, Nov. 1, 2010,
www.space.com/9435-international-space
-station-worth-100-billion.html; NASA Office of Inspector General, "Extending the
Operational Life of the International Space
Station Until 2024," audit report, IG-14-
031, Sept. 18, 2014, oig.nasa.gov/audits/
reports/FY14/ IG-14-031.pdf. (All accessed
Nov. 28, 2017.) European Space Agency,
"International Space Station: How Much
Does It Cost," last update May 14, 2013,
www.esa.int/Our_Activities/Human_Space
flight/International_Space_Station/How
_much_does_it_cost), pegged the total cost
covered by all participants — including
development, assembly, and ten years'
worth of operating costs — at €100B as of
2013, which in 2016 US dollars would be
about $140B. The NASA inspector general's audit report pegs the US contribution
as $43.7 billion for construction and program costs through 2013 plus $30.7 billion
for thirty-seven supporting space shuttle
flights, the last of which took place in July
2011 ("Overview," i).

81. Sheehan, *International Politics of Space,*
176–78; Burrows, *This New Ocean,* 591–
98, 606–609; Johnson-Freese, *Space as a
Strategic Asset,* 177–79, 65–67; European

933

Space Agency, "International Space Station Legal Framework," www.esa.int/Our_Activ ities/Human_Spaceflight/International _Space_Station/International_Space _Station_legal_framework (accessed Mar. 21, 2017).

82. Ker Than, "Nobel Laureate Disses NASA's Manned Spaceflight," Space.com, Sept. 18, 2007, www.space.com/4357-nobel -laureate-disses-nasa-manned-spaceflight .html (accessed Nov. 28, 2017).

83. The authors thank political scientist and space-policy analyst John Logsdon, professor emeritus at George Washington University, for supplying this letter to Tyson.

84. Sheehan calls the United States "the hegemonic partner, in the space station as much as in NATO" (*International Politics of Space,* 178–79).

85. See, e.g., Miriam Kramer, "NASA Suspends Most Cooperation with Russia; Space Station Excepted," Space.com, Apr. 2, 2014, www.space.com/25339-nasa -suspends-russia-cooperation-ukraine.html; Stuart Clark, "Russia Halts Rocket Exports to US, Hitting Space and Military Programmes," *Guardian,* May 15, 2014; Reuters, "Russia to Ban US from Using Space Station over Ukraine Sanctions," *Telegraph,* May 13, 2014; Ralph Vartabedian and W. J. Hennigan, "U.S.-Russia Tension Could Affect Space Station, Satellites," *Los*

Angeles Times, May 16, 2014; "Russia Makes Plans to Kill Space Station in 2020 Due to Sanctions," NBC News, May 13, 2014, www.nbcnews.com/storyline/ukraine-crisis/russia-makes-plans-kill-space-station-2020-due-sanctions-n104531; Irene Klotz, "Atlas V Rocket Launches US Missile-Warning Satellite," Space.com, Jan. 20, 2017, www.space.com/35409-missile-warning-satellite-sbirs-geo-3-launch-success.html; Staff writers, Sputnik, "Why Washington Cannot Ban Russia's RD-180 Rocket Engines," SpaceDaily, May 3, 2016, www.spacedaily.com/reports/Why_Washington_cannot_why_Russias_RD_180_rocket_engines_999.html; "Russia to Supply RD-180 Rocket Engines to US in 2017," TASS, Dec. 1, 2016, tass.com/science/915840; Chris Gebhardt, "U.S. Debates Atlas V RD-180 Engine Ban, ULA's Non-Bid for Military Launch," NASASpaceflight.com, Jan. 29, 2016, www.nasaspaceflight.com/2016/01/u-s-debates-atlas-v-rd-180-ban-ulas-non-bid-military; Phil Plait, "Russian Deputy Prime Minister Threatens to Pull Out of ISS," Bad Astronomy blog, *Slate,* May 14, 2014, www.slate.com/blogs/bad_astronomy/2014/05/14/nasa_and_the_iss_russia_threatens_to_abandon_international_space_effort.html (accessed May 2, 2017).

86. Office of Inspector General, "NASA's

Commercial Crew Program: Update on Development and Certification Efforts," IG-16-028, NASA, Sept. 1, 2016, oig.nasa.gov/audits/reports/FY16/IG-16-028.pdf (accessed May 2, 2017).

87. World Bank, "Gross Domestic Product 2016, PPP," databank.worldbank.org/data/download/GDP_PPP.pdf (accessed Aug. 13, 2017); Joe Rennison and Eric Platt, "China Cuts US Treasury Holdings to Lowest Level Since 2010," *Financial Times,* Jan. 18, 2017; US Census Bureau, "Trade in Goods with China," www.census.gov/foreign-trade/balance/c5700.html (accessed Apr. 6, 2018). See also Central Intelligence Agency, "Country Comparison: GDP (Purchasing Power Parity) — 2016 Est.," The World Factbook, www.cia.gov/library/publications/the-world-factbook/rankorder/2001rank.html; in this ranking the EU, which does not figure in the World Bank's analysis, is second and the US third.

88. Johnson-Freese, *Space as a Strategic Asset,* 223; Sheehan, *International Politics of Space,* 165, 167; Office of the Secretary of Defense, "Annual Report to Congress: Military and Security Developments Involving the People's Republic of China 2016," 117FA69, Apr. 26, 2016, i, 3, www.defense.gov/Portals/1/Documents/pubs/2016%20China%20Military%20Power%20Report

.pdf (accessed May 2, 2017) and "Annual Report to Congress: Military and Security Developments Involving the People's Republic of China 2017," C-B066B88, May 15, 2017, ii, 34–35, 42, www.defense.gov/Portals/1/Documents/pubs/2017_China_Military_Power_Report.PDF?ver=2017-06-06-141328-770 (accessed Aug. 13, 2017); John Costello, "China Finally Centralizes Its Space, Cyber, Information Forces," *The Diplomat,* Jan. 20, 2016, thediplomat.com/2016/01/china-finally-its-centralizes-space-cyber-information-forces/ (accessed May 2, 2017).

89. For the political intricacies of US moves against China in space technology, see especially chap. 6, "The Politicization of the U.S. Aerospace Industry," in Johnson-Freese, *Space as a Strategic Asset,* 141–68. See also Brian Harvey, *China in Space: The Great Leap Forward* (New York: Springer-Praxis, 2013), 12.

90. Select Committee on US National Security and Military/Commercial Concerns with the People's Republic of China, US House of Representatives, "Appendix A: Scope of the Investigation" and "Overview," Report of the Select Committee, Jan. 3, 1999, partly declassified, www.house.gov/coxreport/chapfs/app.html and www.house.gov/coxreport/chapfs/over.html (accessed Mar. 26, 2017); Lowen Liu, "Just the

Wrong Amount of American," *Slate,* Sept. 11, 2016, www.slate.com/ articles/news_and _politics/the_next_20/2016/09/the_case_ of_scientist_wen_ho_lee_and_chinese _americans_under_suspicion_for.html (accessed May 2, 2017); "Statement by Judge in Los Alamos Case, with Apology for Abuse of Power," *New York Times,* Sept. 14, 2000.

91. Moltz, *Asia's Space Race,* 93; Harvey, *China in Space,* 345–46.

92. Sec. 539 of the Consolidated Appropriations Act, 2012, and Sec. 532 of the Consolidated Appropriations Act, 2014, both state that NASA may use no funds provided by these acts "to develop, design, plan, promulgate, implement, or execute a bilateral policy, program, order, or contract of any kind to participate, collaborate, or coordinate bilaterally in any way with China or any Chinese-owned company unless such activities are specifically authorized by a law" unless the activities "pose no risk of resulting in the transfer of technology, data, or other information with national security or economic security implications to China or a Chinese-owned company." Both Acts also stipulate that "official Chinese visitors" may not be hosted at "facilities belonging to or utilized by NASA."

93. Moltz, *Politics of Space Security,* 287; Moltz, *Asia's Space Race,* 95–96; Johnson-

Freese, *Space as a Strategic Asset,* 229; Sheehan, *International Politics of Space,* 167–68; Leonard David, "US–China Cooperation in Space: Is It Possible, and What's in Store?" Space.com, June 16, 2015, www.space.com/29671-china-nasa -space-station-cooperation.html (accessed May 2, 2017).

94. International Astronomical Union, "IAU's Reaction to the Executive Order Banning Access from Seven Countries," announcement, Jan. 30, 2017, www.iau.org/ news/announcements/detail/ann17006/; Royal Astronomical Society, "RAS Responds to the US Executive Order Banning Entry from Seven Countries," news release, Jan. 31, 2017, www.ras.org.uk/news-and -press/2947-ras-response-to-the-us-exec utive-order-banning-entry-from-seven -countries; Multisociety Letter on Immigration, Feb. 10, 2017, mcmprodaaas.s3 .amazonaws.com/s3fs-public/Multisoci ety%20Letter%20on%20Immigration%20 1-31-2017.pdf?utm_medium=email&dm_ i=1ZJN,4QUK6,E29DOV,HT01N,1; William J. Broad, "Top Scientists Urge Trump to Abide by Iran Nuclear Deal," *New York Times,* Jan. 2, 2017, and static01.nyt.com/ packages/pdf/science/03ScientistsLetter .pdf; "March for Science," satellites.march forscience.com; Becky Crystal, "These Washington Restaurants Are Closed for the

'Day Without Immigrants' Protest," *Washington Post,* Feb. 16, 2017. (All accessed May 2, 2017.)

95. Alicia Parlapiano and Gregor Aisch, "Who Wins and Loses in Trump's Proposed Budget," *New York Times,* updated Mar. 16, 2017; Will Thomas, "White House Requesting Immediate $3 Billion Cut to R & D Budgets," American Institute of Physics: FYI Bulletin 40, Mar. 29, 2017; American Institute of Physics, "Congress Stands by Science in Final Budget Deal," FYI Bulletin 53, May 2, 2017; Will Thomas, "Final FY17 Appropriations: NASA," FYI Bulletin 56, May 5, 2017; Associated Press, "Federal Budget Deal Would Spare Arts Agencies," May 1, 2017. Actual 2017 figures are 0.9% president's requested cut; 1.9% increase for NASA overall; 3.1% increase for NASA Science; 5.2% increase for ARPA-E. NEH and NEA each got a 1.3% increase; CPB's funding remained the same as 2016. For 2018 figures, see William Thomas, "Final FY18 Appropriations: Department of Defense," *FYI Bull.* 40, Apr. 5, 2018. See also the American Institute of Physics' continually updated "Federal Science Budget Tracker," www.aip.org/fyi/federal-science -budget-tracker.

9. A Time to Heal

1. Lewis Mumford, "No: 'A Symbolic Act of War . . . ,' " *New York Times,* July 21, 1969, query.nytimes.com/mem/archive/pdf ?res=9804E3DB1738E63ABC4951DFB16 68382679EDE (accessed May 5, 2017); "Reactions to Man's Landing on the Moon Show Broad Variations in Opinions. Some Would Forge Ahead in Space, Others Would Turn to Earth's Affairs," *New York Times,* July 21, 1969, 6–7, timesmachine.nytimes .com/timesmachine/1969/07/21/issue.html (accessed Sept. 18, 2017).
2. See generally Daron Acemoglu, Mikhail Golosov, Aleh Tsyvinski, and Pierre Yared, "A Dynamic Theory of Resource Wars," *Quarterly J. of Economics* (2012), 283–331, economics.mit.edu/files/8041 (accessed Oct. 9, 2017). For recently diminishing supplies of another natural resource, see David Owen, "The End of Sand," *New Yorker,* May 29, 2017, 28–33.
3. Government Accountability Office, *Rare Earth Materials: Developing a Comprehensive Approach Could Help DOD Better Manage National Security Risks in the Supply Chain,* GAO-16-161, Feb. 2016, www.gao.gov/ assets/680/675165.pdf; Lee Simmons, "Rare-Earth Market," *Foreign Policy,* July 12, 2016, foreignpolicy.com/2016/07/12/

decoder-rare-earth-market-tech-defense -clean-energy-china-trade; Lisa Margonelli, "Clean Energy's Dirty Little Secret," *The Atlantic,* May 2009, www.theatlantic.com/ magazine/archive/2009/05/clean-energys -dirty-little-secret/307377; Julie Butters, "This Is Dysprosium — If We Run Out of It, Say Goodbye to Smartphones, MRI Scans and Hybrid Cars," Phys.org, June 6, 2016, phys.org/news/2016-06-dyprosiumif -goodbye-smartphones-mri-scans.html (accessed Apr. 2, 2017).

4. Aluminum constitutes about 8 percent of Earth's crust and is the third most abundant element found there. On the light side of all metals, it has about the same density as quartz. So, along with the silicate rocks, it floated to the top. Never solo on Earth, it is always found combined with other elements, such as oxygen or potassium.

5. Europlanet, "Nanosat Fleet Proposed for Voyage to 300 Asteroids," press release, Sept. 19, 2017, www.europlanet-eu.org/ nanosat-fleet-proposed-to-300-asteroids (accessed Sept. 19, 2017).

6. Tony Judt, *Reappraisals: Reflections on the Forgotten Twentieth Century* (New York: Penguin, 2008), 5–7.

7. "Text: Obama's Speech to the United Nations General Assembly," *New York Times,* Sept. 23, 2009. See, in this context, Joan

Johnson-Freese, *Heavenly Ambitions: America's Quest to Dominate Space* (Philadelphia: University of Pennsylvania Press, 2009): "Our flawed approach to space is the product not of any single source of dysfunction, but a swamplike mixture of partisan politics, bureaucratic gamesmanship, the traditional pressures of the military-industrial complex, and an unfortunate and blissful ignorance on the part of the American public. Worse, this ignorance is wedded to a kind of American exceptionalism that drives Americans — the conquerors of the Moon — to believe that they have an almost inherent right to declare space as their own, the reaction of the rest of the world be damned" (xi–xii).

8. The Paris Agreement was adopted by consensus in December 2015. On April 1, 2016, the presidents of China and the United States issued a joint statement saying that both countries would sign it. The agreement opened for signature on April 22, 2016; 175 nation-states including the European Union signed it that same day. The agreement came into effect seven months later. As of January 2018, of the 197 parties that have signed or otherwise accepted its provisions, 174 parties have ratified it, including the United States. In June 2017, President Trump stated his intention to withdraw the United States

from the agreement. A withdrawal would have to be done within the provisions of international law; a presidential declaration does not constitute withdrawal. (Re the 197 parties: As a body, the United Nations has 193 member states and two "observer" states, Palestine and the Holy See. The other parties to the agreement are two island nations that are not UN member states, Niue and the Cook Islands. The European Union is counted as a member state.) See United Nations Framework Convention on Climate Change, "The Paris Agreement," unfccc.int/paris_agreement/items/9485.php (accessed Jan. 23, 2018).

9. National Nuclear Security Administration, "70 Years of Computing at Los Alamos National Laboratory," www.lanl.gov/asc/_assets/docs/history-computing.pdf; Los Alamos National Laboratory, "Los Alamos' Trinity Supercomputer Lands on Two Top-10 Lists," news release, Nov. 16, 2017, www.lanl.gov/discover/news-release-archive/2017/November/1116-trinity-supercomputer.php (accessed Jan. 25, 2018).

10. National Nuclear Security Administration, "About Us," nnsa.energy.gov/ (accessed Jan. 25, 2018).

11. SAO/NASA Abstract Service, a digital library portal for researchers in astronomy and physics, operated by the Smithsonian Astrophysical Observatory under a grant

from NASA, adsabs.harvard.edu/basic _search.html.

12. Bill Maher, "New Rules" segment, *Real Time with Bill Maher,* HBO, season 7, episode 7, Apr. 3, 2009.

13. Commission to Assess United States National Security Space Management and Organization, "Executive Summary," *Report — Pursuant to Public Law 106-65,* Jan. 11, 2001, 8, fas.org/spp/military/commission/ executive_summary.pdf (accessed Sept. 20, 2017).

14. Scott A. Weston, "Examining Space Warfare: Scenarios, Risks, and US Policy Implications," *Air & Space Power J.* 23:1 (Spring 2009), 75–77.

15. Office of the Assistant Secretary of Defense, "Space Domain Mission Assurance: A Resilience Taxonomy," white paper, Sept. 2015, 1, fas.org/man/eprint/resilience .pdf (accessed May 5, 2017).

16. Sec. 1616, "Organization and Management of National Security Space Activities of the Department of Defense," S. 293 (114th): National Defense Authorization Act for Fiscal 2017, Dec. 13, 2016 (passed Congress/enrolled bill), www.govtrack.us/ congress/bills/114/s2943/text (accessed May 4, 2017).

17. Commission to Assess US National

Security Space Management, *Report,* 17, 13, 33.

18. One notable threat from President Trump, delivered during a news conference at a golf club, was that "North Korea best not make any more threats to the United States. They will be met with fire and fury like the world has never seen" (Peter Baker and Choe Sang-hun, "Trump Threatens 'Fire and Fury' Against North Korea If It Endangers U.S.," *New York Times,* Aug. 8, 2017). Another, delivered the following month from the podium at the UN General Assembly: "No nation on Earth has an interest in seeing this band of criminals arm itself with nuclear weapons and missiles. The United States has great strength and patience, but if it is forced to defend itself or its allies, we will have no choice but to totally destroy North Korea. Rocket Man is on a suicide mission for himself and for his regime" (Peter Baker and Rick Gladstone, "With Combative Style and Epithets, Trump Takes America First to the U.N.," *New York Times,* Sept. 19, 2017).

19. Li Bin, "The Consequences of a Space War," conference paper, Pugwash Workshop on Preserving the Non-Weaponization of Space, Castellon de la Plana, Spain, May 22–24, 2003, www.pugwash.org/reports/nw/space2003-bin.htm (link disabled).

20. Referring to Trump's truculence in early

October 2017, for instance, Senator Bob Corker (R–TN), chair of the Senate Foreign Relations Committee, who had declared he was not running for reelection, said to a reporter, "[W]e could be heading towards World War III with the kinds of comments that he's making." See "Read Excerpts From Senator Bob Corker's Interview With The Times," *New York Times,* Oct. 9, 2017.

21. Johnson-Freese, *Heavenly Ambitions,* 25.

22. A sampling of such statements: An Air Force Space Command master plan describes its long-term strategy as "fielding and deploying space and missile combat forces in depth, allowing us to take the fight to any adversary in, from, and through space, on-demand. . . . The result will be a space combat command that is organized, trained, and equipped to rapidly achieve decisive results on or above the battlefield, anywhere, anytime" (*Strategic Master Plan FY06 and Beyond,* Oct. 1, 2003, 11, www .wslfweb.org/docs/Final%2006%20SMP –Signed!v1.pdf). Another Air Force plan concurs: "A key objective for transformation, therefore, is not only to ensure the U.S. ability to exploit space for military purposes, but also as required to deny an adversary's ability to do so" (HQ USAF/ XPXC, *The U.S. Air Force Transformation Flight Plan 2004,* July 1, 2004, C-10, www

.hsdl.org/?view&did=454273). ISR being indispensable to these goals, the 2006 *Quadrennial Defense Review* speaks of the need "to establish an 'unblinking eye' over the battlespace through persistent surveillance" so as to "support operations against any target, day or night, in any weather, and in denied or contested areas" (US Department of Defense, *Quadrennial Defense Review Report,* Feb. 6, 2006, 55, archive.defense.gov/pubs/pdfs/QDR20060203.pdf). A national security report from the White House describes one of DoD's main thrusts as dealing with "[d]isruptive challenges from state and non-state actors who employ technologies and capabilities (such as biotechnology, cyber and space operations, or directed energy weapons) in new ways to counter military advantages the United States currently enjoys" (President of the United States, *The National Security Strategy of the United States of America,* Mar. 2006, 44, www.state.gov/documents/organization/64884.pdf). (All accessed May 12, 2017.)

23. Two examples from 2016: General James E. Cartwright (USMC, ret.): "The days of 'space dominance' are over and we need to move from thinking of space as a military domain of offense and defense to a more complex environment that needs to

be managed by a wide range of international players" (foreword to Theresa Hitchens and Joan Johnson-Freese, "Toward a New National Security Space Strategy: Time for a Strategic Rebalancing," Atlantic Council Strategy Paper 5, June 2016, i); General John E. Hyten (formerly Commander, USAFSPC): "Space is no longer a sanctuary where the United States or our allies and partners operate with impunity" (Hyten, "Space Mission Force: Developing Space Warfighters for Tomorrow," white paper, US Air Force Space Command, June 29, 2016, 2, www.afspc.af.mil/Portals/3/documents/White%20Paper%20-%20Space%20Mission%20Force/AFSPC%20SMF%20White%20Paper%20-%20FINAL%20-%20AFSPC%20CC%20Approved%20on%20June%2029.pdf?ver=2016-07-19-095254-887 (accessed May 8, 2017). Hitchens and Johnson-Freese contend, however, that in response to recent Chinese and Russian tests of maneuverable satellites and the 2013 launch of a Chinese rocket that nearly reached geostationary orbit, "Defense against counterspace capabilities has taken on a top priority, followed by, in order, a diminished view of space diplomacy, and an increased interest in offensive capabilities. In particular, the increased threat perception was accompanied by more aggressive public diplomacy by the

Pentagon and US Air Force, aimed at making it very clear that the United States would respond to threats in space with the use of force — with rhetoric slipping back toward the 'dominance and control' motif of the Bush administration's space policy" (3).

24. Brian Weeden, "Alternatives to a Space Weapons Treaty," *Bulletin of the Atomic Scientists,* Apr. 17, 2009, thebulletin.org/alternatives-space-weapons-treaty; European Union: External Action, "International Space Code of Conduct — Version Mar. 31, 2014, Draft," eeas.europa.eu/topics/disarmament-non-proliferation-and-arms-export-control/14715_en (accessed May 7, 2017). Organizations focusing on de-escalation in space include, for instance, the Eisenhower Center for Space and Defense Studies, the European Institute for Security Studies, the Federation of American Scientists, GlobalSecurity.org, the Institute of Air and Space Law, the National Security Archive, the Planetary Society, Project Ploughshares, the Secure World Foundation, the Space Policy Institute, the Stimson Center, the Union of Concerned Scientists, and the United Nations Institute for Disarmament Research.

25. Henry R. Hertzfeld, Brian Weeden, and Christopher D. Johnson, "Outer Space: Ungoverned or Lacking Effective Governance?:

New Approaches to Managing Human Activities in Space," *SAIS Review of International Affairs* 36:2 (Summer–Fall 2016), 15–28; Hitchens and Johnson-Freese, "New National Security Space Strategy"; Weeden, "Alternatives to a Space Weapons Treaty"; "Executive Summary," in *Space Security Index 2016,* ed. Jessica West (Waterloo, ON: Project Ploughshares, Sept. 2016), 1.

26. "HST Publication Statistics," Feb. 25, 2017, archive.stsci.edu/hst/bibliography/pubstat.html (accessed Sept. 21, 2017).

27. Arthur S. Eddington (1920), quoted in S. Chandrasekhar, foreword to Eddington, *The Internal Constitution of the Stars* (Cambridge: Cambridge University Press, 1926/1988), x.

28. Eddington, *Internal Constitution,* 301.

29. William A. Fowler, "Formation of the Elements," *Scientific Monthly* 84: 2 (Feb. 1957), 98.

30. Jonathan M. Weisgall, "The Nuclear Nomads of Bikini," *Foreign Policy* 39 (Summer 1980), 83.

31. E. Margaret Burbidge, G. R. Burbidge, William A. Fowler, and F. Hoyle, "Synthesis of the Elements in Stars," *Reviews of Modern Physics* 29:4 (Oct. 1957), 547–650. The US Atomic Energy Commission was a precursor agency to the US Department of Energy.

32. Burbidge et al., "Synthesis," 640. The "*r*-process" refers to free neutrons that march straight into an atomic nucleus and stay there. With their neutral charge, they face no electromagnetic resistance at all. That configuration of particles in the nucleus might be unstable. But in an *r*-process element, a second neutron enters before the nucleus has a chance to decay, creating a stable nucleus.

33. The authors are grateful to economists Mark Harrison of the University of Warwick (editor of *The Economics of World War II: Six Great Powers in International Comparison* [Cambridge, UK: Cambridge University Press, 1998] and co-editor, with Stephen Broadberry, of *The Economics of World War I* [Cambridge, UK: Cambridge University Press, 2005]) and Linda Bilmes of the Harvard Kennedy School (co-author, with Joseph Stiglitz, of *The Three-Trillion Dollar War: The True Cost of the Iraq Conflict* [New York: W. W. Norton, 2008]) for their generosity in helping to clarify the extreme difficulty of calculating the comprehensive costs of war.

34. $3 billion is a high ballpark figure; comprehensive figures are not available. In 2016 NASA's astrophysics budget was $1.35 billion (Astrophysics: $730 million; James Webb Telescope: $620 million). The

2016 budget for the National Science Foundation's Division of Astronomical Sciences was $250 million, including research, education, and facilities. The 2016 budget for the European Space Agency's entire Scientific Programme was $510 million. The India Space Research Organisation budgeted $47 million for space sciences in 2016. About 14 percent, or $180 million, of the Japan Aerospace Exploration Agency budget has been going to space science and exploration, which suggests a space-science portion on the order of $100 million; additional space-science funding is part of JAXA's commitment to the International Space Station. See American Institute of Physics, "Federal Science Budget Tracker," FYI: Science Policy News, www.aip.org/fyi/federal-science-budget-tracker/FY2017; National Science Foundation, Directorate for Mathematical and Physical Sciences: Division of Astronomical Sciences (AST), "AST Funding," *FY2016 Budget Request to Congress,* MPS-12, www.nsf.gov/about/budget/fy2016/pdf/fy2016budget.pdf; European Space Agency, "ESA 2016 Budget by Domain," www.esa.int/spaceinimages/Images/2016/01/ESA_budget_2016_by_domain; Space Foundation, "Exhibits 1s. Indian Space Budgets," "Exhibit 1t. Japanese Space Spending by Agency 2016," *The*

Space Report 2017, 12–13; "FY2015 Annual Budget [JAXA]," reproduced in Chu Ishida, "JAXA Program for Earth Observation Satellites," Japan Aerospace Exploration Agency, Jan. 6, 2016, n.p., www.pco -prime.com/vegetation_lidar2016/pdf/1-3 Ishida_JAXA_EO_program_20160106.pdf (accessed Sept. 24, 2017).

35. Nan Tian, Aude Fleurant, Pieter D. Wezeman, and Siemon T. Wezeman, "Trends in World Military Expenditure, 2016," fact sheet, SIPRI, Apr. 2017, www.sipri.org/ sites/default/files/Trends-world-military -expenditure-2016.pdf; World Bank, "World Development Indicators Database: Gross Domestic Product 2016," Apr. 17, 2017, databank.worldbank.org/data/download/ GDP.pdf (both accessed Oct. 29, 2017). SIPRI's estimate for world military expenditure in 2016 is $1.686 trillion; the World Bank's figure for 2016 global GDP is $75.642 trillion.

36. In 1943, while military spending on the war represented 42 percent of America's income, it represented 43 percent of Japan's, 55 percent of Britain's, and 70 percent of Germany's. See Mark Harrison, "The Economics of World War II: An Overview," chap. 1 of *Economics of World War II,* 34, "Table 1-8: The military burden, 1939–44 (military outlays, per cent of national income)," www2.warwick.ac.uk/

fac/soc/economics/staff/ mharrison/public/ ww2overview1998.pdf (accessed Sept. 23, 2017). "Table 1-11: War losses attributable to physical destruction (per cent of assets)" indicates that Germany lost 17 percent of its industrial assets and Japan 34 percent (42).

37. Stephen Daggett, "Costs of Major U.S. Wars," report, June 29, 2010, Congressional Research Service, 2, fas.org/sgp/crs/natsec/ RS22926.pdf (accessed Sept. 23, 2017). From Dec. 1941 through Sept. 1945, total US military spending was $296 billion in then-current dollars. In a note to "Table 1: Military Costs of Major U.S. Wars, 1775–2010," Daggett specifies that the estimates are based on US government budget data and include the costs of military operations only, not veterans' benefits, interest on war-related debt, or assistance to allies.

Converted to constant 2016 dollars (in terms of simple purchasing power), $296 billion — call it $300 billion — in World War II–era dollars becomes $4 trillion, or $75 billion per year, according to MeasuringWorth, www.measuringworth.com, a public-service website founded by two American professors of economics. However, Harrison emphasizes that a more meaningful conversion should take into account the *rate* of war spending relative to nominal GDP — that is, the change in prices multiplied by the change in output. Between World War II and now, nominal GDP increased ninetyfold.

38. Benito Mussolini, "Plan for the New Italian Economy (1936)," quoted in John Bellamy Foster, "Neofascism in the White House," *Monthly Review* 68:11 (Apr. 2017), monthlyreview.org/2017/04/01/neofascism -in-the-whitehouse (accessed May 9, 2017).

39. "Discretionary Spending 2015: $1.1 Trillion," pie chart, National Priorities Project, www.nationalpriorities.org. Analyses of science-related funding are at, e.g., "Research by Science and Engineering Discipline: Physical Sciences Research Funding, 1978–2014," bar graph, American Association for the Advancement of Science, www .aaas.org/page/research-science-and-en gineering-discipline; "Survey of Federal Funds for Research and Development Fiscal Years 2015–17 — Table 2. Summary of Federal Obligations and Outlays for Research, Development, and R&D Plant, by Type of R&D, Performer, and Field of Science and Engineering: FYs 2014–17," National Science Foundation, ncsesdata.nsf .gov/fedfunds/2015/html/FFS2015_DST _002.html; Jeffrey Mervis, "Data Check: U.S. Government Share of Basic Research Funding Falls Below 50%," *Science,* Mar. 9, 2017, www.sciencemag.org/news/2017/ 03/data-check-us-government-share-basic

-research-funding-falls-below-50. Data on military spending is at "Military Expenditure: World Military Spending in 2016: Military Spending Graphics," SIPRI, www.sipri.org/research/armament-and-disarmament/arms-transfers-and-military-spending/military-expenditure. (All accessed May 9, 2017.)

SELECTED SOURCES

Please see endnotes for a fuller compilation of sources. Sources for which no pagination or details are indicated are accessible online.

2MASS: 2 Micron All Sky Survey. "Introduction: 1. 2MASS Overview." Dec. 20, 2006.

Abrahamson, James A., and Henry F. Cooper. "What Did We Get for Our $30-Billion Investment in SDI/BMD?" National Institute for Public Policy, Sept. 1993.

Air Force Doctrine Documents. GlobalSecurity.org.

Air Force Space Command. "Commander's Strategic Intent." May 6, 2016.

American Presidency Project. University of California, Santa Barbara.

Anson, Peter, and Dennis Cummings. "The First Space War: The Contribution of Satellites to the Gulf War." *RUSI Journal* 136:4 (Winter 1991), 45–53.

Arago, François. "Report" (1839). In Alan Trachtenberg, ed. *Classic Essays in Photog-*

raphy. New Haven: Leete's Island Books, 1981, 15–26.

Arbatov, Alexei, and Vladimir Dvorkin, eds. *Outer Space: Weapons, Diplomacy, and Security.* Washington, DC: Carnegie Endowment for International Peace, 2010.

Arike, Ando. "The Soft-Kill Solution: New Frontiers in Pain Compliance." *Harper's* (Mar. 2010), 38–47.

Augenstein, Bruno W. "Evolution of the U.S. Military Space Program, 1945–1960: Some Key Events in Study, Planning, and Program Development." Paper P-6814. RAND Corporation, Sept. 1982.

Austin, B. A. "Precursors to Radar: The Watson-Watt Memorandum and the Daventry Experiment." *International Journal of Electrical Engineering Education* 36 (1999), 365–72.

Ball, Philip. *Invisible: The Dangerous Allure of the Unseen.* Chicago: University of Chicago Press, 2015.

"Barry Goldwater on Space: GOP Candidate Wants Military, Not Civilians, to Run Space Program." *Science* 145 (July 31, 1964), 470–71.

Bartusiak, Marcia, ed. *Archives of the Universe: A Treasury of Astronomy's Historic Works of Discovery.* New York: Pantheon, 2004.

Barty-King, Hugh. *Eyes Right: The Story of*

Dollond & Aitchison Opticians, 1750–1985. London: Quiller Press, 1986.

Bedini, Silvio A. "Of 'Science and Liberty': The Scientific Instruments of King's College and Eighteenth Century Columbia College in New York." *Annals of Science* 50:3 (May 1993), 201–27.

Bhaskaran, Shyam. "Autonomous Navigation for Deep Space Missions." American Institute of Aeronautics and Astronautics Space-Ops 2012 Conference, Stockholm.

Biagioli, Mario. "Did Galileo Copy the Telescope? A 'New' Letter by Paolo Sarpi." In Albert Van Helden, Sven Dupré, Rob van Gent, and Huib Zuidervaart, eds. *The Origins of the Telescope.* Amsterdam: KNAW Press/Royal Netherlands Academy of Arts and Sciences, 2010, 203–30.

Brown, J. Willard. *The Signal Corps, U.S.A. in the War of the Rebellion.* Boston: US Veteran Signal Corps Association, 1896.

Brown, Louis. *A Radar History of World War II: Technical and Military Imperatives.* Philadelphia: Institute of Physics Publishing, 1999.

Browne, Malcolm W. "New Space Beacons Replace the Compass." *New York Times,* Nov. 8, 1988.

Bruger, Steven J. "Not Ready for the 'First Space War,' What About the Second?" Operations Department, Naval War College, May 17, 1993.

Burbidge, E. Margaret, G. R. Burbidge, William A. Fowler, and F. Hoyle. "Synthesis of the Elements in Stars." *Reviews of Modern Physics* 29:4 (Oct. 1957), 547–650.

Burns, Richard Dean, and Joseph M. Siracusa. *A Global History of the Nuclear Arms Race: Weapons, Strategy, and Politics*, 2 vols. Santa Barbara, CA: Praeger/ABC-CLIO, 2013.

Burrows, William E. *This New Ocean: The Story of the First Space Age.* New York: Random House, 1998.

Butrica, Andrew J. *To See the Unseen: A History of Planetary Radar Astronomy.* NASA History Series: NASA SP-4218. Washington, DC: NASA, 1996.

Cañizares-Esguerra, Jorge. *Nature, Empire, and Nation: Explorations in the History of Science in the Iberian World.* Stanford: Stanford University Press, 2006.

Casson, Lionel. *The Ancient Mariners: Seafarers and Sea Fighters of the Mediterranean in Ancient Times.* Princeton: Princeton University Press, 1991.

Chaisson, Eric J. *The Hubble Wars: Astrophysics Meets Astropolitics in the Two-Billion-Dollar Struggle over the Hubble Space Telescope.* New York: Harper-Collins, 1994.

Chandra, Satish, ed. *The Indian Ocean: Explorations in History, Commerce, and Politics.* New Delhi: Sage, 1987.

Chang, Iris. *Thread of the Silkworm.* New York: Basic Books, 1995.

Clark, Stuart. "Russia Halts Rocket Exports to US, Hitting Space and Military Programmes." *Guardian,* May 15, 2014.

Commission on the Future of the United States Aerospace Industry. *Anyone, Anything, Anywhere, Anytime.* Final report. Nov. 2002.

Commission to Assess United States National Security Space Management and Organization. *Report — Pursuant to Public Law 106-65.* Jan. 11, 2001.

Committee on Aeronautical and Space Sciences, US Senate. *Staff Report: Documents on International Aspects of the Exploration and Use of Outer Space, 1954–1962.* May 9, 1963.

"Cooperation in Space: Agreement Between the United States of America and the Union of Soviet Socialist Republics Concerning Cooperation in the Exploration of the Use of Outer Space for the Peaceful Purposes." May 24, 1972.

Cordesman, Anthony. *The Iraq War: Strategy, Tactics, and Military Lessons.* Westport, CT: Praeger/Center for Strategic and International Studies, 2003.

Costello, John. "China Finally Centralizes Its Space, Cyber, Information Forces." *The Diplomat,* Jan. 20, 2016.

Cotter, Charles H. *A History of Nautical Astronomy.* New York: American Elsevier, 1968.

Covault, Craig. "Desert Storm Reinforces Military Space Directions." *Aviation Week & Space Technology,* Apr. 8, 1991, 42–47.

———. "Recon Satellites Lead Allied Intelligence Effort." *Aviation Week & Space Technology,* Feb. 4, 1991, 25–26.

Covert, Claudia T. "Art at War: Dazzle Camouflage." *Art Documentation: Journal of the Art Libraries Society of North America* 26:2 (Fall 2007), 50–56.

Crawford, Neta C. "US Costs of Wars Through 2016: $4.79 Trillion and Counting: Summary of Costs of the US Wars in Iraq, Syria, Afghanistan, and Pakistan and Homeland Security." Watson Institute, Brown University, Sept. 2016.

C-SPAN. "SDI Debate: Is the Strategic Defense Initiative in the National Interest." Program 532-1. Nov. 18, 1987.

Cunliffe, Barry. *The Extraordinary Voyage of Pytheas the Greek.* New York: Walker, 2002.

Curtis, Heber D. "Optical Glass." *Publications of the Astronomical Society of the Pacific* (Apr. 1919), 77–85.

David, Leonard. "US–China Cooperation in Space: Is It Possible, and What's in Store?" Space.com, June 16, 2015.

Davis, Jason. "What's the Matter with Russia's Rockets?" Blog post. Planetary Society, Dec. 2, 2016.

Day, Dwayne A. "The Flight of the Big Bird (parts 1–4)." *Space Review,* Jan. 17–Mar. 28, 2011.

———, and Robert G. Kennedy III. "Soviet Star Wars." *Air & Space Smithsonian,* Jan. 2010.

Democracy Now! "How the U.S. Narrowly Avoided a Nuclear Holocaust 33 Years Ago, and Still Risks Catastrophe Today." Video and transcript. Sept. 18, 2013.

Deng, Gang. *Chinese Maritime Activities and Socioeconomic Development, c. 2100 BC–1900 AD: Contributions in Economics and Economic History* 188. Westport, CT: Greenwood Press, 1997.

Denny, Neil. "Interview: Eric Schlosser's *Command and Control." Little Atoms* 1, Jan. 17, 2016.

DeVorkin, David. *Science with a Vengeance: How the Military Created the US Space Sciences after World War II.* New York: Springer-Verlag, 1992.

Doel, Ronald E. *Solar System Astronomy in America: Communities, Patronage, and Interdisciplinary Science, 1920–1960.* New York: Cambridge University Press, 1996.

———, and Kristine C. Harper. "Prometheus Unleashed: Science as a Diplomatic Weapon in the Lyndon B. Johnson Administration."

In "Global Power Knowledge: Science and Technology in International Affairs," *Osiris* 21:1 (2006), 66–85.

Dolman, Everett C. *Astropolitik: Classical Geopolitics in the Space Age.* London: Frank Cass, 2002.

Duffner, Robert W. *The Adaptive Optics Revolution: A History.* Albuquerque: University of New Mexico Press, 2009.

Dupuis, Andre. "An Overview of Canadian Military Space in 2014." Pt. 1, Feb. 9, 2015; Pt. 2, Feb. 17, 2015. SpaceRef Canada.

Dyson, George B. *Darwin Among the Machines: The Evolution of Global Intelligence.* Reading, MA: Addison-Wesley Longman, 1997.

Eddington, Arthur S. *The Internal Constitution of the Stars.* Cambridge, UK: Cambridge University Press, 1926/1988.

"Edison and the Unseen Universe." *Scientific American* 39:8, suppl. 138 (Aug. 24, 1878), 112.

Eisenhower, Dwight D. "Annual Message to the Congress on the State of the Union." Jan. 10, 1957. At Gerhard Peters and John T. Woolley, American Presidency Project, University of California, Santa Barbara.

———. "Atoms for Peace: Address to the 470th Plenary Meeting of the UN General

Assembly." Dec. 8, 1953. International Atomic Energy Agency.

———. "Farewell Address: Transcript." Jan. 17, 1961, sec. IV. The Presidency, Miller Center, University of Virginia.

———. "Radio and Television Address to the American People on Science in National Security." Nov. 7, 1957. At Gerhard Peters and John T. Woolley, American Presidency Project, University of California, Santa Barbara.

Ellis, Emma Grey. "Russia's Space Program Is Blowing Up. So Are Its Rockets." *Wired,* Dec. 7, 2016.

Erickson, John. "Radio-location and the Air Defence Problem: The Design and Development of Soviet Radar 1934–40." *Science Studies* 2:3 (July 1972), 241–63.

European Commission. "Space Strategy for Europe." COM(2016) 705. Oct. 26, 2016.

European Council, Council of the European Union. "Implementation Plan on Security and Defence." Nov. 14, 2016.

European Space Agency. "Council Meeting Held at Ministerial Level on 1 and 2 December 2016: Resolutions and Main Decisions." Dec. 2, 2016.

———. "International Space Station Legal Framework." Nov. 19, 2013.

Evangelista, Matthew. "Cooperation Theory and Disarmament Negotiations in the 1950s." *World Politics* 42: 4 (July 1990),

502–28.

Ezell, Edward Clinton, and Linda Neuman Ezell. *The Partnership: A History of the Apollo–Soyuz Test Project.* Washington, DC: NASA, 1978.

Field, Alexander J. "French Optical Telegraphy, 1793–1855: Hardware, Software, Administration." *Technology and Culture* 35:2 (Apr. 1994), 315–47.

Fishel, Edwin C. *The Secret War for the Union: The Untold Story of Military Intelligence in the Civil War.* Boston: Houghton Mifflin, 1996.

Fisher, David E. *A Summer Bright and Terrible: Winston Churchill, Lord Dowding, Radar, and the Impossible Triumph of the Battle of Britain.* Berkeley, CA: Shoemaker & Hoard, 2005.

Fleurant, Aude, Pieter D. Wezeman, Siemon T. Wezeman, and Nan Tian. "Trends in International Arms Transfers, 2016." Fact sheet. SIPRI, Feb. 2017.

Freeth, Tony, and Alexander Jones. "The Cosmos in the Antikythera Mechanism." *ISAW Papers* 4, Feb. 2012.

Galilei, Galileo. *Sidereus Nuncius, or The Sidereal Messenger* (1610). Translated and with commentary by Albert Van Helden. Chicago: University of Chicago Press, 1989/2016.

Garthoff, Raymond L. "Banning the Bomb

in Outer Space." *International Security* 5:3 (Winter 1980–81), 25–40.

Gaskin, Thomas M. "Senator Lyndon B. Johnson, the Eisenhower Administration and U.S. Foreign Policy, 1957–60." *Presidential Studies Quarterly* 24:2 (Spring 1994), 341–61.

Gerovitch, Slava. "Stalin's Rocket Designers' Leap into Space: The Technical Intelligentsia Faces the Thaw." *Osiris* 23:1 (2008), 189–209.

Goodrick-Clarke, Nicholas. *The Occult Roots of Nazism: Secret Aryan Cults and Their Influence on Nazi Ideology: The Ariosophists of Austria and Germany, 1890–1935.* New York: New York University Press, 1992.

GPS.gov. "Selective Availability."

Grafton, Anthony. "Girolamo Cardano and the Tradition of Classical Astrology: The Rothschild Lecture, 1995." *Proceedings of the American Philosophical Society* 142:3 (Sept. 1998), 323–54.

Gray, Colin S. "Clausewitz Rules, OK? The Future Is the Past: With GPS." *Review of International Studies* 25 (Dec. 1999), 161–82.

———. "The Influence of Space Power upon History." *Comparative Strategy* 15:4 (1996), 293–308.

Greely, A. W. "The Signal Corps." In Francis Trevelyan Miller and Robert Sampson Lanier, eds. *Photographic History of The Civil War in Ten Volumes,* vol. 8. New York: Review of Reviews Co., 1912, 312–40.

Greenemeier, Larry. "GPS and the World's First 'Space War.' " *Scientific American,* Feb. 8, 2016.

Grego, Laura, George N. Lewis, and David Wright. *Shielded from Oversight: The Disastrous US Approach to Strategic Missile Defense.* Union of Concerned Scientists, July 2016.

Hagen, Antje. "Export versus Direct Investment in the German Optical Industry: Carl Zeiss, Jena and Glaswerk Schott & Gen. in the UK, from Their Beginnings to 1933." *Business History* 38:4 (1996), 1–20.

Hardy, John W. *Adaptive Optics for Astronomical Telescopes.* New York: Oxford University Press, 1998.

Hartung, William D. *Tangled Web 2005: A Profile of the Missile Defense and Space Weapons Lobbies.* New York: World Policy Institute — Arms Trade Resource Center, 2005.

Harvey, Brian. *China in Space: The Great Leap Forward.* New York: Springer-Praxis, 2013.

Harwit, Martin. *Cosmic Discovery: The Search, Scope, and Heritage of Astronomy,* 1st ed. New York: Basic Books, 1981.

———. "The Early Days of Infrared Space

Astronomy." In J. A. Bleeker, J. Geiss, and M. Huber, eds. *The Century of Space Science.* Dordrecht: Kluwer, 2002, 301–28.

Hawkes, C. F. C. *The Eighth J. L. Myres Memorial Lecture — Pytheas: Europe and the Greek Explorers.* Oxford: Blackwell, 1977.

Heginbotham, Eric, et al. *The U.S.–China Military Scorecard: Forces, Geography, and the Evolving Balance of Power 1996–2017.* Santa Monica, CA: RAND, 2015.

Herschel, William. "Investigation of the Powers of the prismatic Colours to heat and illuminate Objects; with Remarks, that prove the different Refrangibility of radiant Heat . . ." *Philosophical Transactions of the Royal Society* 90 (1800), 255–83.

Hertzfeld, Henry R., Brian Weeden, and Christopher D. Johnson. "Outer Space: Ungoverned or Lacking Effective Governance?: New Approaches to Managing Human Activities in Space." *SAIS Review of International Affairs* 36:2 (Summer–Fall 2016), 15–28.

Hitchens, Theresa, and Joan Johnson-Freese. "Toward a New National Security Space Strategy: Time for a Strategic Rebalancing." Atlantic Council Strategy Paper 5. June 2016.

Holzmann, Gerard J., and Björn Pehrson. *The Early History of Data Networks.* Los Alamitos, CA: IEEE Computer Society Press, 1995.

Howard-Duff, Ian. "Joseph Fraunhofer (1787–1826)." *Journal of the British Astronomical Association* 97:6 (1987), 339–47.

Howe, Ellic. *Astrology and Psychological Warfare During World War II.* London: Rider, 1972.

Hyten, John E., General. "Space Mission Force: Developing Space Warfighters for Tomorrow." White paper. US Air Force Space Command, June 29, 2016.

Jansky, Karl G. "Directional Studies of Atmospherics at High Frequencies." *Proceedings of the IRE* 20 (1932), 1920–32.

———. "Electrical Disturbances Apparently of Extraterrestrial Origin." *Proceedings of the IRE* 21:10 (Oct. 1933), 1387–98.

Johnson, Lyndon Baines. Speech at the opening and dedication of Florida Atlantic University. Oct. 25, 1964.

Johnson-Freese, Joan. *Heavenly Ambitions: America's Quest to Dominate Space.* Philadelphia: University of Pennsylvania Press, 2009.

———. *Space as a Strategic Asset.* New York: Columbia University Press, 2007.

Joint Chiefs of Staff. *Joint Vision 2020: America's Military — Preparing for Tomorrow.* Summer 2000.

———. *Space Operations: Joint Publication*

3-14. May 29, 2013.

Jones, Alexander. "The Antikythera Mechanism and the Public Face of Greek Science." *Proceedings of Science,* PoS(Antikythera & SKA)038, 2012. "From Antikythera to the Square Kilometre Array: Lessons from the Ancients," Kerastari, Greece, June 12–15, 2012.

Jones, Sam. "Satellite Wars." *Financial Times,* Nov. 20, 2015.

Josephson, Paul R. "Atomic-Powered Communism: Nuclear Culture in the Postwar USSR." *Slavic Review* 55:2 (Summer 1996), 297–324.

Kaiser, Walter. "A Case Study in the Relationship of History of Technology and of General History: British Radar Technology and Neville Chamberlain's Appeasement Policy." *Icon* 2 (1996), 29–52.

Kalic, Sean A. *US Presidents and the Militarization of Space, 1946–1967.* College Station: Texas A & M University Press, 2012.

Kallender, Paul. "Japan's New Dual-Use Space Policy: The Long Road to the 21st Century." *Notes de l'Ifri: Asie.Visions* 88 (Nov. 2016).

Katz, Jonathan I. *The Biggest Bangs: The Mystery of Gamma-Ray Bursts, the Most Violent Explosions in the Universe.* New York: Oxford University Press, 2002.

Kennedy, John F. "Address at Rice University

on the Nation's Space Effort." Sept. 12, 1962. John F. Kennedy Presidential Library and Museum.

———. "President Kennedy's Special Message to the Congress on Urgent National Needs." May 25, 1961. John F. Kennedy Presidential Library and Museum.

Klebesadel, Ray W., Ian B. Strong, and Roy A. Olson. "Observations of Gamma-Ray Bursts of Cosmic Origin," *Astrophysical Journal* 182 (June 1, 1973), L84–L88.

Klein, John J. *Space Warfare: Strategy, Principles and Policy.* London and New York: Routledge, 2006.

Kramer, Miriam. "NASA Suspends Most Cooperation with Russia; Space Station Excepted." Space.com, Apr. 2, 2014.

Krepon, Michael. "Not Just Yet for No First Use." Blog post. Arms Control Wonk: Leading Voices on Arms Control, Disarmament and Non-Proliferation, July 31, 2016.

———, with Michael Katz-Hyman. "Space Weapons and Proliferation." *Nonproliferation Revie* w 12:2 (July 2005), 323–41.

LaFeber, Walter. *America, Russia, and the Cold War, 1945–2006,* 10th ed. New York: McGraw Hill, 2006.

Lankford, John. "The Impact of Photography on Astronomy." In Owen Gingerich, ed. *Astrophysics and Twentieth-Century Astronomy to 1950: Part A — The General His-*

tory of Astronomy, vol. 4. Cambridge, UK: Cambridge University Press, 1984, 16–39.

Launius, Roger D. "Historical Dimensions of the Space Age." In Eligar Sadeh, ed., *Space Politics and Policy: An Evolutionary Perspective.* Dordrecht: Springer Netherlands, 2004, 3–25.

Lee, T.-W. *Military Technologies of the World,* vol. 1. Westport, CT: Greenwood/ Praeger Security International, 2009.

Lequeux, James. "Early Infrared Astronomy." *Journal of Astronomical History and Heritage* 12:2 (2009), 125–40.

Lovell, Bernard. "The Cavity Magnetron in World War II: Was the Secrecy Justified?" *Notes and Records of the Royal Society of London* 58:3 (Sept. 2004), 283–94.

———. *The Story of Jodrell Bank.* New York: Harper & Row, 1968.

Lu, Edward T., and Stanley G. Love. "Gravitational Tractor for Towing Asteroids." *Nature* 438 (Nov. 10, 2005), 177–78.

Martin, Andy. "Mentioned in Dispatches: Napoleon, Chappe and Chateaubriand." *Modern & Contemporary France* 8:4 (2000), 445–55.

Mastny, Vojtech. *The Cold War and Soviet Insecurity: The Stalin Years.* New York: Oxford University Press, 1996.

McDougall, Walter A. *The Heavens and the Earth: A Political History of the Space Age.*

Baltimore: Johns Hopkins University Press, 1985/1997.

McNamara, Robert S. "The Military Role of Nuclear Weapons: Perceptions and Misperceptions." *Foreign Affairs* 62:1 (Fall 1983), 59–80.

McNeill, William H. *The Pursuit of Power: Technology, Armed Force, and Society since A.D. 1000*. Chicago: University of Chicago Press, 1982.

Meadows, A. J. "The New Astronomy." In Owen Gingerich, ed. *Astrophysics and Twentieth-Century Astronomy to 1950: Part A — The General History of Astronomy*, vol. 4. Cambridge, UK: Cambridge University Press, 1984, 59–72.

———. "The Origins of Astrophysics." In Owen Gingerich, ed. *Astrophysics and Twentieth-Century Astronomy to 1950: Part A — The General History of Astronomy*, vol. 4. Cambridge, UK: Cambridge University Press, 1984, 3–15.

Menzel, Donald H. "Venus Past, and the Distance of the Sun." *Proceedings of the American Philosophical Society* 113:3 (June 16, 1969), 197–202.

Merton, Robert K. "Science, Technology and Society in Seventeenth Century England." *Osiris* 4 (1938), 360–632.

Moltz, James Clay. *Asia's Space Race: National Motivations, Regional Rivalries, and*

International Risks. New York: Columbia University Press, 2012.

———. *Crowded Orbits: Conflict and Co-operation in Space.* New York: Columbia University Press, 2014.

———. *The Politics of Space Security: Strategic Restraint and the Pursuit of National Interests.* Stanford: Stanford University Press, 2008.

Morgan, J. H. *Assize of Arms: The Disarmament of Germany and Her Rearmament (1919–1939).* New York: Oxford University Press, 1946.

Mumford, Lewis. "No: 'A Symbolic Act of War . . .' " *New York Times,* July 21, 1969.

Myer, Albert J. *A Manual of Signals: For the Use of Signal Officers in the Field, and for Military and Naval Students, Military Schools, etc.* New York: D. Van Nostrand, 1868.

National Aeronautics and Space Act of 1958. Public Law 85-568, 72 Stat., 426. Signed by the President on July 29, 1958.

National Science Board. *S & E Indicators 2016.* Arlington, VA: National Science Foundation, 2016.

———. *Science & Engineering Indicators 2018 Digest.* Arlington, VA: National Science Foundation, Jan. 2018.

National Security Archive. George Washington University.

National Security Council. "National Security Council Report: Statement of Preliminary U.S. Policy on Outer Space." NSC 5814/1. Aug. 18, 1958.

National Security Council — Executive Secretary. "Report to the National Security Council on Basic National Security Policy." NSC 162/2. Oct. 30, 1953.

———. "Report to the National Security Council on United States Objectives and Programs for National Security." NSC 68. Apr. 14, 1950.

Newton, Isaac. "A Serie's of Quere's propounded by Mr. Isaac Newton, to be determin'd by Experiments, positively and directly concluding his new Theory of Light and Colours; and here recommended to the Industry of the Lovers of Experimental Philosophy, as they were generously imparted to the Publisher in a Letter of the said Mr. Newtons of July 8. 1672." *Philosophical Transactions of the Royal Society* 85 (July 15, 1672), 5004–5007.

———. *Opticks: or, A Treatise of the Reflexions, Refractions, Inflexions and Colours of Light,* 4th ed. corr. London: William Innys, 1730. Project Gutenberg.

North Atlantic Military Committee. "Final Decision on MC 14/2 (Revised): A Report by the Military Committee to the North Atlantic Council on Overall Strategic Concept for the Defense of the North Atlantic

Treaty Organization Area" ["Massive Retaliation"]. May 23, 1957. In Gregory W. Pedlow, ed. "NATO Strategy Documents 1949–1969." NATO International Staff Central Archives.

Office of Inspector General. "NASA's Commercial Crew Program: Update on Development and Certification Efforts." IG-16-028. NASA, Sept. 1, 2016.

Office of the Secretary of Defense. "Annual Report to Congress: Military and Security Developments Involving the People's Republic of China 2016." 117FA69. Apr. 26, 2016.

Oliver, Bernard M. "Colloquy 4 — The Rationale for a Preferred Frequency Band: The Water Hole." SP-419. SETI: The Search for Extraterrestrial Intelligence, 1977.

Parker, Geoffrey. *The Military Revolution: Military Innovation and the Rise of the West, 1500–1800,* 2nd ed. Cambridge, UK: Cambridge University Press, 1996.

Parry, J. H. *The Age of Reconnaissance.* London: Phoenix Press, 1963.

Peebles, Curtis. *High Frontier: The U.S. Air Force and the Military Space Program.* Washington, DC: Air Force History and Museums Program, 1997.

Pollpeter, Kevin, Eric Anderson, Jordan Wilson, and Fan Yang. *China Dream, Space*

Dream: China's Progress in Space Technologies and Implications for the United States. Washington, DC: IGCC/US–China Economic and Security Review Commission, 2015.

Portree, David S. F. "NASA's Origins and the Dawn of the Space Age." Monograph 10. NASA History Division, 2005.

President of the United States. "National Space Policy of the United States of America." June 28, 2010.

Preston, Bob, Dana J. Johnson, Sean J. A. Edwards, Michael Miller, and Calvin Shipbaugh. *Space Weapons Earth Wars.* Santa Monica, CA: RAND, 2002.

Price, S. D. *History of Space-Based Infrared Astronomy and the Air Force Infrared Celestial Backgrounds Program.* AFRL-RV-HA-TR-1008-1039. Hanscom AFB, MA: Air Force Research Laboratory — Space Vehicles Directorate, Apr. 2008.

Project Ploughshares. *Space Security Index.* Annual publication, 2003– .

Raines, Rebecca Robbins. *Getting the Message Through: A Branch History of the U.S. Army Signal Corps.* Washington, DC: Center of Military History, US Army, 1996.

Randall, J. T. "Radar and the Magnetron." *Journal of the Royal Society of Arts* 94:4715 (Apr. 12, 1946), 302–23.

Reagan, Ronald. "Address Before a Joint Session of the Congress on the State of the Union." Jan. 25, 1984. At Gerhard Peters and John T. Woolley, American Presidency Project, University of California, Santa Barbara.

————. "Address to the Nation on Defense and National Security." Mar. 23, 1983. Ronald Reagan Presidential Library and Museum.

Reed, Sidney G., Richard H. Van Atta, and Seymour J. Deitchman. *DARPA Technical Accomplishments: An Historical Review of Selected DARPA Projects,* vol. 1. IDA paper P-2192. Institute for Defense Analyses, Nov. 1990.

Regis, Ed. "What Could Go Wrong? The Insane 1950s Plan to Use H-bombs to Make Roads and Redirect Rivers." *Slate,* Sept. 30, 2015.

Reiffel, Leonard. *A Study of Lunar Research Flights,* vol. I. AD 425380/AFSWC TR-59-39. Kirtland AFB, NM: Air Force Special Weapons Center, June 19, 1959.

Richelson, Jeffrey T. *America's Space Sentinels: The History of the DSP and SBIRS Satellite Systems,* 2nd ed. Lawrence: University Press of Kansas, 2012.

————, ed. "U.S. Satellite Imagery, 1960–1999: National Security Archive Electronic Briefing Book No. 13." National Security Archive, George Washington University,

Apr. 1999.

Rogalski, A. "History of Infrared Detectors," *Opto-Electronics Review* 20:3 (2012), 279–308.

Roseman, Christina Horst. *Pytheas of Massalia: On the Ocean — Text, Translation and Commentary.* Chicago: Ares, 1994.

Rowe, A. P. *One Story of Radar.* Cambridge, UK: Cambridge University Press, 1948/2015.

"Russia Makes Plans to Kill Space Station in 2020 Due to Sanctions." NBC News, May 13, 2014.

Sagan, Carl, and Steven J. Ostro. "Dangers of Asteroid Deflection." *Nature* 368 (Apr. 7, 1994), 501.

Sagan, Scott D. "On the Brink: How Safe Are Our Nukes?" *American Scholar,* Sept. 5, 2013.

Sagdeev, Roald. *The Making of a Soviet Scientist: My Adventures in Nuclear Fusion and Space from Stalin to Star Wars.* New York: John Wiley & Sons, 1994.

Sambrook, Stephen C. "The British Armed Forces and Their Acquisition of Optical Technology: Commitment and Reluctance, 1888–1914." In *Year Book of European Administrative History* 20. Baden-Baden: Nomos Verlagsgesellschaft, 2008.

———. "No Gunnery Without Glass: Optical Glass Supply and Production Problems in Britain and the USA, 1914–1918." Working paper. 2001.

———. "The Optical Munitions Industry in Great Britain, 1888–1923." PhD diss. University of Glasgow, 2005.

Sawako, Maeda. "Transformation of Japanese Space Policy: From the 'Peaceful Use of Space' to 'the Basic Law on Space.' " *Asia-Pacific Journal: Japan Focus* 7:44:1 (Nov. 2009), 1–7.

Scheips, Paul J. "Union Signal Communications: Innovation and Conflict." *Civil War History* 9:4 (Dec. 1963), 399–421.

Schilling, Govert. *Flash! The Hunt for the Biggest Explosions in the Universe.* Translated by Naomi Greenberg-Slovin. Cambridge, UK: Cambridge University Press, 2002.

Schlosser, Eric. *Command and Control: Nuclear Weapons, the Damascus Accident, and the Illusion of Safety.* New York: Penguin, 2013.

Schuster, Richard J. *German Disarmament After World War I: The Diplomacy of International Arms Inspection.* London: Routledge, 2006.

Scott, William B., and Craig Covault. "High Ground over Iraq." *Aviation Week & Space Technology* 158:23 (June 9, 2003), 44–48.

Seamans, Robert C., Jr. *Project Apollo: The Tough Decisions.* Monographs in Aerospace History 37, SP-2005-4537. Washington,

DC: NASA History Division, 2007.

Sheehan, Michael. *The International Politics of Space.* London: Routledge, 2007.

Shrock, Keith A. "Space-Based Infrared Technology Center of Excellence." Fact sheet. AFRL Space Vehicles Directorate, Space Technology Division, Infrared Technologies Center of Excellence Branch, Kirtland AFB and Hanscom AFB, Apr. 3, 2007.

Siddiqi, Asif A. "Korolev, Sputnik, and the International Geophysical Year." In Roger D. Launius, John M. Logsdon, and Robert W. Smith, eds. *Reconsidering Sputnik: Forty Years Since the Soviet Satellite.* London: Routledge, 2000, 43–72.

———. *The Red Rockets' Glare: Spaceflight and the Soviet Imagination, 1857–1957.* New York: Cambridge University Press, 2010.

Sluiter, Engel. "The Telescope Before Galileo." *Journal of the History of Astronomy* 28:92 (Aug. 1997), 223–34.

Smith, Marcia S. "Military and Civilian Satellites in Support of Allied Forces in the Persian Gulf War." Congressional Research Service, Feb. 27, 1991.

Sobel, Dava. *Longitude: The True Story of a Lone Genius Who Solved the Greatest Scientific Problem of His Time.* New York: Walker, 2005.

Space Foundation. *The Space Report: The Authoritative Guide to Global Space Activity.*

Annual publication, 2006– .

Stanley-Lockman, Zoe, and Katharina Wolf. "European Defence Spending 2015: The Force Awakens." European Union Institute for Security Studies — *Brief Issue* 10 (Mar. 2016).

State Council Information Office of the People's Republic of China. "China's Space Activities in 2016 — Preamble." *Global Times*/Xinhua, Dec. 27, 2016.

Stine, Deborah D. "U.S. Civilian Space Policy Priorities: Reflections 50 Years After Sputnik." Congressional Research Service, Feb. 2, 2009.

Stone, Richard. "A Renaissance for Russian Space Science." *Science,* Apr. 7, 2016.

Talbot, David. "How Technology Failed in Iraq." *MIT Technology Review,* Nov. 2004.

Taylor, E. G. R. *The Haven-Finding Art: A History of Navigation from Odysseus to Captain Cook.* London: Hollis & Carter, 1956.

Technological Capabilities Panel of the Science Advisory Committee, Office of Defense Mobilization. "Report: Meeting the Threat of Surprise Attack" [Killian Report]. Feb. 14, 1955.

Tester, S. J. *A History of Western Astrology.* Woodbridge, UK: Boydell Press, 1987.

Thompson, George Raynor. "Civil War Signals." *Military Affairs* 18:4 (Winter 1954), 188–201.

Tibbetts, G. R. *Arab Navigation in the Indian Ocean Before the Coming of the Portuguese.* London: Royal Asiatic Society of Great Britain and Ireland, 1971.

Tseng, Lillian Lan-ying. *Picturing Heaven in Early China.* Cambridge, MA: Harvard University Asia Center, 2011.

Turnbull, David. "Cartography and Science in Early Modern Europe: Mapping the Construction of Knowledge Spaces." *Imago Mundi* 48 (1996), 5–24.

Union of Concerned Scientists. UCS Satellite Database.

United Nations General Assembly. Resolutions Adopted by the General Assembly During Its Twelfth Session: 1148 (XII), 1149 (XII). Nov. 14, 1957.

———. Resolution 1884 (XVIII): "Question of General and Complete Disarmament." Oct. 17, 1963.

———. Treaty on Principles Governing the Activities of States in the Exploration and Use of Outer Space, including the Moon and Other Celestial Bodies. Signed in London, Moscow, and Washington, Jan. 27, 1967.

United Nations Office at Geneva: Conference on Disarmament. CD Documents on Prevention of an Arms Race in Outer Space.

United Nations Office for Disarmament Affairs. Treaty Banning Nuclear Weapon Tests in the Atmosphere, in Outer Space, and Under Water. Aug. 5, 1963.

United Nations Office for Outer Space Affairs. Space Law.

United States Air Force. *Counterspace Operations: Air Force Doctrine Document 2-2.1.* Aug. 2, 2004.

———. *Space Operations: Air Force Doctrine Document 3-14.* Nov. 27, 2006. Incorporating Change 1, July 28, 2011. [*AFDD 3-14* is *AFDD 2-2* plus the change.]

United States Army Ordnance Department/ Lt. Col. F. E. Wright. *The Manufacture of Optical Glass and of Optical Systems: A Wartime Problem.* Washington, DC: Government Printing Office, 1921.

United States Central Command. "Operation Desert Shield/Desert Storm: Executive Summary." July 11, 1991.

United States Department of Defense. *Conduct of the Persian Gulf War: Final Report to Congress.* Apr. 1992.

———. *Quadrennial Defense Review 2014.* Mar. 4, 2014.

United States Government Accountability Office. *Operation Desert Storm: Evaluation of the Air Campaign.* GAO/NSAID-97-134. June 1997.

United States Space Command. "Operations Desert Shield and Desert Storm: Assessment." Jan. 1992.

van Creveld, Martin. *Command in War.* Cambridge, MA: Harvard University Press, 1985.

———. *Technology and War, From 2000 B.C. to the Present.* Revised and expanded. New York: Free Press, 1991.

Van Helden, Albert. "The Invention of the Telescope." *Transactions of the American Philosophical Society* 67:4 (June 1977), 1–67.

———. "Telescopes and Authority from Galileo to Cassini." *Osiris* 9 (1994), 8–29.

Vaucouleurs, Gérard de. *Astronomical Photography: From the Daguerreotype to the Electron Camera.* Translated by R. Wright. New York: Macmillan, 1961.

Vice President's Space Policy Advisory Board. "The Future of the U.S. Space Industrial Base: A Task Group Report." Nov. 1992.

Viotti da Costa, Emilia. "The Portuguese–African Slave Trade: A Lesson in Colonialism." *Latin American Perspectives* 12:1 (Winter 1985), 41–61.

Walker, Christopher B. F., ed. *Astronomy Before the Telescope.* London: British Museum Press, 1996.

Walker, Russell G., and Stephan D. Price. *AFCRL Infrared Sky Survey.* Vol. 1, *Catalog of Observations at 4, 11, and 20 Microns.*

ADA 016397. Hanscom AFB, MA: Optical Physics Laboratory, Air Force Cambridge Research Laboratories, July 1975.

Warner, Deborah Jean. *Alvan Clark & Sons: Artists in Optics.* Washington, DC: Smithsonian Institution Press, 1968.

Watson, Fred. *Stargazer: The Life and Times of the Telescope.* Cambridge, MA: Da Capo Press, 2005.

Watson-Watt, Robert. "Radar Defense Today — and Tomorrow." *Foreign Affairs* 32:2 (Jan. 1954), 230–43.

Weekes, Trevor. "Very High Energy Gamma Ray Astronomy 101." Harvard–Smithsonian Center for Astrophysics, June 2012.

Westfall, Richard S. "Science and Patronage: Galileo and the Telescope." *Isis* 76:1 (Mar. 1985), 11–30.

Weston, Scott A. "Examining Space Warfare: Scenarios, Risks, and US Policy Implications." *Air & Space Power Journal* 23:1 (Spring 2009), 73–82.

Wie, Bong. "Optimal Fragmentation and Dispersion of Hazardous Near-Earth Objects: NIAC Phase I Final Report." Asteroid Deflection Research Center, Iowa State University, Sept. 25, 2012.

Williams, J. E. D. *From Sails to Satellites: The Origin and Development of Navigational Science.* Oxford: Oxford University Press, 1992.

Wolter, Detlev. *Common Security in Outer Space and International Law.* Geneva: United Nations Institute for Disarmament Research, 2006.

Wright, David, Laura Grego, and Lisbeth Gronlund. *The Physics of Space Security: A Reference Manual.* Cambridge, MA: American Academy of Arts and Sciences, 2005.

Zak, Anatoly. "A Rare Look at the Russian Side of the Space Station." *Air & Space Smithsonian,* Sept. 2015.

———. "Russia Approves Its 10-Year Space Strategy." Blog post. Planetary Society, Mar. 23, 2016.

Zimmerman, David. *Britain's Shield: Radar and the Defeat of the Luftwaffe.* Stroud, UK: Amberly, 2013.

ABOUT THE AUTHORS

Neil deGrasse Tyson is an astrophysicist with the American Museum of Natural History, director of its world-famous Hayden Planetarium, host of the hit radio and TV show *StarTalk,* and an award-winning author. He lives in New York City.

Avis Lang is a research associate at the American Museum of Natural History's Hayden Planetarium. For half a decade, she edited Tyson's *Natural History* magazine column, Universe, which became the basis for his best-selling *Astrophysics for People in a Hurry,* as well as his anthology, *Space Chronicles: Facing the Ultimate Frontier.* She lives in New York City.